Algal Biotechnology

This book initiates with a general introduction to microalgae and algal biotechnology with subsequent discussions on all the significant aspects of applied biotechnology, bioremediation, nano applications, multidimensional usages of algae as a biofertilizer, and a source of bioactive compounds and phytochemicals. Major themes of the book include algae and the environment, bioremediation using algae, algal-omics and applications, and large-scale bioprocesses for algal cultivation, its constraints and challenges.

Features:

- Focusses on the importance of algae for a sustainable environment.
- Covers algal bioplastics and other commercial products.
- Explores possible utilization of algae in phyco/bioremediation.
- Reviews algae as a biostimulant and biofertilizer.
- Demonstrates challenges during algal cultivation on a large scale.

This book is aimed at graduate students and researchers in biotechnology, bioenergy, renewable energy, energy, bioremediation, fuel and petrochemicals, wastewater, novel technologies, clean technologies, bioremediation environmental, functional foods and nutraceuticals, marine and aquatic science.

Algal Biotechnology

Current Trends, Challenges and Future Prospects for a Sustainable Environment

Edited by
Nivedita Sahu and S. Sridhar

CRC Press
Taylor & Francis Group
Boca Raton London New York

CRC Press is an imprint of the
Taylor & Francis Group, an **informa** business

Designed cover image: Nivedita Sahu

First edition published 2024
by CRC Press
2385 NW Executive Center Drive, Suite 320, Boca Raton FL 33431

and by CRC Press
4 Park Square, Milton Park, Abingdon, Oxon, OX14 4RN

CRC Press is an imprint of Taylor & Francis Group, LLC

ISBN: 978-1-032-11268-8 (hbk)
ISBN: 978-1-032-11269-5 (pbk)
ISBN: 978-1-003-21915-6 (ebk)

DOI: 10.1201/9781003219156

Typeset in Times
by Apex CoVantage, LLC

Dedicated to our families, well-wishers, and
CSIR-IICT's Membrane Separations Laboratory

Specially devoted to the memory of
our student and co-author,
Sajja S Chandrasekhar

Contents

THEME III Algal-Omics and Applications

THEME IV Large-Scale Bioprocesses for Algal Cultivation: Constraints and Challenges

Preface

Algal biotechnology has garnered significant interest recently, owing to its versatile characteristics of harnessing solar energy and CO_2 to generate value-added products, chemicals, oxygen, and biofuels. This innovative research domain can gradually replace or complement conventional processes that operate with heterotrophic compounds using traditional methods. Algal biomass can provide a renewable source of biofuels and high-value chemicals using process optimization, process intensification, and sustainable cultivation. Algal biotechnology is a broad research area that involves integration of both fundamental and applied research. Despite its importance, algae are the most neglected tiny organisms of Mother Nature. It is one of the primitive groups of simplest, photosynthetic organisms with diverse sizes. Its sizes range from micro to macro, growing in discrete environmental conditions varying from freshwater to marine water, municipal sewage water to domestic wastewater, rivers to eutrophicated ponds, etc. The photosynthetic output of these tiny organisms is much higher than other plants as they can trap more solar energy and carbon dioxide from the atmosphere. This fundamental exploration comprises 22 chapters, which involves screening and cultivating suitable algal species that are potentially viable through cytology, stress tolerance, and metabolic engineering. Applied algal research needs to focus on the large-scale production methods of microalgae, unit operations for oil and lipids, economic viability, including capital investment, and cost-lowering components for production. The key bottlenecks in algal biotechnology are the downstream processing methods that involve liquid fuel extraction, which is still in its early stages and not always adequate for industrial production. Significant change and advancement in the domain of algal biotechnology can be brought about through multilateral cooperation between academia, research organizations, governments, stakeholders, and researchers.

This book, ***Algal Biotechnology: Current Trends, Challenges and Future Prospects for a Sustainable Environment*** is contemplated bringing in recent topics of interest that provide the way forward and address some of the critical bottlenecks. The book begins with a general introduction to microalgae and algal biotechnology with subsequent discussions on all the significant aspects of applied biotechnology, bioremediation, nano applications, multidimensional usages of algae as a biofertilizer, and a source of bioactive compounds and phytochemicals. Broadly, the book ***Algal Biotechnology: Current Trends, Challenges and Future Prospects for a Sustainable Environment*** covers the following themes as listed here, with contributions by researchers from across the globe.

Theme I: Algae and the Environment: This theme primarily constitutes algae's role in maintaining and balancing the changes in the environment. The chapters elaborate on the role of algae in both aquatic and terrestrial ecosystems. Additionally, the production of environment-friendly compounds from algae using different economical methods is explained in detail. This theme is intended for the users to understand the importance of algae in the natural ecosystem.

Theme II: Bioremediation Using Algae: Apart from using carbon dioxide through photosynthesis, algae can consume organic matter through heterotrophic mechanisms. Building upon this characteristic, the algae cultures have been used to treat various kinds of waste and wastewater originating from industries and domestic pollutants. Different strategies of using algae to remediate polluted water bodies and remove metal and non-metal contaminants have been elaborated in this section, along with its characteristic property of co-producing value-added products.

Theme III: Algal-omics and Applications: For understanding algae's potential in producing valuable compounds and minimizing environmental impact, various studies have sprung up to improve algae's capabilities. This section reflects algae's algal karyology, metabolic engineering, stress-tolerance, micro-metabolites, and other biotechnological advances. This section helps the audience understand the evolutionary trend, the genetic makeup of algae, and algal nanotechnology.

Theme IV: Large-Scale Bioprocesses for Algal Cultivation: Constraints and Challenges: Though the algal potential is widespread and robust, cultivation needs to be done on a large scale to match the market demand for value-added products obtained from other routes. The various ways of large-scale production and the factors influencing growth in an open environment have been discussed in this section. Overcoming different constraints in harvesting the microalgae, extracting value-added products, and reducing downstream processing costs have also been included.

The chapters describing interventions made in bioplastic, nanotechnology, bioremediation, biofertilizer, and the challenges faced in extensive scale cultivation of algae are the attractive features of this book. Start-ups, environmentalists, and NGOs will find an authoritative and comprehensive reference. The book is meant to be a source of recent developments and a resource for young researchers willing to take up this field of study, besides experienced practitioners and the industry. The editors hope to provide this information in an easily accessible format to enable easy access to a wide range of audiences and help produce breakthroughs in algal biotechnology.

KEY FEATURES

This book describes a unique collection of applied algal research work with particular emphasis on the following topics:

- Explanation of the importance of algae for a sustainable environment
- Focus on algal bioplastic and other commercial products
- Exploration of the possibility of utilizing algae in phyco/bioremediation
- Algae as a bio-stimulant and biofertilizer
- Demonstration of the challenges met during algal cultivation on a large scale

About the Editors

Dr. Nivedita Sahu received her PhD degree on marine algal biotechnology from the Dept. of Botany, University of Delhi in 2004 after her MSc (Algology) from Ranchi University. She has expertise in algal biology, cultivation, and algal polysaccharides. Her research area of interest is applied phycology and industrial microbiology and currently working as principal scientist at CSIR-Indian Institute of Chemical Technology in algae-based membrane integrated technology for water and wastewater treatment.

She has expertise in seaweed (marine algae) research also, which includes seaweed biology and its cultivation on the coast of Odisha, Tamil Nadu, and the Andaman Islands. She is an associate professor at AcSIR (Academy of Scientific and Innovative Research). At present, she is working on developing cost-effective membrane-integrated processes and algal reactors to enhance treatment efficiency and improve % water recovery from domestic and industrial effluents.

To her credits, she has one patent, one book, four book chapters, 10 prestigious awards, and 30 research papers published in various journals. Research scholars and students of MSc/MTech projects worked under her have fetched several awards in various national and international conferences and workshops. She is a member of the Board of Studies at the Department of Microbiology, Osmania University, and St. Ann's College for Women, Hyderabad. She has served as Telangana State President for WICCI-Water Resource Council from 2020-2022 and afterwards as National Vice President for the same. As a part of her research work to collect samples, she has travelled extensively to different coastal areas of India. She has made a substantial contribution to numerous health-related products as a member of Membrane Separation Group, including devices for producing alkaline water, solar-powered AWGs for safe, mineral-rich drinking water, and medical-grade water. She participated in the fabrication of "Saans" masks during the epidemic, which assisted many women from SHGs and NGOs create jobs with a renewal of more than 1.5.cr.

Dr. S. Sridhar is a chemical engineer from College of Technology, Osmania University, Hyderabad, who has been working as a scientist in the area of Membrane Separation Processes at the CSIR-IICT, Hyderabad for the past 24 years. He has developed several technologies for chemical industries besides contributing immensely to rural welfare through water purification projects and academic development via extensive HRD and laboratory development in several schools and colleges.

Major highlights of his career include the commissioning of several membrane pilot plants based on electrodialysis, nanofiltration, and gas permeation of capacities varying from 500–5,000 L/h to facilitate solvent recovery, effluent treatment, and gas purification in pharmaceutical, steel, textile and petrochemical industries, design and installation of 15 model defluoridation plants of 600–4,000 L/h capacity and 25 highly compact low-cost systems of 100–200 L/h capacity for purification of ground and surface water for more than 5 million population affected by fluorosis, gastroenteritis, jaundice, typhoid, and other water-borne diseases in villages of Telangana, Andhra Pradesh, Karnataka, and Tamil Nadu, which is widely appreciated by the press, masses, and His Excellency, The Governor of TS and AP, besides Union Ministers of Science and Technology. A free water camp based on a compact system design was established by Dr. Sridhar on Uppal Road in Hyderabad City and has been providing healthy water to the urban population, including pedestrians, drivers of buses, autos, cars, and two-wheelers, totally free of cost, since April 2016. Similar camps at the All India Industrial Exhibition in Jan.–Feb. 2017 and 2018 and CSIR Science Exhibition have provided free drinking water to 1 million people, including schoolchildren. He has contributed to disaster management by installing hand pump operated hollow fiber membrane systems for the 5 lakh population in response to the recent floods in Kerala, Assam, Karnataka, Maharashtra, and Bihar in 2018 and 2019. Another recent innovation includes a low-cost cascaded RO + Resin system for medical-grade water for dialysis and biochemical applications. More than half a lakh patients suffering from chronic kidney disease have been treated using this dialysis water in Nephroplus hospitals.

Dr. Sridhar has published 172 research papers in reputed international journals, which are widely cited by peers more than 10,400 times in high-impact journals with an h-index of 51. He has 17 foreign patents to his credit, besides 49 book chapters and 300 papers in various symposia/conference proceedings. Dr. Sridhar has trained 400 BTech/MTech and MSc students from different universities and institutes for project work apart from guiding eight scholars for PhD. Dr. Sridhar's students have been awarded 50 prizes for meritorious work and the best oral paper/poster presentation made under his guidance in various conferences/symposia.

Dr. Sridhar is an all-round sportsman who represented Hyderabad State and Osmania University Teams and captained All India CSIR Teams in cricket, lawn tennis, and table tennis with distinction.

Dr. Sridhar is a recipient of more than 65 Prestigious Science Awards, including 35 national awards, such as the IIChE Amar Dye-Chem Award 2003, CSIR Young Scientist Award 2007, Engineer of the Year Award from A.P. State Govt. in 2009, Scopus Young Scientist Award 2011, NASI–Reliance Industries Platinum Jubilee Award 2013, VNMM award from IIT-Roorkee 2015, GYTI Awards for 2015 and 2018, CIPET national awards for 2016, 2017, and 2018, and Nina Saxena Excellence in Technology Award from IIT-Kharagpur in 2017, besides the IIChE Herdillia Award in 2019 and the Jubilant Award from the Indian Institute of Chemical Engineers (IIChE), Kolkata, 2023, ABAP-Senior Scientist Award from Association of Biotechnology and Pharma, 2023. Dr. Sridhar is ranked in the Top 2% Cited Scientists in the World in 2021, 2022, and 2023 by Stanford University and Elsevier, and No. 131 in the Chemistry Field in India by Research.com. He was bestowed the Vocational Excellence-Start Up-Award for Designing, Manufacturing, and Successfully Marketing Meghdoot Atmospheric Water Generator by the Rotary Club of Bombay Sea Face in 2023.

Contributors

G. Ajay
Marine Algal Research Station
CSIR—Central Salt and Marine Chemicals Research Institute
Mandapam Camp—623 519, Mandapam, Dist-Ramanathapuram, Tamil Nadu, India

Ana-Valentina Ardelean
Institute of Biology Bucharest, Romanian Academy, Splaiul Independentei 296, Bucharest 060031, Romania
3Nano-SAE Research Centre, University of Bucharest, Bucharest 077125, Romania

Ioan I. Ardelean
Institute of Biology Bucharest
Romanian Academy, Splaiul Independentei 296
Bucharest 060031, Romania

M. Aruna
Dept. of Botany
Telangana University
Nizamabad
Telangana State

Samadhan Yuvraj Bagul
ICAR—Directorate of Medicinal and Aromatic Plant Research
Boriavi, Anand, Gujarat, India

Chhandashree Behera
Algal Biotechnology and Molecular Systematic Lab
Post Graduate Department of Botany
Berhampur University
Bhanja Bihar, Berhampur, India

Apoorva Bhayani
Division of Applied Phycology and Biotechnology
CSIR—Central Salt and Marine Chemicals Research Institute
Bhavnagar, Gujarat, India
Academy of Scientific and Innovative Research (AcSIR)
CSIR—Central Salt and Marine Chemicals Research Institute
Bhavnagar, India

Prajna Paramita Bhuyan
Algal Biotechnology and Molecular Systematic Lab
Post Graduate Department of Botany
Berhampur University
Bhanja Bihar, Berhampur, India

Sai Kishore Butti
CSIR—Indian Institute of Chemical Technology
Tarnaka, Hyderabad, Telangana, India
Academy of Scientific and Innovative Research (AcSIR)
Gaziabad, Uttar Pradesh, India

Sajja S Chandrasekhar
CSIR—Indian Institute of Chemical Technology
Tarnaka, Hyderabad, Telangana, India
Academy of Scientific and Innovative Research (AcSIR)
Gaziabad, Uttar Pradesh, India

Ayushi Dalmia
The Microalgal Biotechnology Laboratory
The French Associates Institute for Agriculture and Biotechnology
Jacob Blaustein Institute for Desert Research
Ben-Gurion University of the Negev
Sede Boker Campus, Israel-84990

Soumya Ranjan Dash
Algal Biotechnology and Molecular Systematic Lab
Post Graduate Department of Botany
Berhampur University, Bhanja Bihar, Berhampur, India

Dolly Wattal Dhar
Professor, School of Agricultural Sciences
Sharda University
Greater Noida, UP, Uttar Pradesh, India

K Eswaran
Marine Algal Research Station
CSIR-Central Salt and Marine Chemicals Research Institute
Mandapam Camp—623 519, Mandapam, Dist-Ramanathapuram, Tamil Nadu, India

Saeed Fatima
CSIR—Indian Institute of Chemical Technology
Tarnaka, Hyderabad, Telangana, India
Academy of Scientific and Innovative Research (AcSIR)
Gaziabad, Uttar Pradesh, India

Abirami Ramu Ganesan
Group of Fermentation and Distillation
Laimburg Research Centre
Ora (BZ) Aue, Italy

Adrian Ghinea
Institute of Biology Bucharest, Romanian Academy
Splaiul Independentei 296, Bucharest 060031, Romania

Khadke GN
ICAR
Directorate of Medicinal and Aromatic Plant Research
Boriavi, Anand, Gujarat, India

M. Gobalakrishnan
Marine Algal Research Station
CSIR—Central Salt and Marine Chemicals Research Institute
Mandapam Camp—623 519, Mandapam, Dist-Ramanathapuram, Tamil Nadu, India

Bapanipally Govardhan
Indian Institute of Chemical Technology
CSIR
Tarnaka, Hyderabad, Telangana, India
Academy of Scientific and Innovative Research (AcSIR)
Uttar Pradesh, India

P. Gwen Grace
Division of Applied Phycology and Biotechnology
CSIR—Central Salt and Marine Chemicals Research Institute
Bhavnagar, Gujarat—India

S. Hemalatha
National Repository for Microalgae and Cyanobacteria—Freshwater (DBT)
Department of Microbiology
Bharathidasan University
Tiruchirappalli, Tamil Nadu, India

Matei-Tom Iacob
USAMV Bucuresti, Bulevardul Mărăşti 59
Bucureşti 011464, Romania

Mrutyunjay Jena
Algal Biotechnology and Molecular Systematic Lab
Post Graduate Department of Botany
Berhampur University
Bhanja Bihar, Berhampur, India

K. Jothibasu
Department of Microbiology
Karpagam Academy of Higher Education
Coimbatore-641021, Tamil Nadu, India

Pandiyan K
Ginning Training Centre
ICAR–Central Institute for Research on Cotton Technology (CIRCOT)
Nagpur, Maharashtra, India

Jyothi K
Microbiology/Food Science and Technology
GITAM School of Science, GITAM (Deemed to be University)
Gandhinagar, Rushikonda, Vishakhapatnam-530045

Mudassar Anisoddin Kazi
Division of Applied Phycology and Biotechnology
CSIR—Central Salt and Marine Chemicals Research Institute
Gijubhai Badheka Marg, Bhavnagar, India

Krishna Prasad M
Visakhapatnam Department of Chemical Engineering
GMR Institute of Technology, Rajam, Srikakulam

M. Madhumala
CSIR—Indian Institute of Chemical Technology
Tarnaka, Hyderabad, Telangana, India
Academy of Scientific and Innovative Research (AcSIR)
Gaziabad, Uttar Pradesh, India

Sairendri Maharana
Algal Biotechnology and Molecular Systematic Lab
Post Graduate Department of Botany
Berhampur University
Bhanja Bihar, Berhampur-760007, India

Pritikrishna Majhi
Department of Microbiology
CBSH, OUAT, Bhubaneswar, Odisha

J. Malarvizhi
Marine Algal Research Station
CSIR-Central Salt and Marine Chemicals Research Institute
Mandapam Camp—623 519, Mandapam, Dist-Ramanathapuram, Tamil Nadu, India

Subir Kumar Mandal
Applied Phycology and Biotechnology
CSIR—Central Salt and Marine Chemicals Research Institute
G. B. Marg, Bhavnagar-364 002, Gujarat, India

Vaibhav A. Mantri
Division of Applied Phycology and Biotechnology
CSIR—Central Salt and Marine Chemicals Research Institute
Gijubhai Badheka Marg, Bhavnagar, India
Academy of Scientific and Innovative Research (AcSIR)
Ghaziabad, India

Kannan Mohan
PG and Research Department of Zoology
Sri Vasavi College
Erode, Tamil Nadu, India

Cristina Moisescu
Institute of Biology Bucharest, Romanian Academy
Splaiul Independentei 296
Bucharest 060031, Romania

N. Monisha
Marine Algal Research Station
CSIR—Central Salt and Marine Chemicals Research Institute
Mandapam Camp—623 519, Mandapam, Dist-Ramanathapuram, Tamil Nadu, India

D. Mubarak Ali
School of Life Sciences
B.S. Abdur Rahman Crescent Institute of Science and Technology
Chennai, Tamil Nadu, India
National Repository for Microalgae and Cyanobacteria—Freshwater (DBT)
Department of Microbiology
Bharathidasan University
Tiruchirappalli, Tamil Nadu, India

Bidhu Bhusan Mukut
Centre for Energy, Indian Institute of Technology Guwahati
Guwahati, Assam-781039, India

Shanmugam Munisamy
Department Biomarine Resource Valorisation
Division of Food Production and Society
Norway Institute of Bioeconomy Research (NIBIO)
Torggarden, Kudaisveien 6, No-8027
Bodo, Norway

Rabindra Nayak
Algal Biotechnology and Molecular Systematic Lab
Post Graduate Department of Botany
Berhampur University
Bhanja Bihar, Berhampur, India

Srimanta Patra
Algal Biotechnology and Molecular Systematic Lab
Post Graduate Department of Botany
Berhampur University
Bhanja Bihar, Berhampur, India

C. Periyasamy
Department of Botany
Melaneelithanallur Manonmaniam Sundaranar Univlersity
Tirunelveli, Tamil Nadu, India

Biswajita Pradhan
Algal Biotechnology and Molecular Systematic Lab
Post Graduate Department of Botany
Berhampur University
Bhanja Bihar, Berhampur, India

(Late) Prof. (Dr.) Iccha Purak Singh
Department of Botany
Ranchi Women's College
Ranchi, India

Durairaj Karthick Rajan
Centre of Advanced Study in Marine Biology
Faculty of Marine Sciences, Annamalai University
Parangipettai, Tamil Nadu, India

Suchitra Rakesh
Biofuel Research Laboratory
Department of Microbiology
Central University of Tamil Nadu
Thiruvarur, Tamil Nadu, India

Dineshkumar Ramalingam
Division of Applied Phycology and Biotechnology
CSIR—Central Salt and Marine Chemicals Research Institute
Gijubhai Badheka Marg, Bhavnagar, India
Academy of Scientific and Innovative Research (AcSIR)
Ghaziabad, India

G. Rani
SDNB Vaishnav College for Women Chromepet
Chennai, Tamil Nadu, India

Prama Rani
Waste Valorization Research Lab
Lovely Professional University
Phagwara, Punjab, India

B. Sankaran
Presidency College
Chennai-600005

R. Sathya
Department of Biotechnology, School of Life Sciences
H.N.B. Garhwal University (A Central University)
Srinagar, Garhwal, Uttarakhand, India

A. Srinivasa Rao
Department of Botany
Government College (Autonomous)
Rajamundry, A.P., India

P.V. Subba Rao
Aquaculture Foundation of India
Madurai, Tamil Nadu, India

K. Suresh Kumar
Department of Botany
University of Allahabad
Prayagraj, U.P., India

Nivedita Sahu
CSIR—Indian Institute of Chemical Technology
Tarnaka, Hyderabad, Telangana, India
Academy of Scientific and Innovative Research (AcSIR)
Gaziabad, Uttar Pradesh, India

Saubhagya Manjari Samantaray
Department of Microbiology
CBSH, OUAT, Bhubaneswar, Odisha

Jastin Samuel
Waste Valorization Research Lab
Lovely Professional University
Phagwara, Punjab, India

Y. Saurabh
School of Life Sciences
B.S. Abdur Rahman Crescent Institute of Science and Technology
Chennai, India

Rajamohamed Beema Shafreen
Department of Biotechnology
Alagappa University, Science Campus
Karaikudi, Tamil Nadu, India

S. Sridhar
CSIR—Indian Institute of Chemical Technology
Tarnaka, Hyderabad, Telangana, India
Academy of Scientific and Innovative Research (AcSIR)
Gaziabad, Uttar Pradesh, India

Ioan Stamatin
USAMV Bucuresti, Bulevardul Mărăşti 59
Bucureşti 011464, Romania

Serban N. Stamatin
USAMV Bucuresti, Bulevardul Mărăşti 59
Bucureşti 011464, Romania

Ganesh Temkar
Marine Algal Research Station
CSIR—Central Salt and Marine Chemicals Research Institute
Mandapam Camp—623 519, Mandapam, Dist-Ramanathapuram, Tamil Nadu, India

Nooruddin Thajuddin
National Repository for Microalgae and Cyanobacteria—Freshwater (DBT)
Department of Microbiology
Bharathidasan University
Tiruchirappalli, Tamil Nadu, India

J. Tharun Kumar
Biofuel Research Laboratory, Department of Microbiology
Central University of Tamil Nadu
Thiruvarur, Tamil Nadu, India

Ashutosh Tripathy
Waste Valorization Research Lab
Lovely Professional University
Phagwara, Punjab, India
School of Bioengineering and Biosciences
Lovely Professional University
Phagwara, Punjab, India

Ajay W Tumaney
Department of Biochemistry
School of Life Science
University of Hyderabad
Hyderabad—500046, India

Ramachandran Uma
SDNB Vaishnav College for Women Chromepet
Chennai, Tamil Nadu, India

V. Veeragurunathan
Division of Applied Phycology and Biotechnology
CSIR—Central Salt and Marine Chemicals Research Institute
Bhavnagar, Gujarat, India
Academy of Scientific and Innovative Research (AcSIR)
CSIR—Central Salt and Marine Chemicals Research Institute
Bhavnagar, India

M. Vignesh
Division of Applied Phycology and Biotechnology
CSIR—Central Salt and Marine Chemicals Research Institute
Gijubhai Badheka Marg, Bhavnagar, India

Sanda Voinea
USAMV Bucuresti, Bulevardul Mărăşti 59
Bucureşti 011464, Romania

Introduction, Scope, and Significance of Algal Biotechnology

Nivedita Sahu and S. Sridhar

This book on ***Algal Biotechnology: Current Trends, Challenges and Future Prospects for a Sustainable Environment*** is the first volume of this series which enlightens readers on the algae-based unique applications for humankind. Algae, the microgreens, are present in diverse habitats and possess physical, biological, and morphological diversity. They are the primary producers (autotrophs) in aquatic ecosystems. Algae are the fundamental essence of the food web in the marine habitat. Sometimes, the plethora of specific genera of algae can be an indicator of quality and pollution in water bodies. They are one of the leading contributors that reduce the impact of climate change globally. They are used in various food products as gelling, thickening, and stabilizing agents. They are part of many industries like pharmaceutical, cosmetics, fuel, chemicals, etc. The algal biomass was found on earth almost 3.5 million years ago. These organisms were a great source to convert carbon dioxide to organic materials releasing oxygen, which is the utmost need for human survival. Due to the energy crisis of this century, it has gained immense attention from researchers as a feedstock for green renewable fuels. Microalgae have gained attention for their potential utility in various biotechnology applications due to their polytrophic metabolism, including autotrophy, mixotrophy, and heterotrophy, enabling them to take advantage of different energy and carbon sources for growth. These organisms can fix CO_2 with light as an energy source to produce biomass and O_2, thus significantly impacting global CO_2 reduction. They can grow fast, possess high oil content, and their mass cultures can be easily maintained. The growth of microalgae is several times higher compared to any photosynthetic plants, attributing this to the efficient photosynthetic machinery. The growth phase is followed by the stress phase, like nutritional stress or physical stress, which usually induces lipogenesis. Thus, the accumulated carbohydrate in the growth phase is converted into polyunsaturated fatty acids (PUFA) and lipids (triacylglycerides), the precursor molecules for biodiesel production.

Though mixotrophic and heterotrophic cultivation of microalgae is receiving massive attention due to high biomass productivities, there are gaps in deciphering the underlying metabolisms and molecular pathways which microalgae exploit to produce biomass and other metabolites. These techniques can reduce the cost of downstream processing, which may represent a considerable percentage of the total production expense. Biomass from diverse origins, including forestry, agricultural and aquatic sources, has been studied as a raw material for producing a variety of biofuels, including biodiesel, bioethanol, biohydrogen, jet fuel, bio-oil, and biogas. Several studies have shown that some microalgal biomasses are rich sources of omega-3 and omega-6 fatty acids (FAs) and valuable chemicals such as polysaccharides, carotenoids, and phycobilin pigments. Specific microalgae species can be sources for the extraction of different high-value chemical products, such as pigments, antioxidants, polysaccharides, triglycerides, and vitamins. These compounds are mainly employed as bulk commodities or specialties in different industrial sectors, for example, pharmaceuticals, cosmetics, nutraceuticals, functional foods, and aquaculture. They have also found applications in the food and feed industries as fertilizers and growth promoters in agriculture and wastewater treatment.

This book can be used as a textbook by presenting each chapter in one section and a small summary of the various chapters in the next section. The possible usage of the book for introductory and advanced courses is proposed. Each chapter is precise and peculiar, with exhilarating concepts that are not repeated from one chapter to the other. The book describes how these tiny microgreens are ecologically ubiquitous and play a prominent role in bioremediation and recent advances in omics applications. The book also portrays various ways of large-scale production and the factors influencing their growth. The production needs to be done on an enormous scale to match the market demand and the possible routes for value-added products, which have been explained. This book presents extraordinarily unique information pulled collectively from experts working in different parts of the world. The schematic explanations and user-friendly illustrations in the book make it more informative and exciting to the reader with or without a biology background

This book is organized around four themes. Each theme contains new findings on algae's role for different applications. Of course, the idea in each theme is unique and not replicated from one chapter to the other. In fact, each chapter is an independent entity. The readers are entitled to understand the algae's role in maintaining and balancing the changes in the environment from six (06) different chapters on algae and the environment under Theme I.

Moreover, it also gives insights into the algae's role in boosting the blue economy. The aquaculture industry is expanding with algal-derived products due to their immunostimulant properties. Theme II starts with the algal bioremediation and comprises six (06) chapters. It offers the rectification of significant problems associated with the circular economy, wastewater treatment, and the reuse of energy. In this theme, the reader will thus learn how microalgae are receiving attention as one of the future sources of renewable energy. The blooming field of biological fuel cells uses algae to degrade organic matter and generate energy. They can easily overcome the limitations associated with conventional cathodic cells. Phycoremediation is a sustainable alternative to wastewater treatment. Microalgae are very effective in sequestering heavy metals from wastewater as they have a large surface-to-volume ratio. Theme III takes the reader to omics-applications of algae, their nuclear characteristics, with algal nanotechnology and bio-composites for biomedical applications. Also, various groups of algae viz. Cyanophyceae, Chlorophyceae, Bacillariophyceae, Phaeophyceae, and Rhodophyceae show great potential for silver, gold biosynthesis, copper, palladium, and silica nanoparticles. The green synthesis of nanoparticles using algae has attracted the focus of scientists from all fields. The excellent biosynthesis procedures give an idea to develop biocomposites with microalgae for biomedical applications. Biocomposites of microalgae act as a capping agent and have beneficial properties such as biocompatibility and biodegradability. The cultivation of algal biomass on a large scale with its constraints is covered in Theme IV. The thermo physical properties of some algal species are studied with machine learning tools like artificial neural networks (ANN). It also provides insights into the technology developed with scientific efforts for cultivating certain commercially valuable seaweeds in Indian waters. The challenges faced in commercial culturing of microalgae in maritime states in India like temperature, light intensity, culture mixing, pH, salinity, and microbial contamination are major constraints, whereas in seaweeds, a larger quantity of initial seed material, growth cycle variations, grazing and drifting, epiphytism, unawareness, loss of vigor, unfavorable weather conditions are major difficulties highlighted in this part of the book.

Scope and Significance: Microalgae have been an answer for pollution abatement for a long time now. These tiny greens are the most promising source of bioenergy in the search for alternate fuels. Photosynthetic machinery in microalgae is considered efficient and versatile, yielding different products. Microalgal metabolic adaptability enables them to be cultivated in different nutritional modes (autotrophic, heterotrophic, and mixotrophic) with waste as feedstock for the biosynthesis of lipids and other value-added products like pigments, hydrogen, bioelectricity, etc. In fact, they can fulfill the dream of sustainable development.

In this book, the reader will thus learn how to optimize new forms of renewable energy, adding a versatile cap to algal biotechnology. How algae provide high-tech, low-cost, and ecologically responsible solutions to many of our society's current and future requirements is dealt with. Many questions revolving around developing technologies such as synthetic biology, high-throughput phenomics, and the application of Internet of things (IoT) based automation to algae manufacturing technology for enhanced algal biology research are addressed. Ultimately, to enable the formation of an algal-based bio-economy, how algae biofuels have emerged as a clean, environmentally friendly, and cost-effective alternative to traditional fuels in the energy sector, is dealt with. The book has also answered various queries regarding the manufacture of single-cell proteins, pigments, bioactive compounds, medicines, and cosmetics. It sheds light on multiple applications of algae as food and fuel and provides information on various commercially available algae products.

While writing this book, the editors had the opportunity to meet various algologists who have rendered their valuable contributions in the form of chapters. Motivation has been regularly forthcoming from prominent phycologists like Prof Vidyavati, former vice-chancellor, Kakatiya University, Warangal, Telangana, India; Prof Dinabandhu Sahoo, Director, Centre for Himalayan Studies, Univ. of Delhi, and former vice-chancellor, Fakir Mohan University, Odisha; (Late) Prof Ichha Purak Singh, former HOD, Botany Department, Ranchi Women's College, Ranchi, Jharkhand. The editors feel fortunate to have one chapter contributed by (Late) Prof Ichha Purak Singh in this volume.

Theme I

Algae and the Environment

1 A Role of Algae in an Aquatic Ecosystem

P.V. Subba Rao, C. Periyasamy, K. Suresh Kumar, and A. Srinivasa Rao

1.1 INTRODUCTION

An ecosystem results from the integration of all the living and nonliving factors of the environment (Tansley, 1935). Moreover, it is a critical life support system for the planet, delivers essential benefits to humans, and provides stability and resilience after disturbing events through resources and services. It is divided into terrestrial and aquatic ecosystems (Vignieri and Fahrenkamp-Uppenbrink, 2017). The terrestrial ecosystem comprises mainly four components: desert, cropland, forest, and grassland ecosystems, while three major constituents represent aquatic ecosystems *viz.* freshwater, estuarine, and marine ecosystems. Freshwater ecosystems are categorized into lentic ecosystems, with standing waters like ponds, lakes, swamps, and lotic ecosystems having running water like streams, rivers, and springs (Wehr and Sheath, 2015). Estuarine ecosystems represent a transition between fresh and marine waters, influenced by aquatic realms and biologically productive due to special water circulation with plant nutrients stimulating primary production. On the other hand, the marine ecosystem is the largest and is more stable than the freshwater one due to the stability of saline conditions, harboring the most diversified flora and fauna. Important marine ecosystem components are the Coral reef ecosystem, seagrass ecosystem, and seaweed ecosystem (Subba Rao, 2012). Freshwater, terrestrial, and marine (including estuarine) ecosystems represent 0.08%, 28.2%, and 71% of the globe, respectively (Gleick, 1996; Subba Rao, 2012). The uninterrupted water cycles in the aquatic bodies have an immense role in the arrival of minerals and nutrient components from high levels to low-lying areas and eventually to the vast sea and ocean water (Kearns, 2010).

Algae are a highly diverse consortium of primitive, prokaryotic or eukaryotic, photoautotrophic, and thallophytic plants (Lee, 2008). Fossil records of algae date back to three billion years ago, indicating their origin in the prolific Precambrian era (Kolak, 2019). Algae pave the way for the evolution of eukaryotic organisms by producing oxygen. Fossil fuels came from Cretaceous deposits of marine algae. It is an established fact that about 3.5 billion years ago, the origin of blue-green algae paved the way for the beginning of other life-forms by producing oxygen in the earth's atmosphere. Oxygen discharged by algae during photosynthesis accounts for 50% of its total on Earth. They are primary producers in aquatic ecosystems (Chapman, 2013). John (1994) has suggested that there are around 36,000 known species of algae and represent only about 17% of the existing species, and roughly 50% of the global photosynthesis on the plant group is algal derived. Thus every second molecule of oxygen the humans inhale is produced by an alga, and every second molecule of carbon dioxide they exhale is reused by an alga (Melkonian, 1995). Moreover, Dring (1982) has opined that over 90% of the species of marine plants are alga. Algae gave raise to land plants and are directly responsible for all seafood and indirectly responsible for all land food, helping nitrogen fixation to support life on the planet (Chapman, 2013; Ghorbani et al., 2014). Vast numbers of phytoplankton cells are initially food for marine and freshwater webs and expressed as pasturage of the seas. Blue-green algae not only produce oxygen but also play a vital role as primary producers in the food web and also fix atmospheric nitrogen needed for other living organisms (Chapman, 2013). Depending on the magnitude, algae are broadly classified into microscopic and macroscopic forms. Microalgae

DOI: 10.1201/9781003219156-3

are further categorized into two general groups, namely, phytoplankton and periphyton. In lentic and lotic water bodies, phytoplankton live suspended in water column, whereas periphyton live attached to various substrata like rocks, sediment, plant stems, and aquatic organisms. Algal growth (cell numbers) has been affected by season, temperature, amount of sunlight penetrating the water column, amount of available inorganic nutrients, competition from other algae and aquatic plants, and duration of water retention (residence time) in the lake (Simpson, 1991). Benthic algae are the most diverse macrophytes in running water bodies and occur on all possible surfaces along the river and are intimately associated with microbes. In aphotic zones of the rivers, the diatom species occur in large numbers. Macroscopic algae living in seawater are termed as seaweeds. They are generally benthic in nature, but a few like *Sargassum muticum*, *S. fluitans*, and *S. natans* are found to be pelagic (Oyesiku and Egunyomi, 2014). Depending on the dominant pigment in their cells, they vary in color morphology (Imamura et al., 2013; Ghorbani et al., 2014). Marine macroalgae (seaweeds) on the other hand are classified as green algae (Chlorophyta), brown algae (Phaeophyta), and red algae (Rhodophyta) based on the color of pigment (Chapman and Chapman, 1980; Subba Rao et al., 2009, 2018). Algal succession in an ecosystem results due to changes in season, temperature, wind, precipitation patterns, and nutrient cycles (Moore and Thornton, 1988). Abundant algal populations have been observed in spring and early summer when available light and nutrients are high in many lakes (Addy and Green, 1996).

1.2 PRIMARY PRODUCERS

Algae are the source of food and energy base for all living organisms in aquatic ecosystems. They are primary producers known as autotrophs. They harness solar energy, reduce carbon dioxide, and oxidize water molecules to synthesize the cellular carbon, source for entire biota in an aquatic environment. In fact, as of today algae are contributing 50% of the primary produce on the green planet (Falkowski and Knoll, 2007; Field et al., 1998). In general, phytoplankton initiates food chain in aquatic habitats, and they are consumed by zooplankton, fishes, and crustaceans. Algae form the principle basis of food web in the aquatic environment and on the earth as well (Figure 1.1). On the other hand, they liberate oxygen essential for the respiration of heterotrophic organisms through the Chlorophyll—*a* mediated photosynthesis. Periphyton communities are the critical food source for conspicuous benthic invertebrates like snails, caddisflies, crayfish, chironomids, and mayflies in lotic ecosystems (McIntire et al., 1996). In lakes phytoplankton are foodstuff for zooplankton and profundal zoobenthos, but littoral zoobenthos that feed on periphyton are a crucial element of many fish diets (Schindler et al., 1996). Filamentous algae are an important diet item of many minnows in streams (Power et al., 1985). Food webs are principally based on attached algae in very large oligotrophic lakes (James et al., 2000). Fishes in littoral zones ultimately depend on carbon source derived from periphytonas per the evidence obtained from stable isotope analysis (Yoshii, 1999). Vadeboncoeur and Steinman (2002) have recorded an evidence that periphyton is important in small clear lakes, however plays a lesser role in shallow extremely eutrophic lakes, whereas Pardy and Royce (1992) have reported that ascidians of Didenmidae family directly get the fixed carbon from their symbiotic cyanobacteria of the genera *Synechocystis* and *Prochloron*. Certain species of marine algae, benthic diatoms, and phytoplankton in the North Atlantic and Antarctic oceans produce massive quantity of dimethylsulfoniopropionate (DMSP) for the regulation of their internal osmotic processes resulting in the release of dissolved DMSP (DMSPd) into marine environment, which not only serves as dissolved organic carbon to be available for bacteria but also contribute nearly all the sulfur requirement to the microbes in that habitat (Yoch, 2002).

An analytical study by Newell et al. (1995) using stable isotopes has confirmed that benthic microalgae provide a major dietary component for prawns living in tidal creeks, while phytoplankton-based material is the feed for prawns living in off-shore. Chew and Chong (2010) have attributed that phytoplankton-based carbon and energy sources are responsible for the community structure and abundance of copepods in a tropical mangrove estuary. Higher productivity values of

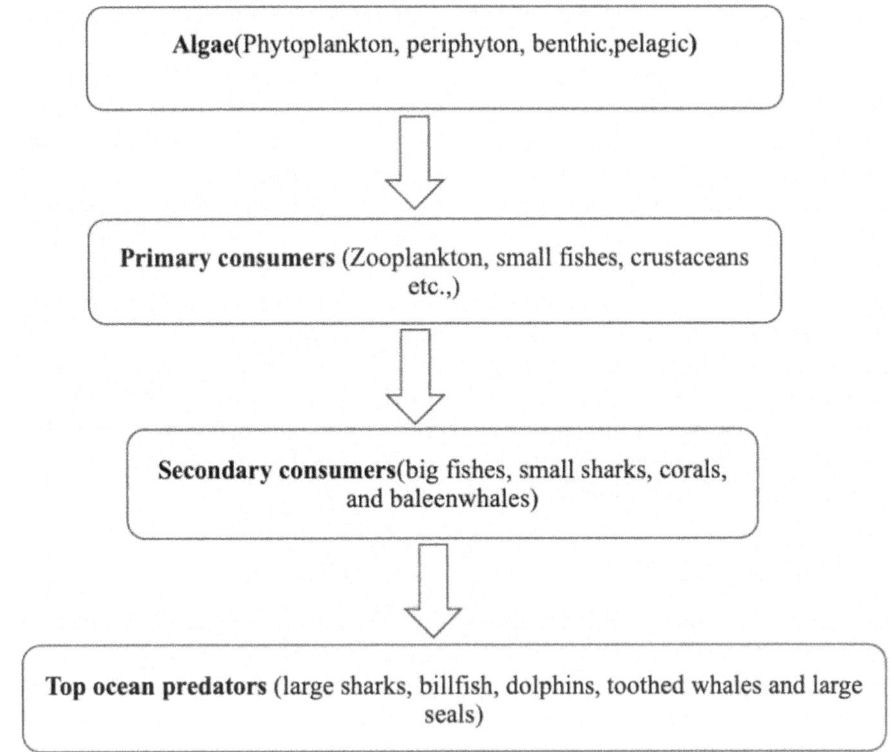

FIGURE 1.1 Algae as primary producer in food web.

phytoplankton such as *Chaetoceros* species, *Chroococcus limneticus*, and *Oscillatoria* species have been reported during wet season, when highest concentration of nutrient levels have been noticed in Mida Creek, Kenya. Abundant supply of plankton-based food has been found to be responsible for larval retention and high productivity in mangrove-lined estuaries at Parangipettai, Southeast coast of India (Rajkumar et al., 2009). Diverse phytoplankton assemblages growing in waters enriched with particulate organic matter (POM) of mangrove estuaries have not only played a significant role in primary productivity of that habitat but also have influenced the fish diversity (Mitra et al., 2004). Viktoria et al. (2011) have observed that the epipsammic communities in the shallow regions have contributed for primary production than the phytoplankton during summer period in Lake Balaton, Hungary, and have reported the importance of increase of the benthic communities with decrease in water level. They have further opined that benthic algae play an important role in the production of organic matter and energy flux in shallow lakes. Khatoon et al. (2010) have stated that phytoplankton having high nutrient content is a valuable food item in the aquatic food chain. Labile Dissolved Organic Carbon (DOC) exuded by benthic marine algae (seaweeds) into the surrounding environment has led to the highest bacterioplankton growth in the reefs of lagoon at Moorea, French Polynesia (Haas et al., 2011). Dittel et al. (2006) have observed a direct trophic relationship between juvenile blue crabs (*Callinectess apidus*) and marine algae (seaweeds) at Delaware Bay in the United States and have reported that the juvenile crabs get refuge from predation and on the other hand consume amphipods that directly graze on macroalgal beds.

1.3 INTERACTIONS AMONG ALGAE

Coccoid cyanobacteria live as endosymbionts in diatoms (*Epithemia* spp. and *Rhopaloidia* spp.) belonging to Rhopalodiaceae family, which can fix nitrogen (Floener and Bothe, 1980), and so those diatoms subsequently has become dominant in nitrogen-limited environment (Peterson and Grimm, 1992). Lucas (1991) has identified three distinct forms of cyanobacteria as symbionts in dinoflagellate genera, namely, *Ornithocercus, Histoneis, Citharistes*, and *Amphisolenia*. He has observed that some dinoflagellates have different provisions outside the cytoplasm proper to hold the symbiotic cyanobacteria. For instance, *Ornithocercus* consists of symbiont externally in the girdle list, whereas *Parahistoneis* has a pocket indentation in the posterior of the singular groove to accommodate the symbiont. In contrast, cyanobacterial symbiont lives in the cytoplasm of *Amphisolenia*. Gordon et al. (1994) have observed the abundance of dinoflagellate genera *Ornithocercus, Histoneis*, and *Citharistes* with cyanobacteria symbionts in the Gulf of Aqaba, Red Sea during autumn when least availability of nitrogenous nutrients in surface waters has been noticed, and the authors have postulated that the host may provide sites for low oxygen concentration for fixing nitrogen by symbiont. Norris (1967) has reported *Synecihocystis consortia* (a cyanobacterium) living symbiotically with the dinoflagellate *Parahistoneis* and a coccoid cyanobacterium as a symbiont within the protoplast of *Dictyocha speculum*, a silico flagellate from Indian Ocean region. Rosenberg and Paerl (1981) have reported a nitrogen-fixing cyanobacteria that live between the utricles of the green macro alga, *Codium* sp. from European and American coastal waters.

1.4 INTERACTIONS BETWEEN ALGAE AND OTHER BIOTA

Croft et al. (2005) have observed a strong mutualistic relation between algae and bacteria and have proved that bacteria have supplied Vitamin B_{12} to algae in exchange for fixed carbon. Kim et al. (2014) have conclusively proved a mutualism between *Rhizobium* sp. and *Chlorella vulgaris* from wastewaters. Cho et al. (2015) have demonstrated that algae supply fixed organic carbon to an artificial consortium of plant growth promoting bacteria (PGPB) or mutualistic bacteria and in turn PGPB supply dissolved inorganic carbon and low molecular organic carbon for algal consumption. An endophytic *Rhodopseudomonas* species isolated from *Caulerpa taxifolia* has been found to be a potential nitrogen fixer (Chisholm et al., 1996). Delbridge et al. (2004) have identified photosynthetic Alphaproteobacteria as endosymbionts in four species of *Caulerpa*. Hollants et al. (2011) have recognized mutual association between bacteria of Flavobacteriaceae family and *Bryopsis* from the Mexican west coast using fluorescence in situ hybridization (FISH) technique. Kazamia et al. (2012) have established experimentally a facultative relation between *Chlamydomonas reinhardtii* and heterotrophic bacteria which delivers Vitamin B_{12}. Algae from estuaries and marine ecosystems produce dimethylsulfoniopropionate (DMSP), a methylated product of this tertiary sulfonium compound that serves as the main source of sulfur for marine bacteria and possibly also for certain eukaryotic organisms (Yoch, 2002).

Schenk (1992) working on submarine as colichens has identified the cyanobacterium *Chroococcus* sp. with *Halographis runica*, the cyanobacterium *Hyellacaespitosa* with *Arthropyrenia halodytes*, and the cyanobacterium *Hyellacaespitosa* with *Arthorpyrenia halodytes*. Twelve different species of cyanobacteria belong to the genera *Aphanocapsa, Synechocystis, Oscillatoria*, and *Phormidium* and have been reported in sponges, and most of them occur extracellularly; however *Aphanocapsa feldmanni* occurs within the specialized amoeboid sponge cells cyanocytes (Adams, 2000). Biomass of cyanobacteria in Indo-Pacific sponges has been recorded as equal to their host species and the sponges derive 50% of their metabolic requirements from the symbionts (Wilkinson, 1983).

Chaves et al. (2013) have proposed that fishes belong to Labridae, Pomacentridae, Acanthuridae, and Haemulidae families get their feed from epiphytic invertebrates associated with high canopy of *Sargassum polyceratium, Dictyopteris delicatula*, and *Canistrocarpus cervicornis*. Tropical seaweed beds have been reported to be important habitats and feeding ground for juvenile fishes that belong to Labridae and Serranidae families (Tano et al., 2017). Horn (1989) has reported that macroalgae are of trivial status as a straight food source for herbivorous fishes, whereas Lewis (1987) has opined that macroalgae provide shelter and lucrative hunting sites for predation of the associated fauna for invertebrate feeders and omnivores. Moreover Carr (1994) has observed a linear relation between recruitment of kelp bass (*Paralabraxclathratus*) and the local abundance of kelp structure (i.e., the number and biomass of overlapping *Macrocystis* blades).

Larval midges (*Cricotopus* sp.) colonize the cyanobacteria *Nostoc parmelioides* (Dodds and Marra, 1989) and *Nostoc verrucosum* (Sabater and Munoz, 2000) and graze the colony from the inside. *N. parmelioides* provides food and protection to the midge, and in reciprocation, midge change cyanobacterium from a spherical to an earlike form, which enhances per biomass exposure to photon and nutrient flux (Brock, 1960; Ward et al., 1985). Zooxanthellae live as symbionts with members of Protozoa, Porifera, Cnidaria, Platyhelminthes, and Mollusca (Trench, 1993). Hard and soft corals get energy-rich food from zooxanthellae, and in turn they provide inorganic carbon and other nutrients to its symbiont (Muscatine, 1990). *Solenicola setigera* (heterotrophic flagellated protozoan) that lives on the frustules of the chain-forming diatom *Leptocylindrus mediterraneus* in oligotrophic open-ocean waters consists of unicellular cyanobacteria of *Synechococcus*as symbiont (Buck and Bentham, 1998). In this case, it is presumed that protozoan feeds on cell contents of the diatom while cyanobacterium fixes the nitrogen. Van Bergeijk and Stal (2001) have reported a symbiotic relationship between a benthic flat worm, *Convolutaros coffensis*, that lives in intertidal mudflats and a green unicellular alga, *Tetraselmis* sp.

Littler and Littler (1984) based on relative dominance model (RDM) have hypothesized that relative abundances of corals, crustose coralline algae, microalgalturfs, and frondose macroalgae on coral reefs often depend on complex interactions of environmental factors (bottom-up controls such as nutrient levels) and biological factors (top-down controls such as grazing). Farming intensity and character of the substratum in *Eucheuma* cultivation have shown an effect on associated fish fauna in terms of abundance, species richness, trophic identity, and fish community composition (Bergmann et al., 2001). Hauxwell et al. (2001) have identified the disappearance of *Zostera marina* (eelgrass) with increase in anthropogenic nitrogen load to estuaries resulting in an increase in macroalgal biomass reaching about 9–12 cm critical canopy height

1.5 ALGAL BLOOMS: CAUSES AND ERADICATION

Solid masses of phytoplankton cells of one or more species in a waterbody are generally referred to as algal blooms (Reynolds and Walsby, 1975). In other words, an outburst of phytoplankton cells well above the usual quantity for a given region or water body is termed as an algal bloom. Less than 5% of phytoplankton species used to form algal blooms in lotic, lentic, and marine waters (Assmy and Smetacek, 2009). Phytoplankton are often seen when massive bloom like red tides and other harmful algal blooms used to appear changing the color of the ocean for miles (Paerl, 1988). Moreover, blue-green algae also form blooms that color water often reddish due to red phycoerythrin pigment, which masks blue-green phycocyanin pigment. Blue-green algal blooms could develop during the spring when higher water temperature with increased light is noticed, and they generally do not persist in temperate regions through winter months due to low water temperatures suggesting that higher water temperatures observed in tropical regions could be the cause to persist throughout the year. In summer though enough sunlight is available, the amount of phosphorus in the lake often controls the abundance of algae, and so phosphorus is considered the limiting nutrient in most freshwater bodies. However, adaptation of exact environmental conditions like nutrient ratios or light conditions for bloom formation has not been established yet as they are often formed and usually dispersed throughout their array (Paerl et al., 2018; Griffith and Gobler, 2020). Bloom-forming species are diverse in shape, size, and behavior due to their unlike phylogenetic clusters, pigment composition, and biochemistry. These algal blooms have very dissimilar impacts on the ecosystem and biogeochemistry as their appearance or reappearance differs. In the same water body, a species may bloom every year as an integral part of its seasonal cycle. Conversely, another species forms bloom only in some years but always in the same season. On the other hand, a species may form blooms in a sporadic way. Therefore, it is not possible to make a general statement about bloom-forming algae. The physical factors accountable for algal blooms are depth of the surface mixed layer (SML) of a water body, wind force and fetch, halocline and pycnocline, and column of euphotic zone. The depth and vertical mixing of SML are of critical importance for bloom formation, whereas the rate of bloom growth depends on the relative depth of the euphotic zone and the SML (Assmy and Smetacek, 2009). Major chemical-essential elements responsible for algal blooms are nitrogen, phosphorus, and iron. Pathogens and grazers are the two important biological factors that determine the bloom. *Microcystis, Synechococcus, Nodularia, Aphanizomenon*, and *Anabaena* are the major bloom-forming cyanobacteria and pose a severe problem in reservoirs and lakes (Paerl et al., 2001). *Botryococcus* chlophycean algae are highly buoyant surface dwelling and accumulate as scums in freshwater bodies (Nakamoto, 1975). Oligotrophic lake viz Lake Taupo in New Zealand has blooms with species of *Sphaerocestis, Dictosphaerium, Scenedesmus*, and *Chlorococcus* (Paerl et al., 2001). *Trichodesmium* blue-greenalgal blooms are regularly encountered in the seas and oceans (Sarangi, Chauhan, and Nayak, 2005; Sarangi, Chauhan, Nayak, and Shreedhar, 2005). Species of *Skeletonema, Thalassiosira*, and *Chaetoceros* are the recurrent bloom-forming diatoms in coastal waters (Degerlund and Eilertsen, 2010).

Algal blooms are a natural phenomenon throughout the coastal region of Indian seas (Devassy and Nair, 1987), and on the west coast of India, they are caused by blue-green alga *Trichodesmium* sp. (Sarangi, Chauhan, and Nayak, 2005; Sarangi, Chauhan, Nayak, and Shreedhar, 2005) and the occurrence of the diatom *Hemidiscus hardmannianus* in bloom conditions has also been reported (Subramaniam and Purushothaman, 1985). Most estuaries in India are moderately to highly eutrophic with freshwater discharge during the spring season enhancing the nutrient load from the drainage freshwater and thereby intensifying the phytoplankton bloom (Admiraal et al., 1990). *Coscinodiscus radiata*, a centric diatom, created a severe bloom at a confined region of Sagar Island, the western part of Indian Sundarbans, along with two other diatoms *viz. Thalassiothrix frauenfeldii*, which is common in both temperate and tropical regions (Sahu et al., 2012), and *Chaetoceros lorenzianus*, a chain-forming diatom encountered in warm, temperate waters. Complete discoloration (deep greenish color) of the water appeared where *C. radiatus* was predominant, and the bloom was categorized as a periodic event (Takahashi et al., 1977).

1.6 HARMFUL ALGAL BLOOMS AND THEIR TOXINS

Harmful algal blooms are formed due to abnormal, unnatural, and excessive growth of algae (nuisance algal blooms) interfering with the utilization of aquatic resources and creating adverse impact in an aquatic environment where dead algae decomposes and decreases dissolved oxygen causing anoxia, increases pH level affecting the existence of other organisms by releasing toxins. Diatoms and dinoflagellates are abundant causing HABs and can exhibit bio luminescence playing a role of symbionts in corals. Elevated nutrient loading due to application of fertilizers (anthrapogenitic activity) has been advocated as primary reason for increasing HBAs, discolor the water, float on surface in scums, cover beaches with biomass, or exudates (foam) and deplete oxygen levels through excessive respiration or decomposition. Further some of these HBAs through their synthesized toxins could alter cellular processes from plankton to humans and as well could cause mortalities, loss of coastal resources, submerged aquatic vegetation, and benthic epi- and in-fauna (Sellner et al., 2003; Paerl et al., 2001).

These blooms prior to becoming harmful ought to be eradicated for proper functioning of the ecosystem. Mechanical removal is one aspect, and the other one is chemical treatment, that is, use of herbicides like copper sulphate mostly is found to be toxic or detrimental. Calcium compounds are sometimes added to bind phosphorus present in the water. Biological control is another option introducing predators to feed on blooms. The best way to control algal blooms is limiting the growth of blooms before reaching the excessive growth by controlling the nutrients entering into the system, besides removing manually. Moreover, aeration especially in freshwater ecosystem is helpful as oxygen inactivates phosphorus to reduce the effect of algal blooms (Ghorbani et al., 2014). HBAs have direct impact on human health and negative influences on human well-being mainly through consequences to coastal ecosystem services—fisheries, tourism and recreation, and other marine organisms and environment (Berdalet et al., 2016).

Blooms in fresh marine and brackish waters constitute largely cyanobacteria producing multiple toxins like liver, nerve, and skin toxins affecting human and animal health. Most common toxin is microcystin, a group of liver toxins causing intestinal illness in humans, mortality in pets, livestock, and wildlife. Dinoflagellates and diatoms (phytoplankton) are common in marine and brackish waters, including estuaries discoloring water into red/blue shades and a few bioluminescent. Some notorious freshwater blue-green algal HBAs are toxic, surface dwelling, and scum forming, and they include genera like *Anabena, Aphanizomenon, Lyngbya, Nodularia, Microcystis, Oscillatoria* (Reynolds and Walsby, 1975; Paerl, 1988).

Control and management of cyanobacterial and another phytoplankton blooms invariably include nutrient input constraints like nitrogen and phosphorus The types and amount of nutrient input constraints depend on hydrologic, climatic, geographic and geological factors interacting with anthropogenic and natural nutrient regimes. Single nutrient input constraint may be effective in some waterbodies and dual N and P input reductions are completely required for effective long term control and management of HBAs (Paerl et al., 2018; Berdalet et al., 2016).

1.7 EUTROPHICATION

Eutrophication is an ecological process where waters are increasingly enriched with organic matter. Hypertrophic water bodies with excessive enrichment of nutrients (phosphorus and nitrogen) due to anthropogenic activities are termed as eutrophication and are identified as the main cause of impaired surface water quality (Schindler, 1977). Eutrophication restricts use of water for fisheries, recreation industry, and drinking due to growth of undesirable algae, mainly blue-green algae and also causes oxygen depletion by their death and decomposition (Nasir Khan and Mohammad, 2014). Excess nutrients come from two types of sources *viz.*, (1) point sources like sewage and industrial discharges and (2) non-point sources from excessive application of fertilizer in agriculture, construction sites, and urban areas. Non-point sources replaced point sources to be the cause of eutrophication in many regions (Carpenter et al., 1998). Eutrophication in lakes and rivers involves algal blooms and fish kills and is accelerated by nutrient pollution of phosphorus and nitrogen from stormwater and fossil fuel combustion besides sewage and agriculture. Increase in population and intensification of land use have accelerated eutrophication of water bodies from the Great Lakes of North America to Lake Tai of China to Lake Victoria in Africa (Kleinman and Sharpley, 2012). All blooms cause deoxygenation and water toxicity disrupting normal ecosystem functioning (Sharabian et al., 2018). Anthropogenic (agriculture, urban, and industrial) activities have increased aquatic nitrogen and phosphorus pollution leading to eutrophication, thereby threatening water quality and biotic integrity from up streams to coastal areas worldwide. Eutrophication reduces fish and shellfish production, creates harmful algal blooms threatening safety of drinking water and adequate food supplies, as well as stimulation of greenhouse gases. Comprehensive strategies are required to curb eutrophication. A variety of watershed programs have yielded success, but they are outnumbered by the ever-expanding number of lakes and rivers undergoing eutrophication (Duncan et al., 2012; Schindler, 2006).

1.8 PHYCOREMEDIATION

Microalgae are the suitable and potential candidates for the development of innovative mass production with inexpensive growth requirements (solar, light, and CO_2) and advantage of being utilized simultaneously for multiple technologies like carbon mitigation, biofuel production, and bioremediation (Suresh Kumar et al., 2015). Phycoremediation is potential tool employing algae viz microalgae, macroalgae, and cyanobacteria to treat wastes or wastewaters resulting in clean water and also production of useful biomass to be used as food feed fertilizer pharmaceuticals and biofuel. Algae also remove carbon dioxide from the wastewaters or polluted waters through photosynthesis and thus acting as carbon-reducing agents, removing toxicants, xenobiotics, and heavy metals. It is eco-friendly and solar-driven technology as well as cost-effective over conventional treatment processes, which require chemical load and energy inputs (Brar et al., 2017; Olguin and Sanchez-Galvan, 2012). On the other hand, bioremediation is a process exploring the potential of microorganisms to clean up contaminated sites, namely, wastewater, ground or surface water, soils, sediments, and air in environment (Boopathy, 2000). A heavy metal (HM) is a metal or metalloid element causing environmental pollution without having any vital function, and on the other hand, it is toxic at low concentrations (e.g., Pb and Hg), and when it has vital function, it is harmful to organisms at higher concentrations (e.g., Cu and Mo) (Herrera-Estrella and Guevara-Garcia, 2009). Some of the heavy metals are micronutrients (e.g., Zn, Cu, Mn, Ni, and Co), and others have unknown biological function and are toxic (e.g., Cd, Pb, and Hg) (Gaur and Adholeya, 2004). Most common heavy metal contaminants are Cd, Cr, Hg, Pb, and Zn. In fact, HMs removal from environment is a challenging task as they are indestructible as well as cannot be biologically or chemically degraded. HMs discharged into the water bodies through wastes pose a threat, besides having great impact on aquatic ecosystem destroying its self-purification ability (Khan et al., 2008). Algal species may bind up to 10% of their biomass as metals. This process involves adsorption of metals on the cell surface and independent of the cell (Ahmad et al., 2020). Hanumantha Rao et al. (2011) have employed a green alga *Chlorella vulgaris* to treat the solid waste generated from the leather processing industry and has reported a significant reduction in calcium (63%) and magnesium (50%) along with reductions of free ammonia, nitrite, BOD, and COD by 80%, 89%, 22%, and 38% respectively, besides removing considerable quantity (14%) of total dissolved solids. *Chlorococcum humicola*, a green alga, has shown a reduction of sludge, BOD, total suspended solids, and TDS by 47.75%, 93.20%, 60.83%, and 80.79% respectively, when employed in an oil-drilling effluent plant (Sivasubramanian and Muthukumaran, 2012). In a laboratory study, a brown color form of *Kappaphycus alvarezii*, a red alga, has been found to be an effective bio-sorbent to remove heavy metals like cadmium, cobalt, and chromium by 3.064 mg.g FW^{-1}; 3.365 mg.g FW^{-1}, and 2.799 mg.g FW^{-1} respectively (Suresh Kumar et al., 2007).

1.9 BIOMASS PRODUCTION

Aquatic ecosystem is a habitat for primary producers *viz*. phytoplankton, microalgae, and macroalgae, as well as for their biomass production. Large-scale production of *Spirulina* mostly for food purpose for human beings is being cultivated in freshwater bodies viz ponds and race ways with a total annual production of 860 tons (dry) in six countries viz., Mexico, USA, Thailand, Taiwan, Japan, and Hawaii (Vonshak and Richmond, 1988). Nowadays marine algae viz. *Laminaria japonica* and *Undaria pinnatifida*—brown algae; *Porphyra, Eucheuma, Kappaphycus*, and *Gracilaria*—red algae; *Monostroma* and *Enteromorpha*—green algae are intensively cultivated in different countries to meet the industrial demand, besides two microalgae—*Dunaliella salina* and *Spirulina platensis* (Wikfors and Ohno, 2001). Global production of marine algae (seaweeds) through aquaculture has been arrived at 11.66, 16.83, and 19.9 million tons fresh during 2002, 2008, and 2010 respectively (FAO, 2012; Paul and Tseng, 2012), whereas the same has been estimated 23.78 million tons fresh in 2012 (FAO, 2014). The top 10 countries producing seaweeds through aquaculture in the world are China, Korea, Japan, Philippines, Indonesia, Chile, Taiwan, Vietnam, Russia, and Italy (Bixler and Porse, 2011). On the other hand in 2016 global production of seaweeds has been estimated at 30.1 million wet tons with 95% from culture and the remaining 5% from natural beds with the top seven countries that account for global production are China, Japan, Korea, Indonesia, Philippines, Malaysia, and Vietnam (FAO, 2018). Marine red alga, *Kappaphycus alvarezii*, has been cultivated in Indian coastal waters for industrial chemicals, carrageenan, semi-refined carrageenan, and liquid fertilizer since 2008 (Subba Rao and Periyasamy, 2019), and in 2010 its production from aquaculture in the world has been reported as 1, 83, 000 tons (dry) (Bixler and Porse, 2011), while the same is found to be 1,490 tons fresh in India (Krishnan and Narayankumar, 2013).

1.10 CONCLUSION

Aquatic ecosystems (fresh, estuarine, and marine) harbor different algae and phytoplankton besides various fauna. Algae along with phytoplankton in these ecosystems are the primary producers. Algal blooms are formed when nutrients' availability is in excess of the requirement for growth. Eutrophication of waters results due to various contaminants from sewage, industrial effluents, and excess use of fertilizer in agriculture. Further, it envisages algal blooms, including Harmful Algal Blooms (HABs). HBAs produce toxins and kill fish, affect recreation and human health. Phycoremediation employing algae is the potential tool to remove excess nutrients as well as toxic heavy metals from eutrophicated waters which ultimately controls the formation of Algal blooms or HBAs. Remedial measures have been mentioned in this review to contain formation of HBAs and to avoid eutrophication of water bodies. Aquatic ecosystem is habitat for large-scale production of biomass to be used for food, fodder, fertilizer, for production chemicals, pharmaceuticals, and biofuel.

Acknowledgment: This review article is dedicated to the late Prof. V. Krishnamurthy, a walking encyclopedia of algae during his time, and there is no one parallel to him even as of today, and he is also the Father of Seaweed Cultivation in India. He was instrumental for training many algologists in Tamil Nadu as well as in CSIR Central Salt and Marine Chemicals Research Institute, Bhavnagar, Gujarat. During his tenure as head of the Division of Marine Algae, his vision of establishing Marine Algal Research Station, Mandapam, has been a living memory forever.

REFERENCES

Adams, D. G. 2000. Symbiotic Interactions. In: B. A. Whitton and M. Potts (Eds.), *The Ecology of Cyanobacteria*. Kluwer Academic Publishers, Dordrecht, pp. 523–561.

Addy, K. and L. Green. 1996. *Phosphorus and Lake Aging*. University of Phode Island: Natural Resources Facts; South Kingstown Report No 96–2.1p.

Admiraal, W., P. Breugem, D. M. L. H. A Jacobs and E. D. De R. V. Steveninck. 1990. Fixation of dissolved silicate and sedimentation of biogenic silicate in the lower river Rhine during diatom blooms. *Biogeochemistry* 9: 175–185.

Ahmad, S., A. Pandey, V. V. Pathak, V. V. Tyagi and R. Kothari. 2020. Phycoremediation: Algae as eco-friendly tools for removal of heavy metals. In: R. N. Bharagava and G. SAxena (Eds.), *Bioremediation of Industrial Waste for Environmental Safety*. https://doi.org/10.1007/978.981-13-3428-9-3.

Assmy, P. and V. Smetacek. 2009. Algal blooms. In: M. Schaechter (Ed.), *Encyclopedia of Microbiology*. Elsevier, Oxford, pp. 27–41.

Berdalet, E., L. E. Fleming, R. Growen, K. Devidson, P. Hess, L. C. Baker, S. K. Moore, P. Hoagland and H. Enevoldsen. 2016. Marine harmful algal blooms, human health and wellbeing: Challenges and Opportunities in 21st century. *Journal of Marine Biological Association of United Kingdom* 96(1): 61–91.

Bergman, K. C., S. Svensson and M. C. Ohman. 2001. Influence of algal farming on fish assemblages. *Marine Pollution Bulletin* 42: 1379–1389.

Bixler, H. J. and H. Porse. 2011. A decade of change in the seaweed hydrocolloids industry. *Journal of Applied Phycology* 23: 321–335.

Boopathy, R. 2000. Factors limiting bioremediation technologies. *Bioresource Technology* 74. 63–67.

Brar, A., M. Kumar, V. Vivekanad and N. Pareek. 2017. Photoautotrophic microorganisms and bioremediation of industrial effluents: Current status and Future prospects. *Biotech* 3: 7–18. DOI: 10.1007/s 13205–017–0600.5.

Brock, E. M. 1960. Mutualism between the Midge Cricotopus and the Alga Nostoc. *Ecology, Ecological Society of America* 41(3): 474–483.

Buck, K. and W. N. Bentham. 1998. A novel symbiosis between a cyanobacterium, *Synechococcus* sp., an aplastidicprotist, *Solenicola setigera*, and a diatom, *Leptocylindrus mediterraneus*, in the open ocean. *Marine Biology* 132: 349–355.

Carpenter S. R., N. F. Caraco, D. L. Correll, R. W. Howarth, A. N. Sharpley and V. H. Smith. 1998. Nonpoint pollution of surface waters with phosphorus and nitrogen. *Ecological Applications* 8(3): 559–568.

Carr, M. H. 1994. Effects of macroalgal dynamics on recruitment of a temperate reef fish. *Ecology* 75: 1320–1333.

Chapman, R. 2013. Algae: The world's most important "plants"—an introduction. *Mitigation and Adaptation Strategies for Global Change* 18: 5–12.

Chapman, V. J. and D. J. Chapman. 1980. *Seaweeds and Their Uses*. Chapman and Hall. London.

Chaves LT C., P H C Pereira and J L L Feitosa. 2013. Coral reef fish association with macroalgal beds on a tropical reef system in North-eastern Brazil. *Marine and Freshwater Research 64*(12): 1101–1111.

Chew, L. L. and V. C. Chong. 2010. Copepod community structure and abundance in a tropical mangrove estuary, with comparisons to coastal waters. *Hydro* 666: 127–143.

Chisholm, J. R. M., C. Dauga, E. Ageron, P. A. D. Grimont and J. M. Jaubert. 1996. 'Roots'inmixotrophic algae. *Nature* 381: 382–382.

Cho, D. H., R. Ramanan, J. Heo, J. Lee, B. H. Kim, H. M. Oh and H. S. Kim. 2015. Enhancing microalgal biomass productivity by engineering a microalgal–bacterial community. *Bioresource Technology* 175: 578–585.

Croft M. T., A. D. Lawrence, E. Raux-Deery, M. J. Warren and A. G. Smith. 2005. Algae acquire vitamin B12 through a symbiotic relationship with bacteria. *Nature* 438(7064): 90–93.

Degerlund, M. and H. C. Eilertsen. 2010. Main species charictriristics of phytoplankton spring blooms in NE Atlantic and Arctic waters (68–80 ^{0} N). *Estuaries and Coasts* 33(2): 242–269.

Delbridge L., J. Coulburn, W. Fagerberg and L. S. Tisa. 2004. Community profiles of bacterial endosymbionts in four species of *Caulerpa*. Symbiosis 37: 335–344.

Devassy, V. P. and S. R. Nair. 1987. Discoloration of water and its effect on fisheries along the Goa coast. Mahasagar Bulletin. *National Institute of Oceanography* 20: 121–128.

Dittel, A. I, C. E. Epifanioand and M. L. Fogel. 2006. Trophic relationships of juvenile blue crabs (*Callinectes sapidus*) in estuarine habitats. *Hydrobiologia* 568: 379–390.

Dodds, W. K. and J. L. Marra. 1989. Behaviors of the midge, *Cricotopus* (Diptera, Chironomidae) related to mutualism with *Nostoc parmelioides* (Cyanobacteria). *Aquatic Insects* 11: 201–208.

Dring, M. J. 1982. *The Biology of Marine Plants*. Edward Arnold, London.

Duncan, E., P. J. A. Kleinman and A. N. Sharpley. 2012 Eutrophication of lakes and rivers. In: *eLS*. John Wiley & Sons Ltd, Chichester. www.els.net doi:10.1002/9780470015902.a0003249.pub2.

Falkowski, P. G. and A. H. Knoll. 2007. The evolutionary transition from anoxygenic to oxygenic photosynthesis. In: P. G. Falkowski and A. H. Knoll (Eds.), *Evolution of Primary Producers in the Sea*. Elsevier, Boston.

FAO. 2012. The state of world fisheries and aquaculture. ISBN: 978-92-5-107225-7.

FAO. 2014. Fish statj—software for fishery statistical time series. www.fao.org/fishery/statistics/software/fishstatj/en.

FAO. 2018. *The State of World Fisheries and Aquaculture-Meeting the Sustainable Development Goals*. FAO, Rome, Italy.

Field, C. B., M. J. Behrenfeld, J. T. Randerson and P. Falkowski. 1998. Primary production of the biosphere: Integrating terrestrial and oceanic components. *Science* 281: 237–240.

Floener, L. and H. Bothe. 1980. Nitrogen fixation in *Rhopalodiagibba*, a diatom containing blue-greenish inclusions symbiotically. In: W. Schwemmler and H. E. A. Schenk (Eds.), *Endocytobiology, Endosymbiosis, and Cell Biology*. Walter de Gruyter and Co., Berlin, Germany, pp. 541–552.

Gaur, A. and A. Adholeya. 2004. Prospects of arbuscular mycorrhizal fungiin phytoremediation of heavy metal contaminated soils. *Current Science* 86(4): 528–534.

Ghorbani, M., S. A. Mirbagheri, A. H. Hassani, J. Nouri and S. M. Monavari. 2014. Algal bloom in aquatic ecosystems—an overview. *Current World Environment* 9(1): 105–108.

Gleick, P. H. M. 1996. Basic water requirements for human activities: Meeting basic needs. *Water International* 21: 83–92.

Gordon, N., D. L. Angel, A. Neori, N. Kress and B. Kimor. 1994. Heterotrophic dinoflagellates with symbiotic cyanobacteria and nitrogen limitation in the Gulf of Aqaba. *Marine Ecology Progress Series* 107: 83–88.

Griffith, A. W. and C. J. Gobler. 2020. Harmful algal blooms: A climate change co-streessor in marine and freshwater ecosystems. *Harmful Algae* 91: 101590.

Haas, A. F., C. E. Nelson, L. W. Kelly, C. A. Carlson, F. Rohwer, J. J. Leichter, A. Wyatt and J. E. Smith. 2011. Effects of coral reef benthic primary producers on dissolved organic carbon and microbial activity. *PLoS One* 6(11): e27973. DOI:10.1371/journal.pone.0027973.

Hanumantha Rao, P., R. Ranjith Kumar, B. Raghavan, V. Subramanian and V. Sivasubramanian. 2011. Application of phycoremediation technology in the treatment of wastewater from a leather-processing chemical manufacturing facility. *Water SA* 37(1): 7–14. https://doi.org/10.4314/wsa.v37i1.64099.

Hauxwell, J., J. Cebrián, C. Furlong and I. Valiela. 2001. Macroalgal canopies contribute to eelgrass (*Zostera marina*) decline in temperate estuarine ecosystems. *Ecology* 82(4): 1007–1022.

Herrera-Estrella, L. R. and A. A. Guevara-Garcia. 2009. Heavy metal adaptation. In: *eLS Encyclopaedia of Life Sciences*. John Wily & Sons Ltd, Chichester, pp. 1–9. http://doi.org/10.1002/9780470015902.a0001318.pub2.

Hollants J, O. Leroux, F. Leliaert, H. Decleyre, O. De Clerck and A. Willems. 2011. Who is in there? Exploration of endophytic bacteria within the siphonous green seaweed Bryopsis (Bryopsidales, Chlorophyta). *PLoS One* 6(10): e26458.

Horn, M. H. 1989. Biology of marine herbivorous fishes. *Oceanography and Marine Biology: An Annual Review* 27: 167–272.

Imamura, S., A. Ishiwata, S. Watanabe, H. Yoshikawa and K. Tanaka. 2013. Expression of budding yeast FKBP12 confers rapamycin susceptibility to the unicellular red alga *Cyanidioschyzon merolae*. *Biochem Biophys Res. Commun* 439(2): 264–269.

James, M. R., I. Hawes, M. Wheatherfield, C. Stanger and M. Gibbs. 2000. Carbon flow in the littoral food web of an oligotrophic lake. *Hydrobiologia* 441: 93–106.

John, D. M. 1994. Biodiversity and conservation: An algal perspective. *The Phycologist* 38: 3–15.

Kazamia, E., H. Czesnick, T. T. V. Nguyen, M. T. Croft, E. Sherwood, S. Sasso, S. J. Hodson, M. J. Warren and A. G. Smith. 2012. Mutualistic interactions between vitamin B_{12}-dependent algae and heterotrophic bacteria exhibit regulation. *Environmental Microbiology* 14(6): 1466–1476.

Kearns, D. B. 2010. A field guide to bacterial swarming motility. *Nature Reviews Microbiology* 8: 634–644.

Khan, M. A., R. A. K. Rao and M. Ajmal. 2008. Heavy metal pollution and its control through nonconventional adsorbents (1998–2007). *International Environmental Application and Science* 3(2): 101–141.

Khatoon, H., S. Banarjee, F. M. Yusoff and M. Shariff. 2010. Effects of salinity on the growth and proximate composition of selected tropical marine periphytic diatoms and cyanobacteria. *Aqua Res* 41: 1348–1355.

Kim, B. H., R. Ramanan, D. H. Cho, H. M. Oh and H. S. Kim. 2014. Role of Rhizobium, a plant growth promoting bacterium in enhancing algal biomass through mutualistic interaction. *Biomass and Bioenergy* 69: 95–105.

Kleinman, P. J. A. and A. N. Sharpley. 2012. Eutrophication of lakes and rivers. http://doi.org/10.1002/9780470015902.a0003249.pub2

Kolak, M. 2019. Ecological importance of algae. https://sciencing.com/ecological-importance-algae-8655847.html

Krishnan, M., S. and R. Narayanakumar. 2013. Social and economic dimensions of carrageenan seaweed farming: In India. In: D. Valderram, J. Cai, N. Hishamunda and N. Ridler (Eds.), *Social and Economic Dimensions of Carrageenan Seaweed Farming*. Fisheries and Aquaculture technical paper No. 580. FAO, Rome, pp. 163–184.

Lee, R. E. 2008. *Phycology*. 4th edition. Cambridge University Press, Cambridge, p. 561. ISBN-13-978-0-511-38669-5, eBook (EBL)

Lewis, F. G. 1987. Crustacean epifauna of seagrass and macroalgae in Apalachee Bay, Florida, USA. *Marine Biology* 94: 219–229.

Littler, M. M. and D. S. Littler. 1984. Models of tropical reef biogenesis: The contribution of algae. In: F. E. Round and D. J. Chapman (Eds.), *Progress in Phycological Research*, Vol. 3. Biopress, Bristol and London, pp. 323–364.

Lucas, I. A. N. 1991. Symbionts of the tropical Dinophysiales (Dinophyceae). *Ophelia* 33: 213–224.

McIntire, C. D., S. V. Gregory, A. D. Steinman and G. A. Lamberti. 1996. Modeling benthic algal communities: An example from stream ecology. In: R. J. Stevenson, M. L. Bothwell and R. Lowe (Eds.), *Algal Ecology: Freshwater Benthic Ecosystems*. Academic Press, San Diego, pp. 669–704.

Melkonian, M. 1995. Introduction: XI—XIII. In: W. Wiessner, E. Schnepf and R. C. Starr (Eds.), *Algae, Environment and Human Affairs*. Biopress Ltd., Bristol, p. 258.

Mitra A., K. Banarjee and A. Gangopadhyay. 2004. *Introduction to Marine Plankton*. Daya Publishing House, New Delhi.

Moore, L. and Thornton, K. W. 1988. *Lake and Reservoir Restoration Guidance Manual*. United States Environmental Protection Agency EPA-440/5–88–002, Washington, DC.

Muscatine, L. 1990. The role of symbiotic algae in carbon and energy flux in coral reefs. In: Z. Dubinsky (Ed.), *Ecosystems of the World, Vol. 25, Coral Reefs*. Elsevier, Amsterdam, pp. 75–87.

Nakamoto, N. 1975. A freshwater red tide on a water reservoir. *Japanese Journal of Limnology* 36: 55–64.

Nasir Khan, M. and F. Mohammad. 2014. Eutrophication: Challenges and solutions. In: A. A Ansari and S. S. Gill (Eds.), *Eutrophication: Causes, Consequences and Control*. Spriger Science, pp. 1–15. DOI:10.1007/978-94-007-7814-6_1.

Newell, R. I. E., N. Marshall, A. Sasekumar and V. C. Chong. 1995. Relative importance of benthic microalgae, phytoplankton, and mangroves as sources of nutrition for penaeid prawns and other coastal invertebrates from Malaysia. *Marine Biology* 123: 595–606.

Norris, R. E. 1967. Algal consortisms in marine plankton. In: V. Krishnamurty (Ed.), *Proceedings of the Seminar on Sea, Salt and Plants, Central Salt and Marine Chemicals*. Research Institute, Bhavnagar (India), pp. 178–189.

Olguin, E. J. and G. Sanchez-Gauan. 2012. Heavy metal removal in phytofiltration and phycoremediation, the to differentiate bioadsorption and bioaccumulation. *New Biotechnology* 30: 3–8.

Oyesiku, O. and A. Egunyomi. 2014. Identification and chemical studies of pelagic masses of *Sargassumnatans* (Linnaeus) Gaillon and *S. fluitans* (Borgessen) Borgesen (brown algae), found offshore in Ondo State, Nigeria. *African Journal of Biotechnology* 13: 1188–1193.

Paerl, H. W. 1988. Nuisance phytoplankton blooms in coastal, estuarine and inland waters. *Limnology and Oceanography* 33: 823–847.

Paerl, H. W., R. S. Fulton, P. H. Moisander and J. Dyble. 2001. Harmful freshwater algal blooms, with an emphasis on cyanobacteria. *The Scientific World* 1: 76–113. ISSN 1532–2246. DOI:10.1100/tsw.2001.16.

Paerl, H. W., T. G. Otten and R. Kudela. 2018. Mitigating the expansion of harmful algal blooms across the freshwater-to-marine continuum. *Environmental Science and Technology* 52: 5519–5529.

Pardy, R. L. and C. L. Royce. 1992. Ascidians with algal symbionts. In: W. Reisser (Ed.), *Algae and Symbioses, Plants, Animals, Fungi, Viruses, Interactions Explored*. Biopress Ltd, England, pp. 215–230.

Paul, N. A. and C. K. Tseng. 2012. Seaweed and microalgae aquaculture, Farming aquatic animals and plant. In: S. John and C. Lucas (Eds.), *South Gate*. 2nd edition. Blackwell Publishing Ltd. Hoboken, United States.

Peterson, C. G. and N. B. Grimm. 1992. Temporal variation in enrichment effects during periphyton succession in a nitrogen limited desert stream ecosystem. *Journal of the North American Benthological Society* 11: 20–36.

Power, M. E., M. J. Mathews and A. J. Stewart. 1985. Grazing minnows, piscivorous bass and stream algae: Dynamics of a strong interaction. *Ecology* 66: 1448–1456.

Rajkumar. M., P. Perumal, V. A. Prabu, N. V. Perumal and T. Rajasekar. 2009. Phytoplankton diversity in Pichavaram mangrove waters from the south-east coast of India. *Journal of Environmental Biology* 30(4): 489–498.

Reynolds, C. S. and A. E. Walsby. 1975. Water blooms. *Biological Reviews* 50: 437–481.

Rosenberg, G. and H. W. Paerl. 1981. Nitrogen fixation by bluegreen algae associated with the siphonous green seaweed *Codium decorticatum*: Effects on ammonium uptake. *Marine Biology* 61: 151–158.

Sabater, S. and I. Munoz. 2000. *Nostoc verrucosum* (Cyanobacteria) colonized by a chironomid larva in a Mediterranean stream. *Journal of Phycology* 36(1): 59–61. DOI:10.1046/j.1529–8817.2000.99100.x.

Sahu, G., K. K. Satpathy, A. K. Mohanthy and S. K. Sarkar. 2012. Variations in community structure of phytoplankton in relation to physicochemical properties of coastal waters, southeast coast of India. *Journal Geo-Marine Sciences* 43(3): 223–241.

Sarangi, R. K., P. Chauhan and S. R. Nayak. 2005. Inter-annualvisibility of phytoplankton blooms in northern Arabian sea during winter monsoon period (February-March) using IRS P4 OCM data. *IJMS* 32(2): 163–173.

Sarangi, R. K., P. Chauhan, S. R. Nayak and U. Shreedhar. 2005. Remote sensing of *Trichodesmium* blooms in the coastal waters off Gujarat, India using IRS-P4 OCM. *International Journal of Remote Sensing* 26(9): 1777–1780.

Schenk, H. E. A. 1992. Cyanobacterial symbioses. In: A. Balows, H. G. Trüper, M. Dworkin, W. Harder and K. H. Schleifer (Eds.), *The Prokaryotes Vol. IV*. Springer-Verlag, New York, pp. 3819–3854.

Schindler, D. E., S. R. Carpenter, K. L. Cottingham, X. He, J. R. Hodgson, J. F. Kitchell and P. A. Soranno. 1996. Food web structure and littoral zone coupling to pelagic trophic cascades. In: G. A. Polis and K. O. Winemiller (Eds.), *Food Webs: Integration of Patterns and Dynamics*. Chapman and Hall, New York.

Schindler, D. W. 1977. Evolution of phosphorus limitation in lakes. *Science* 195: 260–262.

Schindler, D. W. 2006. Recent advances in the understanding and management of eutrophication. *Limnology and Oceanography* 5: 356–363.

Sellner, K. G., J. Doucette and G. J. Kirkpatrick. 2003. Harmful algal blooms: Causes, impacts and detection. *Journal of Industrial Microbiology and Biotechnology* 30(7): 383–406.

Sharabian, M. N., S. Ahmad and M. Karakouzin. 2018. Climate change and eutrophication: A short review. *Engineering, Technology and Applied Science Research* 8(6): 3668–3672.

Simpson. 1991. Volunteer lake monitoring: A methods manual. EPA440/4–91–002.

Sivasubramanian, V. and M. Muthukumaran. 2012. Large scale phycoremediation of oil drilling effluent. *Journal of Algal Biomass Utilization* 3(4): 5–17.

Subba Rao, P. V. 2012. Seaweed biodiversity and conservation. In: R. Varatharajan (Ed.), *Proceedings of Biodiversity Status and Conservation Strategies with Special Reference to Northeast India*. Publication of CAS in Life Sciences, Manipur University, Imphal—795 003, pp. 1–8 (Invited Articles).

Subba Rao, P. V., K. Ganesan and K. Suresh Kumar. 2009. Seaweeds: A survey of research and utilization. In: J. I. S. Khattar, D. P. Singh and G. Kaur (Eds.), *Algal Biology and Biotechnology*. I.K. International Publishing house, PVT Ltd., New Delhi, Banglore, pp. 165–178.

Subba Rao, P. V. and C. Periyasamy. 2019. *Kappaphycus farming* for Socio-economic development of coastal people in India. In: G. A. Ravisankar and A. Ranga Rao (Eds.), *Hand Book of Algol Technologies and Phytochemicals, Volume II: Phycoremediation, Biofuels and Global Biomass Production*, Publishers CRC Press, Tayllor and Francis Group, Boca Raton, London and New York, pp. 145–153.

Subba Rao, P. V., C. Periyasamy, K. Suresh Kumar, A. Srinivasa Rao and P. Anantharaman. 2018. Seaweeds: Distribution, production and uses. In: M. N. Noor, S. K. Bhatnagar and ShaShi Kr Sinha (Eds.), *Bioprospecting of Algae-Prof. J. P. Sinha Memorial Volume*. Published by Society for Plant Research, New Delhi India, pp. 59–78.

Subramanian, A. and A. Purushothaman. 1985. *Hemidiscus hardmanianus* bloom and fish mortality. *Limnology and Oceanography* 30(4): 910–911.

Suresh Kumar, K., K. Ganesan and P. V. Subba Rao. 2007. Phycoremediation of heavy metals by *Kappaphycus alvarezii*. *Journal of Hazardous Materials* 143: 590–592.

Suresh Kumar, K., H. Uwe Dahms, E. Ji Won, J. Seong Lee and K. Hoon Shin. 2015. Microalgae—A promising tool for heavy metal remediation. *Ecotoxicology and Environmental Safety* 113: 329–352.

Takahashi, M., D. L. Seibert and W. H. Thomas.1977. Occasional blooms of phytoplankton during summer in Saanich Inlet, B.C., Canada. *Deep Sea Research* 24(8): 775–780.

Tano, S., M. Eggertsen, S. Wikstrm, C. Berkström, A. Buriyo and C. Halling. 2017. Tropical seaweed beds as important habitats for juvenile fish. *Marine and Freshwater Research* 68. 10.1071/MF16153.

Tansley, A. G. 1935. The use and abuse of vegetational terms and concepts. *Ecology* 16(3): 284–307.

Trench, R. K. 1993. Microalgal-invertibratesymbiosis: A review. *Endocyt Cell Research* 9: 135–175.

Vadeboncoeur, Y. and A. D. Steinman. 2002. Periphyton function in lake ecosystems. *The Scientific World Journal* 2: 1449–1468.

Van Bergeijk, S. A. and L. J. Stal. 2001. Dimethylsulfoniopropionate and dimethylsulfide in the marine flatworm *Convoluta roscoffensis* and its algal symbiont. *Marine Biology* 138: 209–216.

Vignieri, S. and J. Fahrenkamp-Uppenbrink. 2017. Ecosystem earth. *Science* 356(6335): 258–259.

ViktóriaÜveges, L., J. P. Vörös and A. Kovács. 2011. Primary production of epipsammic algal communities in Lake Balaton (Hungary). *Hydrobiologia* 660: 17–27.

Vonshak, A. and A. Richmond. 1988. Mass production of blue green alga, *Spirulina*: An overview. *Biomass* 15: 233–247.

Ward, A. K., C. N. Dahm., and K. W. Cummins. 1985. Nostoc (Cyanophyta) Productivity in Oregon stream ecosystems invertebrate influences and differences between morphological types. *J Phycology* 21(2) 223–227.

Wehr, J. D. and G. Robert Sheath. 2015. Habitats of freshwater algae. In: D. John, G. Wehr, J. S. Robert and P. Kociolek (Eds.), *Freshwater Algae of North America*. Academic Press, USA. pp. 13–74.

Wikfors, G. H. and M. Ohno. 2001. Impact of algal research in aquaculture. *Journal of Applied Phycology* 11: 968–974.

Wilkinson, C. R. 1983. Net primary productivity in coral reef sponges. *Science* 219: 410–412.

Yoch, D. C. 2002. Dimethylsulfoniopropionate: Its sources, role in the marine food web, and biological degradation to dimethylsulfide. *Applied and Environmental Microbiology* 68: 5804–5815.

Yoshii, K. 1999. Stable isotope analyses of benthic organisms in Lake Baikal. *Hydrobiologia* 411: 145–159.

2 Phytoplankton as Bioindicators

M. Aruna

ABBREVIATIONS

%	Percentage
CPCB	Central Pollution Control Board
Et al.	et alia (and others)
p., pp.	Page, pages
pH	Potential of Hydrogen
sp	Species

2.1 INTRODUCTION

India has most varied inland water resources, which are considered to be one of the richest in the World. About 73% of Earth is covered with marine and fresh water, which is present in oceans, rivers, lakes, ponds, glaciers, and on mountains with ice caps.

Natural lakes are largely confined to the Himalayan region, while the rest of the subcontinent is dominated by river systems and manmade impoundments. The surface fresh water in the form of lakes and rivers is hardly 0.01% of total water available on the Earth. Demand for fresh water has increased markedly in recent years. Freshwater ecosystem comprises environment exhibiting enormous diversity in terms of nutritional, physiochemical, and biological characteristics. Assessment of physicochemical factors is more important to understand any ecosystem, especially that of aquatic ecosystem.

There is need to assess the quality of water as chemistry of water reveals much about the metabolism of the ecosystem and explains the general hydrobiological interrelationship. The life first originated in water, and first organisms were also aquatic, where water was principle external as well as internal milieus for organisms. Thus, water is a valuable natural resource, very important for sustaining all life on planet Earth, and is in regulation continuously studied as hydrological cycle. We depend on water for drinking, domestic needs, and irrigation. Normally water in nature is never pure in a chemical sense. It contains impurities of various kinds, such as dissolved gases, dissolved minerals, and suspended matters. These natural impurities are in very low amounts.

Due to anthropogenic activities and urbanization, many unwanted substances are introduced into water, and it is polluted. Polluted water is turbid, bad smelling, unpleasant, and unfit for drinking as well as domestic purpose and cause many diseases and is harmful to human beings. Major Indian rivers, such as Ganga, Yamuna, Tapti, Narmada, Chambal, Damodar, Krishna, Cauvery, Brahmaputra, Mahi, and other rivers are severely polluted. According to the Central Pollution Control Board (CPCB), 90% of the water supplied to the towns and cities is polluted. Out of which only 1.6% get treated. It is evident from literature that the general effect of pollution is that it leads to decrease in oxygen, along with increase in content of organic matter like, NH3, PO_4, Cl- and total solids with a change in color and odor of water. The presence of certain species of algae in abundance will serve as indicators of water quality and pollution.

 DOI: 10.1201/9781003219156-4

2.1.1 Some Representative Algae Collected from Different Lentic Water Bodies of Telangana State

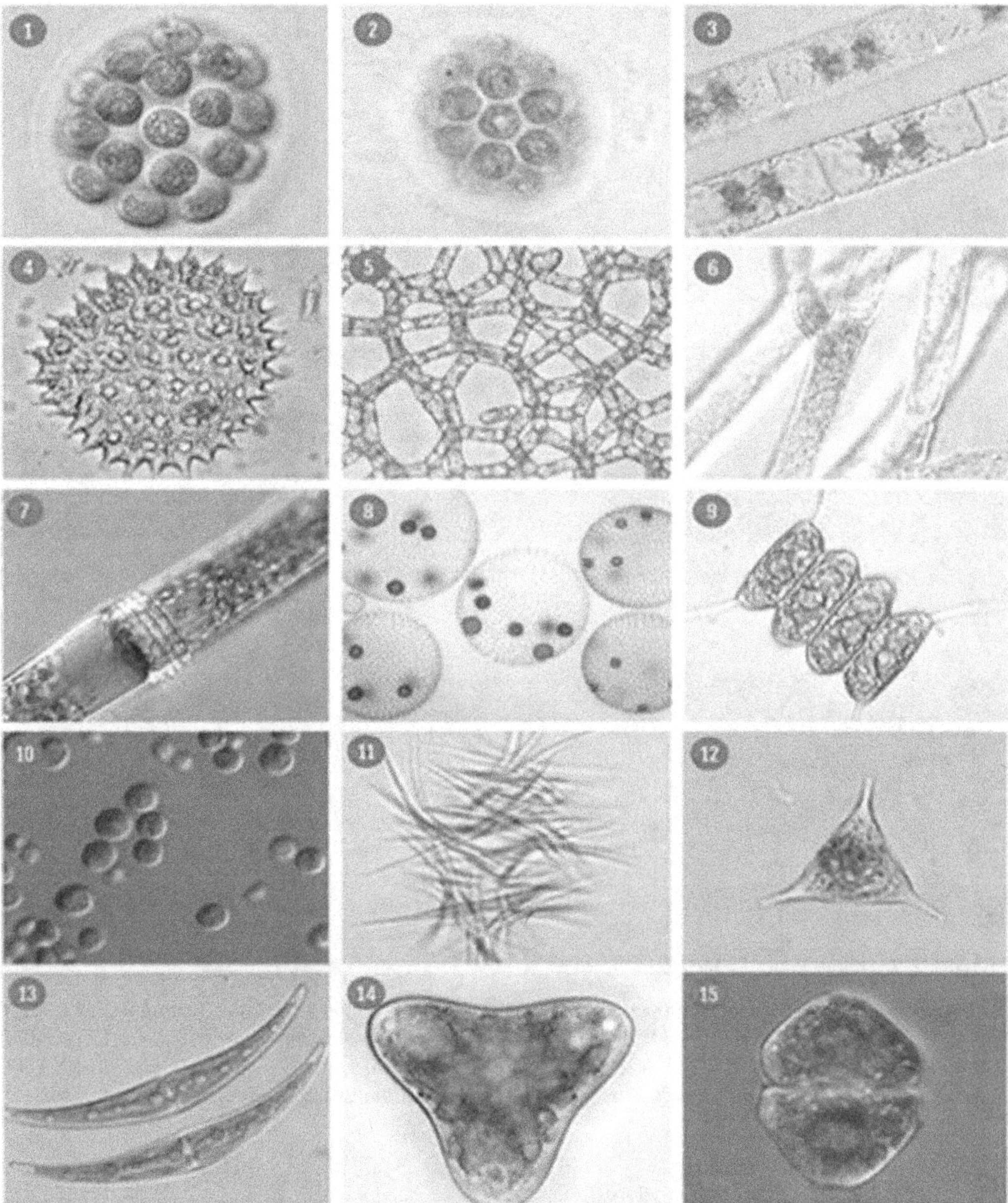

FIGURE 2.1 Photo Plate I: Chlorophyceae.

1. *Pandorina cylindricum* Iyengar X 500 2. *Eudorina elegans* Ehr X 500 3. *Zygnema pectinatum* voucher X 250 4. *Pediastrum duplex* Meyen X 500 5. *Hydrodictyon reticulatum* L. Lagerheim X 500 6. Cladophora glomerata (L.) Kutz X 500 7. *Oedogonium giganteum* Knitzing X 400 8. *Volvox aureus* Ehr 9. *Scenedesmous quadricauda* (Turpin) Brebisson X 1000 10. *Chlorella vulgaris* Beyerinck X 400 11. *Ankistrodesmous falcatus* (Corda) Ralfs X 400 12. *Tetraedron bifercatum* Langerheim X 400 13. *Closterium dianae Ralfs* ex Ehrenb X 400 14. *Straurastrum bieneanum* WB Turner X 250 15. *Cosmarium granalum* Brb. & Ralfs X 400

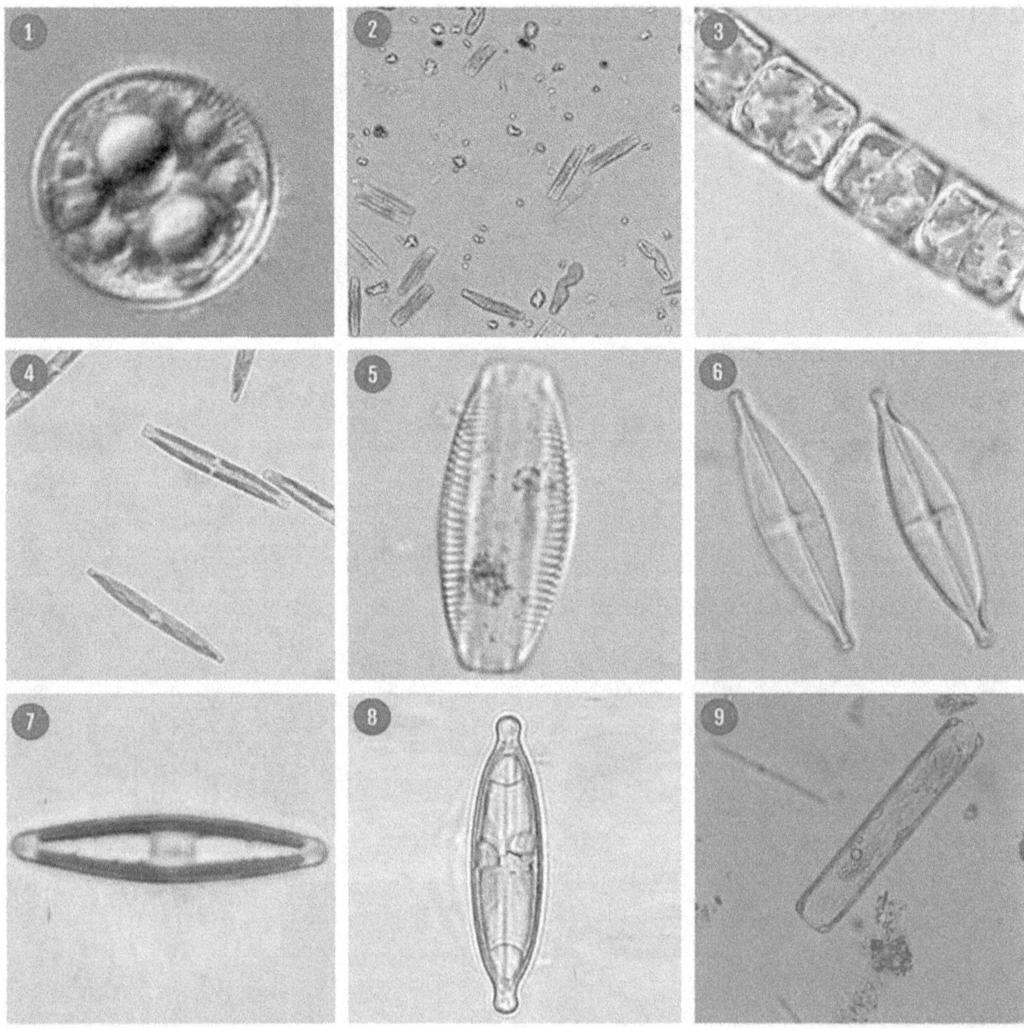

FIGURE 2.2 Photo Plate II: Bacillariophyceae.

1. *Cyclotella glomerata* Bachm X 1000 2. *Synendra rumpens* kuetz X 250 3. *Melosira granulate* Ehr Ralfs X 250 4. *Nitzschia palea* (Kutzing) X 400 5. *Amphora veneta* Kutz X 400 6. *Stauronensis anceps* Ehr X 1000 7. *Navicula lanceolata* Kutz X 400 8. *N. Rhyncocephala* Kutz X 500 9. *Pinnularia graciloides* Hust X 1000 *Staurastrum bieneanum* WB Turner X 250 15. *Cosmarium granatum* Brb. & Ralfs X 400

2.2 BIOLOGICAL MONITORING

Biological monitoring is a laboratory technique applicable to evaluate the quality of water. The application of biological data assessing the quality of water was first reported by Kolkwitz and Marsson (1909) in Saprobian system. Biological monitoring is a useful water quality management tool because the structure of aquatic community is a function of environmental condition. The process of biological monitoring is very important because community structure can be altered with any change in their environment by the effect of pollution, which could easily be missed by conventional and physical indicators. Biological monitoring aims to monitor ambient water quality by the application of one biological technique, i.e., the examination of algal community. Recently it has been reported by many workers that algal communities are used in bio monitoring pollution.

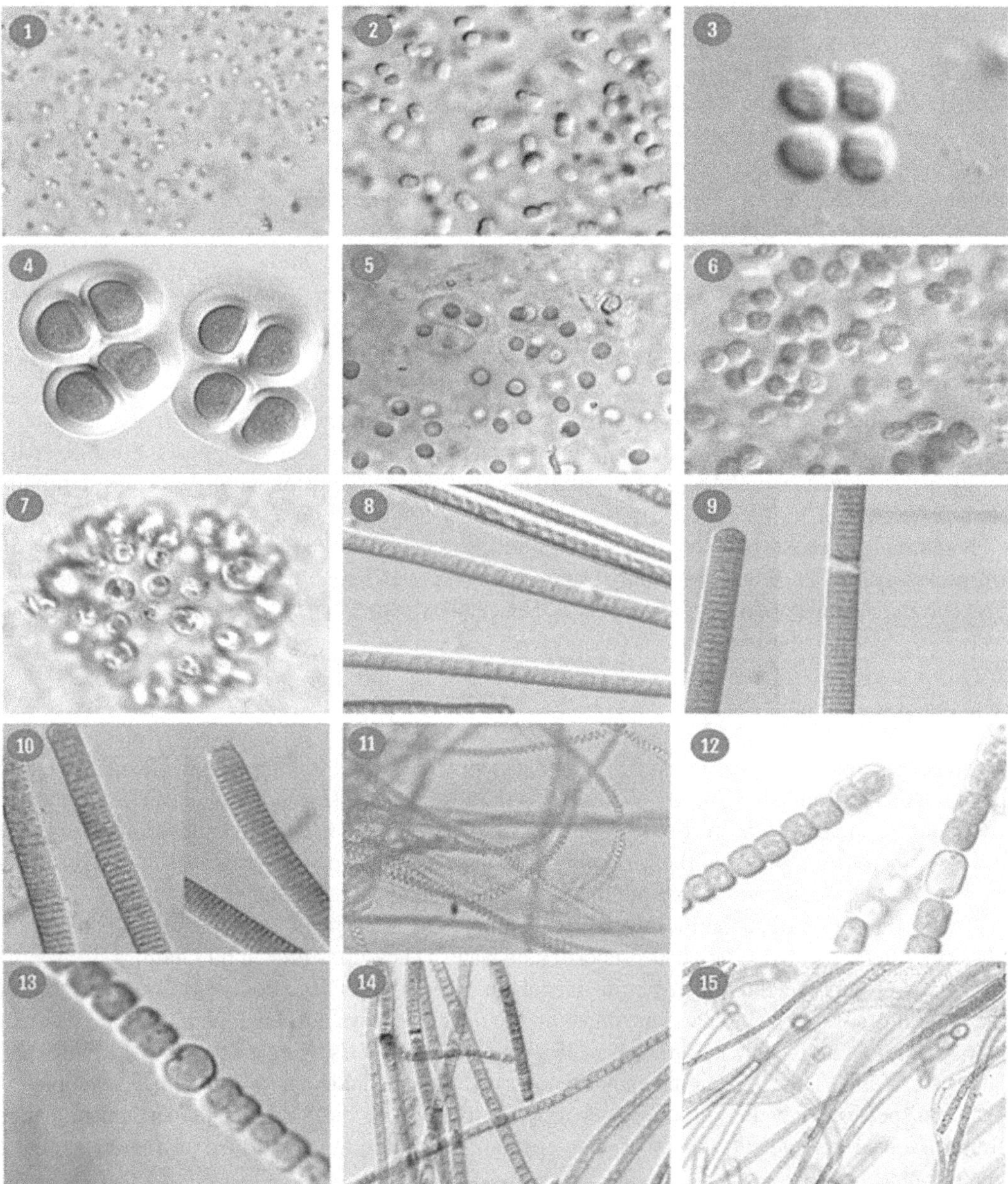

FIGURE 2.3 Photo Plate III: Cyanophyceae.

1. *Aphanocapsa brunnea* Naeg X 1000 2. *Aphanothece bullosa* Rabenh X 1000 3. *Chroococcus minor* Kutz X 1000 4. *Chroococcus turgidus* Kutz X 400 5. *Gloeocapsa nigrescens* Nag X 1000 6. *Microcystis marginata* Kutz X 1000 7. *M. aerugniosa* Kutz X 500 8. *Lyngbya majuscule* Harvey X 400 9. *Oscillatoria acuta* Bruhl X 400 10. *O.* curviceps Ag X 400 11. *Spirulina major* Kutzing X 1000 12. *Anabaena azollae* Strasb X 400 13. *A. oryzae* X 400 14. *Scytonema subtile* Moeb X 400 15. *Gloeotrichia ghosei* Singh X 400

The importance of lagoon as science was recognized in early decades. Wetzel (1975a) has defined Lake Ecosystem. Every lake in the country is associated with problems which are needed to be addressed. Water quality is an important factor and the primary causes of deterioration of surface water quality are municipal and domestic wastewater, industrial and agricultural waste, and solid and semi-solid refuge. The materials which lead to pollution include inorganic salts, acids and alkalis, organic matter, color, toxic chemicals, and radioactive materials.

Algae are involved in water pollution in a number of significant ways. Certain algae are to flourish in water polluted with organic wastes and to play an important role in "self purification" of the water. The selective type of algae that exists in polluted water also is being used as indicators of collection of animal drinking the water or living in it. Since algae constitute part of a chain of aquatic life in the water, whatever alters the number and kind of algae affects all of the other organisms, including fish. Thus it requires a continuous monitoring and study of algae existing in waters of various quality in order to determine what control, what changes, or what uses be instituted for the benefit of man and for conservation of water and desirable aquatic life.

In freshwater lakes and ponds, green filaments such as *Oedogonium, Spirogyra, Microspora, Ulothrix, Cladophora* and *Pithophora* serve directly as fish food or consumed by midge fly larvae and other insects, which in their turn are eaten by the fish. Algae are found in successful ecological niches from fresh water to extreme saline water from the warmth of natural hot springs to the frozen reaches of Antarctic.

For a number of years there has been a series of proposals indicating that one or more algae could be used as indicators of clean water or polluted water. Clean water would support a great diversity of organisms, whereas polluted water would yield just a few organisms with one or few dominant forms.

Planktonic algae are widely used for monitoring the intracellular concentration of metals in freshwater system since they are totally immersed in water and are efficient in absorbing almost all concentrated metal ions from ambient waters. Hence, the relationship between metal absorption and its toxicity to microalgae became very important.

2.3 PHYTOPLANKTON

The term phytoplankton was derived from an ancient Greek word Phyton, which means plant, and plankton means wanderer (Thurman, 2007). Planktonic organisms move freely in surface waters of rivers, lakes, and oceans in the direction of wind and water current. The phytoplanktons are a large group of microscopic photosynthetic organisms found growing in diverse habitats. These organisms survive in water bodies both fresh and marine wetlands in and around the globe.

The composition, distribution, diversity, and abundance of phytoplankton are dependent on physical, chemical, and biological properties of water. They constitute food chain, food web of an aquatic ecosystem, and play a significant role in production of organic matter. Their absence in the natural water bodies indicate imbalance in the aquatic ecosystem. Phytoplanktons are autotrophic in nature, that is, they can fix solar energy by photosynthesis using carbon dioxide, nutrients, and trace metals. The chief pigments are chlorophylls, carotenoids, and phycobilins. Some phytoplankton species can be temporarily heterotrophic, that is, they build up organic particulate matter from dissolved organic substances or even particulate organic matter termed as osmotrophy and phototrophy.

The main component of phytoplankton in rivers, lakes, and ponds are diatoms, desmids, dinoflagellates, blue-green algae, green algae, and some other flagellates. The blue-green algae are very much abundant in freshwater ponds, lakes, and springs. The blue-green algae which are primitive prokaryotes tend to concentrate towards the surface waters. Within the phytoplankton, there is a large variation in cell size and growth rates, the species ranging from smaller prokaryotic to macroscopic eukaryotic forms sometimes equivalent in size to that of bacteria to those of largest dinoflagellates, which are visible to the naked eye. Blue-green algae are a very important group of phytoplankton which play a significant role in assessment of water quality and also occur as water blooms. Diatoms are common in both marine and fresh water, while dinoflagellates are more widespread in oceans. The diversity of green and blue-green algae recorded was found to be highest in fresh waters, and they tend to concentrate towards surface waters.

Most phytoplankton organisms have a density greater than that of water, and the higher density in part may be due to skeletal portion of silica, calcium carbonate, and cellulose. Water turbulence combined with other factors, such as shape or physiological state, reduces the sinking rate of non-motile organisms such as diatoms. There is evidence that the setting of phytoplankton to the bottom, at least in neritic zone is not uniform but occurs at irregular intervals. Motile phytoplankton like dinoflagellates may actively swim to compensate against activity of benthos organisms.

Diversity of phytoplankton is based on spatial and physical factors such as light, temperature, and availability of nutrients. Biological variables such as grazing pressure and competition for habitation are also significant (Roy and Chattopadhyay, 2006a, 2006b). The study of aquatic ecosystem communities and the factors affecting their strength are the essential challenges of ecology because water bodies serve as resource for mankind for various aspects including food. The probable natural complications of nutrients were strengthening and interruption, which include modification of the normal phytoplankton population.

Physical characters play a major role in determining the inimitability of any water body with reference to organic matter. Physical factors such as light, transparency, temperature, wave stroke, total abundance of the inorganic nutrients play a vital role in making us understand the phytoplankton dynamics. Phytoplankton population existing in a community gets declined due to overgrazing and competition.

Phytoplankton populations further undergo changes susceptible to grazing by aquatic fauna, which show fluctuations in the aquatic environment. With gradual changes in physical, chemical, and biological factors, several species disappear from the water body. The reason for disappearance of species is strongly because they cannot tolerate the new environmental conditions and withstand competing with the new species in new environment. These changes correspond to an essential ecological phenomenon known as Phytoplankton succession.

Members of Bacillariophyceae usually show dominance in winter season when compared to members of Cyanophyceae and Chlorophyceae (Pearsall, 1924). A major number of reports stressed the significance of temperature as a factor in determining the periodicity of Bacillariophyceae members. Low temperature and availability of silicates are conducive to the growth and development of members of Bacillariophyceae (Allen, 1934; Patrick, 1948; Iyengar and Venkatraman, 1951; Roy, 1955; Han Cock, 1973; Milliman and Boyle, 1975; Agarwal and Kumar, 1978).

Chlorophyceae members occur in summer when the NO_3 and PO_4 are low and the Cyanophyceae have a general association with the time of year when the organic matter is abundant and are able to grow profusely in minimal amount of NO_3 and PO_4. Many of the Cyanophyceae members which exist in eutrophic waters in summer were found to be nitrogen fixers, and diatoms were abundant during winter. The physiology and biochemistry of the phytoplankton has been well reviewed in a number of standard works (Lewin, 1962; Morries, 1980).

In spite of the number of pervasive generalizations about the common seasonal succession in fresh waters, close inspection of existing data shows a great diversity pattern. Nevertheless, simplifying briefly the temperate pattern of succession involves minimum of small flagellates adapted to low light and temperature in winter, burst of diatom activity, and biomass in spring followed rapidly by a smaller development of green algae and a transitional state between spring and summer. Summer populations vary in relation to the trophic status of water bodies but can include either another diatom development in less productive lakes by late summer and early autumn or increase in nitrogen-fixing blue-green algae in eutrophic lakes.

The widely held assumption that winter productivity is not universally valid, and rates of primary production under ice cover can constitute a very significant portion of the total annual primary productivity of the phytoplankton. Increase in light and decrease in temperature are the central factors contributing to the outburst of diatoms in spring. Few of the dominant species of these diatoms are *Asterionella, Cymbella*, and *Stephanodiscus*.

As silica concentrations are reduced in productive lakes, diatoms populations are often succeeded by preponderance of green algae first and later by blue-green algae. Growth in these eutrophic lakes can be so intense that combined nitrogen (NO_3, NH_4^+) sources were reduced to be lower than detectable in the trophogenic zone. In epilimnetic zones, at warmer temperature blue-green algae occurred, which were capable of fixing molecular nitrogen. Finally, it was learned that a heavy load of phosphorus was required for sustainability of lakes.

In addition to the physical and chemical parameters, biological parameters are equally important in monitoring the water quality, as reported first by Kolkwitz and Marsson (1909). Many certain algal forms which grow in the special type of polluted water serve as indicators of particular environment. *Microcystis aeruginosa* is the best solitary indicator of organic contamination (Prescott, 1951; Singh, 1953). *Euglena*, *Oscillatoria*, and *Phacus* are the reliable indicator of eutrophication (Patrick, 1961; Palmer, 1969).

Seven genera most tolerant to organic pollution are Euglena, *Oscillatoria*, *Chlamydomonas*, *Scenedesmus*, *Chlorella*, *Nitzschia*, and *Navicula* (Gunnale and Bala Krishna, 1979; Yadava and Pandey, 1980). Presence of highest amount of phosphate and nitrogen in chemical wastes affect growth of the phytoplanktons. In the presence of excess nutrients, the phytoplankton are capable of growing rapidly and increase their number. If the conditions are not restricted, it may lead to modifications in the population, dominance, or algal bloom, which are indicators of the deterioration of water quality.

Among aquatic microorganisms, desmids stand at high level, particularly in assessment of water quality and conserve nature. Phytoplankton form good indicators of water quality as they have rapid turnover time and are sensitive indicators of environmental stresses. Phytoplankton survey thus helps to find trophic status and the organic pollution in the ecosystem (Ramachandra and Solanki, 2007).

On the other hand, phytoplankton become the major food source for aquatic fauna and thus can also be used as indicators of the trophic status of water body. Thus, phytoplanktons play a significant role in transforming energy from atmosphere to aquatic ecosystem. According to Saladia (1997), phytoplankton are very sensitive to pollution and make supplementary changes in aquatic ecosystem thus act as biological indicator of water pollution. He observed that the clean water supports the phytoplankton diversity, while very few organisms survive in contaminated water in dominant form.

Wetzel (2001) stated that phytoplankton represent the basis of nutrient cycle of an aquatic ecosystem; therefore, they play a vital role in maintaining equilibrium between biotic and abiotic factors. The study of phytoplankton is also important to estimate the fresh ecosystem. Nair et al. (1983) illustrated that the density and species richness of phytoplankton are restricted by several physicochemical factors of water body. Hence phytoplankton was used frequently as indicators to examine and understand changes in the hydrobiology and limnology under climatic and seasonal diversity. Similar work made by several authors such as Beaugarand and Reid (2000) and Steinacher et al. (2010). Kurano and Miyachi (2004) opined that they play a vital role in solving several environmental issues, understanding water ecosystem, and also the productivity of aquatic ecosystem.

Species diversity indices:

Species diversity of phytoplankton is determined by using following indices.

1. *Margalef Diversity Index* (d)
2. *Shannon Diversity Index* (H')
3. *Simpson Diversity Index* (1-Δ)

2.4 TROPHIC STATUS AND EUTROPHICATION

Nygaard (1949) has projected five indices for biological monitoring of aquatic resides based on the individual populations of Cyanophyceae, Chlorophyceae, Bacillariophyceae, Euglenophyceae, and a combination of all the populations together. For finding if the organic pollution of the water sample was high or low, the Algal Generic Pollution Index of Palmer (1969) was followed.

Any water body is usually classified as being in one of three possible classes, such as oligotrophic, mesotrophic, and eutrophic. Water bodies with extreme trophic indices may also be considered hyper oligotrophic or hyper eutrophic. They are categorized on the basis of their richness of nutrients, which typically affects the plant growth. Oligotrophic lakes are nutrient-poor lakes and are generally clear having low concentration of plant life. Mesotrophic lakes have a good clarity and an average level of nutrients. Eutrophic lakes initiate as nutrient deficient, hence unproductive and detritus-free systems, and develop once with increasing amount of nutrients, thus leading to the organic materials. Senescent lakes are the oldest developmental stage lakes, which consist of their layer of organic matter.

Eutrophication is the enrichment of water bodies with plant nutrients, particularly phosphorous and nitrogen compounds, and it is a natural phenomenon that normally occurs during the life of an impoundment of lakes and can take thousands of years to occur without the influence of man. The chief cause of eutrophication is draining of sewage and agricultural fertilizers. A substantial fraction of the nitrate content of rivers comes from land drainage, but this does not apply to phosphates, which mostly or entirely come from sewage and detergents. This is because unlike nitrate, phosphate does reach off easily from soil or land.

Round (1981) stated that eutrophication results in increase in growth-promoting substances. He concluded that first major source of artificial eutrophication was due to the flow of excess fertilizers of agricultural land, and secondarily it was due to addition of sewage of organic waste. As a resultant of eutrophication algal blooms were formed, which were either temporary or permanent. It is believed that high pH is related to heavy blooms of phytoplankton (Das and Srivatsava, 1956; Sreenivasan, 1963).

Increase in the dissolved oxygen may also increase in algal blooms (Das and Srivastava, 1956). Huber Pestalozzi (1938) listed 40 Cyanophycean forms which produce algal blooms. In India *Microcystis aeruginoosa* and *Trichodesmium sps.* are common blooms-forming algae. Desikachary (1959) and Iyengar (1938) observed permanent water blooms in many temple tanks of South India at Madras, *Microcystis aeruginoosa*. Bharati and Hosmani (1973) investigated the cause of water bloom occurrence. It was noticed that increase in phosphates, calcium oxidizable organic matter, ammonia nitrate, carbohydrate, with more number of hours of bright sunlight, low pH and oxygen with higher degree of algal forms had a prolific effect in accelerating the blooms.

The genus *Anabaena* with its 11 species was the most common bloom-forming BGA (Desikachary, 1959; Whitton, 1973) both in India and abroad.

Some bloom-forming Cyanophycean members are releasing algal toxins into water. A few lakes have been plagued by races or strains of blue-green algae, six genera such as *Microcystis* sp., *Nodularia* sp., *Coelospaerium* sp., *Gloeotrichia* sp., *Anabaena* sp., and *Aphanizomenon* sp. have been implicated for the killing of fishes and death of horses, cattle, dogs, chicken, and even human beings (Heise, 1951; Ingram and Prescott, 1954; Gorham, 1964). An increased nutrient input causes a stepwise change in productivity, with phytoplankton biomass increase being one of the most obvious effects (Skei et al., 2000).

Although most eutrophic lakes are dominated by Cyanophycean blooms, some are dominated by the Chlorococcales; for instance, Lake Arreso in Denmark is highly eutrophic. The diatoms and other chlorophyll-c containing algae are, in general, photosynthetically less efficient than many smaller green algae, such as *Chlorella*, especially under low light intensities. The diatom *Turbullaria* and *Melosira* almost invariably reduce the length of filter runs. The brown flagellate *Synendra* even in small numbers is a notorious alga giving bad taste and producing bad odor.

Low concentrations of most of the algae however are often an asset rather than a liability in raw waters. Although many environmental factors, relatively in smaller amount, tend to keep the phytoplankton population stable, other factors will equally influence the growth and relative abundance of various genera and species.

2.5 CONCLUSION AND FUTURE PERSPECTIVES

Reservoir and irrigation tanks are one of the most important sources of water. Almost all the tanks in India are polluted. India and China are counted among the water hot spots because of their large populations that have to be provided with food and drinking water. In India villages do not have a source of drinking water. In addition, millions of people are annually affected by poor quality of water; around 80% of the total diseases are water borne.

In the field of agriculture, to get better yield, man had used a number of agrochemicals that had to be indiscriminately employed to keep the balanced environment required for growth and reproduction of plants (Round, 1979; Singh, 1961; Thamos, 1973; Venkataramana, 1957; Kumar 1993). After using the agrochemicals in agricultural fields, runoff involving various chemicals brought major changes in the natural ecosystem, affecting the aquatic flora.

Abiotic environment of freshwater ecosystem affects the biotic component of the ecosystem. If any change occurs in the physicochemical characteristics of water, it causes a direct impact upon the biota. Therefore, the knowledge of physicochemical characteristics of water is essential for proper exploitations of aquatic environment. So the main objective of physicochemical analysis of water is to determine the nutrient status of the medium (APHA, 2005). Since the water contains dissolved and suspended constituents in varying proportions, it has different chemical and physical properties. Water may be affected in various ways by pollution. Physicochemical and biological characteristics help evaluate the habitat, thus the concentration of water pollution can be assessed. Most of the fresh waters are alkaline, and alkalinity of fresh water is mainly due to carbonates, bicarbonates, sulfates, and chlorides. The total ionic salinity of water is constituted by the concentration of major anions like CO_3, HCo_3^-, Cl-, and SO_{4-} cations like Na^+, K^+, Ca^+, and Mg^+ (Wetzel, 1975b). The concentrations of ionized components, iron, and numerous minor elements are of immense biological importance. The physicochemical and biological characteristics of water bodies together account for the trophic status. In lentic waters, physical and chemical parameters keep fluctuating by consequent cycling of water through evaporation and its subsequent return as rain. There will be considerable seasonal changes in these parameters. These seasonal changes in addition to anthropogenic activities affect the biological environment.

Our soils, fresh water, and marine sediments are some of the least known biological habitats or domains on earth. These subsurface domains have been perceived as dark and lifeless. However, it is predicted that the subsurface portion contains higher species diversity than at ground level all over the world (Schulze and Mooney, 1994). Importantly, the functioning and linkages among these habitats is essential to sustain life on Earth (Daily, 1997).

We are only beginning to realize the extent of the interdependence of biotic interactions in subsurface habitats.

The increasing perturbation of the natural ecosystem, including habit loss, pollution, and overharvesting, highlights the necessity of developing methodologies for the rapid detection of ecosystem change. It is also necessary to evaluate sustainable utility of different components of biodiversity. Increase in population and rapid urbanization results in expansion of cities, rise in the number of industries, which in turn create many environmental problems hazardous to man. Because of these hazardous health effects, man should learn and gain knowledge and skills to conserve the basic requirements of life such as air, soil, and water to lead a better life in society.

A primary goal of reflecting on past directions and accomplishments is to learn from previous progress and improve slowly. There is an urgent need to inventory the world's algae, conserve species in culture collections and genetic stocks in seed banks, as well as to preserve natural communities by protecting sites of phycological importance.

ACKNOWLEDGMENT

I am thankful to the department for providing facilities and grateful to **Prof. Vidyavati,** former vice-chancellor of Kakatiya University, Warangal, for her constant encouragement.

REFERENCES

Agarwal M and Kumar HD (1978) Physico-chemical and Phycology assessment of two mercury polluted effluents. *Indian J Environ Health* 20(2): 141–155.

Allen E (1934) Neglected factors in the development of thermal springs. *US Nat Acad Sci Proc* 20: 345–349.

APHA (2005) Standard methods for the examination of water and waste water. 21st edn. APHA, AWWA, WPCF, Washington, DC.

Beaugarand, GFI and Reid, PC (2000) Spatial, seasonal and long term fluctuations of plankton in relation to hydroclimatic features in the English Channel, Celtic Sea and Bay of Biscay. *Mar Ecol Prog Ser* 200: 1813–1819.

Daily GC (1997) Nature's services: Societal dependence on natural ecosystems. Island Press, Washinton, DC. p. 392.

Das SM and Srivasthava VK (1956) Some new observation on plankton from fresh waterponds and tanks of Lucknow. *Ind Sci Cult* 21(8): 466–467.

Desikachary TV (1959) Cyanophyta. Indian Council of Agriculture Research, New Delhi.

Gorham PR (1964) Toxic algae in Algae and man. Jackson DF (ed). Plenum Press, New York. pp. 307–366.

Gunale VR and Balakrishnan MS (1979) Schizomeris leibleninii kuetz. as an indicator of eutrophication. *Biovigyanam* 5(2): 171–172.

Hancock FD (1973) The ecology of the diatoms of the Klip River, Southern Transval. *Hydrobiologia* 42(2–3): 243–284.

Huber PG (1938) Das, Phytoplankton des Susswasers. Systematik und Biologie. Thienemann A (ed), Die Binnengewasser. E Schweizerbart'sche Verlag, Stuttgart, Germany. pp. 367–549.

Ingram WM and Prescott GW (1954) toxic fresh water Alge Annar. *Midil Net* 52: 75–87.

Iyenger MOP (1938) The vegetation of Madras and its environs. Scientific survey of Madras and its environs. University Madras. pp. 52–59.

Iyengar MOP and Venkataraman GS (1951) The ecology and seasonal succession of theriver Cooum at Madras with special reference to the Diatomaceae. *J Madras Univ* 21: 140–192.

Kolkwitz R and Marsson M (1909) Okolodieder Herische Seprobien. Beitrage Zurlehrevonder biologische Gewasser burteilung. *Inst Revuc Gos Hydrobiologia* 2: 126–152.

Kumar S (1993) Correlation among water quality parameter of ground water in Balmerdistrict. *Indian J Environ Prot* 13: 487–489.

Kurano N and Miyachi S (2004) Microalgal studies for the 21st century. *Hydrobiologia* 512: 27–32.

Lewin RA (1962) Physiology and biochemistry of algae. Academic Press, New York.

Milliman JD and Boyle E (1975) Biological up take of dissolved silica in the Amazon. *Riverestury Sci* (Wash D.O) 189(4207): 995–997.

Morries IE (1980) Physiological ecology of phytoplankton (Studies in ecology 7). Blackwell, Oxford.

Nair NB, Dharmaraj K, Abdul Aziz PK, Arunachalam M, Krishna Kumar K and Balsubramaniam NB (1983) Nature of primary production in a tropical backwater of south- west Coast of India. *Proc. The Indian Nation Sci Academy* 49(6): 521–597.

Nygaard G (1949) Hydrobiological studies on some Danish ponds and lakes II. Thequotient hypothesis and some new or little known phytoplankton organisms. *Dat KurgeDanske Vid Sel Biol Skr* 7: 1–293.

Palmer CM (1969) A composite rating of algae tolerating organic pollution. *J Phycol* 5: 78–82.

Patrick R (1948) Factors affecting the distribution of Diatoms. *Bot Rev* 14(8): 437–524.

Patrick R (1961) A study of the number and kinds of species found in rivers in Eastern United States. *Proc Acad Nat Sci* 113: 215–258.

Pearsall WH (1924) A suggestion as to the factors influencing the distribution of free-floating vegetation. *J Ecol* 19(2): 241–262.

Prescott GW (1951) Algae of western great lake areas. Wm. C. Brown Co. Publishers Dubuque, Iowa.

Ramachandra TV and Solanki M (2007) Ecological assessment of lentic water bodies of Bangalore. ENVIS Technical Report 25. Centre for Ecological Sciences, Indian Institute of Science, Bangalore.

Round FE (1979) The ecology of the river Hoogly at Palta, West Bengal. *J Ecol* 36(2): 169–175.

Round FE (1981) The ecology of algae. Cambridge University Press, London, New York and Cambridge. p. 629.

Roy JC (1955) Periodicity of the plankton diatoms of the Chilka lake for the years 1950–1951. *J Bomb Nat Hist Soc* 52: 112–123.

Roy S and Chattopadhyay J (2006a) Enrichment and stability: A phenomenological coupling of energy value and carrying capacity. *BioSystems* 90(2): 371–378.

Roy S and Chattopadhyay J (2006b) Enrichment and ecosystem stability: Effect of toxicfood. *BioSystems* 90(1): 151–160.

Saladia PK (1997) Hydrobiological studies of Jait Sagar Lake, Bunch (Rajastan). Thesis submitted to MDS University, Ajmer, India.

Schulze E and Mooney H (1994) Ecosystem function and biodiversity: A summary in biodiversity and ecosystem function. Schulze MH (ed). Springer-Verlag, Berlin. pp. 497–570.

Singh RN (1953) Limnological relations of Inland waters with special reference to waterblooms. *Verh Int ver Ther A new Limnol* 12: 831–836.

Singh RN (1961) The role of BGA in nitrogen economy of Indian agriculture. ICAR, New Delhi, India. pp. 112–142.

Skei J, Larson P, Rosenberg R, Jonsson P, Olsson M and Broman D (2000) Eutrophication and contaminats in Aquati ecosystem. *Ambio* 29(4–5): 184–194.

Sreenivasn A (1963) Primary production in the upland lakes of Madras state, India. *Current Science* 32(3): 130–131.

Steinacher M, Joos F, Frolicher TL, Bopp L, Cadule P, Cocco V, Doney S C, Gehlen, M, Lindsay K, Moore JK, Schneider B and Segschneider J (2010) Projected 21st century decrease in marine productivity: A multi-model analysis. *Biogeosciences* 7(3): 979–1005.

Thomas EA (1973) Phosphorus and eutrophication. Griffith EJ, Beeton A, Spencer JM and Mitchell DT (eds), Environmental phosphorus handbook. Wiley, New York.

Thuruman HV and Trujillo AP (1988) Introductory oceanography. 5th edn. Merrill Publishing Company, a Bell & Howell Information Company, Columbus, OH. p. 515.

Venkataraman GS (1957) The algal flora of the ponds and puddles inside the Banaras Hindu University Ground, India. *J Bomb Nat Hist Soc* 54: 908–914.

Wetzel RG (1975a) Limnology. W.B. Saunders Co., Philadelphia, London, and Toronto. p. xii + p. 743.

Wetzel RG (1975b) Primary production. Whitton BA (ed), River ecology. Cambridge University Press, Cambridge, pp. 230–247.

Wetzel RG (2001) Limnology: Lake and river ecosystems. Gulf Professional Publishing. p. 1006.

Whitton BA (1973) Fresh water plankton in cars. Whitton BA (ed). *The Biology of Blue Green Algae* 676: 352–367.

Yadava RN and Pandey DC (1980) Certain cultural observations on Schizomeris leibleniniikuetz. *Phykos* 19: 204–209.

3 Bioplastic Production from Algae and Impact on Environment

Ashutosh Tripathy, Prama Rani, and Jastin Samuel

3.1 INTRODUCTION

Plastic pollution is currently one of the world's major problems. We are using plastics extensively in our day-to-day life and is an intricate part of our life and one of the most important materials. It is found that a very large percentage of plastics are used in applications related to packaging, agriculture, and the medical sector. Especially in India, the use of plastic is increasing because of its low cost and better qualities in terms of strength and longevity compared to other materials like paper and glass (Andrady and Neal, 2009). Hence, it is very important to consider these factors while finding an alternative to nondegradable or single-use plastics. It is estimated that global plastic production will touch a value of 1,100 metric tons by 2050; out of which considering the current plastic discard rate, which is 76%, it would generate around 33,000 metric tons of plastic waste. It raises a greater concern as only 10% of the plastic waste is recycled, and 14% of the waste is incinerated, and it will strongly affect our environment, increasing pollution at all levels of the ecosystem (Geyer, 2020). It is estimated that mismanaged plastic waste (MPW) will have a potential impact on Asian and African countries as it would raise to the value of 155–265 metric tons per year by 2060 (Lebreton and Andrady, 2019). Microplastics of a size less than 5 mm is also a potential source of pollution as it affects both marine and even remote ecosystems like tropical coral reefs and the deep sea. Plastic mulch used in agriculture, plastics disposal, landfills of plastic waste create a lot of concern as it affects the soil ecosystem and also can affect the overall soil productivity (Horton and Barnes, 2020; Chae and An, 2018). Hence it is important to find an alternative that can not only fulfill the needs of everyone as well as will not affect our environment at the same time. Bioplastics are mainly produced from a biological resource such as renewable biomass (corn, sugar, and potato) or microorganism (bacteria and algae) (Saharan and Sharma, 2012; Ezgi, 2015). Nowadays cost-effective agricultural residues are also used as feedstock to produce bioplastics but require pretreatment before use and sometimes require other nutrients for growth (Getachew and Woldesenbet, 2016). The main concern is their production at the industrial level is still immature. It is also suggested that if we can screen, select, and isolate appropriate organisms and use genetic engineering and metabolic engineering tools, then we can explore the capacity of different microorganisms to produce bioplastics (Luengo et al., 2003). According to European Bioplastic, as compared to other bioplastics, PLA (polylactic acid) and PHA (polyhydroxyalkanoates) are manufactured with a high percentage, and they are biobased and biodegradable (European Bioplastic, 2018). Microalgae can be a potential candidate to produce bioplastics. It is found that bioplastic produced from microalgae is biodegradable and environment-friendly (direct carbon capture from the environment), and different studies have been conducted from direct use of microalgae biomass as a bioplastic, intermediate processing, and blended microalgae with plasticizer and conventional plastic (Abdo et al., 2019). This chapter discusses the bioplastic production from microalgae and their impact on the environment.

DOI: 10.1201/9781003219156-5

3.2 ALGAE AS A SUSTAINABLE FEEDSTOCK FOR BIOPLASTIC FORMATION

Algae is a photosynthetic, autotrophic aquatic organism. It is microscopic/macroscopic, unicellular/multicellular, and is devoid of many advanced and complex structures in terms of cell alignment like in higher plants (Sudhakar and Viswanaathan, 2019). By June 2012 about 72,500 algal species are estimated to exist, as per the records in the algae-based database, which although is not an exhaustive list but can explain how vast the world of algae is (Guiry, 2012). Algae are mainly classified into green algae, or Chlorophyta; red algae, or Rhodophyta; and brown algae called Phaeophyta. They are also classified in terms of their size, that is, macroalgae or seaweeds, which are visible and large algae, and multicellular and microalgae, which are microscopic and include prokaryotes like cyanobacteria and eukaryotes like green algae (Khan et al., 2018). Algae are a great source of many nutrients like proteins, carbohydrates, fatty acids, and source of many vitamins and minerals and have their application in cosmetics, dairy, food, and pharmaceutical industries. Algae and algae products have important therapeutic properties and have anticancer, antiviral, and antioxidant properties. Complex algal protein products can also be used as dilatory agents, anticoagulants, insecticides, cholesterol reducers, antihypertensive agents (Goswami, 2015; Sudhakar and Viswanaathan, 2019). Algae can also act as a sustainable feedstock for bioplastic production. Bioplastics can be produced from algae in three different ways: (1) direct conversion, (2) intermediate processing, (3) blending with petroleum plastics (Rahman and Miller, 2017). Bioplastic can be manufactured from different microalgae species such as *Spirulina* sp., *Nostoc* sp., *Calothrix* sp., *Cyanobacteria, Synechocystis* sp., etc.

Johnssom and Steuer observed that *C. sorokiniana* contains 38% starch in dried form and can be used to produce starch granules, which can be further used as raw material for bioplastic production (Johnsson and Steuer, 2018). Poly(butylene adipate-co-terephthalate) (PBAT) is a by-product during biodiesel production. Torres et al. used microalgae (*Nannochloropsis gaditana*) and PBAT as a plasticizer for bioplastic production and observed that the tensile strength and impact factor of the end product is improved after the addition of PBAT (Torro et al., 2015). Tharani et al. produced bioplastic (PHA) from microalgae (*Chlorella pyrenoids*) and used glycerol as a plasticizer and observed improved plasticity property (Tharani and Ananthasubramanian, 2020). Onen et al. manufactured bioplastic by using microalgae (*Spirulina and S. plantensis*), mixed with different plasticizers and compatibilizers such as polybutylene succinate (PBS), ultra-high molecular weight PE (UHMW-PE), maleic anhydride-grafted PBS, polyethylene, etc. and observed that biocomposite produce is improved in properties like tensile strength, lower elongation, and smooth surface area (Onen et al., 2020).

Researchers are manipulating microalgae strain genetically to increase production and improve certain characters (Hlavova et al., 2015). Osnani et al. observed that by altering the carbon metabolism and overexpression of sigma factor SigE during nitrogen starvation, the PHA production from microalgae (*Synechocystis* sp. PCC 6803) is increased (Osnani et al., 2013). Rahman and Miller used gene incorporation (phbA, phbB, phbC) to produce bioplastic (PHB) from microalgae (Rahman and Miller, 2017). Kaparapu used advanced techniques such as clustered regularly interspaced short palindromic repeats (CRISPR) and Cas9 (an RNA guided cleaving enzyme) to produce bioplastic (PHA) from microalgae (Kaparapu, 2018).

Using algae to produce bioplastic rather than using bacteria can reduce the costs of production as algae can be grown in a much simpler way as they can use CO_2 and sunlight to produce useful products and for biomass growth, which are a cheaper alternative to costly media used to grow bacteria. According to recent reports, microalgae can be produced using wastewater as a substrate and can produce bioplastic also by applying biorefinery models such as wet lipid extraction procedure (WLEP) etc. through mixotrophic cultivation and cost-effective also (Zhang et al., 2019). Hence algae can be considered as a sustainable feedstock for bioplastic production.

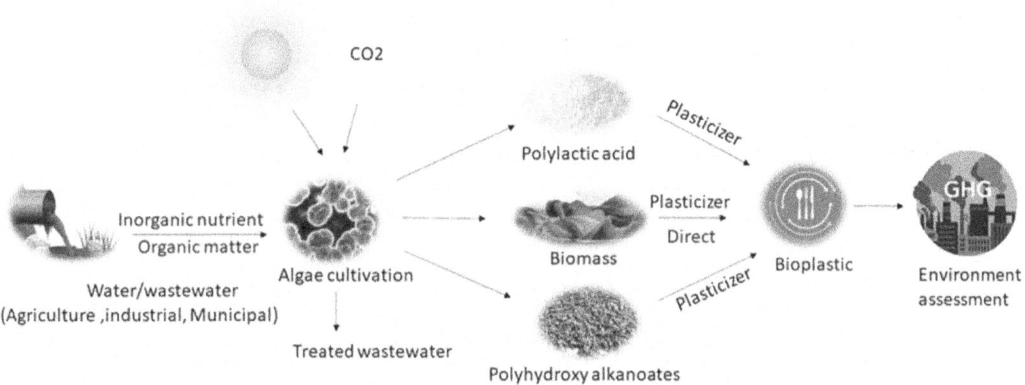

FIGURE 3.1 Pictorial flowchart of the bioplastic production process from algae.

3.3 ALGAE CULTIVATION

Microalgae belong to heterogeneous groups of microorganisms photosynthetic, eukaryotic/prokaryotic, gram-negative, and colored due to the presence of photosynthetic pigment and can be developed rapidly under adverse conditions due to simple structure and are photosynthetic microorganisms that use light and inorganic nutrient for growth. To cultivate microalgae in an artificial environment, resource input should match the natural environment. There is a different mode of microalgae cultivation such as (a) heterotrophic cultivation, (b) photoautotrophic cultivation, and (c) mixotrophic cultivation is shown in Table 3.1.

3.3.1 Heterotrophic Cultivation

In heterotrophic cultivation algae, growth is independent for light energy and has a higher growth rate, easy to scale-up, and low harvesting cost due to high cell density, but require more energy as compared to photosynthesis because requires an organic source for the initiation of the process. Microalgae cultivation through heterotrophic results in higher biomass content (Miao and Wu, 2006).

3.3.2 Mixotrophic Cultivation

Some microalgae species can use both photosynthesis and organic substrate for growth and metabolism, but growth is influenced by the dark and light phase and for maximum growth of biomass integration of both photo and hetero can be investigated (Andrade and Costa, 2007).

3.3.3 Photoautotrophic Cultivation

Photoautotrophic cultivation is the cheapest mode of microalgae cultivation. It can be carried out in open ponds as well as closed bioreactors also. Open ponds are cheaper as compared to bioreactors, but only limited species can be cultivated in the open pond (Masanne et al., 2012).

3.3.4 Photoautotrophic Open Cultivation System

The open pond is mostly rectangular with the flow from one end to another, and length, width, and depth are important parameters. Depth and length are mainly depending upon the amount of culture

TABLE 3.1
Different Culture Strategies for Algal Cultivation

S. No.	Cultivation System		Advantages	Disadvantages	References
1)	Heterotrophic	Utilize only organic compound as carbon and energy source.	• Higher growth rate and biomass density • Cheaper and simple • Easier scale-up	• Require more energy • Limited number of microalgae species can grow	Mia and Wu, 2006
2)	Mixotrophic	Both organic and CO_2 essential Perform photosynthesis as main energy source	• Higher growth rate and biomass production • Prolonged exponential growth phase	• Limited number of microalgae species can grow	Andrade and Costa, 2007
3)	Photoautotrophic	Light is used as sole energy source Convert light energy to chemical energy through photosynthetic reaction	**Open** • Installation and operating cost low • Low energy consumption • External cooling system is not required • The utilization of sunlight and atmospheric air as their surviving source	• Limited type of algae species cultivates. • High chance of contamination • Low energy consumption	Chisti, 2007 Posten, 2009
			Closed • Good control on cultivation parameter • High productivity • Reduced contamination risk • Less CO_2 losses	• Cooling system required • Cost is high • Scale-up is difficult	

used and light. The open system mainly includes big shallow ponds, circular, and raceways ponds. It includes both natural water (lake, pond, and lagoon) and artificial water (stored ground tanks with large surface area and shallow depth). In open cultivation, mixing is generally done by aeration, stirrer, rotating agitator (circular pond), and paddlewheel in raceway pond. The open system is cost-effective, sustainable because it requires less energy, is easy to maintain, and net production is also high. A major limitation in the open system is water loss, lower light utilization, limited type of algae production, and insufficient mixing (Chisti, 2007; Ugwu et al., 2008; Hallmann, 2016).

3.3.5 Photoautotrophic Closed Cultivation System

Bioreactors are mostly close system to produce microalgae and where energy is provided through an artificial source of light, offer uniform distribution, and efficient utilization of light distribution (Carvalho et al., 2006). The closed system contains four phases (solid, liquid, gas, and light phase). Microalgae as solid phase, growth media as liquid phase, CO_2 and O_2 as gas phase, and light phase. There are different types of photobioreactor such as flat plate (Banerjee and Ramaswamy, 2019), vertical column (Diaz et al., 2019), bubble column (Lopez and Rosales et al., 2017), and stirred (Guler et al., 2020) mentioned in Table 3.2. An ideal closed system would have a highly transparent

TABLE 3.2
Advantages and Disadvantages of Different Photobioreactor

S No	Photobioreactor	Advantages	Disadvantages	References
1	Vertical	Low shear force, cheaper, wall growth is absent, high CO_2 use efficiency, good mixing, high mass transfer, easy to sterile, reduced photo-oxidation, high photosynthetic efficiency	Small illumination surface	Diaz et al., 2019
2	Bubble	Good heat and mass transfer, low operating and maintenance cost	In reactor back mixing is happening, which adversely affect product conversion	Lopez-Rosales et al., 2017
3	Flat	Easy to clean, high biomass yield, large illumination surface area, good light pattern	Difficult to scale up/ expensive, wall growth, low photosynthetic efficiency, shear damage from aeration	Banerjee and Ramaswamy, 2019
4	Horizontal/Tubular	Relatively cheap, large illumination surface area, good biomass, suitable for outdoor	Some degree of wall growth, require large space, photoinhibition, dissolved oxygen CO_2 along the tube	Posten, 2009
5	Stirred tank	High-value product (Pharma), precise control over culture, small illumination surface area	Limited to heterotrophic algae	Guler et al., 2020

surface, high minimum illuminated parts, and consider best for the cultivation of specific species in a controlled environment (Ugwu et al., 2008; Posten, 2009).

During bioplastic production, main concern is the scale-up of bioreactors as it creates light and dark regions, which decreases light distribution efficiency due to the increase in the diameter of the tubes and thickness of the plates. As we go for mass production of algae in an open system or close system, it requires a large area for their cultivation. In outdoor conditions, it is very difficult to provide required light conditions as natural light may be affected by local weather. Although the outdoor system of algae cultivation can be a possible solution to decrease demand and supply gaps but are not quite successful as poor sterile conditions, improper light illumination, and improper mixing and requires a large area for cultivation (Xu et al., 2009) (Carvalho et al., 2006).

3.4 MICROALGAE HARVESTING

Microalgae is small, 3–30 um, and the harvesting process of algal biomass mostly carried out by chemical, mechanical, and biological methods but complicate downstream processing, and these methods investigate the reduction of harvesting cost (Kumar et al., 1981).

3.4.1 Centrifugation

Centrifugation is the most efficient method for microalgae recovery, and centrifugal force is used to separate the broth. This process (centrifugal force) accelerates the separation of particles based on density difference. This method is fast, preferred over gravitational sedimentation, and offers biomass recovery 89–90% in 2–3 min, feasible, easy to clean, and low risk of contamination. But

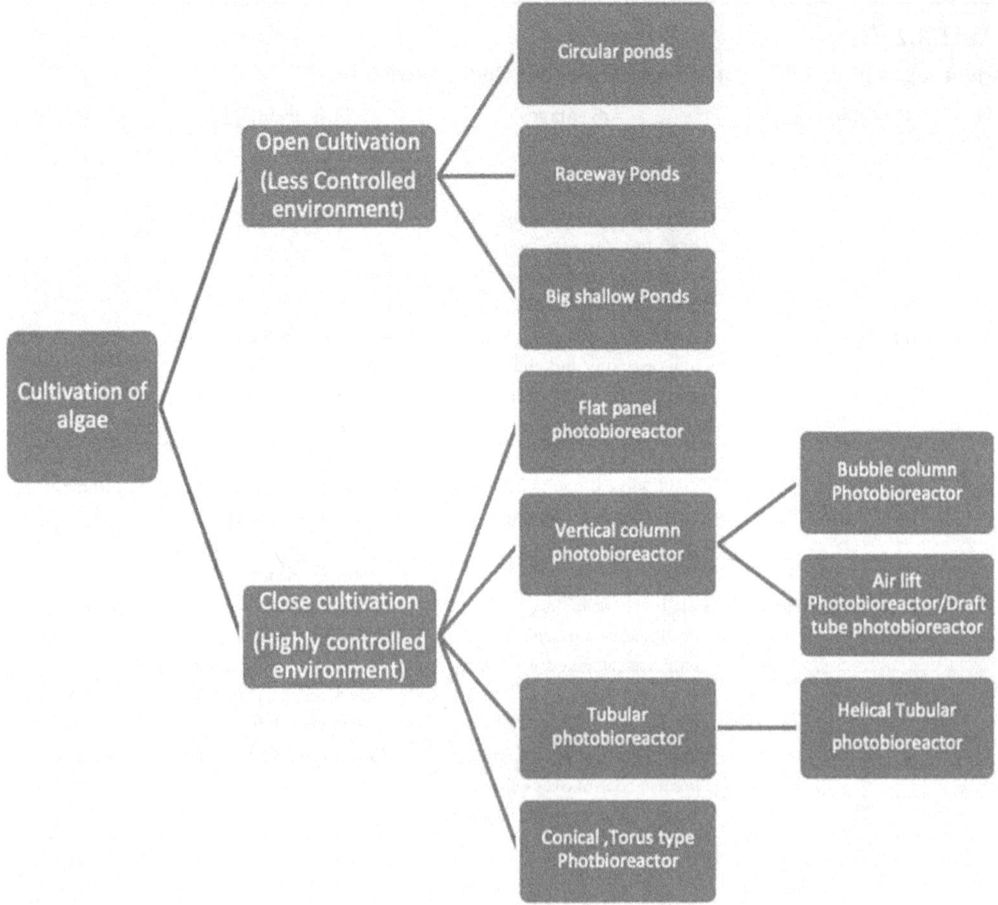

FIGURE 3.2 Different cultivation systems for algae biomass production.

the main fact, or that affects biomass recovery is cell depth, residence time, and cell settling, etc. Disc centrifuge is used for all microalgae types, sterile, and clean but the centrifugal pump requires high energy input for operation and maintenance, therefore, limits its application and not ideal for large-scale microalgae harvesting (Shah et al., 2014; Aramaki et al., 2020).

3.4.2 Filtration

Membrane technology is a common method for microalgae harvesting and does not require additional chemical, coagulants, and energy-intensive due to isothermal behavior. First-time membrane filtration was reported in 1995. It consists of recovering the solid phase from the liquid phase through meshes of a filter or membrane. Conventional filtration processes are dead-end filtration, microfiltration, ultrafiltration, vacuum filtration, and pressure filtration. Ultrafiltration and microfiltration techniques are used to recover biomass in smaller microalgae, more appropriate as compared to simple filtration (Rossi et al., 2004). The main cost of this process is associated with frequent replacement of filtration membrane due to clogging and energy consumption of the pumping system, but when compared to other harvesting methods, filtration is effective and economic competition. It was observed that through this process biomass recovery is up to 70–80%. However, the main challenge is membrane fouling for that frequent cleaning is required to result in increased operating costs (Aramaki et al., 2020). Different material is studied for microalgae filtration, and hydrophilic membrane showed a more efficient result as compared to other easy-to-clean due to negatively charged membrane fouling and cake formation.

3.4.3 Flocculation

Flocculation can be used as a previous step to biomass recovery. Flocculation is seen as a promising low-cost harvesting method for primary concentration that can facilitate and increase the efficiency of the subsequent process. Through this method, algal cells can aggregate by the addition of some coagulants and flocculants, resulting in rapid settling or floatation. The repulsive charge between microalgae is neutralized or reduced by using flocculants such as aluminum sulfate salt ($AL2(so_4)_3$), iron chloride ($FeCl_3$), and iron sulfate ($Fe(so_4)_3$) among others (Aramaki et al., 2020). Some microalgae species may naturally flocculate from environmental stimuli, such as changes in the culture medium pH, and can naturally cluster by establishing a well-defined sedimentation rate. Other microalgae species can synthesize chemicals excreted in the outer cells producing a colloidal suspension that make them difficult to settle. Besides, some species are mobile and do not settle naturally. However, this approach is limited to preconcentration, and chemicals may negatively affect the downstream process (Li et al., 2020).

Due to energy-intensive and costly harvesting techniques, addition of flocculant to promote microalgae cell aggregation for high-density floc formation started to hit the dewatering trends three decades back. Microalgae cells carry predominantly negative charge due to the ionized functional group that presents on their surface and also adsorption of ion from organic matter, causing cell-cell repulsion. Combining flocculation with a mechanical harvesting technique can improve biomass recovery. Flotation is another simple method used to facilitate biomass harvesting. It uses air microbubbles dispersed in an aqueous medium and microalgae cells are absorbed and dragged out of the aqueous medium or removed (Zhao et al., 2020; Li et al., 2020).

3.4.4 Drying

Drying is an essential part of microalgae biomass production, although harvesting results in higher biomass concentration; drying is considered for the thermochemical energy conversion platform, such as food supplement, pigment extraction, etc. Drying removes both water and other material, which can become a barrier between solvent and salute. The process requires a lot of energy (to evaporate 1kg of water at least 800 kcal of energy) to provide a final biomass concentration of at least 90%. Thus, the drying step can be an economic bottleneck of the entire process due to high energy consumption and the following advantages: it prolongs shell life, and it helps better transportation (Chen et al., 2015).

The most common drying methods are freeze-drying, drum drying, spray drying, and sun drying. However, the main drying method used for many microalgae genera is spray drying. Spray drying can yield a powdered product with better quality, such as high versatility, the possibility to pack directly, the powder is produced without the need for any grinding process, and the ease of processing control, allowing the product quality to remain constant during processing. However, its main disadvantage is the high capital and energy/operating cost. Modeling of the algal drying process and system would be a useful tool to assess the energy requirement and efficiency of the current drying method and develop novel, low-cost eco-friendly drying systems (Hosseinizand et al., 2018).

3.5 BIOPLASTIC PRODUCTION FROM ALGAE

3.5.1 Direct

Zeller et al. used *Chlorella* and *Spirulina* to develop algae-based bioplastic and grow on rice-wastewater and plasticized with glycerol in 0–30 wt% with algae:glycerol ratio being 80:20 to produce the best material property. All samples were processed via thermomechanical molding for 20 min at 150°C to make the ASTM standard for mechanical testing. The processing temperature determined by thermogravimetric analysis (TGA), and differential scanning calorimetry (DSA) and found that above 170°C algal biomass is degraded, and *spirulina* derived bioplastic is better than *Chlorella* (Zeller et al., 2013).

Kato et al. produced bioplastic beads from microalgae (*Chlamydomonas reinhardtii*). The triacylglycerol produced from *Chlamydomonas reinhardtii* without any purification and extraction is directly molded into a 7 mm bead and withstand compressive stress up to 1.7 megapascals (Kato, 2019).

3.5.2 PHA

Polyhydroxyalkanoates (PHA) are polyesters produced by different microorganisms such as bacteria, algae, etc. PHA is used in different sectors, such as packaging, lamination, pharmaceutical, agriculture, etc., due to its excellent biocompatibility and biodegradability.

Roja et al. manufactured PHA from four algal species (*Chlorella* sp., *Oscillatoria salina.*, *Leptolyngbya valderiana*, and *Synechococcus elongatus*). Analyzed the higher PHA production in all algal species and observed that 2.75 g/L reached in *Leptolyngbya valderiana* and *Chlorella* sp., whereas *Oscillatoria salina* and *Synechococcus elongatus* showed low dry cell weight 0.99 g/L and 0.33 g/L. Also, observe their thermal stability. PHA produce from different are thermally stable and below 260°C, but as compared to all PHA produced from *Leptolyngbya valderiana* is more thermally stable (Roja et al., 2019).

Nematollahi et al. produced bioplastic from microalgae (*Arthrospira platensis*) and observe carbon and nutrient effect on their production. Microalgae growth can be enhanced by supplementing extra carbon sources and limited nutrient supply. Polyhydroxybutyrate (PHB) produced from *Arthrospira platensis* MURI126 under the photoautotrophic and mixotrophic condition with and without nutrient supply. It was observed that PHB (33%) and biomass concentration (23%) increased in photoautotrophic conditions, with nutrient depletion and extra $CO_{2.}$ As compared to photoautotrophic in mixotrophic conditions, PHB production is less with nutrient depletion required extra carbon for growth (Nematollahi et al., 2020).

Costa et al. used microalgae species (*Chlorella minutissima, Synechoccus subsalaus,* and *Spirulina* sp.) and observed nitrogen, cell growth, biomass production of polyhydroxyalkanoates during the cultivation process and observed that microalgae growth reduced with limited use of nitrogen source, and biochemical composition is altered observed in degradation processes such as protein, chlorophyll, and accumulation of carbonaceous storage molecule also. PHA obtained is a long chain monomer with 14 to 18 carbon atoms and the highest yield got in 15 days dry mass concentration is 16% (Costa et al., 2018).

3.5.3 PLA

Polylactic acid (PLA) is generally produced from renewable material and different microorganisms. It has the second-highest consumption in the world. PLA is used in several sectors such as packaging, medical implants, and used as feedstock in electronics.

Bulota and Budtova prepared bioplastic from PLA and used algae as a filler and prepared by melt mixing. Algae size was below 50 μm and between 200 to 400 um and concentration from (2 to 40 wt%) varied. PLA and algae flakes dried at 80°C mixed at various ratios (40g final mass) at 180°C. Observed tensile strength and young modulus at (20% GA 80% PLA) 45 MPa, and when algae concentration increased, tensile strength decreased and young modulus increased (Buloto and Budtovo, 2015).

Arrieta et al. used microalgae (*D. antarctica*) as filler for PLA with triethyl citrate (TEC) and polyvinyl alcohol (PVA) to synthesize bioplastic. It was observed that after the addition of microalgae and electrospun PVA fiber, the oxygen barrier, crystallinity, antioxidant property, and mechanical performance of PLA are enhanced (Arrita et al., 2018).

Wu et al. used microalgae power (MAP) as filler to improve PLA property and glycidyl methacrylate (GMA) used as compatibilizer. The PLA performance was studied with and without compatibilizer. It was analyzed that as compared to PLA-MAP, PLA-g-GMA/MAP exhibit higher mechanical property due to high compatibility between polymer and MAP. Also, observe cytocompatibility by seeding of human foreskin fibroblast (FBs) in each biocomposite. Through cytotoxicity assay, it was analyzed that the compatibility of PLA-MAP is greater as compared to PLA-g-GMA/MAP but as compared to pure PLA both PLA-MAP and PLA-g-GMA/MAP were more compatible (Wu et al., 2015).

3.5.4 Others

Dianursanti et al. manufactured bioplastic from microalgae (*Spirulina plantensis*), also, observed the effect of different compatibilizer ratios (0–6%). Bioplastics produce from polyvinyl alcohol (PVC), microalgae, and compatibilizer (maleic anhydride). It was observed that the best quality of bioplastic is obtained from a 6% compatibilizer. Tensile strength reached up to 28.26 Kgf/cm^2 and elongation break reached 59.17%. The surface area of bioplastic with compatibilizer more uniform as compared to without compatibilizer (Dianursanti et al., 2019).

Torro et al. manufactured bioplastic (polybutylene succinate) (PBS) with microalgae (MA) as filler by injection molding. MA was used at different concentrations (0, 20, 30) and observed that tensile strength reached 21.6 MPa at ratio 20/80 and 18.7 30/70, the rigidity of bioplastic also improved (Torro et al., 2013).

Zhang et al. manufactured bioplastic (polypropylene) (PP) with *Chlorella* microalgae by a melt-mixing method with compatibilizer (maleic anhydride)-modified propylene. Bioplastics were produced from polypropylene-*Chlorella* (PP-*Chlorella*) and maleic anhydride-modified polypropylene and *Chlorella* (MPP-*Chlorella*). It was observed that bioplastic produced from MPP-*Chlorella* has higher tensile strength and young modulus as compared to PP-chlorella because there is stronger interaction between MPP and *Chlorella* (Zhang et al., 2000a).

Zhang et al. manufactured PVC-*Chlorella* composite material (PCCM) by molding method. PCCM shows lower elongation, more brittle as compared to pure PVC, the tensile strength of PCCM reached 30 MPa meets the requirement of rigid PVC. It was observed that as filler (*Chlorella*) concentration is increasing, bioplastic is less effective. For better result, filler concentration should be less than 20% (Zhang et al., 2000b).

3.6 LIFE CYCLE ASSESSMENT (LCA)

Abdul-Latif et al. examined carbon reduction through Carbon Emission Pith Analysis (CEPA) in microalgae-based plastic as compared to conventional plastic. It was observed that currently conventional plastic contributing 400 MT carbon and reduced carbon production required an alternative such as bioplastic production from algae (Abdul-Latif et al., 2020). According to Viswanaathan et al. microalgae can cultivate in wastewater also and offer several advantages such as does not require land, soil, and biomass generate can be transformed into different end products such as bioplastic, biofuel, food, and lipid, etc. It can valorize waste and simultaneously mitigate carbon dioxide from the environment (Viswanaathan and Sudhakar, 2019). It is examined that 1kg of microalgae biomass can consume 1.83 Kg of carbon dioxide. It can meet our market demand and at the same time reduce the carbon footprint also (Shahid et al., 2020).

Bussa et al. compared PLA produced from microalgae and plant-based routes and observed significant environmental improvement potential in terms of land use and terrestrial ecotoxicity through the microalgae route. Through the LCA study result, it was concluded that microalgae-based bioplastic is advantageous to the environment due to lower eutrophication, acidification, and

TABLE 3.3
Bioplastic Production from Different Microalgae Species

S. No	Species	Percentage of Material	Type of product	Product	Reference
1	*Chlorella/Spirulina*	Glycerol (0–30%) by weight	100% algal based	Biofilm	Zeller et al., 2013
2	*Chlamdomonas reinhardtii*	Triacylglycerol from *Chlamdomonas*	100% algal based	Beads from triacylglycerol	Kato et al., 2019
3	*Chlorella, Oscillatoria, Leptolyngbya and Synechococcus*	*Chlorella/Leptolyngbya* (2.75 g/L) *Oscillatoria* 0.99 g/L *Synechococcus* 0.33 g/L	PHA	Biofilm	Roja et al., 2019
4	*Arthrospira platensis*	NA	PHB (33%)	Biofilm	Nematollahi et al., 2020
5	*Chlorella, Synechoccus, spirulina*	NA	PHB (16%)	Biofilm	Costa et al., 2018
6	*(Mixed culture) Red, Green, Brown algae*	MA as filler (2–40%) by weight	PLA/MA	Biofilm	Bulota and Budtova, 2015
7	*D. antarctica*	PLA Triethyl citrate (TEC) Polyvinyl alcohol (PVA)	MA/PLA/TEC/ PVA	Biofilm	Arrieta et al., 2018
8	*Spirulina*	Maleic anhydride (MA) 0–6%	MA/MA/polyvinyl alcohol (PVC)	Biofilm	Dianursanti et al., 2019
9	*B. braunii*	MA as filler (0–20%)	Polybutylene succinate (PBS)/ MA	Biofilm	Torro et al., 2013
10	*Chlorella*	*Chlorella*/PVC composite	PVC/*Chlorella*	Biofilm	Zhang et al., 2000b

human toxicity. But microalgae might worsen the performance of PLA concerning global warming, energy demand, increase in depletion of material. But to date, the effect of microalgae on PLA blends has not been studied properly. There is a need for proper additive and LCA studies to understand their mechanism and get more insight to result in a better end product (Bussa et al., 2019).

Beckstrom et al. studied the environmental impact of microalgae-based bioplastic as compared to conventional-based plastic. They include nine different production models and observe mass, energy balance, and cost and their impact. Compared to greenhouse gas intensities and observed that bioplastic produced from microalgae reduced 67 to 116% GHG as compared to conventional plastic, they are economically stable and sustainable (Beckstrom et al., 2020).

Posda et al. presented ten different biorefinery and nine different final products compared. Also observe different technologies for algal cultivation, harvesting, cell disruption, and lipid extraction. Cultivation on wastewater was considered as a particular case scenario analysis for the various confirmation. The result of this study however does not provide insight into the performance of microalgae-based plastic as compared to conventional plastic (Posda et al., 2016).

3.7 CONCLUSION

Microalgae species have been studied for several years for various applications. These microorganisms do not require land and can be cultured naturally in an open environment such as a pond, but also, they can be grown under specific conditions for commercial purposes. In recent decades, the investigation has revealed that they are used as a valuable resource in various fields of cosmetics, biofuel, and bioplastic production. In this chapter the current situation of bioplastic from microalgae is examined and their effect on the environment. PHA/PLA production from microalgae is already reported but as compared to PHA, PLA is less studied. According to literature study, it was analyzed that there is still need of new development of technologies for bioplastic production from microalgae to overcome economic feasibility. The main problem is industrial scale-up, which prevents the wider usage of bioplastic product from microalgae in the market. As the demand for bioplastic increases, there will be a place for microalgae-derived bioplastic. Furthermore, microalgae feedstock, i.e., grown on the waste stream will provide the lowest cost for the biorefinery system.

REFERENCES

Abdo, S.M. and Ali, G.H., 2019. Analysis of polyhydroxybutrate and bioplastic production from microalgae. *Bulletin of the National Research Centre*, *43*(1).

Abdul-Latif, N.I.S., Ong, M.Y., Nomanbhay, S., Salman, B. and Show, P.L., 2020. Estimation of carbon dioxide (CO2) reduction by utilization of algal biomass bioplastic in Malaysia using carbon emission pinch analysis (CEPA). *Bioengineered*, *11*(1), pp. 154–164.

Andrade, M.R. and Costa, J.A., 2007. Mixotrophic cultivation of microalga Spirulina platensis using molasses as organic substrate. *Aquaculture*, *264*(1–4), pp. 130–134.

Andrady, A.L. and Neal, M.A., 2009. Applications and societal benefits of plastics. *Philosophical Transactions of the Royal Society B: Biological Sciences*, *364*(1526), pp. 1977–1984.

Aramaki, T., Watanabe, M.M., Nakajima, M. and Ichikawa, S., 2020. Bench-scale dehydration of a native microalgae culture by centrifugation, flocculation and filtration in Minamisoma city, Fukushima, Japan. *Bioresource Technology Reports*, p. 100414.

Arrieta, M.P., de Dicastillo, C.L., Garrido, L., Roa, K. and Galotto, M.J., 2018. Electrospun PVA fibers loaded with antioxidant fillers extracted from Durvillaea antarctica algae and their effect on plasticized PLA bio nanocomposites. *European Polymer Journal*, *103*, pp. 145–157.

Banerjee, S. and Ramaswamy, S., 2019. Dynamic process model and economic analysis of microalgae cultivation in flat panel photobioreactors. *Algal Research*, *39*, p. 101445.

Beckstrom, B.D., Wilson, M.H., Crocker, M. and Quinn, J.C., 2020. Bioplastic feedstock production from microalgae with fuel co-products: A techno-economic and life cycle impact assessment. *Algal Research*, *46*, p. 101769.

Bulota, M. and Budtova, T., 2015. PLA/algae composites: Morphology and mechanical properties. *Composites Part A: Applied Science and Manufacturing*, *73*, pp. 109–115.

Bussa, M., Eisen, A., Zollfrank, C. and Röder, H., 2019. Life cycle assessment of microalgae products: State of the art and their potential for the production of polylactic acid. *Journal of Cleaner Production*, *213*, pp. 1299–1312.

Carvalho, A.P., Meireles, L.A. and Malcata, F.X., 2006. Microalgal reactors: A review of enclosed system designs and performances. *Biotechnology Progress*, *22*(6), pp. 1490–1506.

Chae, Y. and An, Y.-J., 2018. Current research trends on plastic pollution and ecological impacts on the soil ecosystem: A review. *Environmental Pollution*, *240*, pp. 387–395.

Chen, C.L., Chang, J.S. and Lee, D.J., 2015. Dewatering and drying methods for microalgae. *Drying Technology*, *33*(4), pp. 443–454.

Chisti, Y., 2007. Biodiesel from microalgae. *Biotechnology Advances*, *25*(3), pp. 294–306.

Costa, S.S., Miranda, A.L., Andrade, B.B., de Jesus Assis, D., Souza, C.O., de Morais, M.G., Costa, J.A.V. and Druzian, J.I., 2018. Influence of nitrogen on growth, biomass composition, production, and properties of polyhydroxyalkanoates (PHAs) by microalgae. *International Journal of Biological Macromolecules*, *116*, pp. 552–562.

Dianursanti, Noviasari, C., Windiani, L. and Gozan, M., 2019, April. Effect of compatibilizer addition in Spirulina platensis based bioplastic production. In *AIP Conference Proceedings* (Vol. 2092, No. 1, p. 030012). AIP Publishing LLC.

Díaz, J.P., Inostroza, C. and Fernández, F.A., 2019. Fibonacci-type tubular photobioreactor for the production of microalgae. *Process Biochemistry*, *86*, pp. 1–8.

European Bioplastic, 2018. New market data: The positive trend for the bioplastics industry remains stable. www.european-bioplastics.org/new-market-data-the-positive-trend-for-the-bioplastics-industry-remains-stable/.

Ezgi, B.A.H.D., 2015. A review: Investigation of bioplastics. *Journal of Civil Engineering and Architecture*, *9*, pp. 188–192.

Getachew, A. and Woldesenbet, F., 2016. Production of biodegradable plastic by polyhydroxybutyrate (PHB) accumulating bacteria using low cost agricultural waste material. *BMC Research Notes*, *9*(1), pp. 1–9.

Geyer, R., 2020. Production, use, and the fate of synthetic polymers. In *Plastic Waste and Recycling* (pp. 13–32). Academic Press.

Goswami, G., Bang, V. and Agarwal, S., 2015. Diverse applications of algae. *International Journal of Advance Research in Science and Engineering*, *4*(1), pp. 1102–1109.

Guiry, M.D., 2012. How many species of algae are there? *Journal of Phycology*, *48*(5), pp. 1057–1063.

Guler, B.A., Deniz, I., Demirel, Z., Oncel, S.S. and Imamoglu, E., 2020. Computational fluid dynamics modelling of stirred tank photobioreactor for Haematococcus pluvialis production: Hydrodynamics and mixing conditions. *Algal Research*, *47*, p. 101854.

Hallmann, A., 2016. Algae biotechnology—green cell-factories on the rise. *Current Biotechnology*, *4*(4), pp. 389–415.

Hlavova, M., Turoczy, Z. and Bisova, K., 2015. Improving microalgae for biotechnology—From genetics to synthetic biology. *Biotechnology Advances*, *33*(6), pp. 1194–1203.

Horton, A.A. and Barnes, D.K., 2020. Microplastic pollution in a rapidly changing world: Implications for remote and vulnerable marine ecosystems. *Science of the Total Environment*, *738*, p. 140349.

Hosseinizand, H., Sokhansanj, S. and Lim, C.J., 2018. Studying the drying mechanism of microalgae Chlorella vulgaris and the optimum drying temperature to preserve quality characteristics. *Drying Technology*, *36*(9), pp. 1049–1060.

Johnsson, N., and Steuer, F. (2018). Bioplastic material from microalgae : Extraction of starch and PHA from microalgae to create a bioplastic material (Dissertation).

Kaparapu, J., 2018. Polyhydroxyalkanoate (PHA) production by genetically engineered microalgae: A review. *Journal on New Biological Reports*, *7*, pp. 68–73.

Kato, N., 2019. Production of crude bioplastic-beads with microalgae: proof-of-concept. *Bioresource Technology Reports, 6*, pp. 81–84.

Khan, M.I., Shin, J.H. and Kim, J.D., 2018. The promising future of microalgae: Current status, challenges, and optimization of a sustainable and renewable industry for biofuels, feed, and other products. *Microbial Cell Factories*, *17*(1), p. 36.

Kumar, H.D., Yadava, P.K. and Gaur, J.P., 1981. Electrical flocculation of the unicellular green alga Chlorella vulgaris Beijerinck. *Aquatic Botany*, *11*, pp. 187–195.

Lebreton, L. and Andrady, A., 2019. Future scenarios of global plastic waste generation and disposal. *Palgrave Communications*, *5*(1), pp. 1–11.

Li, S., Hu, T., Xu, Y., Wang, J., Chu, R., Yin, Z., Mo, F. and Zhu, L., 2020. A review on flocculation as an efficient method to harvest energy microalgae: Mechanisms, performances, influencing factors and perspectives. *Renewable and Sustainable Energy Reviews*, *131*, p. 110005.

López-Rosales, L., García-Camacho, F., Sánchez-Mirón, A., Contreras-Gómez, A. and Molina-Grima, E., 2017. Modeling shear-sensitive dinoflagellate microalgae growth inbubble column photobioreactors. *Bioresource Technology*, *245*, pp. 250–257.

Luengo, J.M., García, B., Sandoval, A., Naharro, G. and Olivera, E.R., 2003. Bioplastics from microorganisms. *Current Opinion in Microbiology*, *6*(3), pp. 251–260.

Miao, X. and Wu, Q., 2006. Biodiesel production from heterotrophic microalgal oil. *Bioresource Technology*, *97*(6), pp. 841–846.

Msanne, J., Xu, D., Konda, A.R., Casas-Mollano, J.A., Awada, T., Cahoon, E.B. and Cerutti, H., 2012. Metabolic and gene expression changes triggered by nitrogen deprivationin the photoautotrophically grown microalgae *Chlamydomonas reinhardtii* and *Coccomyxa* sp. C-169. *Phytochemistry*, *75*, pp. 50–59.

Nematollahi, M.A., Laird, D.W., Hughes, L.J., Raeissosadati, M. and Moheimani, N.R., 2020. Effect of organic carbon source and nutrient depletion on the simultaneous production of a high value bioplastic and a specialty pigment by *Arthrospira platensis*. *Algal Research*, *47*, p. 101844.

Onen Cinar, S., Chong, Z.K., Kucuker, M.A., Wieczorek, N., Cengiz, U. and Kuchta, K., 2020. Bioplastic production from microalgae: A review. *International Journal of Environmental Research and Public Health*, *17*(11), p. 3842.

Osanai, T., Numata, K., Oikawa, A., Kuwahara, A., Iijima, H., Doi, Y., Tanaka, K., Saito, K. and Hirai, M.Y., 2013. Increased bioplastic production with an RNA polymerase sigma factor SigE during nitrogen starvation in Synechocystis sp. PCC 6803. *DNA Research*, *20*(6), pp. 525–535.

Posada, J.A., Brentner, L.B., Ramirez, A. and Patel, M.K., 2016. Conceptual design of sustainable integrated microalgae biorefineries: Parametric analysis of energy use, greenhouse gas emissions and techno-economics. *Algal Research*, *17*, pp. 113–131.

Posten, C., 2009. Design principles of photo-bioreactors for cultivation of microalgae. *Engineering in Life Sciences*, *9*(3), pp. 165–177.

Rahman, A. and Miller, C.D., 2017. Microalgae as a Source of Bioplastics. *Algal Green Chemistry*, pp. 121–138.

Roja, K., Sudhakar, D.R., Anto, S. and Mathimani, T., 2019. Extraction and characterization of Polyhydroxyalkanoates from marine green alga and cyanobacteria. *Biocatalysis and Agricultural Biotechnology*, *22*, p. 101358.

Rossi, N., Jaouen, P., Legentilhomme, P. and Petit, I., 2004. Harvesting of cyanobacterium *Arthrospira platensis* using organic filtration membranes. *Food and Bioproducts Processing*, *82*(3), pp. 244–250.

Saharan, B. and Sharma, D., 2012. Bioplastics-for sustainable development: A review. *International Journal of Microbial Resource Technology*, *11*, pp. 11–23.

Shahid, A., Malik, S., Zhu, H., Xu, J., Nawaz, M.Z., Nawaz, S., Alam, M.A. and Mehmood, M.A., 2020. Cultivating microalgae in wastewater for biomass production, pollutant removal, and atmospheric carbon mitigation; A review. *Science of the Total Environment*, *704*, p. 135303.

Sudhakar, M.P. and Viswanaathan, S., 2019. Algae as a Sustainable and Renewable Bioresource for Bio-Fuel Production. *New and Future Developments in Microbial Biotechnology and Bioengineering*, pp. 77–84.

Tharani, D. and Ananthasubramanian, M., 2020. Microalgae as sustainable producers of bioplastic. *Microalgae Biotechnology for Food, Health and High Value Products*, pp. 373–396.

Toro, C., Reddy, M.M., Navia, R., Rivas, M., Misra, M. and Mohanty, A.K., 2013. Characterization and application in bio composites of residual microalgal biomass generated in third generation biodiesel. *Journal of Polymers and the Environment*, *21*(4), pp. 944–951.

Torres, S., Navia, R., Campbell Murdy, R., Cooke, P., Misra, M. and Mohanty, A.K., 2015. Green composites from residual microalgae biomass and poly (butylene adipate-co-terephthalate): Processing and plasticization. *ACS Sustainable Chemistry & Engineering*, *3*(4), pp. 614–624.

Ugwu, C.U., Aoyagi, H. and Uchiyama, H., 2008. Photobioreactors for mass cultivation of algae. *Bioresource Technology*, *99*(10), pp. 4021–4028.

Viswanaathan, S. and Sudhakar, M.P., 2019. Microalgae: Potential agents for CO2 mitigation and bioremediation of wastewaters. In *New and Future Developments in Microbial Biotechnology and Bioengineering* (pp. 129–148). Elsevier.

Wu, T.Y., Yang, M.C. and Hsu, Y.C., 2015. Improvement of cytocompatibility of polylactide by filling with marine algae powder. *Materials Science and Engineering: C*, *50*, pp. 309–316.

Xu, L., Weathers, P.J., Xiong, X.-R. and Liu, C.-Z., 2009. Microalgal bioreactors: Challenges and opportunities. *Engineering in Life Sciences*, *9*(3), pp. 178–189.

Zeller, M.A., Hunt, R., Jones, A. and Sharma, S., 2013. Bioplastics and their thermoplastic blends from Spirulina and Chlorella microalgae. *Journal of Applied Polymer Science*, *130*(5), pp. 3263–3275.

Zhang, C., Show, P.-L. and Ho, S.-H., 2019. Progress and perspective on algal plastics—A critical review. *Bioresource Technology*, *289*, p. 121700.

Zhang, F., Endo, T., Kitagawa, R., Kabeya, H. and Hirotsu, T., 2000a. Synthesis and characterization of a novel blend of polypropylene with Chlorella. *Journal of Materials Chemistry*, *10*(12), pp. 2666–2672.

Zhang, F., Kabeya, H., Kitagawa, R., Hirotsu, T., Yamashita, M. and Otsuki, T., 2000b. An exploratory research of PVC-Chlorella composite material (PCCM) as effective utilization of Chlorella biologically fixing CO2. *Journal of Materials Science*, *35*(10), pp. 2603–2609.

Zhao, Z., Li, Y., Muylaert, K. and Vankelecom, I.F., 2020. Synergy between membrane filtration and flocculation for harvesting microalgae. *Separation and Purification Technology*, *240*, p. 116603.

4 Economic Potential of Algae for Enhancing Blue Economy

Ramachandran Uma and G. Rani

ABBREVIATIONS

SDG	Sustainable Development Goals
BE	Blue Economy
CC	Climate Change
UNCLOS	United Nations Convention on the Law of the Sea
CBD	Convention on Biodiversity
NBA	National Biodiversity Authority
FAO	
WIO	Western Indian Ocean
LC-PUFA	Long Chain-Poly Unsaturated Fatty Acids
GOLD	Genome On Line Database

4.1 INTRODUCTION

Adopted in January 2016, the Sustainable Development Goals (SDGs) are a universal call to action to end poverty, protect the planet, and ensure that all people enjoy peace and prosperity. SDG 14 "Life Below Water" aims to Conserve and sustainably use oceans, seas, and marine resources and provides a framework of targets to guide strategic actions by stakeholders.

"Blue Economy" (BE) simultaneously promotes economic growth, environmental sustainability, social inclusion, and helps strengthen the marine ecosystems. This involves using smart shipping to lessen impacts on environment, inclusive and improved lives for all, harnesses renewable energy, creates jobs, reduces poverty, takes action against illegal fishing, conserves marine life and oceans, protects coastal communities from the impacts of CC, and also tackles marine litter and ocean pollution issues. The key ingredients of BE are: (a) 10th COP to Convention on Biodiversity (CBD), (b) UNFCCC including international policy instruments under Island Ocean States (IOS) that encompasses Large Island Developing States (LIDS) and Small Island Developing States (SIDS). IOS face issues like limited space of land and sea, having national and international dependencies for food, livelihoods, transport, etc., and (c) few SDGs. Several studies have been conducted on blue economy, and literature review reveals that BE has links not only with SDG-14 but also SDGs 15, 16, and 17 (Life on land; Peace, justice, and strong institutions; Partnerships for goals), while several stakeholders preferred SDGs 3 and 4 (Good health and well-being; Decent work and economic growth) in the context of BE, reiterating the importance of stakeholders. Therefore, steps to achieve BE seem to be multi-stakeholder partnerships at national, regional, and local levels (Uma Ramachandran, 2020).

The BE concept is subject to a multilayer regulatory framework under the United Nations Convention on the Law of the Sea (UNCLOS) and other national, regional, multilateral as well as sectoral governance regimes which has also been formed in India. BE can address some of the concerns associated with economic and environmental vulnerability, such as those associated with remoteness, by fostering international and regional cooperation under an **"ocean space approach"**. For instance, a recent initiative of the government of India—the *Sagarmala Project* is meant for

DOI: 10.1201/9781003219156-6

enhancing the performance of the country's logistics sector. The program envisages unlocking the potential of waterways and coastline to minimize infrastructural investments required to meet these targets, and one of the interests is marine spatial planning.

An ocean space approach requires the development of a more coherent, integrated, and structured framework that considers economic potential of all marine natural resources, including seaways and energy sources from the oceans. As mentioned previously, BE offers significant development challenges and opportunities for vulnerable coastal areas in sectors such as sustainable fisheries and aquaculture, renewable marine energy, marine bio-prospecting, maritime transport, and marine and coastal tourism. Many of the vulnerable coastal areas could explore ways to mainstream generation of renewable energy into their national and regional planning and energy mix. Similarly, a great potential exists to increase offshore wind-generated electricity and use of algae biomass in the production of fuel. Bio-prospecting of marine genetic resources offers interesting opportunities for benefit-sharing and creation of scientific capacities in vulnerable coastal regions, especially in relation to **pharmaceuticals, cosmetics, and food products** development.

Therefore, there is a need to mainstream BE to address SDGs. Consideration should be given to a comprehensive goal focusing on use of marine ecosystems and resources within ecological limits. Further, it is suggested that there is a need to find opportunities to engage in the process of global reporting and assessments of the State of the Marine Environment, including socioeconomic aspects, under national coastal development program. This process will lead to key findings and conclusions that can shape the future governance of the oceans and seas. In conclusion, BE conceptualizes **Oceans** as **Development Spaces.** As we acknowledge that we receive multiple benefits from marine habitats such as coastal protection, fish production, tourism, etc., it is only by valuing **Blue Capital** in the decision-making process, several options will exist to strengthen the BE. Developing BE sustainably and responsibly is key for a country like India with a large coastline and home to a significant share of the country's coastal population. While the importance of a sustainable BE has been recognized by several countries including India, the issue has still not garnered enough attention at the regional level. Thus, there is a need to examine the barriers and enablers to the development of BE in further detail and identify strategies to promote its sustainable development in the long-term.

The objective of this chapter aims to nurture an ecosystem-based approach to safeguard the ecology of marine resources, while exploring a window of opportunities to improve local economies significantly, also considering social benefits to a host of coastal population as well as a range of stakeholders. While there are numerous opportunities, this chapter specifically deals with some key economic opportunities associated with marine algal resources in the context of BE.

4.2 ALGAL RESOURCE POTENTIAL

Algae are present in water bodies, common in terrestrial environments, and are also found on snow and ice. Seaweeds grow mostly in shallow marine waters, below 100 m depth. Few species such as *Navicula pennata* have been recorded to a depth of 360 m (Round, 1981). An alga, *Ancyloneman ordenskioeldii*, was found in Greenland in the "Dark Zone", causing an increase in the rate of melting ice sheet (Space.com, 2018) and also were found in the Italian Alps, after pink ice appeared on parts of the Presena glacier (Sky News, 2020). These could be due to climate change.

Algae play significant roles in aquatic ecology. Marine algae form the base of most aquatic food chains, or rather ecological pyramid, are important in biogeochemical cycling and serve as habitats for many organisms in aquatic ecosystems (Carpenter and Kitchel, 1993; Wetzel R, 1996; De Clerck and Coppejans, 2002). Marine algae are either prokaryotic (unicellular organisms that lack membrane-bound structures, including nucleus) or eukaryotic (organisms whose cells have a nucleus and other organelles enclosed by a plasma membrane). They belong to two main subgroups: macroalgae and microalgae (Figure 4.1). Microscopic forms called phytoplanktons live suspended in the water column and provide the food base for most of the marine food chains. In very

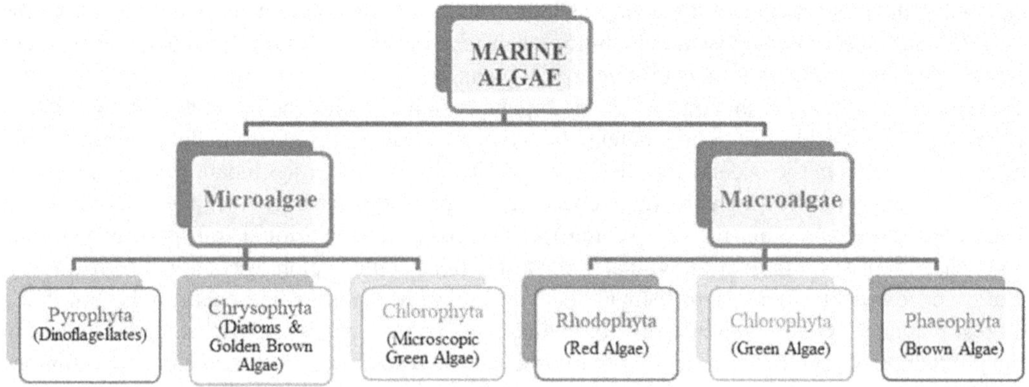

FIGURE 4.1 Classification of marine algae.

high densities "algal blooms" are formed that may discolor the water, poison, or asphyxiate other life-forms. Algae can be used as indicator organisms (bioindicators) to monitor pollution in various aquatic systems (Omar and Wan, 2010). Algal metabolism is sensitive to various pollutants. Due to this, the species composition of algal populations may shift in the presence of chemical pollutants (Necchi, 2016). To detect these changes, algae can be sampled and maintained in laboratories with relative ease (Necchi, 2016).

4.2.1 Global Overview

Globally algae have been estimated to range from 30,000 to more than 1 million species. Several regional estimates of the algal resources exist. Few of them are the National Museum of Natural History consisting of approximately 3,200,500 dried specimens. UK Biodiversity Steering Group Report estimated around 20,000 algal species in the UK, and around 500 to 5,500 species of red algae have been estimated worldwide. An online taxonomic database for algae, namely AlgaeBase (www.algaebase.org), is an attempt to arrive at a more accurate estimate using species numbers in phyla and classes. There are uncertainties regarding what organisms need to be classified as algae and what a species is in the context of the various algal phyla and classes. A conservative approach made by AlgaeBase has resulted in an estimate of 72,500 algal species. Names for 44,000 of them were already published and available in public domain. Besides these, several more names have been processed by AlgaeBase. Some published estimates of diatom numbers are of over 200,000 species, which would result in four to five diatom species for every other algal species (Guiry, 2012).

4.2.2 Indian Overview

Total number of known species of algae both with global and Indian figures is represented in the next section (Figure 4.2). Number of endemic algae in India is 1,924 (Singh and Dash, 2018), while number of threatened algal species are not known.

The 2018 UNEP-WCMC-IUCN-NGS report revealed that India's land area represents only 2–4% of the world but interestingly comprises 7–8% of total known plant and animal taxa globally. The Biological Diversity Act enacted in 2002 in the country aimed in encouraging conservation of living resources, and a National Biodiversity Authority (NBA) was established in 2003. This is due to the country's high degree of endemism established by the fact that it ranks fourth out of 34 biodiversity-rich hotspots in the world, as stated in the 2019 Sixth National Report for the CBD (CBD-CHM, 2018). The biogeographic zones encompass high degree of variation in ecological environments, ranging from grasslands, alpine forests, and desert to wetlands and coastal and marine habitats.

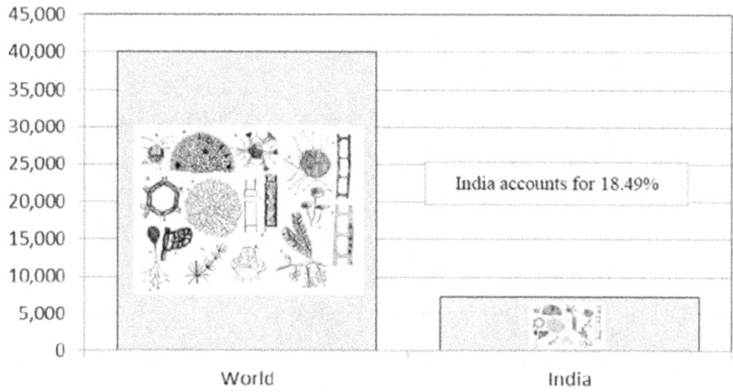

FIGURE 4.2 Total number of known algae.

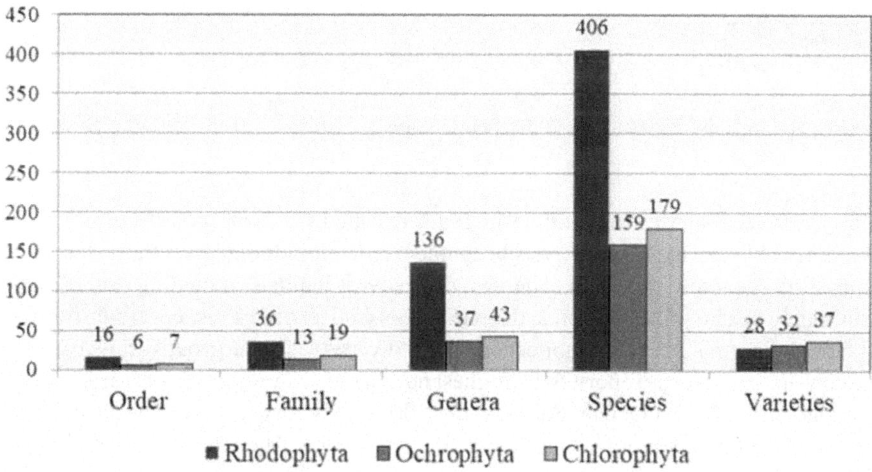

FIGURE 4.3 Seaweed resources of India.

The country is bestowed with a long coastline of approximately 7,500 km with an exclusive economic zone of 2.5 million km^2 and a vast shelf area of 0.13 million km^2. Despite rich availability of seaweeds, a renewable marine resource, they have been underutilized essentially due to the lack of realization of its economic potential. Vast and diverse habitats, including estuarine mangrove vegetation, sandy beaches, rocky shores, deep tide pools, cliffs and caves, coral substrates, and artificial offshore structures, provide abundant support for diverse seaweed groups to thrive in intertidal and subtidal waters. India has reported the highest number of seaweed taxa compared to all the nations bordering the Indian Ocean (Chitale *et al.*, 2014; Sahoo *et al.*, 2001). India with its long coastlines is blessed with several algal resources and have been put to use for multiple purposes. A qualitative survey undertaken by Subba Rao and Mantri (2006) revealed the species composition of marine algae along the Indian coast (Figure 4.3). Detailed survey in the Sunderbans by Sarkar and Sekh (2014) revealed that the region hoards a rich diversity of marine algae mostly contributed by red, brown, and green algae also in the Indian coast. This has been attributed to the complex habitats created by the dominant benthic primary producers along with the phytoplankton of the productive shallow waters, further stimulated by terrestrial inputs and upwelling incidents. The paper suggests that energy and habitats play an important role in determining the present-day distributional patterns observed for marine algae along with the role of historical processes.

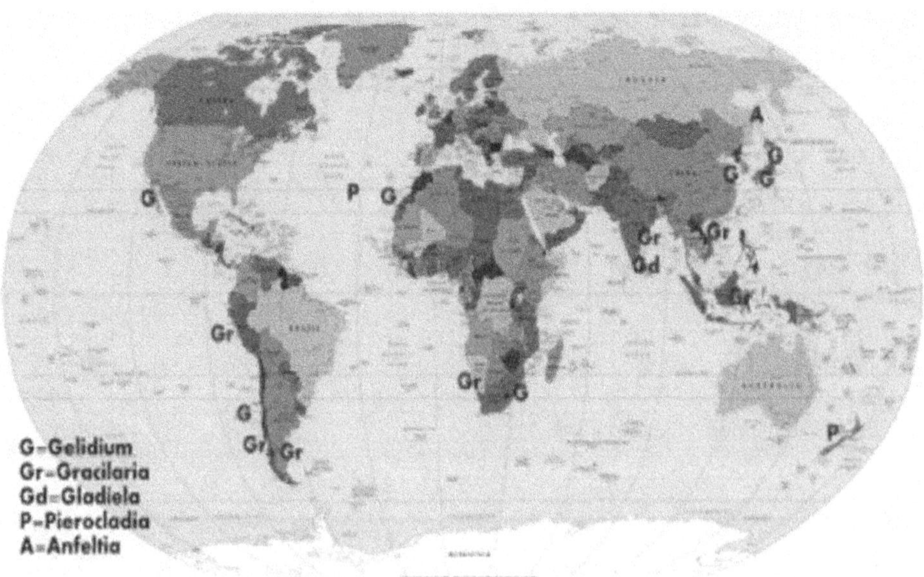

FIGURE 4.4 Agarophyte seaweed: world distribution map.

Similarly, seaweed survey undertaken in India revealed rich varieties (Figure 4.4). Seaweed, which is a renewable marine resource, has been underutilized in India, essentially due to the lack of realization of its economic potential. The vast diversity of habitats, including estuarine mangrove vegetation, sandy beaches, rocky shores, deep tide pools, cliffs and caves, coral substrates, and artificial offshore structures provide ample support for diverse seaweed groups to thrive in intertidal and subtidal waters. India has reported the highest number of seaweed taxa compared to all the other nations bordering the Indian Ocean (Sahoo *et al.*, 2001). This current trend reveals that the number of species described by taxonomists are increasing worldwide. The most comprehensive estimates for Indian waters reported that there are 841 seaweed species belonging to 216 genera of 68 families present in Indian waters as depicted in Figure 4.3 (Guiry, 2012). An estimate by Kaliaperumal (2017) shows 871 species from Indian waters. The more recent data (Mantri *et al.*, 2020) provides the phylum-wise distribution of algae, as shown in the next section (Figure 4.3), which has been consolidated from works of several researchers studying seaweed biodiversity. Therefore, Mantri *et al.* (2020) aimed to consolidate and report all information related to seaweed biodiversity, identified gaps, and recommend future research directions with respect to algae and highlighting the importance of maintaining a database that could be integrated into AlgaeBase.

Subba Rao and Mantri (2006) reported the estimated availability of seaweed biomass for the entire coast to be ranging from 677,308.87 to 682,758.87 tons (fresh weight) accounting for nearly 3.4% of the global seaweed resource. Figure 4.5 depicts updated details of algae/seaweed production, volume of algae produced globally as well as its major utilization. The paper concluded that the outcome of such an exercise would not only be helpful in discovering new taxa but would also essentially help in developing conservation policies.

According to FAO (2014, 2016) major seaweed-producing countries are China, Indonesia, and Philippines, which cultivated greatest diversity of seaweed species (7, 6, and 4, respectively). Of the top seven most cultivated seaweeds, three are used mainly for hydrocolloid extraction, namely, *Eucheuma* spp. and *Kappaphycus alvarezii* for carrageenans; *Gracilaria* spp. for agar; *Saccharina japonica*, *Undaria pinnatifida*, *Pyropia* spp., and *Sargassum fusiforme* were most important in human food usage. Seaweed consumption in Southeast Asia has been common, traditional, and depended on taste and price. In non-Asian countries, like European and USA markets, its use as food considered

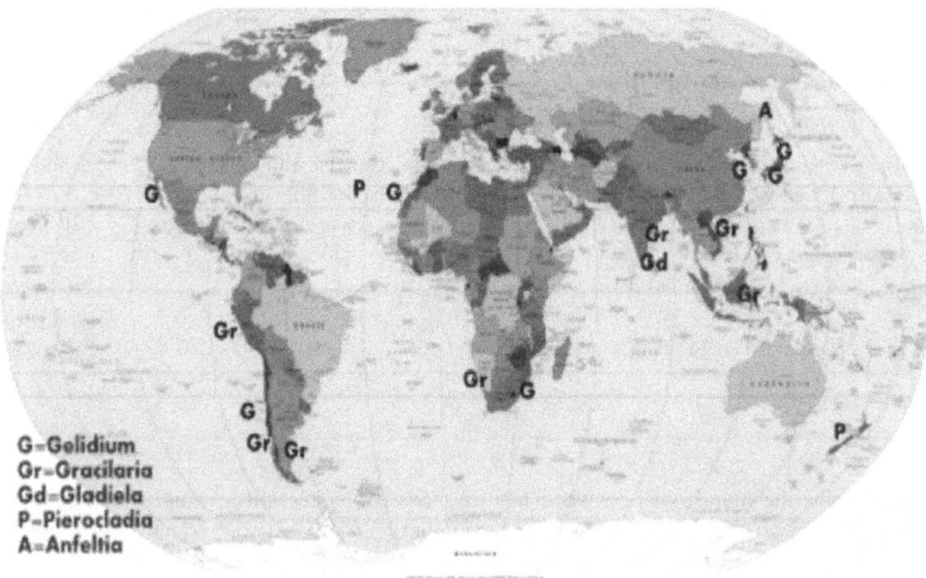

FIGURE 4.5 a. Major producers of seaweed across the world (%); b. Algae Market Value Share by Region 2018 (in Mn US$).

additional parameters such as nutritional value and "health food" with strong consumer preference towards organic, sustainable, and fair trade products. This represented low impacts on both environment and biodiversity (Chapman *et al.*, 2015; Gomez Pinchetti and Martel Quintana, 2016). However, the global market share of seaweed farming production used for food and "other uses", that is, other than for hydrocolloids, is still below 1% of the total biomass production (Figure 4.6a, b).

4.2.2.1 Global Algae Market

There has been a gradual growth in the global algae market. This is due to many algae oil production facilities that are cropping up but yet to be fully commercialized. In this era of climate change there has been serious concerns revolving around emission of greenhouse gases (GHGs), which has led to shift in the focus towards adopting renewable energy sources. Algae are becoming prominent and are being targeted. Apart from a rising demand for algae-based biofuels in road, marine, and aviation applications, plastic industries have also increased its demand for algae to produce biodegradable plastics. In the year 2018, global algae market was valued at US$ 717.14 Mn and is projected to touch US$ 1,365.8 Mn by 2027, at a CAGR of 7.42%. In terms of volume, the market is projected to expand at a 5.35% CAGR between 2019 and 2027 (Transparency Market Research, 2020).

4.3 EXPLORING OPPORTUNITIES AND SUITABLE TECHNOLOGIES UNDER BLUE ECONOMY

This section describes different technologies using marine algae, such as study of the metabolites of interest for subsequent commercial/industrial applications in the context of blue economy both and covers these aspects across the world as well as in India.

4.3.1 Women's Role and Employment in Seaweed Aquaculture

Flower *et al.* (2017) studied in detail the role of women in seaweed aquaculture in the Western Indian Ocean and Southeast Asia. According to him in most developing countries, involvement

TABLE 4.1
Type of Plastics from Marine Algae

Procedure Followed	Plastic Type
Adding denatured algal biomass (for e.g., filamentous macroalga Cladophora) to petroleum-based plastics (like polyurethane and polyethylene) as filler to increase their biodegradability	Hybrid plastics
Made from cellulose component of the algal biomass left after the extraction of algal oil (about 30% of the total algal biomass)	Cellulose-based plastics
Produced by the bacterial fermentation of algal biomass	Poly-lactic acid (PLA)
Derived from the bacterial digestion of algal biomass	Bio-polyethylene

(*Source:* Mc Hugh, 2003)

of women forms the majority of seaweed farming. Several successful examples and case studies have widely demonstrated and evaluated their contribution to this sector and have made significant advances in the sustainability of seaweed farming for more than four decades. The roles of women were found to be complex, including hands-on farming activities as well as small-scale processing units to produce value-added products based on the cultivated seaweeds. In Africa, there are significantly more women than men employed, and their roles are more varied. In Southeast Asia, men and women are involved in almost equal numbers at the various levels of the seaweed industry, where it is carried out by family-owned businesses involving all working age members of the family, irrespective of nuclear or extended families, as compared with the WIO region. The paper highlights five case studies of individual women to show how they have been and continue to be the pillars of the seaweed farming industry. Economic gains from seaweed farming in both regions have provided positive and favorable changes in the quality of life (e.g., food, shelter, clothing, health care, and social acceptance) of the family members involved. The case studies exhibit commitment of women as the driving force of this industry, adding value to seaweeds, especially in the WIO. The benefits will continue to be shared with whole families as well as other community members as mothers are strong anchors of the families in many communities of the developing countries.

Rafael *et al.* (2018) reviewed biotechnology applications of microalgae in the context of EU "Blue Growth" Initiatives. Many sectors such as aquaculture, improving alga culture conditions, biofuels, abatement of pollution, genetic engineering, etc., have been highlighted. In addition to the ecosystems services, globally marine algae and their derivatives are gaining increasing attention in several other sectors too both at the molecular and organismal levels such as food (Kovac *et al.*, 2013; Rajauria *et al.*, 2015) and nutraceuticals (Suleria *et al.*, 2015), animal and fish feed (Rajauria *et al.*, 2015; Maisashvili *et al.*, 2015), biofertilizer (Priyadarshini and Rath, 2012), bioplastics (Gade *et al.*, 2013), pharmaceuticals (Martins *et al.*, 2014), cosmeceuticals (Martins *et al.*, 2014; Thomas and Kim, 2013; Wang *et al.*, 2015), fluorophores [Glazer, 1994], food colorants and textile dyes (Khattar *et al.*, 2015; Sudhakar *et al.*, 2015), biofuels (Beetul *et al.*, 2014) as well as for phycoremediation (Jeyakumar and Chandrasekaran, 2014; Imani *et al.*, 2011; Mata *et al.*, 2009). Polysaccharides from macroalgae—carrageenan, agar, and alginate can be used to make bioplastics. Various types of plastics derived from marine algal feedstock are shown in Table 4.1.

4.3.2 Furthering Culture Conditions

Many researchers are attempting to tap microalgae to increase the fraction of diverse compounds and components of possible commercial interest. Apart from total biomass, which is of research interest, other valuable fractions are chlorophyll, PUFA, and carbohydrates. Literature review shows means of promoting chlorophyll content in microalgae through variation in light intensity, culture agitation, temperature, and nutrient availability, including nitrogen, phosphorus, carbon, micronutrients, and other parameters. Studies show the importance of identifying optimal chlorophyll-inducing conditions in the species of interest (Ferreira and Anna, 2017). Investigations of optimal culture parameters to obtain both exopolysaccharides and biomass from microalgae were made (Delattre *et al.*, 2016). It was found that the culture condition in which microalgae accumulate large amounts of exopolysaccharides does not coincide with best conditions for cultivating microalgae biomass. A greater proportion of exopolysaccharides could be segregated as PFA when the microalgae are under stressed conditions. Currently cultivating microalgae using urban or industrial wastewater as both the medium and the source of nutrients, effectively obtaining biomass and purification treatment at the same time, is being tried for making it more cost-effective (Gupta *et al.*, 2016). It has been demonstrated that environmental conditions, particularly temperature, light intensity, photoperiod, and other physicochemical factors influences biomass growth and composition (Chang *et al.*, 2017). Different lipid content was obtained in microalgae by modifying these conditions, but best conditions for obtaining maximum lipid content were not the same.

4.4 GENETIC ENGINEERING AND BIOTECHNOLOGY TOOLS TO EXPLORE HIDDEN PROPERTIES AND THE "OMICS" APPLICATIONS OF MARINE ALGAE

In the context of genetic engineering and molecular techniques, it is essential to know that several properties of marine algae have the potential to produce several biological products of interest.

4.4.1 Genomics: This is the first "omics" technology. It defines native biosynthetic and metabolic capacities of an organism as a potential microbial cell factory and provides a blueprint for engineering and optimizing productivity. Early genomics studies focused mainly on bacterial and mammalian organisms for biomedical applications. Subsequently genomics became an important technology for improvement in human health sector, involving several steps of unlocking the biocatalytic potential of marine algae.

Next-generation sequencing technologies like sequencing by litigation, pyrosequencing, and real-time sequencing has revolutionized the field of genomics (Metzker, 2010) and helped in the rapid, reliable, and accurate sequencing of a number of marine algal species at a comparative cost (Grossman, 2007; Parker and Thomas, 2004). After over a decade of the human genome project, scientists across the world have sequenced many different organisms, including marine algae (Jamers *et al.*, 2009). About 8,000 organisms' genomes from different kingdoms have been completely sequenced and published with several more in the pipeline in the "Genome On Line Database (GOLD)". The first marine alga to be sequenced was *Guillardiathet*, followed by *Thalassiosira pseudonana, Phaeodactylum tricornutum, Aureococcus anophagefferens, Emilianiahuxleyi, Ostreococcus tauri, O. lucimarinus*, and *Micromonas pusilla* strains.

Genome sequencing provides a sequence of nucleotides that should be assembled and analyzed for gene annotation. It indicates genes encoding proteins and functional RNAs available to the cell along with their associated regulatory elements (Joyce and Palsson, 2006). It also allows identification of protein-coding genes in the nuclear genome, understanding of the metabolic pathways of the pigment biosynthesis and photosynthesis, understanding of evolutionary hypothesis of the carbon storage, and cell wall biosynthesis metabolic processes (Cock and Coelho, 2011; Michel *et al.*, 2010a, 2010b; Chan *et al.*, 2012). Furthermore, DNA barcoding is another important aspect of taxonomic investigation and the economics of the marine algae because it allows the identification of strains of interest for the study of evolution as well as commercial or industrial applications (Qin *et al.*, 2004; Saunders, 2005). DNA barcoding uses molecular markers for characterizing marine algae of interest. Other technologies include DNA-DNA hybridization, restriction fragment length polymorphism (RFLP), amplified fragment length polymorphism (AFLP), and random amplified polymorphic DNA (RAPD) (Qin *et al.*, 2004). In addition, marine algal genomics encourages the understanding by serving as model organisms (Cock and Coelho, 2011).

4.4.2 Transcriptomics refers to the study of transcriptome, or the whole set of transcribed RNAs, during a certain period of development as well as under specific biological conditions. It gives insights into genome expression that lends a view on gene structure, gene expression regulation, gene product function, and the dynamics of the genome. Techniques used for transcriptome analysis over the years have evolved from the initial expression sequencing tag (EST) strategy to gene chips, and now the RNA-seq and bioinformatics analysis. Dong and Chen (2013) and Morozova *et al.* (2009) studied and provided an in-depth review of transcriptomics techniques.

4.4.3 Proteomics is defined as the study of proteins coupled with transcriptomics and genomics as they are complementary (Graves and Haystead, 2002). A cell's proteome or the whole cell protein is dynamic. Proteins extracted and studied at a particular point of time and under certain physiological condition(s) represent a cell's immediate response to its environment or transcriptome. However, the aim remains to study a cell's proteome—protein-protein interaction, protein modifications, protein function, and location of proteins etc. Proteomics involves different separation techniques, multiple analyses, and different identification tools. Graves and Haystead, 2002 reviewed proteomics techniques in-depth. In the past decade, several novel marine algal proteins have been identified by two-dimensional electrophoresis (DE) and mass spectrometry (MS), including proteins from the macroalga *Gracilaria changii* (Wong *et al.*, 2006) and microalgae *Dunaliella bardawil* (Katz *et al.*, 1995) and *Nannochloropsis oculata* (Kim *et al.*, 2008). In addition to proteomics leading to particular protein identification, it also helps in revealing underlying functions including its role in evolution as well as taxonomic studies and biochemical pathways (Le Bihan *et al.*, 2011; Kim *et al.*, 2008; Nunn *et al.*, 2009).

According to Carrasco *et al.* (2018), transforming the proteome in information with high value for all investigations in proteomics research can design best strategies in genetic engineering (Vaudel *et al.*, 2016). The most efficient strategy to obtain the information of how microalgae work is the proteomics, giving information of the "software" of the Microalgae supported by the "hardware" genomics (Acero *et al.*, 2011). Proteomic studies are a fundamental tool to verify the expression of the genes of interest sought in genomics, corroborating the presence of by-products from microalgae. A microalga accumulates more PUFA under culture conditions or multiplies rapidly in other culture conditions due to the set of proteins that are expressed differently. Therefore, it helps in understanding the behavior and relationships of microalgae with the environment. Further, it produces huge amounts of information on biological molecules of interest for many fields of research. For instance, proteins are responsible for PUFA accumulation, restoring contaminated water, producing anti-cancer protein, or accumulating greater quantities of pigments of interest such as phycocyanin. Proteomics have revolutionized the way of understanding the behavior of organisms, based not only on genomics, the expression of this genes depends on the proteome, in certain conditions (Aussant *et al.*, 2018; Savchenko *et al.*, 2003). Additionally, proteomic is used for investigating biosynthesis mechanism of harmful marine algae particularly in relation to human health and safety (Wang *et al.*, 2013). Saxitoxin, for instance, is associated with paralytic shellfish poisoning. The biosynthetic pathway of saxitoxin synthesis was not known for long, but studies are now revealing the enzymes implicated (Wang *et al.*, 2007). Marine algal toxins are being considered for potential commercial applications, such as the use of saxitoxin and tetrodotoxin as an anaesthetic (Camacho *et al.*, 2007), and insights of their production would be an advantage for the industry. It is in this way that the true value of microalgae can be obtained, remarking these studies and important information obtained from using "omics" techniques. However, proteomic studies on algae are still relatively limited compared to higher plants. So far, it is the freshwater microalga—*Chlamydomonas reinhardtii*—that has been thoroughly studied from various omics-angle. *O. tauri* is now being considered as a research model (Gutman and Niyogi, 2004).

Modern experimentation tools help in closing research gaps and provide a better understanding of genomics-based approaches in microalgae (Anand *et al.*, 2017). **Synthetic biology** is the readapting of biological systems for objectives and applications of interest. Through coordinated and balanced expression of genes, resources within industries can be used for commercial production of high-value substances from both native and introduced organisms. This interdisciplinary field has the potential to revolutionize natural product discovery, by providing a diverse array of tools, technologies, and strategies for exploring the large chemically complex space of natural plant-based products using unicellular organisms (Moses *et al.*, 2017). The application of molecular techniques to microalgae will undoubtedly allow selection of desirable characteristics and discarding undesirable ones (Scafie and Smith, 2016), to obtain microalgae that are custom-designed to the specification required by a sector of industry.

4.4.4 Metabolomics refers to the study of the complete set of metabolites of an organism or metabolome. Metabolites of an organism can be categorized as primary (fundamental for cell development and continually produced, e.g., amino acids and polysaccharides) or secondary (produced in response to a stimulus such as an environmental distress, e.g., sterols). Secondary metabolites of marine algae are of major interest. However, not all can be expressed at all times as their synthesis must be triggered by either physiological and/or environmental stimulus. While proteomics used the evaluation of phenotypic responses of organisms, it gradually took the backseat after the advent of metabolomics. Verpoorte *et al.* (2008) stated that the main aim of metabolomics is the qualitative and quantitative analysis of all the metabolites present in an organism. The study mentioned five major approaches, including high-performance liquid chromatography/thin layer chromatography—ultraviolet (HPLC/TLC–UV), gas chromatography—MS (GC-MS), LC-MS, MS and nuclear magnetic resonance (NMR) spectrometry. According to Roessner and Bowne (2009), metabolomics approaches could be summarized as target analysis, metabolite profiling, metabolomics, and metabolic fingerprinting.

Since metabolomics depicts physiological states of any organism, research database of this omics outshines proteomics and transcriptomics in the area of functional genomics. The exometabolome and endometabolome of both marine microalgae and macroalgae have been thoroughly studied for multiple purposes, including ecology (Barofsky *et al.*, 2009), physiological states (Vidoudez and Pohnert, 2012), and applications in multiple sectors such as health (Lin *et al.*, 2010; Kubanek *et al.*, 2005) and energy (Beetul *et al.*, 2014).

4.5 BARRIERS TO SUSTAINABLE UTILIZATION OF ALGAL BIO-RESOURCES AND OPPORTUNITIES

There are a range of available technologies to produce seaweed, but optimization and more efficient developments are still required. There are fundamental and very significant hurdles to overcome in order to achieve the potential contributions of seaweed cultivation. There are critical aspects, like improving the value of seaweed biomass, along with a proper consideration of the ecosystem services that seaweed farming can provide, for example, a reduction in coastal nutrient loads. Additional considerations are environmental risks associated with climate change, pathogens, epibionts, and grazers, as well as the preservation of the genetic diversity of cultivated seaweeds. Several technological options exist for sustainable utilization of algal resources. Nevertheless, there are challenges and various constrains to their commercialization.

Emily *et al.* (2015) reports that large-scale cultivation of algae, or algaculture, has existed for over half a century. Large-scale cultivation of algae is a mix of fundamental aspects of traditional agricultural farming and aquaculture. Despite this overlap, alga culture and the benefits associated with it has not yet been positioned within agriculture. Various federal/state agricultural support and assistance programs are currently adopted for crops, but extension to algal biomass is uncertain. In India, Mantri *et al.* (2019) examined the existing policies and legislations acting as constrains for the agarophyte trade in India and products which would provide a framework for improving the overall prospects of agar producing red seaweeds.

Olaizola (2003) emphasizes that marine algal-omics studies have given rise to new opportunities and subsequent industrial applications. While the prospects are huge, there are multiple challenges that this sector faces and available opportunities, which are listed in the next section.

4.5.1 Marine Algae Taxonomic Classification

Taxonomic classification of marine algae has been tedious for the past decades. Morphological identification (using both microscopic and macroscopic features) of algae causes various limitations, such as change in morphology due to environmental factors (Luo *et al.*, 2006), presence of similar morphotypes, complex cellular structure (Santhi *et al.*, 2013), lack of characteristics, morphological features of the organisms as well as time-consuming and expertise-requiring in this field (Krietnitz and Bock, 2012). Taxonomic classification of algae is therefore being reviewed through phylogenetic studies supported by analyses of cell division processes, physiological products (pigments and oils produced), and genetic characterization (Lewin, 2015). Phylogenetic analyses help in categorizing species difficult to identify. However, from a taxonomic point of view, DNA sequence information without other evidences can never be used as an indicator for species delimitation (DeSalle, 2006). Combining molecular phylogenetic studies could address such issues. To enhance and make taxonomic exploration of marine algae more reliable, it is thought to be best if different tools are used and sorting out of all lineages of algae is done.

Marine algal-omics suffer similar challenges like other microbial "omics" technologies. Graves and Haystead (2002) state that the foremost challenge of proteomic studies is the analysis of low-abundance proteins, and criticized archaic methods used to study proteins, including bioinformatics tools for data interpretation. The study advocates urgency for new algorithms and technologies involving large-scale analyses rather than mundane technologies. In the case of transcriptomics being complex to decipher, Zhi Cheng and Yan (2013) summarized the challenges as being experimental, technological, and at the level of data interpretation. While advanced high-throughput technologies of metabolomics is gaining thrust, several data interpretation difficulties and identification of an array of unknown metabolites and transformative agents of metabolism, and data mining exist (Mendes, 2006). Importance of using bioinformatics tools such as Algal Functional Annotation Tool (Lopez *et al.*, 2011) and Green Cut2 (Grossman *et al.*, 2010) is emphasized. In the future proteomics studies of marine algae are expected to increase by multiple folds owing to their commercial/industrial implication and the will of many countries to develop their blue economy. Such "omics" studies are opening up opportunities for several algal species in multiple fields such as *D. salina* for carbon capture to address climate change.

However, as listed in Table 4.2, marine algal applications face many challenges that need to be addressed prior to commercialization.

Other challenges while considering marine algae are shown in Table 4.2.

TABLE 4.2
Challenges in Commercialization and Industrial Application of Marine Algae

Challenge	Reference
Internal Factors Affecting Production	
Biomass production: High production cost due to significant cost Implication with respect to resource supply: water, CO_2, and nutrients	Chisti (2013)
In-sea cultivation: Alterations to biosynthesis of molecules of interest due to spatial and seasonal variations, location and depth	Hafting *et al.* (2015)
Biomass harvest: Energy intensive process: centrifugation oil extraction; use of petroleum derivatives	Beetul *et al.* (2014) Hannon *et al.* (2010) Hannon *et al.* (2010)
Algal strain: In-breeding leads to negative alteration of trait of economic importance such as decline in yield and quality	Chisti (2013) Hafting *et al.* (2015) Hannon *et al.* (2010)
Algal diseases: Affects candidature of strain to be considered as a feedstock	Hafting *et al.* (2015)

TABLE 4.2 (*Continued*)
Challenges in Commercialization and Industrial Application of Marine Algae

Challenge	Reference
External Factors Affecting Production	
Impact of microbial interactions with marine algae on bioactivity	Hafting *et al.* (2015)
Land use: Nonarable land hosting marine algal mass production increases proportionately with increase in demand	Wei *et al.* (2013) Hannon *et al.* (2010)
CO_2 input: Contribute to high production cost due to purchased CO_2	Chisti (2013)
Nutrient supply (especially nitrogen and phosphorus): Limited nutrient supply limits algal mass production	
Dehydration is energy intensive	Chisti (2013) Kleivdal *et al.* (2013)
Water use: Limited by fresh water	Chisti (2013) Hannon *et al.* (2010)
Light: Optimal production limited by shading as well as photoinhibition [Applicable to marine microalgae only]	Chisti (2013)

4.5.2 Future Prospects for Integration of Algal Resources in the Context of Blue Economy

For the challenges faced by marine algae associated industries, future directions for dovetailing both research and commercialization/industrialization arenas into blue economy are required. At the global level, biorefinery approach is proposed as a sustainable strategy. An outline for future needs is provided by Buschmann *et al.* (2017) in the anticipation that phycologists around the world will rise to the challenge, such that the potential to be derived from seaweed biomass becomes a reality. The future can look at increasing the efficiency and economic viability of microalgae production by (1) improving culture conditions so that the quantity of microalgae biomass produced is increased; (2) improving culture conditions so that one or more fractions of interest are obtained more cost-effectively; and (3) developing new or improved cultivation, processing, and extraction methods to increase efficiency and/or reduce cost.

1. As a whole, the numerous challenges mentioned in Table 4.2 can be addressed by genetically enhancing algal strains, improving production of high-value products, and reducing production cost. Hybridization of marine macroalgae (Hafting *et al.*, 2015) and genetic engineering (Chisti, 2013) of macroalgae and microalgae are promising tools for important traits such as disease resistance and overproduction of specific compounds of interest. The compounds of interest may be targeted molecules as well as by-products.
2. In-sea culture exercise should be carried out along a standard for cultivation to circumvent issues like inconsistent depth, environmental changes, and others which may impact the production of bioactive compounds and composition (Hafting *et al.*, 2015).
3. In view of lowering production cost of marine algae, the by-products of macroalgae biomass production such as protein, alginates, and phenolic compounds should be considered as an integral part of commercialization to enhance the economic value of marine algal biofuel production process. Chisti (2013) suggests that nutrients can be recycled while wastewater can also be used to some extent.
4. The last few decades have witnessed several developments in biotechnological applications of marine algae on three scales: research, commercial, and industrial. Many challenges hindering the full-fledged application of marine algae are being progressively addressed, and opportunities are being generated through "marine algal omics" with a determinant role in the marine algal economy.

5. Remya and Radhika Rajasree (2016) state that there is evidence of several studies in Tamil Nadu, India, on seaweed distribution, resources, taxonomy, culture, harvest, industry, utilization, postharvest technology, and utilization. However, it was noted that intensive studies are required in the industrial and pharmaceutical applications in the future.
6. **Marine Energy:** US Department of Energy (2019) discussed in their findings that the power requirements for large-scale cultivation and harvesting operations of macroalgae at sea are not well understood but resembles aquaculture operations, including power for safety, navigation lights, maintenance equipment, pumps for nutrients and ballast control, refrigeration and ice production, drying operations, marine sensors, recharging of autonomous underwater vehicles, and transport vessels. Huge potential of integrating marine energy systems co-developed with algal cultivation and harvesting systems exists. By replacing fossil fuels with marine renewable energy, biofuels industry could reduce harm to air and water quality, supply chain and transport risks, and also potentially reduce operational costs.
7. **Cost-effectiveness:** Projected costs for marine algae are at present several times higher than terrestrial biomass, but improvements in yields, scale, and operations can make algae cost competitive with terrestrial crops (National Renewable Energy Laboratory, 2017). Rapid growth in seaweed farming exists in about 50 countries, and 27.3 million tons of aquatic plants including seaweeds were harvested in 2014, totaling $5.6 billion (FAO, 2016). Larger marine farms proposed for biofuel production requires enormous energy for harvesting, drying, monitoring, maintenance activities, maneuvering and buoyancy controls. This could be met wholly/partially through energy generated from marine energy devices by designing marine energy systems for growing and harvesting to provide off-grid power. Marine energy is less geographically limited at high latitudes where some macroalgae thrive and also provide shelter to more exposed sites by decreasing wave action while power is generated simultaneously.
8. **Linkage of products and coproducts with marine energy markets:** It is predicted that increased demand for cleaner fuels, including air-quality mandates and petroleum spill protections, will spur biofuel markets. High-value coproducts, including complex polysaccharides like algin, laminarian, mannitol, fucoidan, and agar can be extracted from macroalgae, leaving the residue for animal feed. These products and coproducts will trigger marine energy markets. The state of technology for algal biofuel production is continuously maturing with ongoing investments. Additional research, development, and demonstration are needed to achieve extensive placement of affordable, scalable, and sustainable algae-based biofuels.
9. **Blue economy driving forces:** Lu Wenhai *et al.* (2019) elaborated many successful examples of blue economy emphasizing on international perspectives, careful definition, and illustrative case studies as being fundamental in developing a blue economy. As importance of blue research expands, policy makers, and research institutions worldwide are demanding further improved analysis of the blue economy. Particularly, in terms of the management, data access, monitoring, and product development, countries are making decisions according to their own needs, and there is a lack of consensus. Future prospects would be proposals for development of blue economy, including shouldering global responsibilities to protect marine ecological environment, strengthening international communication, sharing development achievements, and promoting the establishment of global blue partnerships with scope and depth of our collective understanding and analysis. The major suggestions are the following:
   a. Understanding, utilizing, and protecting oceans and create shared goals and responsibilities for all human being to achieve future sustainable marine development. At

present, blue economy, viewed as "blue engine", is becoming an important driving force to achieve global sustainable development.

b. Environmental observations play a powerful technical supporting role in realizing benefits from blue economy but require joint efforts, pushing development between blue economy and global economy, society, and ecosystem in the next decade.

c. It is critical to shoulder global responsibilities, foster deep sea environmental management, understand effects of human and climate on deep-sea creatures' diversity along with ecosystem health, control micro-plastics in global oceans, strive to establish a responsible community of marine ecological protection and environment governance, and establish a community of shared future.

d. Development achievements need to be shared. It is important to play a key role in verifying data, make data-driven decisions under blue economy sectors, strengthen international communications in terms of technology, human talents and information, and jointly design and produce science-based products through collaborative public/private partnerships (government/university/enterprise, and society), provide members with a platform to share policies, markets, and growth.

e. It is critical to establish blue partnerships across the globe, take mutual efforts to explore new markets, generate new growth, co-establish service platforms, and provide industrial service platform to achieve global blue economy development, connecting technologies and markets, and also linking enterprises with finance.

10. Emily *et al.* (2015) emphasized that it is essential for nascent industries to encourage investment, build infrastructure, disseminate technical experience and information, and create markets. Potential agricultural policies and programs could support algal biomass cultivation, and remove the barriers to the expansion of these programs to algae. The US government has supported R&D of alga culture for biofuels over last few decades, but policies must anticipate and stimulate the evolution of the industry to next level. Recognition of alga culture within agriculture under USDA at national, regional, and local levels will expand agricultural support and assistance programs for algae cultivation and progress of the industry. Currently, they are world leaders in algal biomass technology hosting a number of companies devoted to the industry.
11. **Mantri *et al.* (2019)** proposed a framework for improving agar trade: CSIR-Central Salt and Marine Chemicals Research Institute (CSMCRI), Bhavnagar along with 264 Technology Information, Forecasting and Assessment Council (TIFAC), Department of Science and Technology, New Delhi, conducted a meeting of seaweed researchers, industry experts, and other stakeholders in November 2017. Indian Chambers of Commerce in collaboration with CSMCRI conducted "Indian Seaweed Summit" in February 2018. These two important events took stock of the issues confronting the sector and called for taking up commercial cultivation and processing of seaweeds on priority at pan Indian level, under a mission mode program. Many recommendations were put forward.

a. Have a common platform consisting of academia, industry, and government representatives that will largely handle three domains—farming, marketing, and industry.

b. Farming sector will liaise with Indian Chambers of Commerce through regional seaweed produce committee (for biomass production and product) facilitating permits required for commercial farming, linking seaweed growers to government schemes and incentives, disseminating R&D information and other details that would improve farming techniques.

c. Marketing domain will have a strategic market analysis group to conduct in-depth studies pertaining to market assessment, import export opportunities, and novel applications to boost farming and products.

   d. Third domain consisting of industry partners would provide crucial support to overall development of trade infrastructure, production requirements, advocate green processes and technologies, facilitate development of new standards a range of agar products. This will add greater value of the produce thereby facilitating positive gains. These three domains are interlinked with each other and shall work together in tandem, improving the growth of agar trade in India.
12. **Employment opportunities of coastal communities:** Immanuel and Sathiadhas (2004) worked on the employment potential of fisher women in seaweed sector. Seaweeds, a renewable marine resources forms the primary raw material for the *agar* and *algin* industries. They are mostly exploited from the southeast coast of India, Tamil Nadu (Mandapam to Kanyakumari), west coast (Gujarat), Lakshadweep Islands, and Andaman and Nicobar Islands. Although substantial resources are available, they are neither fully harvested, nor are efforts up to the mark. Seaweed collection renders extensive employment to coastal fisherfolk, but estimation of their resources indicated that quantity harvested is very limited. At present around 5,000 women depend on seaweed industries for their livelihood. If the available resources are harvested to its optimal level, it can provide employment to another 20,000 coastal fisherfolk in harvesting and an equal number in postharvesting activities. Since this domain is dominated by women, special efforts should be taken for its optimum exploitation and market expansion through diversified product development and popularization. Since cultivation has to be in open sea, it is necessary to foster suitable measures. Women involved in seaweed collection are not fully aware about its demand and value. This requires capacity building on its nutritive value, marketing, sales promotion, and popularization of edible seaweeds. Diversified utilization of seaweed in other areas like manure or as feed need to be enhanced.
13. **Policies and regulations:** Mantri *et al.* (2019) analyzed the legislation and policy background in India relevant to agarophyte trade. Legislation and policies pertaining to commercial cultivation, processing, and trade of agarophytes in India assume considerable significance, as currently only 30% of agar requirement is being produced indigenously (Subba Rao and Mantri, 2006) and requires expansion of biomass production and agar, given the potential and globally expanding seaweed market. The key recommendations of ICAR-Central Marine Fisheries Research Institute (CMFRI) pertain to various dimensions of seaweed resource management, cultivation, trade, and regulation. Government of India has identified seaweed culture as one of the components under blue revolution scheme and National Fisheries Development Board (NFDB) provided guidelines for seaweed cultivation and specific programs for its promotion in India (NFDB, 2015). They are as follows:
   a. Selected seaweeds known to co-occur in the reef areas are protected under different regulatory regimes. Various stretches of coastal areas comprising of coral reefs, including seaweed beds, have been declared as ecologically sensitive areas (ESAs) or ecologically sensitive zones (ESZ) under the provisions of Environment (Protection) Act (MoEF, 1986). Similar zones are established as per the Island (Protection) Zone notification (IPZ, 2011), having spatial jurisdiction limited to island territories of India. Coastal Regulation Zone notification given by the Ministry of Environment Forest ans Climate Change in 2011, does not classify the seaweed beds existing (312) along the coastal as ESAs.
   b. The National Policy on Marine Fisheries, 2017, called for periodic review and evaluation of existing spatial conservation measures for legislative support to protect the rights of traditional fishermen and their livelihoods (MoAFW, 2017). It highlights evolving of leasing policies and guidelines for spatial planning, environmental and social impacts assessment as essential for promoting mariculture.

c. Section 4 of CRZ notification 2011 provides demarcation of sensitive coastal ecosystems as critically vulnerable coastal areas (CVCA), which in turn can be managed as per the integrated management plans (IMP), prepared involving local communities. The seaweed beds and farming areas can be demarcated as CVCAs.

d. The Biological Diversity Act 2002 provided equitable sharing of benefits arising from the use of traditional biological resources which also includes seaweeds. It also provides for constitution of local-level Biodiversity Management Committees (BMC) that empowers local traditional dwellers for sustainable use of seaweeds.

e. The Coastal Aquaculture Authority (CAA) under the Ministry of Agriculture is mandated to regulate all coastal aquaculture efforts in India including seaweed farming (Coastal Aquaculture Authority, 2005). However, this sector in the coastal regions of India has not witnessed growth, commensurate with the efforts, attributed to inconsistencies in the regulatory framework, weak institutional mechanisms, and poor awareness among the community.

f. Penetration of agarophyte farming technology in coastal rural areas is crucial to meet existing domestic demand of agar. Cultivated seaweed shall be treated as agricultural product, enabling seaweed farmers quick and timely access to affordable credit through schemes such as Kisan Credit Card (KCC) and insurance for their product.

g. Product development is critical in the seaweed value chain and requires innovative support schemes with clear specifications for gel strength and various other quality criteria for economic competitiveness.

h. Linkage between research organizations and industry are to be strengthened by easing the procedures for exchange of knowledge and manpower, so as to meet the targeted production.

i. Government can set up seed-banks to ensure supply of quality seed material of agarophytes in a sustainable manner to seaweed cultivators.

j. Government may promote implementation of agarophyte farming projects through Corporate Social Responsibility (CSR) funding, for social equilibrium.

k. Novel applications of functionally modified agar and agarose shall pave the avenues of growth. Application of matrigel–agarose hybrid hydrogels in 3D printing of biological architectures mimicking the structural and functional features of *in vivo* tissues (Fan *et al.*, 2016) will aid in tissue engineering and development of transplantable organ constructs. The government shall consider instituting a "Technology Development Fund", with active participation of agar manufactures association, to promote more start-ups in the domain.

l. Since majority of the Indian seaweed cultivators are women, expansion of seaweed farming through appropriate promotional policies and schemes would aid in meeting various societal goals viz. rural employment, alleviating poverty, social equity and empowerment, that would contribute to improve their living standards such as education, health, access to better amenities (Mantri *et al.*, 2015). National Mariculture Policy 2018, Government of India will enhance suitable steps towards various goals.

m. Following issues crucial for creating a policy framework and some recommendations are proposed (Mantri *et al.*, 2019) for consolidating agar trade in India:

  i. Mapping the agarophytes along Indian coast and pertinent water bodies to understand current availability status and also bring new species under the umbrella of commercial farming.

  ii. Collaborating with appropriate international agencies to fine-tune existing processes and adopting new technologies for agar production.

iii. Ensuring crop insurance programs against disease outbreak, epiphytic infestation, natural calamity, and climate change related disasters for end user protection and sustained supply of agarophytic raw materials.
iv. Intensifying skill development and conducting feasibility studies for pilot-scale farming for scaling up geographically.
v. Ensuring regular income by engaging government agencies and leading seaweed industries by fixing minimum assured price for agarophytes like most agricultural crops.
vi. Integrating farming and processing to empower grass-root level farmers for accessing direct benefits.
vii. Granting export license to all those engaged in agarophyte farming as it directly helps in improving livelihood of local communities.

14. Rafael *et al.* (2018) Microalgae are positioned as an important future food for humans due to its composition and advantages it offers as a cultivated crop. It can help to counter global warming and restore the atmosphere and water resources of the planet, does not require large areas of scarce land, and helps protect the natural environment. Microalgae are currently source of many interesting products in biomedicine, healthy food, and technological applications. Emphasis is required to exchange of research results between various fields of investigation leading to expansion in terms of new applications in all the sectors of industry in the future.

4.5.3 The Way Forward

Blue economy will be compliant with SDG-14, with the attribute focused on conserving and sustainably using the oceans, seas, and marine resources. The core is to realize social economic development and dynamic balance of resources and environment. The approaches to adopt blue economy is consistent with the core contents of RIOC20 Summit 2011. Based on the analysis of marine industrial activities and the health of marine ecosystem, the goal is to maintain a healthy marine and land ecosystem, solve pollution such as marine transport waste and plastic litter and microplastic, mitigate global climate change effects, and construct a blue economy sustainable management model based on maintaining a healthy ecosystem. The outcome of such steps will lead to the following:

- **Catalysing Partnerships:** Create a regional network of partners encompassing the Tamil Nadu coast to enhance ecosystem-based management and facilitate regional learning exchanges and outreach/training events among marine ecosystem project initiatives on a regular basis
- **Knowledge and Resource Mapping:** Create a mapping of available sources of enhanced tools and/or evolve new specific tools for marine ecosystem-based conservation and protection, and capture best governance practices to enhance the management effectiveness to integrate issues across pertaining to oceans/seas and institutional scales
- **Co-creating a Policy Agenda:** Bring out a customized ***Charter for Sustainable Blue Economy*** which will have key action points for stakeholders as given under future prospects section

Success can be viewed in four major dimensions, as depicted here.

Dimension 1: Blue economy business opportunities using marine algae can look at the following:

- Measures which blue economy business opportunities can offer to overcome the problem of growing unemployment in vulnerable coastal areas in algal sector

- Opportunities for environmentally friendly investments in bio energy in coastal areas
- Opportunities for sustainable growth and expansion of SMEs in algal cultivation and processing

Dimension 2: SMEs, resource efficiency, productivity, and resilience can look at the following:

- Ways to achieve resource use efficiency in fishery, agriculture, and livestock in rural areas and its impact on job creation
- Dovetailing suitable policy options for increasing farm-level and aggregate productivity for rural green SMEs
- Ways of resource use efficiency and resilience in agriculture, mining, fisheries, and livestock contribute to improve labor and capital productivity
- Means of SMEs contribution to and benefits from environmentally friendly, quality infrastructure including transport, energy, and shelter

Dimension 3: Rural SMEs, biodiversity, and ecosystem services

- Ingredients for an attractive business and investment environment for rural SMEs specializing in ecosystem services such as fish breeding and coastal protection of coral reefs, water filtering, soil protection, and carbon sequestration
- Leveraging the potential of ecosystem services in rural coastal areas to appeal to SMEs in the area of land and natural resources management, including agriculture, forestry, water, and mining
- Channels/mechanisms for commercial ecosystem management contributing to inclusive growth
- Green/blue entrepreneurship fostering innovation in fishery and livestock management
- Opportunities by overcoming challenges to public-private partnerships for ecosystem management and support benefits of carbon stocking and sequestration

Diemnsion 4: Institutions and governance for environmentally friendly business practices

- Ingredients for an optimal institutional framework to promote labor-intensive green business activities
- Ingredients for an optimal institutional and legal framework to foster an insurance market around rural ecosystem management businesses and activities
- Ingredients of the optimal incentive framework that promotes maximum productivity in land, natural resources, and infrastructure management businesses while minimizing their negative impacts on environment

REFERENCES

Acero, F.J.F., Carbú, M., El-Akhal, M.R., Garrido, C. and González-Rodríguez, V.E. (2011) Development of proteomics-based fungicides: New strategies for environmentally friendly control of fungal plant diseases. International Journal of Molecular Sciences. 12: 795–816 pp.

Anand, V., Singh, P.K., Banerjee, C. and Shukla, P. (2017) Proteomic approaches in microalgae: Perspectives and applications. 3 Biotech. 7: 197 p.

Aussant, J., Guihéneuf, F. and Stengel, D.B. (2018) Impact of temperature onfatty acid composition and nutritional value in eight species of microalgae. Applied Microbiology and Biotechnology. 102: 5279–5297 pp.

Barofsky, A., Vidoudez, C. and Pohnert, G. (2009) Metabolic profiling reveals growth stage variability in diatom exudates. Limnology and Oceanography. 7(6): 382–390 pp.

Beetul, K., Sadally, S.B., Taleb-Hossenkhan, N., Bhagooli, R. and Puchooa, D. (2014) An investigation of biodiesel production from microalgae found in Mauritian waters. Biofuel Research Journal. 1(2): 58–64 pp.

Buschmann, A.H., Camus, C., Infante, J., Neori, A., Israel, Á., Hernández-González, M.C., Pereda, S.V., Gomez-Pinchetti, J.L., Golberg, A., Tadmor-Shalev, N. and Critchley, A.T. (2017) Seaweed production: Overview of the global state of exploitation, farming and emerging research activity. European Journal of Phycology. 52(4): 391–406 pp. DOI: 10.1080/09670262.2017.1365175.

Camacho, F.G., Rodríguez, J.G., Mirón, A.S., García, M.C.C., Belarbi, E.H. and Chisti, Y. (2007) Biotechnological significance of toxic marine dinoflagellates. Biotechnology Advances. 25(2): 176–194 pp.

Carpenter, S. and Kitchel, J. (1993) The trophic cascade in lakes. Cambridge University Press, Cambridge.

Carrasco, R., Fajardo, C., Guarnizo, P., Vallejo, R.A. and Fernandez-Acero, F.J. (2018) Review: Biotechnology applications of microalgae in the context of EU "blue growth" initiatives. Journal of Microbiology and Genetics. DOI: 10.29011/2574-7371. 2018(1): 1–14 pp

CBD-CHM. (2018) Sixth national report for the convention of biological diversity. https://chm.cbd.int/database/record?documentID=241351

Chan, C.X., Blouin, N.A., Zhuang, Y., Zauner, S., Prochnik, S.E. and Lindquist, E. (2012) *Porphyra* (bangiophyceae) transcriptomes provide insights into red algal development and metabolism. Journal of Phycology. 48: 1328–1342 pp.

Chang, J., Le, K., Song, X., Jiao, K. and Zeng, X. (2017) Scale-up cultivation enhanced arachidonic acid accumulation by red microalgae *Porphyridium purpureum*. Bioprocess and Biosystems Engineering 40: 1763–1773 pp.

Chisti, Y. (2013) Constraints to commercialization of algal fuels. Journal of Biotechnology. 167: 201–214 pp.

Chitale, V.S., Behera, M.D. and Roy, P.S. (2014) Future of endemic flora of biodiversity hotspots in India. *PLoS One*. 9: e115264 p.

Coastal Aquaculture Authority. (2005) Guideline for regulating coastal aquaculture. http://caa.gov.in/uploaded/doc/Guidelines-Englishnew.pdf

Cock, J.M. and Coelho, S.M. (2011) Algal models in plant biology. Journal of Experimental Botany. 62(8): 2425–2430 pp.

De Clerck, O. and Coppejans, E. (2002) Marine macroalgae. In Richmond, H.D. (ed.), A field guide to: The seashore of Eastern Africa and Western Indian Ocean Islands. Swedish International Development Cooperation Agency; Swedish Agency for Research Cooperation; University of Dar es Salaam, Sweden.

Delattre, C., Pierre, G., Laroche, C. and Michaud, P. (2016) Production, extraction and characterization of microalgal and cyanobacterial exopolysaccharides. Biotechnology Advances 34: 1159–1179 pp.

DeSalle, R. (2006) Species discovery versus species identification in DNA barcoding efforts: Response to rubinoff. Conservation Biology. 20(5): 1545–1547 pp.

Dong, Z.C. and Chen, Y. (2013) Transcriptomics: Advances and approaches. Science China Life Sciences. 56(10): 960–967 pp.

Emily, M., Trentacoste, A., Martinez, M. and Zenk, T. (2015) The place of algae in agriculture: Policies for algal biomass production. Photosynth Research. 123: 305–315 pp.

Fan, R., Piou, M., Darling, E., Cormier, D., Sun, J. and Wan, J. (2016) Bio-printing cell-laden matrigel–agarose constructs. Journal of Biomaterials Applications. 31(5): 684–692 pp.

FAO. (2016) The state of world fisheries and aquaculture 2016: Contributing to food security and nutrition for all. Rome. 200 p. www.fao.org/3/a-i5555e.pdf.

Ferreira, D.S.V. and Anna, S.C. (2017) Impact of culture conditions on thechlorophyll content of microalgae for biotechnological applications. World Journal of Microbiology and Biotechnology. 33: 20 p.

Flower, E., Msuya, E. and Hurtado, A.Q. (2017) The role of women in seaweed aquaculture in the Western Indian Ocean and South-East Asia. European Journal of Phycology. 52: 4, 482–494 pp.

Gade, R., Tulasi, M.S. and Bhai, V.A. (2013) Seaweeds: A novel material. International Journal of Pharmacy and Pharmaceutical Sciences. 5(2): 40–44 pp.

Glazer, A.N. (1994) Phycobiliproteins—a family of valuable, widely used fluorophores. Journal of Applied Phycology. 6(2): 105–112 pp.

Gomez Pinchetti, J.L. and Martel Quintana. (2016) Algae production and their potential contribution to a nutritional sustainability. Journal of Environmental Health Sciences. 2(3): 1–3 pp.

Graves, P.R. and Haystead, T.A.J. (2002) Molecular biologist's guide to proteomics. Microbiology and Molecular Biology Reviews. 66(1): 39–63 pp.

Grossman, A.R. (2007) In the grip of algal genomics. In León, R., Galván, A. and Fernández, E. (eds.), Transgenic microalgae as green cell factories. Springer, New York: 54–76 pp.

Grossman, A.R., Karpowicz, S.J., Heinnickel, M., Dewez, D., Hamel, B. and Dent, R. (2010) Phylogenomic analysis of the Chlamydomonas genome unmasks proteins potentially involved in photosynthetic function and regulation. Photosynthesis Research. 106(1): 2–17 pp.

Guiry, M.D. (2012) How many species of algae are there? Journal of Phycology. 48: 1057–1063 pp.

Gupta, P.L., Lee, S.M. and Choi, H.J. (2016) Integration of microalgal cultivation system for wastewater remediation and sustainable biomass production. World Journal of Microbiology and Biotechnology. 32: 139 pp.

Gutman, B.L. and Niyogi, K.K. (2004) *Chlamydomonas* and *Arabidopsis*. A dynamic duo. Plant Physiology. 135(2): 607–610 pp.

Hafting, J.T., Craigie, J.S., Stengel, D.B., Loureiro, R.R., Buschmann, A.H. and Yarish, C. (2015) Prospects and challenges for industrial production of seaweed bioactives. Journal of Phycology. 51(5): 821–837 pp.

Hannon, M., Gimpel, J., Tran, M., Rasala, B. and Mayfield, S. (2010) Biofuels from algae: Challenges and potential. Biofuels. 1(5): 763–784 pp.

Imani, S., Rezaei-Zarchi, S., Hashemi, M., Borna, H., Zand, A.M. and Abarghouei, H.B. (2011) Hg, Cd and Pb heavy metal bioremediation by *Dunaliella* alga. Journal of Medicinal Plants Research. 5(13): 2775–2780 pp.

Immanuel, S. and Sathiadhas, R. (2004) Employment potential of fisherwomen in the collection and post-harvest operations of seaweeds in India. Seaweed Research and Utilisation. 26(1&2): 209–215 pp.

Jamers, A., Blust, R. and De Coen, W. (2009) Omics in algae: Paving the way for a systems biological understanding of algal stress phenomena? Aquatic Toxicology. 92: 114–121 pp.

Jeyakumar, S. and Chandrasekaran, V. (2014) Adsorption of lead(II) ions by activated carbons prepared from marine green algae: Equilibrium and kinetics studies. International Journal of Industrial Chemistry. 5(1): 1–10 pp.

Joyce, A.R. and Palsson, B.Ø. (2006) The model organism as a system: Integrating omic' data sets. Nature Reviews Molecular Cell Biology. 7(3): 198–210 pp.

Kaliaperumal, N. (2017) Studies on phycocolloids from Indian marine algae—A review. Seaweed Res Utilin. 39, 1–8 pp.

Katz, A., Jiménez, C. and Pick, U. (1995) lsolation and characterization of a protein associated with carotene globules in the alga *Dunaliellabardawil*. Plant Physiology. 108(4): 1657–1664 pp.

Khattar, J.I.S., Kaur, S., Kaushal, S., Singh, Y., Singh, D.P. and Rana, S. (2015) Hyperproduction of phycobiliproteins by the cyanobacterium *Anabaena fertilissima* PUPCCC 410.5 under optimized culture conditions. Algal Research. 12: 463–469 pp.

Kim, G.H., Shim, J.B., Klochkova, T.A., West, J.A. and Zuccarello, G.C. (2008) The utility of proteomics in algal taxonomy: *Bostrychia radicans-B. moritziana* (Rhodomelaceae, Rhodophyta) as a model study. Journal of Phycology. 44(6): 1519–1528 pp.

Kleivdal, H., Chauton, M.S. and Reitan, K.I. (2013) ProAlgae–industrial production of marine microalgae as a source of EPA and DHA rich raw material in fish feed: Basis, knowledge status and possibilities. www.indbiotech.no/content/report-proalgae-final-report-april-2013.

Kovač, D.J., Simeunovič, J.B., Babič, O.B., Mišan, A.Č. and Milovanovič, I.L. (2013) Algae in food and feed. Food and Feed Research. 40(1): 21–31 pp.

Krietnitz, L and Bock, C. (2012) Present state of the systematics of planktonic coccoid green algae of inland waters. Hydrobiologia. 698: 295–326 pp.

Kubanek, J., Prusak, A.C., Snell, T.W., Hardcastle, K.I., Giese, R.A. and Fairchild, C.R. (2005) Antineoplastic diterpene–benzoate macrolides from the fijian red alga *Callophycusserratus*. Organic Letters. 7(23): 5261–5264 pp.

Le Bihan, T., Martin, S.F., Chirnside, E.S., van Ooijen, G., Barrios-Llerena, M.E. and O'Neill, J.S. (2011) Shotgun proteomic analysis of the unicellular alga *Ostreococcus tauri*. Journal of Proteomics. 74(10): 2060–2070 pp.

Lewin, R.A. (2015) Encyclopædia Britannica. www.britannica.com/science/algae.

Lin, A.S., Stout, E.P., Prudhomme, J., Roch, K.L., Fairchild, C.R. and Franzblau, S.G. (2010) Bioactive bromophycolides R–U from the Fijian Red Alga *Callophycus serratus*. Journal of Natural Products. 73(2): 275–278 pp.

Lopez, D., Casero, D., Cokus, S.J., Merchant, S.S. and Pellegrini, M. (2011) Algal Functional Annotation Tool: A web-based analysis suite to functionally interpret large gene lists using integrated annotation and expression data. BMC Bioinformatics. 12(1): 282 p.

Lu Wenhai, Caroline Cusack, Maria Baker, Wang Tao, Chen Mingbao, Kelli Paige, Zhang Xiaofan, Lisa Levin, Elva Escobar, Diva Amon, Yin Yue, Anja Reitz, Antonio Augusto Sepp Neves, Eleanor O'Rourke, GianandreaMannarini, Jay Pearlman, Jonathan Tinker, Kevin, J. Horsburgh, Patrick Lehodey, Sylvie Pouliquen, Trine Dale, Zhao Peng and Yang Yufeng. (2019) Successful blue economy examples with an emphasis on international perspectives frontiers in marine science. www.frontiersin.org. 6, Article 261, 1–14pp.

Luo, W., Pflugmacher, S., Pröschold, T., Walz, N., Krienitz, L. (2006) Genotype versus phenotype variability in *Chlorella* and *Micractinium* (Chlorophyta, Trebouxiophyceae), Protist. 157(3): 315–333 pp.

Maisashvili, A., Bryant, H., Richardson, J., Anderson, D., Wickersham, T. and Drewery, M. (2015) The values of whole algae and lipid extracted algae meal for aquaculture. Algal Research. 9: 133–142 pp.

Mantri, V. A., K.S., Ashok, K.R., Saminathan, J., Rajasankar, P. and Harikrishna (2015) Concept of triangular raft design: Achieving higher yield in *Gracilaria edulis*. Aquacultural Engineering. 69: 1–6 pp.

Mantri, V.A., Ganesan, M., Gupta, V., Krishnan, P. and Siddhanta, A.K. (2019) An overview on agarophyte trade in India and need for policy interventions. Journal of Applied Phycology. 1–54 pp.

Martins, A., Vieira, H., Gaspar, H. and Santos, S. (2014) Marketed marine natural products in the pharmaceutical and cosmeceutical industries: Tips for success. Marine Drugs. 12(2):1066–1101 pp.

Mata, T.M., Martins, A.A. and Caetano, N.S. (2009) Microalgae for biodiesel production and other applications: A review. Renewable and Sustainable Energy Reviews. 14(1): 217–232 pp.

McHugh, D.J. (2003) A guide to the seaweed industry. Food and Agriculture Organisation of the United Nations, Rome, Italy.

Mendes, P. (2006) Metabolomics and the challenges ahead. Briefings in Bioinformatics.7(2):127p.

Metzker, M.L. (2010) Sequencing technologies—the next generation. Nature Reviews Genetics. 11(1): 31–46 pp.

Michel, G., Tonon, T., Scornet, D., Cock, J.M. and Kloareg, B. (2010a) Central and storage carbon metabolism of the brown alga *Ectocarpussiliculosus*: Insights into the origin and evolution of storage carbohydrates in Eukaryotes. New Phytologist. 188(1): 67–81 pp.

Michel, G., Tonon, T., Scornet, D., Cock, J.M. and Kloareg, B. (2010b) The cell wall polysaccharide metabolism of the brown alga *Ectocarpussiliculosus*. Insights into the evolution of extracellular matrix polysaccharides in Eukaryotes. New Phytologist. 188(1): 82–97 pp.

Ministry of Agriculture and Farmers Welfare (MoAFW). (2017) National policy for marine fisheries. http://dahd.nic.in/news/notification-national-policy-marine-fisheries-2017

Ministry of Environment and Forest, Government of India (MoEF). (1986) The environmental protection act. http://www.moef.nic.in/sites/default/files/eprotect_act_1986.pdf

Morozova, O., Hirst, M. and Marra, M.A. (2009) Applications of new sequencing technologies fortranscriptome analysis. Annual Review of Genomics and Human Genetics. 10: 135–151 pp.

Moses, T., Mehrshahi, P., Smith, A.G. and Goossens, A. (2017) Synthetic biology approaches for the production of plant metabolites in unicellular organisms. Journal of Experimental Botany. 68: 4057–4074 pp.

National Fisheries Development Board (NFDB). (2015) Guidelines for seaweed cultivation. http://nfdb.gov.in/

National Renewable Energy Laboratory. (2017) 2015 Bioenergy market report. www.nrel.gov/docs/fy17osti/66995.pdf

Necchi, J.R.O. (2016) River algae. Springer, River Algae. ISBN 9783319319841.

Neera Sen, S. and Sanoyaz, S. (2014) Algal resources in the Indian coastal and marine systems with special reference to Sundarbans. Research and Reviews: Journal of Botany. 3(2): 11–19 pp. ISSN: 2278–2222.

Nunn, B.L., Aker, J.R., Shaffer, S.A., Tsai, S., Strzepek, R.F. and Boyd, P.W. (2009) Deciphering diatom biochemical pathways via whole-cell proteomics. Aquatic Microbial Ecology. 55(3): 241–253 pp.

Olaizola, M. (2003) Commercial development of microalgal biotechnology: From the test tube to the marketplace. Biomolecular Engineering. 20: 459–466 pp.

Omar, W. and Maznah, W. (2010) Perspectives on the use of algae as biological indicators for monitoring and protecting aquatic environments, with special reference to Malaysian freshwater ecosystems. Tropical Life Sciences Research. 21(2): 51–67 pp. PMC 3819078, PMID 24575199.

Parker, M.S. and Thomas, M.A. (2004) Genomic insights into marine microalgae. Annual Review of Genetics. 42: 616–645 pp.

Priyadarshani, I. and Rath, B. (2012) Commercial and industrial applications of micro algae—A review. Journal of Algal Biomass Utilization. 3: 89–100 pp.

Qin, S., Jiang, P. and Tseng, C.K. (2004) Molecular biotechnology of marine algae in China. In Ang, Jr., P.O. (ed.), Asian Pacific phycology in the 21st century: Prospects and challenges. Kluwer Academic Publishers, Dordrecht, The Netherlands: 21–26 pp.

Rafael, C., Fajardo, C., Guarnizo, P., Roberto, A.V., Francisco, J. and Fernandez, A. (2018) Review: Biotechnology applications of microalgae in the context of EU "blue growth" initiatives. Journal of Microbiology and Genetics. JMGE. 118(1): 1–14 pp.

Rajauria, G., Cornish, L., Ometto, F., Msuya, F.E. and Villa, R. (2015) Chapter 12: Identification and selection of algae for food, feed, and fuel applications. In Tiwari, B.K. and Troy, D.J. (eds.), Seaweed sustainability—food and non-food applications (1st ed.). Academic Press, Elsevier: Amsterdam, 315–345 pp.

Remya, R.R. and Radhika Rajasree, S.R. (2016) Biodiversity, distribution and utilization of brown algae in Tamil Nadu coastline. Journal of Advanced Research in Geo Sciences & Remote Sensing. 2(3&4).

Roessner, U. and Bowne, J. (2009) What is metabolomics all about? Biotechniques. 46: 363–365 pp.

Round, F.E. (1981) Dispersal, continuity and phytogeography. In The ecology of algae: 357–361 pp. ISBN 978052169063 via Google Books.

Sahoo, D., Sahu, N. and Sahoo, D. (2001) Seaweeds of Indian coast. A.P.H. Publication, New Delhi, India: 283 pp.

Santhi, N., Pradeepa, C., Subashini, P. and Kalaiselvi, S. (2013) Automatic identification of algalcommunity from microscopic images. Bioinformatics and Biology Insights. 7: 327–334 pp.

Saunders, G.W. (2005) Applying DNA barcoding to red macroalgae: A preliminary appraisal holds promise for future applications. Philosophical Transactions of the Royal Society B: Biological Sciences. 360(1462): 1879–1888 pp.

Savchenko, A., Yee, A., Khachatryan, A., Skarina, T. and Evdokimova, E. (2003) Strategies for structural proteomics of prokaryotes: Quantifying the advantages of studying orthologous proteins and of using both NMR and X-ray crystallography approaches. Proteins: Structure, Function and Bioinformatics. 50: 392–399 pp.

Scaife, M.A. and Smith, A.G. (2016) Towards developing algal synthetic biology. Biochemical Society Transactions. 44: 716–722 pp.

Singh, P. and Dash, S.S. (2018) Plant discoveries 2017–new genera, species and new records. Botanical Survey of India, Kolkata.

Sky News. (2020) Alpine glacier turning pink due to algae that accelerates climate change, scientists say. Sky News. July 6.

Space.com. (2018) Greenland has a mysterious 'dark zone'—and it's getting even darker. Space.com. April 10.

Subba Rao, P.V. and Mantri, V. (2006) Indian seaweed resources and sustainable utilization: Scenario at the dawn of a new century. Current Science. 91(2).

Sudhakar, M.P., Jagatheesan, A., Perumal, K. and Arunkumar, K. (2015) Methods of phycobiliprotein extraction from *Gracilariacrassa* and its applications in food colorants. Algal Research. 12: 115–120 pp.

Suleria, H.A.R., Osborne, S., Masci, P. and Gobe, G. (2015) Marine-based nutraceuticals: An innovative trend in the food and supplement industries. Marine Drugs. 13(10): 6336–6351 pp.

Thomas, N.V. and Kim, S. (2013) Beneficial effects of marine algal compounds in cosmeceuticals. Marine Drugs. 11(1): 146–164 pp.

Transparency Market Research (2020) Seaweed Commercial Market (www.transparencymarketresearch.com).

Uma R. (2020) Linking blue economy for sustainability of marine ecosystems. In The blue economy initiative: Blue carbon asset management webinar series-1. Core Carbon X Solutions Private Limited, Hyderabad, India: 9–12 pp. https://www.linkedin.com/posts/core-carbonx-sols-pvt-ltd_the-blue-economy-initiative-blue-carbon-activity-6698532879229698048-q3gL

UNEP-WCMC, IUCN and NGS. (2018) Protected planet report 2018. UNEP-WCMC, IUCN and NGS, Cambridge, UK, Gland, Switzerland and Washington, DC, USA.

U.S Department of Energy-Office of Energy Efficiency and Renewable Energy. (2019) 5-Marine algae: Powering the blue economy: Exploring opportunities for marine renewable energy in maritime markets. U.S. Department of Energy.

Vaudel, M., Verheggen, K., Csordas, A., Raeder, H. and Berven, F.S. (2016) Exploring the potential of public proteomics data. Proteomics. 16: 214–225 pp.

Verpoorte, R., Choi, Y.H., Mustafa, N.R. and Kim, H.K. (2008) Metabolomics: Back to basics. Phytochemistry Reviews. 7(3): 525–537 pp.

Vidoudez, C. and Pohnert, G. (2012) Comparative metabolomics of the diatom *Skeletonema marinoi* in different growth phases. Metabolomics. 8(4): 654–669.

Wang, D.Z., Gao, Y., Lin, L. and Hong, H.S. (2013) Comparative proteomic analysis reveals proteins putatively involved in toxin biosynthesis in the marine dinoflagellate *Alexandriumcatenella*. Marine Drugs. 13; 11(1): 213–232 pp.

Wang, D.Z., Zhang, S. and Hong, H. (2007) A sulfotransferase specific to N-21 of gonyautoxin 2/3 from crude enzyme extraction of toxic dinoflagellate *Alexandrium tamarense* CI01. Chinese Journal of Oceanology and Limnology. 25(2): 227–234 pp.

Wang, H.M.D., Chen, C.C., Huynh, P. and Chang, J.S. (2015) Exploring the potential of using algae in cosmetics. Bioresource Technology. 184: 355–362 pp.

Wei, N., Quaterman, J. and Jin, Y. (2013) Marine macroalgae: An untapped resource for producing fuels and chemicals. Trends in Biotechnology. 31(2): 70–77 pp.

Wetzel, R. (1996) Benthic algae and nutrient cycling in lentic freshwater ecosystems. Algal Ecology: Freshwater Benthic Ecosystems: 641–667 pp.

Wong, P.F., Tan, L.J., Nawi, H. and AbuBakar, S. (2006) Proteomics of the red alga, *Gracilaria changii* (Gracilariales, Rhodophyta). Journal of Phycology. 42: 113–120 pp.

Zhi Cheng, D. and Yan, C. (2013) Transcriptomics: Advances and approaches. Science China Life Sciences. 56(10): 960–967 pp.

5 Mass Mortality of *Kappaphycus alvarezii* and Its Subsequent Impact on Marine Phytoplankton Diversity

A Consequence of Climate Change

Subir Kumar Mandal, Ganesh Temkar, J. Malarvizhi, N. Monisha, G. Ajay, M. Gobalakrishnan, and K. Eswaran

ABBREVIATIONS

°C	Degree Centigrade
SST	Sea Surface Temperature
SHGs	Self Help Groups
H'	Shanon-Weiner Index
DGR	Daily Growth Rate
NW/NE	Northwest/Northeast
SW/SE	Southwest/Southeast
BDL	Below Detection Level
μM	Micro Molar

5.1 INTRODUCTION

Kappaphycus alvarezii belongs to the red algae (Rhodophyceae). It yields κ-carrageenan, a commercially important polysaccharide (Bixler, 1996). This carrageenan is used as a gelling, thickening, and stabilizing agent, especially in food products such as frozen desserts, chocolate milk, cheese, cream, instant products, jellies, pet foods sauces. Apart from that, it is also used in the pharmaceutical and cosmetics industries. *K. alvarezii* cultivation was initiated in the Philippines in 1960 (Parker, 1974). After that, other countries like Japan, Indonesia, Tanzania, Fiji, Hawaii, and South Africa have started cultivating this species on a large-scale basis (Hurtado *et al.*, 2001). Carrageenan containing seaweed production increased from 2 million wet tones in 2000 to 9 million wet tones in 2010, and the value of the total production also increased from USD 72 million to USD 1.4 billion. Indonesia, Philippines, Tanzania, Malaysia, and China are the major carrageenan-producing seaweeds (Cai *et al.*, 2013). In India, cultivation of this seaweed emanated at Mandapam (Tamil Nadu), the southeast coast of India, during 2000 by several self-help groups (SHGs) based on CSMCRI cultivation technology with an unprecedented interest for large-scale cultivation of *Kappaphycus* in India (Mandal *et al.*, 2010).

A few severe problems are associated with *K. alvarezii* cultivation, such as epiphytes, grazing, and diseases (Johnson and Gopakumar, 2011). Apart from that grazing, there are several reports on the occurrence of "*ice-ice*" disease and mass mortality of *K. alvarezii* in several countries. The first report on the havoc wreck on *K. alvarezii* production was published in Tawi-Tawi, the Philippines, during 1974 (Barraca and Neish, 1978; Uyengco *et al.*, 1981; Largo *et al.*, 1995) and after that in

DOI: 10.1201/9781003219156-7

Indonesia during 2009 caused by pathogenic bacteria like *Cytophage* sp. and *Vibrio* sp. (Largo *et al.,* 1995). The outbreak of the "*ice ice*" disease of *K. alvarezii* completely damaged all the cultivation sites situated at Mandapam and Rameswaram coast, Tamil Nadu, India, during 2013.

Even though several authors contributed to the "*ice ice*" disease of *K. alvarezii* and tried to find out its causative factors, it remained clueless. There are publications from (Largo *et al.*, 1995, 1999) where it has been shown that *Cytopgaha* sp. and *Vibrio* sp. might be responsible for the mass mortality of the cultivated *K. alvarezii.* However, physiochemical properties and changes in other associated microalgal populations were not given enough attention (personal communication with Prof. Largo). Therefore, this present manuscript has described the probable cause of the massive outbreak of the *Kappaphycus* cultivation at Mandapam and Rameswaram coast, Tamil Nadu, India, and its consecutive impact on the diversity of marine phytoplankton in and around the cultivation sites at southeast coast of India.

5.2 MATERIALS AND METHODS

5.2.1 Study Area

The outbreak occurred in the Palk Bay area, the extreme southeast coast of India. The Palk Bay is a shallow basin with an average depth of 9 m (max. 13.5 m depth), mainly with sandy and a muddy bottom at shore regions. The eastern part of the bay is connected with Sri Lanka, whereas the western part of the bay is the Indian subcontinent. Seawater samples and marine phytoplankton samples were collected from eight different sampling sites, where the *K. alvarezii* cultivation was going on, and subsequently, the massive outbreak was observed along the Palk Bay (Figure 5.1). The specific location of the cultivation sites are as follows: Munnaikadu (N 09°17.269', E 079°07.977'), T. Nagar (N 09°17.486', E 079°08.589'), Thoniturai (N 09°16.981', E 079°10.995'), Pinnaikulam (N 09°18.869', E 079°17.086'), Sangumal (N 09°17.936', E 079°19.620'), and Lighthouse (N 09°19.062', E 079°19.875'). The cultivators of Munnaikadu, T. Nagar, Thonithurai, and Pinnaikulam, used to

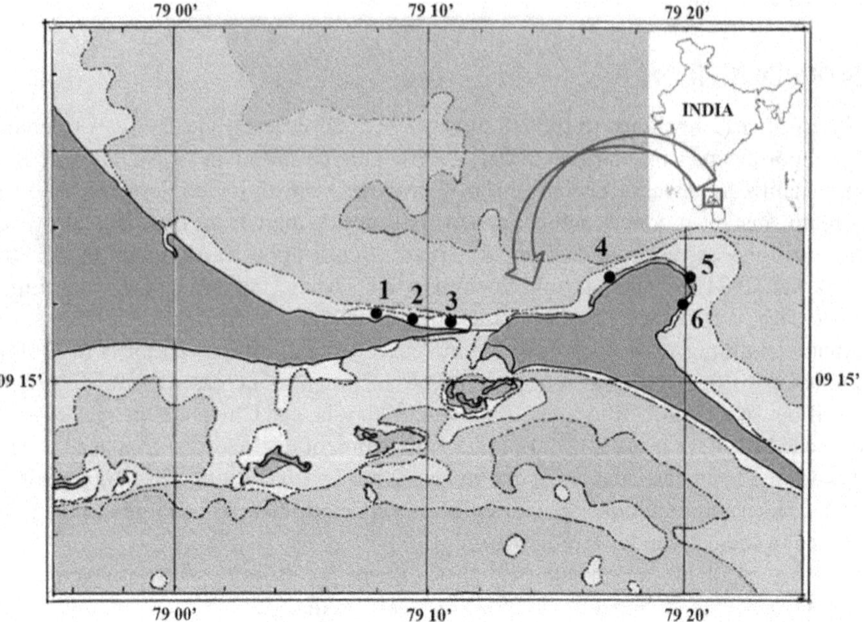

FIGURE 5.1 Map of the study area: (1) Munaikadu, (2) T. Nagar, (3) Thonithurai, (4) Pillaikulam, (5) Lighthouse, (6) Sangumal.

cultivate *K. alvarezii* using rafts methods. In the Sangumal and Lighthouse area, the cultivators used the monoline method.

5.2.2 Environmental Anomalies

5.2.2.1 Air Temperature during the Mass Mortality

Daily temperature data (high and low) in degree centigrade of the month of May, June, and July for 15 years (1997 to 2013) were collected from the field research station of Central Electrical and Corrosion Research Institute, Kadaikudi, Tamil Nadu (Figure 5.2).

5.2.2.2 Physiochemical Parameters of Seawater

Environmental parameters of seawater were carried out by following standard protocols described by Strickland and Parsons (1972) and Grasshoff *et al.* (1998). To analyze nutrient parameters, surface seawater samples were collected in clean high-density polypropylene bottles, kept within an icebox in field conditions, and transported immediately to the laboratory. For example, DO, and BOD_3 were estimated by the modified Winkler's method, and all other nutrients parameters (NO_2-N, NO_3-N, PO_4-P, and SiO_3-Si) were carried out by adopting the standard methods. Estimation of nutrients parameters completed within six hours of collecting samples. Wind parameters have been received by the cup counter mechanical anemometer (Model No. IS: 5192–1970, Sr. No. 25012/2000).

5.2.3 Phytoplankton Community Study

Phytoplankton samples were collected, filtering 30 liters of seawater using a plankton net (Model No. 23.000, Den Mark) (mesh size of 60 µm). Samples were collected from the central point of each cultivation raft, and two samples were collected from two opposite sides of the cultivated raft, on the raft (OR) and out of the raft (100 m away). Samples were collected from above the monoline in a few stations, where cultivation practices were conducted using monoline methods and 100 m away from the monoline. The collected phytoplankton samples were immediately preserved in a

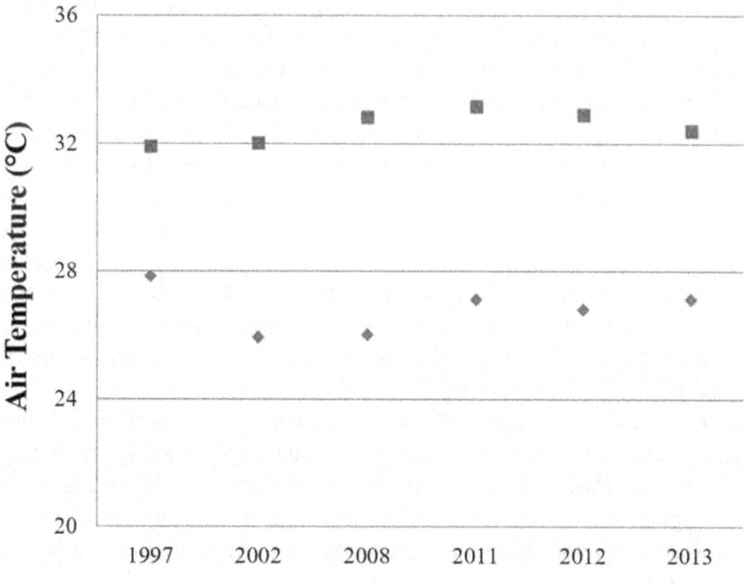

FIGURE 5.2 Increasing air temperature in the southeast coast of the Palk Bay region. Redrawn from Temkar *et al.*, 2014.

4% formaldehyde-seawater solution to avoid damage by bacterial action and autolysis in the collection bottles. Phytoplankton species are identified up to the possible taxonomic level (genus/species) by using "Identifying Marine Phytoplankton" (Thomas, 1996), "Illustration of the Plankton of the Kuroshio-waters", "Marine Phytoplankton Atlas of Kuwait's Waters" (Al-Kandari *et al.*, 2009) and "Phytoplankton of the Indian Seas" (Santhanam *et al.*, 1987). The marine phytoplankton community's qualitative and quantitative studies were carried out using the Sedgwick Rafter Counting Chamber and presented the total number of phytoplankton cells present within 1 liter of seawater.

The diversity of phytoplankton along *K. alvarezii* cultivation sites was expressed cells/L. The species diversity indices were calculated using Shanon-Weiner Index (H') as per the given formula (Shannon and Wiener, 1949).

Shanon-Weiner Index (H')

$$H' = -\sum_{i=1}^{s} \frac{ni}{n} \times \ln \frac{ni}{n}$$

Where H' = the sample diversity
S = number of species.

5.3 RESULTS AND DISCUSSION

5.3.1 Environmental Anomalies and Their Impact

The global average temperature is being increased by ~0.2°C per decade, and presently, the global average sea surface temperature has already been elevated to 0.58°C above the average global temperature recorded for the 20th century (Guldberg and Bruno, 2010). The climatic conditions of the south Palk Bay area, southeast India, are also changing very fast. The monthly average of daily high and low air temperatures have been increased @ 0.21 °C/5 years and @ 0.03 °C/5 years respectively during the last 15 years in the Palk Bay region (Temkar *et al.*, 2014). The atmospheric condition of the Mandapam, as well as the Rameswaram area, is arid. It is observed that the three monthly (May to July) average highest temperature during daytime and average lowest temperature during night time in the last 15 years are increasing (Figure 5.2). In August and September 2013, most of the days, specifically during the mass mortality, it was observed that the air temperature was 33°C and above (Table 5.1) (Source: CECRI Field station at Mandapam). Generally, *K. alvarezii* plant cannot tolerate above 33°C. The high atmospheric temperature (33°C) for a week created low pressure, and results of that the wind speed was very low (<2 km/h) during the occurrence of the mass mortality of *K. alvarezii* (Figure 5.3). In addition to the low wind flow or not at all wind flow, the *K. alvarezii* plants were already under stress conditions due to low nutrient contents in the seawater, as wave actions were absent during that period. The NO_3-N was detected below the detection limit (BDL) in all the stations at Palk Bay from 27 August to 6 September 2013. The recorded precipitation was also increased with time. During 2002, there was no record of rain from June to August months. Recently (2013), the recorded rainfall was 36.2 mm, the highest among the last five years. In 2011, the highest rainfall (10.2 mm) was recorded, and the lowest (9.0 mm) was in 2008. After that, a rainfall (10 mm) was recorded on 28 August 2013. Along with the changing climatic conditions, the cultivation activity such as *K. alvarezii* was also increased in the Palk Bay region. It was initiated with a yield of 21 MT (dry wt.) during the year 2001 to 1450 MT (dry wt.) in 2013 (Temkar *et al.*, 2014).

Due to these geographical locations, the surface seawater of Palk Bay moved away from the coast during the southwest wind, and the bottom enriched seawater came out and nourished the massively dense *K. alvarezii* cultivation sites due to the coastal upwelling. Before the massive death of *K. alvarezii*, environmental parameters' inaptness was recorded. Generally, in August–September month, the wind speed of this region and direction remain 10–18 km/h SW/SE (24 h average).

TABLE 5.1
Wind-Speed and the Direction of the Wind-Flow at the Mandapam and Rameswaram Coast during Mass Mortality of *K. alvarezii*

Date	Temperature Max (°C)	24 h Average Min (°C)	Wind Velocity Km./h.	Direction	Date	Temperature Max (°C)	24 h Average Min (°C)	Wind Velocity Km./h.	Direction
01–08–2013	32	27	16.3	SE/SW	01–09–2013	33	27	10.3	SE/SW
02–08–2013	32	27	9.2	SE/SW	02–09–2013	31	26	15.9	SE/SW
03–08–2013	32	26	9.2	SE/SW	03–09–2013	32	27	18.8	SE/SW
04–08–2013	32	25	9.2	SE/SW	04–09–2013	32	26	17.7	SE/SW
05–08–2013	32	25	9.2	SE/SW	05–09–2013	31	26	11.6	SE/SW
06–08–2013	32	25	9.2	SE/SW	06–09–2013	31	26	11.3	SE/SW
07–08–2013	32	27	9.2	SE/SW	07–09–2013	30	26	11.3	SE/SW
08–08–2013	33	27	9.2	SE/SW	08–09–2013	31	26	11.3	SE/SW
09–08–2013	32	27	9.2	SE/SW	09–09–2013	32	27	11.3	SE/SW
10–08–2013	33	28	9.2	SE/SW	10–09–2013	32	27	11.3	SE/SW
11–08–2013	33	26	9.2	SE/SW	11–09–2013	32	27	11.3	SE/SW
12–08–2013	32	26	9	SE/SW	12–09–2013	31	26	11.3	SE/SW
13–08–2013	32	26	7.7	SE/SW	13–09–2013	32	27	11.3	SE/SW
14–08–2013	32	27	13.7	SE/SW	14–09–2013	32	27	11.3	SE/SW
15–08–2013	33	27	13.7	SE/SW	15–09–2013	32	26	11.3	SE/SW
16–08–2013	32	27	13.7	SE/SW	16–09–2013	32	26	18.3	SE/SW
17–08–2013	33	27	13.7	SE/SW	17–09–2013	32	26	14.6	SE/SW
18–08–2013	32	27	13.7	SE/SW	18–09–2013	31	26	14.6	SE/SW
19–08–2013	31	26	13.9	SE/SW	19–09–2013	32	26	17.3	SE/SW
20–08–2013	31	26	15.7	SE/SW	20–09–2013	32	26	15.1	SE/SW
21–08–2013	31	26	13	SE/SW	21–09–2013	33	27	15.1	SE/SW
22–08–2013	31	27	11	SE/SW	22–09–2013	33	27	15.1	SE/SW
23–08–2013	32	27	10.5	SE/SW	23–09–2013	32	26	14.2	SE/SW
24–08–2013	32	27	10.5	SE/SW	24–09–2013	32	26	13.3	SE/SW
25–08–2013	32	27	10.5	SE/SW	25–09–2013	32	26	14.5	SE/SW
26–08–2013	33	27	10.5	SE/SW	26–09–2013	32	27	14.5	SE/SW
27–08–2013	33	27	10.5	SE/SW	27–09–2013	32	27	13.4	SE/SW
28–08–2013	32	26	10.5	SE/SW with rain (10 mm.)	28–09–2013	33	27	13.4	SE/SW
29–08–2013	33	26	5.9	NE/NW	29–09–2013	33	27	13.4	SE/SW
30–08–2013	33	27	10.3	SE/SW	30–09–2013	33	27	13.4	SE/SW
31–08–2013	33	27	10.3	SE/SW					

Furthermore, the wind direction changed from SW/SE to NW/NE during the late October month, but during this anomaly, the wind direction suddenly changed from SE/SW to NE/NW at the end of August, along with very low wind speed, that is, 5.9 km/h (24 h average). Additionally, there was rain (10 mm), and the havoc wreck occurred at the *Kappaphycus* cultivation sites. The rainfall might have decreased the salinity (below 20‰) suddenly as freshwater flux due to rain could not mix with the seawater due to the highly dense *Kappaphycus* plant in commercial cultivation area and low wind blow (Figure 5.3).

A sudden salinity change disrupts active K-uptake by the *Kappaphycus* plant, and plants thrive under environmental stress, leading to the release of secondary metabolites, which might have

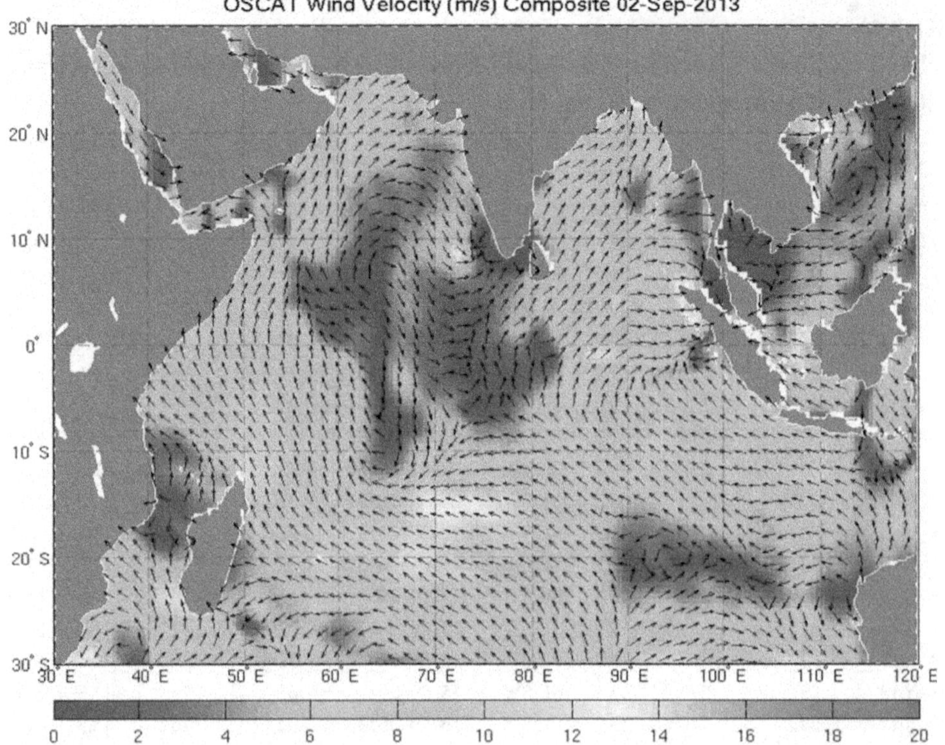

FIGURE 5.3 The environmental anomaly, that is, high air temperature, seawater temperature, low wind speed, and wind direction, might have played as primarily responsible factors for the massive death of *K. alvarezii*. For example, the air temperature was very high (33°C), which was more than the average air temperature during August (32.19°C). Due to low wave action, the seawater temperature reached 34.5°C in the shallow coastal region. The low wind pressure was generated from 30 August to 4 September 2013 (Temkar *et al.*, 2014); a feeble or nil wind blow was recorded during the incident. The setting point of *kappa, Iota*, and *lambda* carrageenan is 30–50°C, and the melting point is 50–70°C (Source: www.agargel.com.br/carrageenan-tec.html visited 12 December 2013). Above the 33°C temperature, the plants become fragile due to substantial adequate energy (vigor) loss. According to Ohno *et al.* (1996), the plants become weak, with decreased DGR at temperatures higher than 33°C. As all the cultivation sites are situated in the north and northeast part of the mainland and as the submerged vegetation, like other macroalgal vegetation, seagrasses are massively dense in these regions, so that natural nutrient concentration remains comparatively low (Kumar and Manivannan, 2001; Sridhar *et al.*, 2006, 2010; Anantharaj *et al.*, 2013; Govindasamy *et al.*, 2012).

helped grow marine bacteria surface seawater rapidly. Those bacterial loads might have influenced the "*ice–ice*". Doty (1987) noted that ice-ice was seasonal and correlated with monsoon season (Doty and Alvarez, 1975).

The other associated abiotic factors (air temp., water temp., S‰, and pH), including nutrient parameters (NO_2-N, NO_3-N, PO_4-P, and SiO_3-Si) on the day of 6 September, are depicted in Table 5.2. The atmospheric temperature varied from 26.5°C to 30.0°C (avg. 29°C), and seawater temperature ranged from 26.5°C to 29.5°C (avg. 27.83°C). The average salinity (‰), as well as pH, varied from 34.0 to 36.0 (Mean 35.33‰) and 7.93 to 8.33 (Mean 8.13), respectively. The minimum values for salinity (34.0‰) and pH (7.93) have been recorded in Pinnakulam. NO_2-N varied from 0.23 to 5.82 µM (avg. 1.91 µM) on the affected raft (OR) and the same varied from 0.41 to 3.67 µM (avg. 1.47 µM) at away from the affected raft (AFR). NO_3-N was recorded as below detection level (BDL) at all the affected sites. The complete absence of NO_3-N concentration for more than 10 days, that is, from 27 August to 6 September 2013 as per the record, created detrimental stress on

TABLE 5.2
Physiochemical Parameters during the Mass Mortality Occurred

Places	GPS Locations	Air Temp °C	Water Temp °C	Salinity (%)	pH	NO_2-N (μM)		NO_3-N (μM)		PO_4-P (μM)		SiO_4-Si (μM)	
						OR	AFR	OR	AFR	OR	AFR	OR	AFR
Munaikadu	N09°17.269, E079°07.977	29.5	27.5	36	8.17	2.75	2.25	BDL	BDL	0.12	0.16	27.95	11.88
T. Nagar	N09°17.486, E079°07.589	26.5	27.5	36	8.26	1.05	0.78	BDL	BDL	1.15	0.13	1284	18.06
Thonithurai	N09°16.981, E079°10.995	28	26.5	35	8.04	0.96	0.64	BDL	BDL	0.08	0.07	18.88	23.07
Pinnakulam	N09°18.869, E079°17.086	30	28	34	7.93	0.23	0.41	BDL	BDL	0.13	0.20	18.68	20.26
Sangulam	N09°17.936, E079°19.620	30	28	35	8.27	5.82	3.67	BDL	BDL	0.07	0.12	21.15	23.35
Lighthouse	N09°19.062, E079°19.875	30	29.5	36	8.13	0.64	1.05	BDL	BDL	0.07	0.06	10.85	20.74
	AVG	29.00	27.83	35.33	8.13	1.91	1.47	BDL	BDL	0.27	0.12	18.39	19.56

Note: AFR = Away from raft, OR = On raft. Air and water temperature recorded just after the massive death of *K. alvarezii* occurred on 6 September 2013.

the full-grown *K. alvarezii* plant (>600kg/raft). This low nutrient condition of NO_3-N concentration might have led to *K. alvarezii* mass mortality. PO_4-P showed variations from 0.07 to 1.15 μM (avg. 0.27 μM) on the affected raft and recorded as higher than the away from the affected raft, that is, 0.06 to o.20 μM (avg. 0.12 μM). The SiO_3-Si value was registered as a slightly lower value on the affected raft and varied from 10.85 μM to 27.95 μM with an average value of 18.39 μM as compared to away from the raft, which ranged from 11.88 to 23.35 μM with an average value of 19.56 μM (Table 5.2).

Inorganic reactive nitrogen in different bioavailable forms, such as NO_2-N, NO_3-N, and NH_4-N are essential nutrients for *K. alvarezii* species (Mairh *et al.*, 1995; Tewari *et al.*, 2006). Red algae have two kinds of accessory pigments, that is, phycoerythrin and phycocyanin (called phycobiliprotein) and Chlorophyll-a as a primary photosynthetic pigment (Veeragurunathan, 2006). Phycobiliproteins are synthesized in high nitrogen availability and are considered an essential nitrogen reserve by Bogorad (1975) and Bird *et al.* (1982). Lapointe (1981) and Gantt (1981) observed that the phycoerythrin has an additional role in storing and/or utilizing bioavailable nitrogen rapidly and growing fast, even in the dark, and primary functions of the light-harvesting ability suggest that it in storing. Fujita (1985) and Duke *et al.* (1989) described that 60% of the total soluble protein in red algae are phycobiliproteins and represent a large nitrogen reserve. From 4.5% to 8.8%, DGR is reported for *K. alvarezii* (Eswaran *et al.*, 2006), so a considerable amount of nitrogen might also be utilized by the *K. alvarezii* plant every day. Here it is observed that during the massive death of *K. alvarezii*, the NO_3-N found below the detection limit might be due to rapid consumption, and as there was significantly less wind blow and changed to the opposite direction within 48 h, which might have reduced air-sea inter-surface interaction and decreased NO_3-N level in seawater. Just 15 days before the incidence, NO_3-N concentration was recorded within a range of 5.43 to 8.24 μM. In earlier reports, the same was varied from 2.15 to 8.28 μM (Sridhar *et al.*, 2006) and 3.81 to 18.42 μM (Anatharaj *et al.*, 2013). The biomass load was high in most of the cultivation points. It varied from 800 rafts (Vadakadu) to 8,750 rafts (Lighthouse) with very high density at all the cultivation points (supplementary Table 5.1). NO_3-N stress occurred due to very low wind blow and no surface

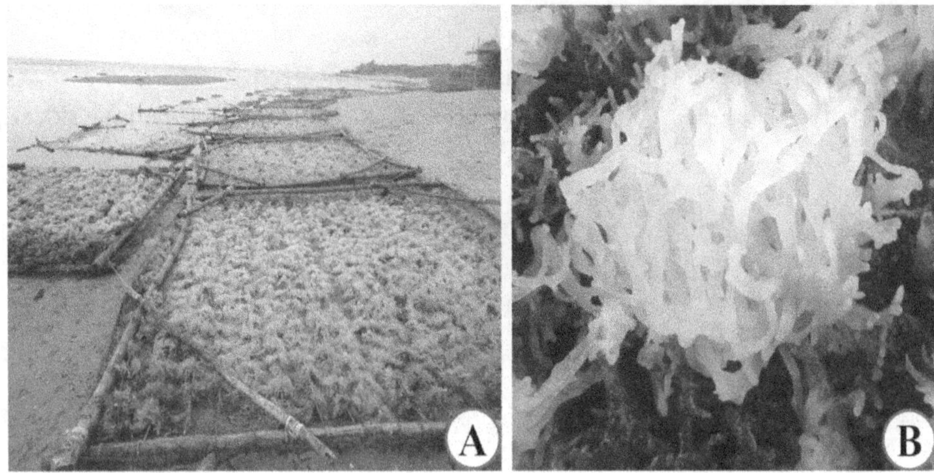

FIGURE 5.4 Mass mortality of *K. alvarezii* throughout the Mandapam-Rameswaram coast. A) Dead biomass of *K. alvarezii* at Munaikadu, Palk Bay coast. B) Close view of completely bleached *K. alvarezii* thalli on an "ice ice" disease-affected raft at Munaikadu, Palk Bay.

water movement, and also high uptake by *K. alvarezii* might have triggered phycobiliproteins degradation very fast, and those plants became bleached. PO_4-P also reduced from 5.29 to 0.22 µM at Thonithurai, Palk Bay, and 5.52 to 1.13 µM at the Munaikadu area within 15 days.

On the other hand, the SiO_3-Si concentration was increased within 15 days of the interval from 8.49 to 12.71 µM at Thonithurai, Palk Bay, and 9.10 to 24.19 µM at Munaikadu area, as *K. alvarezii* does not uptake SiO_3-Si more as compared to other nutrients, and also it uptakes less in higher temperature (30–35°C). The same becomes excess in the water column and helps diatoms grow (Mandal *et al.*, 2014). *Kappaphycus* needs at least 5 km/h wind blow to grow (Castelar *et al.*, 2009). So very slow wind blow and very high seawater temperature reached 34.5°C, which crossed the critical temperature of 33°C and low light intensity due to the cloudy environment might have caused autolysis/bleaching of the phycobiliproteins and massive death of *K. alvarezii* (Figure 5.4). Loss of plant vigor due to the repeated use of the same seed stocks, year after year, might also be an essential factor along with other environmental factors. However, the loss of plant vigor may not be the only cause for mass mortality of the *K. alvarezii*. The combined effect of high temperature (33°C), low wind pressure (<2km/h), and sudden rainfall (10mm) caused the mass mortality of *K. alvarezii* (Figure 5.4) along the Mandapam-Rameswaram coast, the southeast coast of India. This lower concentration of SiO_3-Si on the cultivated raft indicates that the available bio-silica on the cultivated raft induces the diatoms (Bacillaria) growth on the cultivated raft of *K. alvarezii*. Therefore, the phytoplankton community's diversity on cultivation raft and away from raft were also significantly different.

5.3.2 Phytoplankton Abundances

A total of 46 phytoplankton genera were recorded during the mass mortality study along the Mandapam-Rameswaram coast of India. Out of 46 phytoplankton genera, 39 genera were recorded on the affected cultivation raft compared to away from the raft (35 genera). On average, the number of genera (17 genera) was more on the affected cultivation raft along the Mandapam-Rameswaram coast than 100 m away from the raft (13 genera). The generic diversity of dinoflagellates was also high (4 genera) on the affected *K. alvarezii* rafts compared to 100 m away from the raft (1 genus) only. In the earlier days, Sridhar *et al.* (2006) reported 45 species (43 diatoms and 2 cyanobacterial species) comprising total cell count varied from 18×10^3 to 34×10^3 cells L^{-1} from the

Kattumavadi, north Palk Bay region, and 21.040 × 10^3 to 30.971 × 10^3 cells L^{-1} from Kadalur region, Palk Bay (Prabhahar *et al.*, 2011). In the present study, the total phytoplankton count varied approximately from 1.099 × 10^3 no. of cells L^{-1} in Munaikadu and T. Nagar to 13.737 × 10^3 no. of cells L^{-1} in Pinnakulam on the affected cultivation raft. The average phytoplankton (4.113 × 10^3 no. cells L^{-1}) was observed on the affected cultivation rafts along the Mandapam-Rameswaram coast. Whereas, the same ranged from 0.387 × 10^3 no. of cells L^{-1} (T. Nagar) to 2.578 × 10^3 no. of cells L^{-1} (Pinnakulam) in the 100 m away from the raft. The average phytoplankton count (4.113.33 × 10^3 no. cells L^{-1}) on the "*ice ice*" affected raft was more than 100 m away from the affected raft area (1.335 × 10^3 no. cells L^{-1}). The phytoplankton community's overall abundances were low compared to earlier records by Sridhar *et al.* (2006) and Prabhahar *et al.* (2011). Anantharaj *et al.* (2013) reported that 32 genera of phytoplankton were reported from the Palk Bay area. Amongst them, 21 genera were diatoms, and 11 genera belonged to blue-green algae (Cyanophyceaae). *Fragillaria* sp., *Navicula* sp., *Nitzschia* sp., and *Pluerosigma* sp., are the most abundant forms. It is vital to note that the dinoflagellate species were not reported in earlier studies by Sridhar *et al.* (2010) and Anantharaj *et al.* (2013). Diatoms community was found dominant in both areas, that is, 96.69% and 95.57% on the affected raft area and away from the raft area, respectively. *Trichodesmium erythrum* was detected both on the affected raft and away from the raft, but quantitatively, the average *T. erythrum* was more away from the raft (10 × 103 no. of cells L^{-1}) than on the raft affected raft (1 × 10^3 no. of cells L^{-1}). In some cases, Shanon-Weiner Diversity Index (SWDI) was higher on the affected cultivation raft, for example, Munaikadu (2.55), Sangumal (2.81), and Lighthouse (2.43), as compared to away from the raft and some cases. The SWDI values were higher on the away from the raft, for example, T. Nagar (2.58), Thonithurai (2.07), and Pinnaikulam (2.70), as compared to the affected raft (Table 5.3).

TABLE 5.3
Qualitative and Quantitative Studies of Marine Phytoplankton Were Studied in the "*Ice Ice*" Disease-Affected Area of *K. alvarezii* Cultivation Sites

		On Raft/Monoline						Away from Raft/Monoline					
No.	Species	MKD	T. NAG	T.T.	PNK	SAN	LH	MKD	T. NAG	T.T.	PNK	SAN	LH
1	***Amphiprora alata***	79	50	69	381	–	81	60	–	88	–	50	–
2	***Amphora* sp.**	–	–	412	–	–	–	–	–	–	–	–	–
3	***Amphora sulcata***	–	–	–	–	–	–	–	–	–	–	–	97
4	***Asterionella glacialis***	–	–	–	191	–	–	–	–	–	–	–	–
5	***Bacillaria paradoxa***	43	91	–	763	–	803	42	–	–	91	–	–
6	***Biddulphia mobiliensis***	–	–	–	–	–	–	–	–	–	39	–	–
7	***Biddulphia pulchella***	–	–	–	–	–	–	–	–	–	–	–	23
8	***Cheatoceros affinis***	–	–	–	–	–	–	–	–	184	–	–	–
9	***Cheatoceros curvisetus***	–	116	–	–	–	–	114	16	–	–	–	–
10	***Climacosphenia elongata***	82	–	52	–	87	1099	–	–	–	–	–	143
11	***Cocconeis* sp.**	–	–	–	173	–	86	–	44	–	26	91	–
12	***Cymbella* sp.**	48	–	–	–	57	46	21	–	–	74	29	73
13	***Dactyliosolen antarcticus***	–	–	–	78	–	–	–	–	–	–	–	–

(*Continued*)

TABLE 5.3 (*Continued*)
Qualitative and Quantitative Studies of Marine Phytoplankton Were Studied in the "*Ice Ice*" Disease-Affected Area of *K. alvarezii* Cultivation Sites

		On Raft/Monoline						Away from Raft/Monoline					
No.	Species	MKD	T. NAG	T.T.	PNK	SAN	LH	MKD	T. NAG	T.T.	PNK	SAN	LH
14	*Diploneis splendica*	–	–	147	–	–	–	–	–	132	–	–	–
15	*Gunardia delicatula*	–	–	–	–	–	–	–	–	–	132	–	–
16	*Gyrosigma* sp.	–	–	–	–	47	–	–	–	–	–	–	–
17	*Leptocylindrus danicus*	–	–	–	633	93	1183	–	–	–	132	–	187
18	*Licmophora nubecula*	65	66	–	139	–	132	63	27	68	56	–	–
19	*Licmophora* sp.	–	–	–	2721	–	122	–	–	–	95		–
20	*Mastagloia* sp.	–	–	–	–	27	–	–	–	–	–	70	–
21	*Navicula blanda*	–	124	–	295	–	–	–	–	–	48	–	–
22	*Navicula robertsiana*	–	–	–	615	–	–	–	–	–	–	–	–
23	*Navicula* sp. (Giant)	28	–	–	–	–	–	–	–	–	–	–	–
24	*Navicula* sp. 1	119	–	–	–	–	–	27	31	304	204	124	243
25	*Navicula* sp. 2	–	–	–	–	147	–	–	–	–	–	–	93
26	*Navicula* sp. 3	–	116	–	156	–	200	–	–	–	–	–	–
27	*Nitzschia cloasterium*	204	79	321	1066	170	89	81	24	136	156	–	693
28	*Nitzschia sigma*	–	–	–	199	–	–	–	–	–	277	–	–
29	*Nitzschia* sp. 1	–	–	22	61	87	195	48	19	–	74	45	57
30	*Nitzschia* sp. 2	–	128	230	217	23	238	–	36	208	117	136	110
31	*Oscillatoria* sp.	31	33	56	485	37	117	30	29	–	100	–	107
32	*Pleurosigma aestuarii*	57	41	321	381	67	476	–	37	–	–	211	110
33	*Pleurosigma angulatum*	–	–	–	–	37	–	–	–	–	–	–	–
34	*Pleurosigma normannii*	62	–	–	147	73	–	–	29	228	–	–	–
35	*Pseudonitzschia* sp.	40	58	–	199	163	410	84	21	–	–	–	207
36	*Rhabdonema* sp.	–	–	–	–	–	–	–	–	–	–	–	17
37	*Rhizosolenia imbricata*	–	–	13	–	–	–	–	–	–	–	45	–
38	*Rhizosolenia shrubsolei*	–	–	–	–	–	122	–	–	–	39	–	47
39	*Thalassiosira subtilis*	–	–	–	–	–	–	–	41	–	–	–	–
40	*Thalassiothrix fraunfeldii*	133	112	13	2011	87	–	–	21	–	529	–	–
41	*Thalassiothrix longissima*	–	–	–	2522	70	99	–	–	–	165	–	–
	Total	**992**	**1013**	**1655**	**13433**	**1270**	**5497**	**570**	**376**	**1348**	**2353**	**802**	**2207**
	Cyanophyceae												
42	*Trichodesmium erythreaum*	–	–	9	–	–	–	–	–	–	30	29	–

TABLE 5.3 (*Continued*)
Qualitative and Quantitative Studies of Marine Phytoplankton Were Studied in the "*Ice Ice*" Disease-Affected Area of *K. alvarezii* Cultivation Sites

		On Raft/Monoline						Away from Raft/Monoline					
No.	**Species**	**MKD**	**T. NAG**	**T.T.**	**PNK**	**SAN**	**LH**	**MKD**	**T. NAG**	**T.T.**	**PNK**	**SAN**	**LH**
	Total	**–**	**–**	**9**	**–**	**–**	**–**	**–**	**–**	**–**	**30**	**29**	**–**
	Dinophyceae				–								
43	***Allexandrium* sp.**	62	87	–	208	23	–	–	–	–	–	–	40
44	***Ceratium furca***	–	–	–	–	20	–	–	–	–	–	–	–
45	***Procentrum* sp.**	–	–	–	–	30	–	–	–	–	–	–	–
46	***Protoperidinium cerasus***	–	–	–	–	–	58	–	–	–	–	–	–
	Total	**62**	**87**	**–**	**208**	**73**	**58**	**–**	**–**	**–**	**–**	**–**	**40**
	Unknown Species	**45**	–	17	95	97	68	–	11	52	195	–	–
	Total Count	**1099**	**1099**	**1681**	**13737**	**1440**	**5624**	**570**	**387**	**1400**	**2578**	**831**	**2247**
	Highest	204	128	412	2721	170	1183	114	44	304	529	211	693
	No. of Genera	15	13	13	23	20	19	10	14	9	20	10	16
	Species Indices												
	Shanon-Weiner Index (H')	**2.55**	**2.49**	**2.02**	**2.55**	**2.81**	**2.43**	**2.18**	**2.58**	**2.07**	**2.70**	**2.10**	**2.35**

Note: MKD = Munaikadu, T. Nagar = Tata Nagar, T.T. = Thonithurai, PNK = Pinnaikulam, SAN = Sangumal, and LH = Lighthouse.

A few diatoms include *Amphora sulcata*, *Biddulphia mobiliensis*, *B. pulchella*, *Chaetoceros affinis*, *Guinardia delicatula*, and *Rhabdonema* sp., and *Thalassiosira subtilis* were only available away from the raft area. In contrast, *Amphora* sp., *Asterionella glacialis, Dactyliosolen antarcticas, Gyrosima* sp., *Navicula robertsiana, Navicula* sp., and *Pleurosigma angulatum* were present only on the affected raft (Table 5.3). Some species were exclusively dense on the affected raft area, for example, *Climacosphenia elongata* (1.099 $\times 10^3$ no. of cells L^{-1}) and *Leptocylindrus danicus* (1.183 $\times 10^3$ no. of cells L^{-1}) at the Lighthouse area. *Licmophora abbreviate* (2.721 $\times 10^3$ no. of cells L^{-1}), *Nitzschia cloasterium* (1.066 $\times 10^3$ no. of cells L^{-1}), *Thalassiothrix fraunfeldii* (2.011 $\times 10^3$ no. of cells L^{-1}), and *T. longissimi* (2.522 $\times 10^3$ no. of cells L^{-1}) at Pinnakulam area.

Some dinoflagellates genera, for example, *Ceratium*, *Procentrum*, and *Protoperidinium*, were found only on affected raft areas. In contrast, *Allexandrium* sp. was reported from both the sites, that is, on most of the affected raft and varied from 23 no. of cells L^{-1} at Sangumal to 208 no. of cells L^{-1} at Pinnaikulam, and only in one case, the same was found (40 no. of cells L^{-1}) away from the monoline in Lighthouse area.

5.4 CONCLUSION

The consequences of climate change on the world's largest ecosystem, i.e., the marine ecosystems globally and regional, are the most complicated ones. The large-scale cultivation of commercially important seaweeds, for example, *K. alvarezii* in the marine environment, face frequent mass mortality, called "ice ice" disease due to environmental anomaly. This "ice ice" disease is mainly influenced by environmental factors like high temperature, low nutrients (NO_3-N) concentration, and sudden rainfall that reduce the seawater's salinity on the cultivation raft. These primary factors might have triggered the rapid growth of secondary elements like pathogenic and decomposer microbes, which might have competed with the *K. alvarezii* for nutrients and invaded into the outer cell of *K. alvarezii* plant. After that it degrades the plant's vigor and destroys the canopy. During this mass mortality, the secondary metabolites released from the stressed *K. alvarezii* in the seawater support heterotropic phytoplankton's growth and change the phytoplankton community composition in and around the cultivation sites.

ACKNOWLEDGMENT

The authors are thankful to Dr. Pushpito Kumar Ghosh, ex. director, CSIR-CSMCRI, for his constant support during this work, and Dr. V. A. Mantri, principal scientist, CSIR-CSMCRI, for his encouragement. The authors are also thankful to Dr. Eswar, ex scientist in charge, Central Electrochemical Research Institute (Field station), Mandapam Camp, Tamil Nadu, for providing the rainfall, wind speed, and wind direction data. This work has been supported by the different project grants from CSIR, New Delhi (PSC 0105, and K-TEN). This paper is communicated by PRIS approval, and the code is CSIR-CSMCRI—047/2014.

REFERENCES

Al-Kandari M, Al-Yamani F, Al-Raifaie K (2009) Marine Phytoplankton Atlas of Kuwait's Waters. Kuwait Institute for Scientific Research, Safat, Kuwait, p. 344.

Anantharaj K, Anantharaj, Packiyalakshmi P, Ganesh J (2013) Studies on the physico chemical status of Kattumavadi coastal region, Southeast Coast of India. Int J Res Mar Sci 2(2): 45–49.

Barraca RT, Neish IC (1978) A survey of *Eucheuma* farming practices in Tawi-Tawi. I. The Sitangkai, Sibutu, Tumindao Region. A report on Marine Colloids (Phils.) Inc. 65 pp.

Bird KT, Habig C, DeBusk T (1982) Nitrogen allocation and storage patterns in *Gracilaria tikvahiae* (Rhodophyta) J Phycol 18: 344–348.

Bixler HJ (1996) Recent developments in manufacturing and marketing carrageenan. Hydrobiologia 326/327: 35–57.

Bogorad L (1975) Phycobiliproteins and complementary chromatic adaptation. Annu Rev Plant Physiol 26: 369–401.

Cai J, Hishamunda N, Ridler N (2013) Social and Economic Dimensions of Carrageenan Seaweed Farming: A Global Synthesis. FAO Fisheries and Aquaculture Technical Paper. FAO, Viale delle Terme di Caracalla, 00153 Rome, Italy, p. 580.

Castelar B, Reis RP, Moura AL, Kirk R (2009). Invasive potential of *Kappaphycusalvarezii* off the south coast of Rio de Janeiro state, Brazil, contributes to environmentally secure cultivation in the tropics. Bot. Mar 52: 283–289.

Doty MS (1987) The production and use of *Eucheuma*. Case studies of seven commercial seaweed resources. FAO Fish Tech 281: 123–161.

Doty MS, Alvarez VB (1975) Status, problems, advances, and economics of *Eucheuma* farms. Mar Technol Soc J 9: 30–35.

Duke CS, Cezeaux A, Allen MM (1989) Changes in polypeptide composition of *Synechocystis* sp. strain 6308 phycobilisomes induced by nitrogen starvation. J Bacteriol 171: 1960–1966.

Eswaran K, Ghosh PK, Mairh OP (2006) Environmental impact assessment for *Kappaphycus alvarezii* cultivation along the coast of Mandapam, Tamil Nadu. Submitted to the Department of Environment, Government of Tamil Nadu.

Fujita MR (1985) The role of nitrogen status in regulating transient ammonium uptake and nitrogen storage by macroalgae. J Exp Mar Biol Ecol 92: 283–301.

Gantt E (1981) Phycobilosomes. Annu Rev Plant Physiol 32: 327–347.

Govindasamy C, Arulpriya M, Ruban P, Meenakshi RV (2012) Hydro-chemical evolution of Palk strait region, Bay of Bengal. J Trop Life Sci 2(1): 1–5.

Grasshoff K, Erhardt M, Kremling K (1998) Methods of Seawater Analysis. Verlag Chemie Weinheim, Deerfield Beach, FL, p. 419.

Guldberg OH, Bruno JF (2010) The impact of climatic change on the world's marine ecosystems. Science 328: 1523–1528.

Hurtado AQ, Agbayani RF, Sanares R, Castro-Mallare MTR (2001) The seasonality and economic feasibility of cultivating *Kappaphycus alvarezii* Panagatan Cays, Caluya, Antique Philippines. Aquacult 199: 295–310.

Johnson B, Gopakumar G. (2011). Farming of the seaweed *Kappaphycus alvarezii* in Tamil Nadu coast—status and constraints. Marine Fisheries Information Service T&E Ser 208: 1–5.

Kumar V, Manivannan V (2001) Benthic foraminifer's responses to bottom Water characteristics in the Palk Strait off Rameswaram, South-east coast of India. Ind J Mar Sci 30: 173–179.

Lapointe BE (1981) The effects of light and nitrogen on growth, pigment content, and biochemical composition of *Gracilaria foliifera v. angustissima* (Gigartinales, Rhodophyta). J Phycol 17: 90–95.

Largo DB, Fukami K, Nishijima T (1995) Occasional pathogenic bacteria promoting *ice-ice* disease in the carrageenan-producing red algae *Kappaphycus alvarezii* and *Eucheuma denticulatum* (Solieriaceae, Gigartinales, Rhodophyta). J Appl Phycol 7: 545–554.

Largo DB, Fukami K, Nishijima T (1999) The time-dependent attachment mechanism of the bacterial pathogen during ice-ice infection in *Kappaphycus alvarezii* (Gigartinales, Rhodophyta). J Appl Phycol 11: 129–136.

Mairh OP, Zodape ST, Tewari A, Rajyaguru MR (1995) Culture of marine red alga *Kappaphycus striatum* (Schmitz) Doty on the Saurashtra region, west coast of India. Indian J Marine Sci 24: 24–31.

Mandal SK, Ajay G, Monisha N, Malarvizhi J, Temkar G, Mantri VA (2014) Differential responses of varying temperature and salinity regimes on nutrient uptake of drifting fragments of *Kappaphycus alvarezii*: Implication on survival and growth. J Appl Phycol 27(4): 1571–1581.

Mandal SK, Mantri VA, Haldar S, Eswaran K, Ganesan K (2010) Invasion potential of *Kapphycus alvarezii* on coral at Kurusadai Island, Gulf of Mannar, India. Algae 25(4): 205–216.

Ohno M, Nang HQ, Hirase S (1996) Cultivation and carrageenan yield and quality of *Kappaphycus alvarezii* in the waters of Vietnam. J Appl Phycol 6: 431–437.

Parker HS (1974) The culture of the red algal genus Eucheuma in the Philippines. Aquacult 3: 425–439.

Prabhahar C, Saleshani K, Enbarasan R (2011) Studies on the ecology and distribution of phytoplankton biomass in Kadalur coastal zone Tamil Nadu, India. Curr Bot 2(3): 26–30.

Santhanam R, Ramanathan N, Venkataramanujam K, Jegatheesan G (1987) Phytoplankton of the Indian Seas. Daya Publishing House, New Delhi.

Shannon CE, Wiener W (1949) The Mathematical Theory of Communication. University of Illinois Press, Urbana, p. 177.

Sridhar R, Thangaradjou T, Kannan L (2010) Spatial and temporal variations in phytoplankton in coral reef and seagrass ecosystems of the Palk Bay, south-east coast of India. J Environ Biol 31(5): 765–771.

Sridhar R, Thangaradjou T, Kumar SS, Kannan L (2006) Water quality and phytoplankton characteristics in the Palk Bay, south-east coast of India. J Environ Biol 27(3): 561–566.

Strickland JDH, Parsons TR (1972) A Practical Handbook for Seawater Analysis. Fisheries Research Board of Canada, Canada, pp. 167–311.

Temkar G, Mandal SK, Monisha N, Malarvizi J, Ajay G, Jenifer, R (2014) Dominance of *Rhizosolenia* in palk bay: Does *Kappaphycus* cultivation play any role? Proceedings of an International Symposium on "Marine Ecosystems Challenges & Opportunities" organized by Marine Biological Association of India on 2–5 December 2014 at Kochi, India, p. 316.

Tewari A, Basha S, Trivedi RH, Raghunathan C, Sravan Kumar VG, Khambhaty Y, Joshi HV, Kotiwar OS (2006) Environmental impact assessment of *Kappaphycus* cultivation in India in context to global scenario. Recent Advances in Applied Aspects of Indian Marine Algae with Reference to Global Scenario. Central Salt and Marine Chemicals Research Institute Publication, vol. 1, pp. 262–287.

Thomas CR (1996) Identifying Marine Phytoplankton (eds. Hasle R, Syvertsen EE). Academic Press, Harcourt Brace & Company, San Diego. Marine Diatoms, pp. 5–361.

Uyengco FR, Saniel LS, Jacinto GS (1981) The 'ice-ice' problem in seaweed farming. Proc. Inter. Seaweed Symp 10: 625–630.

Veeragurunathan V (2006) A Ph. D. thesis on "Studies of uptake and assimilation of nitrate and nitrite in agarophyte *Gracilaria edulis* (GMEL) SILVA under in vitro condition". The thesis was submitted to Madurai Kamraj University, Madurai, India.

6 Algal-Derived (By) Products as an Immunostimulant in the Aquaculture Industry

Kannan Mohan, Durairaj Karthick Rajan, Abirami Ramu Ganesan, and Shanmugam Munisamy

6.1 INTRODUCTION

Algal species are known to be photosynthetic organisms with marine and freshwater habitats. These are the fastest-growing organisms and have a short doubling time. The atmospheric CO2 is fixed by different autotrophic pathways, and it utilizes the nutrients for the conversion of biomass. The most common products involved in algal metabolism are carbohydrates, protein, pigments, and lipids (Radmer, 1996). Functional foods, feed additives, nutraceuticals, and feed materials are some of the commercial products obtained from various algal sources. The most widely used microalgae in human and animal nutrition are *Spirulina* and *Chlorella*. The predominant products from algal species include hydrocolloids, carrageenan, fucoidans, agars, and alginates. These obtained products are used as gelling agents, healthcare products, cosmetics, and other purposes. The pigments β-carotene, astaxanthin, and phycobiliproteins are served as a rich source of antioxidants and are used as natural food colorants, feed additives, and nutraceuticals (Pulz and Gross, 2004; McClure et al., 2019; Gautam and Mannan, 2020). There is a steady increase in the production of algae for human consumption and animal nutraceutical preparations. Globally, the algal market value is projected to reach 6.5 billion and 700 million for aquaculture (Yaakob et al., 2014; Hamed, 2016; Mobin and Alam, 2017). The production of algal by-products can be achieved from low-tech ocean farming to high-tech farming through bioprocess engineering and by using photo bioreactors. Due to its high cost, the closed photo bioreactors are used in pharmaceutical sectors (Radmer, 1996; Wijffels, 2008; Depra et al., 2019).

In aquaculture industries, steps have been taken to prevent the culturable aquatic organisms from infectious disease and morbidity (Liu et al., 2017). There are several major issues such as lack of advanced facility, misuse of antibiotics, and improper disposal of effluents was noticed. The antibiotics residues present in the aquaculture waste may discharge into the environment and nearby water bodies and cause problems to the living organism in the environment. In China, a variety of antibiotics was mixed in the coastal waters of Bohai City (Liu et al., 2016). The antibiotics mixed in the water reservoirs may decrease after a certain time period (Zhang et al., 2013). However, antibiotic residues present in the water may cause several effects to the organism in the aquatic environment, and finally, it reaches humans via bioaccumulation (Boonsaner and Hawker, 2013).

In recent years, algal biotechnology gained much attention and found to solve several problems in traditional aquaculture (Cho et al., 2015; Halfhide et al., 2014). There are several algal species which is cultivated for their nutritional value, high energy, essential amino acids, pigments, polyunsaturated fatty acids, and other value-added products used to improve the growth of aquatic animals (Lu et al., 2017). It has been reported that the proteins in microalgae serve as good protein sources for aquatic animals and the amino acid profiles of microalgae proteins are similar to that of fish protein (Becker, 2013). The natural pigments obtained from the microalgae including carotene and astaxanthin can be used as a feed additive for immunity stimulation and the excess usage of antibiotics can be prevented (Singh et al., 2017). Hence, the usage of algal-derived natural

DOI: 10.1201/9781003219156-8

immune-stimulants in aquaculture is one of the safety measures to avoid the excess or misuse of antibiotics in the future. The main goal of this chapter was to narrate the role of some algal-derived products. A major concern and special emphasis were made to highlight the promising features of algal-derived polysaccharides, pigments, and single-cell proteins as a possible option for the development of successful and eco-friendly aquaculture.

6.2 SOURCES OF ALGAL-DERIVED PRODUCTS

Currently, various algal-derived compounds have been studied in China, Taiwan, Thailand, South Korea, and India for their beneficial role in improving the health status and disease management in aquaculture given in Table 6.1. The most studied marine algal polysaccharides are fucoidan extracted from brown algae (Phaeophyceae), such as *Cladosiphon okamuranus*, *Saccharina japonica*, *Fucus vesiculosus*, *Undaria pinnatifda*, *Sargassum fusiforme*, carrageenan isolated from red algae (Rhodophyta) like *Chondrus crispus*, *Kappaphycus alvarezii*, and *Eucheuma cottonii*, agar isolated from red seaweed genera of *Gelidium* and *Gracilaria* species, microalgal pigments

TABLE 6.1
Traditional Application of Seaweeds in Fish Farming

Seaweed	Active Substance	Fish Species Tested	Dose Tested	Expected Health Benefits	References
Ulva rigida	Protein	*Dicentrarchus labrax*	10% DM	An increase of protein content, lipid and ash in the internal organs was noted	Batista et al., 2020
Gracilaria bursa-pastoris	Protein	*Dicentrarchus labrax*	10% DM	An increase of protein content, lipid, and minerals in the fillet was noted	Peixoto et al., 2019
Gracilaria cornea	Protein	*Dicentrarchus labrax*	10% DM	An increase of protein content, lipid, and ash in the internal organs was noted	Valente et al., 2006
Cladophora glomerata	Protein	*Sarotherodon niloticus*	5, 10, 15, 20, 25% DM	With the increase of algal content in the feed, increased the feed conversion rate, however, decreased weight gain. An increase of total protein content in meat of fish was observed.	Dewi et al., 2014
Cladophora glomerata	Carotenoids	*Oncorhynchus Mykiss*	0.00045 and 0.0009 % DM	Increased content of carotenoids in meat.	Aquaculture International 2001, 9, 87–93.

(*Continued*)

TABLE 6.1 (*Continued*)
Traditional Application of Seaweeds in Fish Farming

Seaweed	Active Substance	Fish Species Tested	Dose Tested	Expected Health Benefits	References
Ascophyllum	Biomass	*Pagrus major*	2.5 and 5% DM	The increase of the content of total protein in fish feed with algal additives.	Mustafa, 1994
Ulva pertusa	Biomass	*Pagrus major*	5% DM	Addition of macroalgae to the fish diet did not affect the growth rate and feed efficiency. The decrease of fatty acids, lipids, and sugars content in serum was observed.	Yone et al., 1986
Ascophyllum nodosum	Fatty acids	*Chrysophrys major*	5 and 10% DM	The highest growth rate and feed efficiency were observed when adding 5% of the *U. penatifida*, 5% A. nodosum, 10% of *U. penatifida*, and finally 10% of A. nodosum. For the first three additives, increased lipid content in muscles was observed.	Yasuo et al., 1986
Undariapenatifida	Fatty acids	*Chrysophrys major*	5 and 10% DM	The highest growth rate and feed efficiency were observed when adding 5% of the *U. penatifida*, 5% A. nodosum, 10% of *U. penatifida*, and finally 10% of A. nodosum. For the first thee additives, increased lipid content in muscles was observed.	Yone et al., 1986
***Enteromorpha* sp.**	Biomass	*Siganuscanaliculatus*	10, 20, 30% DM	Fresh biomass of macroalgae positively influenced survival, weight gain, feed consumption,	Pulz and Gross, 2004

TABLE 6.1 (*Continued*)
Traditional Application of Seaweeds in Fish Farming

Seaweed	Active Substance	Fish Species Tested	Dose Tested	Expected Health Benefits	References
				increase of the content of crude protein and fat in fish.	
Ulva clathrata	Biomass	Shrimp *Litopenaeus vannamei brood stock*	3.4%DM	Improve the quality of the broodstock and increase the reproductive capacity of the females.	Corral-Rosales et al., 2019
Porphyrayezoensis	Dry Biomass	Red sea bream	5.0% DM	Increased growth, feed efficiency, and protein deposition. Elevated liver glycogen and triglyceride accumulation in muscle.	Kalla et al., 2008
Porphyrayezoensis	Dry Biomass	Yellowtail	2.0% DM	Improved flesh quality.	Morioka et al., 2008
Porphyra spheroplasts	Dry Biomass	Red sea bream	5.0% DM	Survival, growth, and nutrient retention significantly higher than control.	Kalla et al., 2008
Ascophyllum nodosum	Dry Biomass	Red sea bream	5.0% 10%	Improved growth and feed efficiency at 5% inclusion level.	Yone et al., 1986
Ascophyllum nodosum	Dry Biomass	Red sea bream	5.0%	Delayed absorption of dietary carbohydrate and protein. The dietary nutrients are utilized effectively by this delaying effect of the seaweed; thus the growth and feed efficiency of red sea bream are improved.	Yone et al., 1986
Ascophyllum nodosum	Dry Biomass	Red sea bream	5.0%	Elevated growth rates; improved feed conversion, protein efficiency, and muscle protein deposition.	Mustafa, 1994

(*Continued*)

TABLE 6.1 (*Continued*)
Traditional Application of Seaweeds in Fish Farming

Seaweed	Active Substance	Fish Species Tested	Dose Tested	Expected Health Benefits	References
Ascophyllum nodosum	Dry Biomass	Red sea bream	5.0%	Increased growth, feed efficiency, and protein deposition. Elevated liver glycogen and triglyceride accumulation in muscle.	Mustafa et al., (1994)
Ascophyllum nodosum	Dry Biomass	Yellowtail	0.50%	Prevented a nutritional disease that causes retardation of growth and high mortality.	Nakagawa et al., 1986
Undaria pinnatifida	Dry Biomass	Rockfish	5.0%	Showed prominent physiological effects on haematocrit value and red blood cell number.	Yi et al., 1994
Undaria pinnatifida	Dry Biomass	Red sea bream	5.0 and 10%	Improved growth and feed efficiency, and higher muscle lipid deposition at 5% level of inclusion.	Yone et al., 1986
Ulva conglobata	Dry Biomass	Nibbler	5.0%	Improved growth.	Nakazoe, 1986
Ulva pertusa	Dry Biomass	Black sea bream	2.5, 5.0, 10, and 15%	Ulva meal diets repressed lipid accumulation in intraperitoneal body fat without loss of growth and feed efficiency. Fish fed 2.5%, 5%, and 10% Ulva meal did not show significant body weight loss during wintering. During starvation, lipid reserves were preferentially mobilized for energy.	Nakagawa, 1993
Ulva pertusa	Extract	Black sea bream	10.0%	Improved tolerance to hypoxia.	Nakagawa et al., 1984

TABLE 6.1 (*Continued*)
Traditional Application of Seaweeds in Fish Farming

Seaweed	Active Substance	Fish Species Tested	Dose Tested	Expected Health Benefits	References
Ulva pertusa	Dry Biomass	Red sea bream	5.0%	Activated lipid mobilization and suppressed protein breakdown observed during starvation for fish fed Ulva meal supplemented diet before starvation. Preferential use of glycogen observed.	Nakagawa et al., 1986
Ulva pertusa	Dry Biomass	Red sea bream	5.0%	Demonstrated a decrease in susceptibility to *Pasteurella piscicida*, an elevation of phagocytosis and spontaneous haemolytic and bactericidal activity.	
Ulva pertusa	Dry Biomass	Red sea bream	5.0%	Increased growth, feed efficiency, and protein deposition. Elevated liver glycogen and triglyceride accumulation in muscle.	Mustafa et al., 1995
Laminaria* sp., *Kelp	Dry meal	Atlantic salmon (*Salmo salar*)	3, 6, 10%	10% inclusion increased food intake and improved growth. Antioxidant properties improved resistance to temperature stress.	Kamunde et al., 2019
Cladophora	Fermented Cladophora	Tilapia (*Oreochromis* sp.)	10, 20, 30, and 40	10% inclusion increased growth rate, feed efficiency, protein digestibility, and essential amino acid index.	Dewi et al., 2014

(*Continued*)

TABLE 6.1 (*Continued*)
Traditional Application of Seaweeds in Fish Farming

Seaweed	Active Substance	Fish Species Tested	Dose Tested	Expected Health Benefits	References
Asparagopsistaxiformis	Ethanol extract	Sea bass (*Dicentrachus labrax*)	1%	Antibacterial activity against *Aeromonas*, *Vibrio*, and *Photobacteriaum* spp.	Dewi et al., 2014
	Alga	"	10%	"	Marino et al., 2016
	Ethanol extract	Sea bream (*Sparus aurata*)	1%	"	"
	Alga	"	10%	"	"
Ulva rigida	Dry meal	Tilapia (*Oreochromis niloticus*)	9, 18, 27	Good nutritional profile for the growth of tilapia.	"
Ulva rigida	Dry meal	Senegalese sole (*Solea senegalensis*)	10%	Found valid ingredient to replace soyabean and wheat meal.	Mensi et al., 2005
Undaria pinnatifida	Dry meal	"	10%	Growth impairment associated with reduced protein gain and reduced intestinal villi width.	Moutinho et al., 2018

(astaxanthin) extracted from *Haematococcus* sp. and algal protein (single-cell protein) extracted from *Spirulina, Chlorella, Senedessmus, Dunaliella*, diatoms, and *Porphyra* species.

6.3 COMMERCIAL PRODUCTION OF ALGAL-DERIVED PRODUCTS

6.3.1 Algal Polysaccharides: Fucoidan, Carrageenan, and Agar

Fucoidans refer to sulfated fucose-rich polymers found in brown macroalgae. The presence of sulfate groups is attributed to many of its bioactive properties (Berteau and Mulloy, 2003). There are numerous reports available for the structural properties and biological functions of fucoidan (Costa et al., 2011; Fitton et al., 2015). Fucoidan has been found to gain approval for its broad biological activities as a functional food for cancer prevention. Fucoidans are commercially available as a standard chemical by manufacturers of Sigma Aldrich with a specific molecular weight. The process of removing external impurities from fucoidan is quite a challenging process during the ethanol precipitation step (Yang et al., 2017). The biological functions of fucoidan purely depend on the chemical structure and fucose content. The commercial production of fucoidan always requires pretreatment, extraction, and purification stages, as shown in Figure 6.1.

Carrageenan is a sulfated polysaccharide found to occur in the cell walls of some red algal species. They are widely used as gelling agents in food industries and aquaculture feed preparations. There are several steps involved in the isolation of carrageenans using alkali, acids, and salts with heating to obtain pure carrageenan from red algal biomass. Based on the presence of sulfated galactan, the structure of carrageenan is classified as κ-, ι-, and λ-carrageenans. Carrageenan has

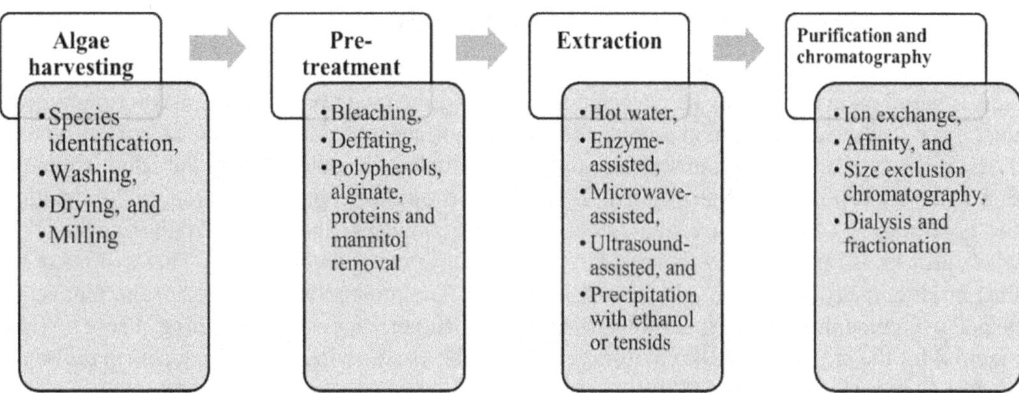

FIGURE 6.1 Required downstream processes including steps in each process for fucoidan production (Zayed and Ulber, 2020).

been used in various industrial sectors, including food industries, pharmaceutical industries, and cosmetic industries due to its antioxidant nature (Ahmad et al., 2014). It has a high market value due to its multifaceted role in various sectors, and the total market value was more than 300 million USD (McHugh, 2003).

Similar to carrageenan, agar is also obtained from red algae. It occurs as a combination of agarose and agaropectins. The agar structure mainly consists of β-(1–3)-D and α-(1–4)-L-linked galactose residues. In Asia-Pacific regions, the estimated production of agar was around 10,600 tons annually (Rhein-Knudsen et al., 2015). Agar has been found to exhibit good gelling properties, and it has been used in various fields, including food, biotechnology, and pharmaceutical sectors as stabilizing agents, thickener, and texture modifier (Armisen and Gaiatas, 2009). Globally, companies involved in large-scale commercialization of agar include MSC in Korea and Huey Shyang Seaweed Industrial Company in China. The *Gracilaria* species are one of the most preferred sources for extracting agar. Among the other seaweed-derived hydrocolloids, agar has a huge market value and 18 USD per kg as of 2009 when compared with carrageenan and alginates respectively (Bixler and Porse, 2011). There are several factors required for successful agar production, including raw material source, an eco-friendly approach, and technically skilled persons.

6.4 ALGAL PIGMENT: ASTAXANTHIN

The pigment astaxanthin (3, 3'-dihydroxy-β, β'-carotene-4, 4'-dione) is a carotenoid, produced from *Haematococcus pluvialis*, *Chlorella zofingiensis*, *Chlorococcum*, and *Phaffiarhodozyma* species for various potential health-promoting effects (Abati et al., 2014). Several companies are engaged in the production of astaxanthin throughout the world and most of the companies apply a two-step process to cultivate the *Haematococcus* for astaxanthin production (Han et al., 2013; Liu and Huang, 2016). The production process of astaxanthin was achieved based on the two stages of the life cycle green and red stage. The reproduction happens at the first stage of the life cycle followed by loss of mobility and loss of reproduction at the red stage. They are capable of accumulating about 5% of the astaxanthin dry weight of biomass (Kang et al., 2005). For industrial production and large-scale commercialization process, more than ten companies are engaged in the production of astaxanthin using *Haematococcus* species. Among that, Cyanotech Inc., USA; Mera Pharmaceuticals Inc., USA; Alga Technologies Inc., Israel; Fuji Chemical Industries Co. Ltd., Japan; and Parry Pharmaceuticals Inc., India, have established the production units in the 1990 and 2000 (Del Campo et al., 2007).

6.5 ALGAL PROTEIN: SINGLE-CELL PROTEIN (SCP)

Fermentation plays a major role in the production of SCP (Figure 6.2). The selection of microorganisms is based on the situation. The selected strains are grown with proper substrates for multiplication, and it leads to an increase in growth and biomass followed by a separation process (Figure 6.3). The screening of microbes from soil, air, and water samples was optimized by the selection, mutation, and genetic protocols (Nasseri et al., 2011). Then, the culture optimization required for cultivation is achieved by cell structure and various metabolic pathways. The technical performance of the whole process is influenced by the apparatus, technology, and processing time. This will make the final product ready for use on a large technical scale. The protection and safety of the innovation help to overcome important aspects regarding product authorizations and operating. These are also required for the protection of the environment due to the use of living microorganisms in the whole process (Anupama Ravindra, 2000).

6.6 ALGAL-DERIVED POLYSACCHARIDES AS IMMUNE-STIMULANTS IN AQUACULTURE

There are several naturally derived immune-stimulants like prebiotic, symbiotic, probiotics, yeast, amino acids, antioxidants, enzymes, minerals, vitamins, hormones, polysaccharides, and plant extracts from various algae that have been utilized as functional feed additives in aquatic animals to promote growth and survival (Goda, 2008). The algal-derived polysaccharides as growth promoting agents have been used as an immune-stimulant, and it is shown in Table 6.2. The dietary supplementation of fucoidan on *O. niloticus* significantly improved the weight, SGR, FCR, and stimulates the immune parameters (Hb, RBC, WBC, Total protein, ALU, SOD, CAT, and PHA) that have been reported (Fabrini et al., 2017; Abdel-Daim et al., 2020; Abdel-Warith et al., 2021). In another study, a diet supplemented with fucoidan increased the weight, SGR, FCR, and PER was observed on *L. rohita* (Mir et al., 2017). The oral administration of (0.4%) fucoidan showed significant elevation in LYZ, and NBT was observed on *P. major* (Sony et al., 2020). The fucoidan supplemented diets on

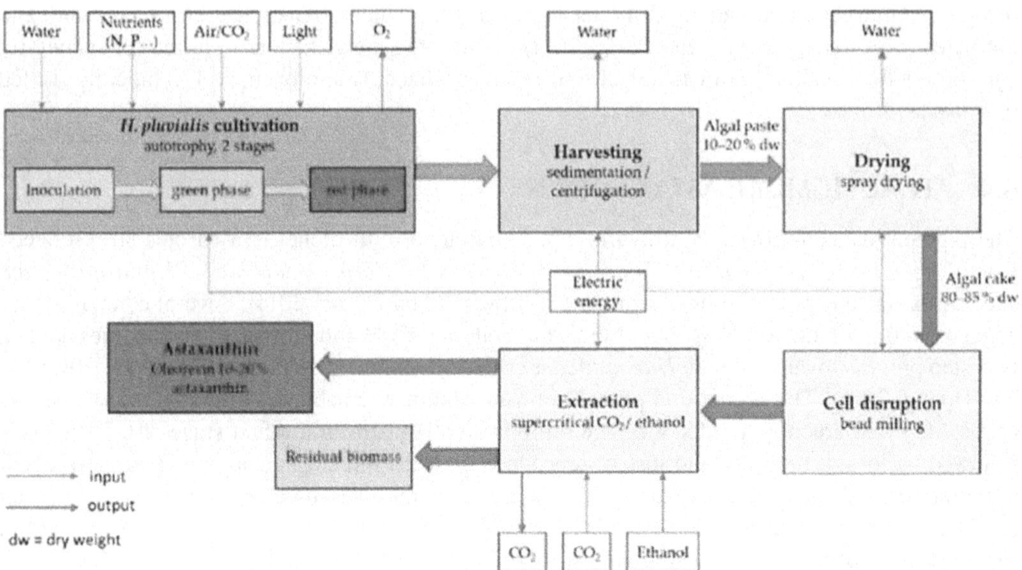

FIGURE 6.2 Flowchart of a possible astaxanthin production process. Reprinted with permission from Algal Research (Onorato and Rosch, 2020), copyright 2020 Elsevier.

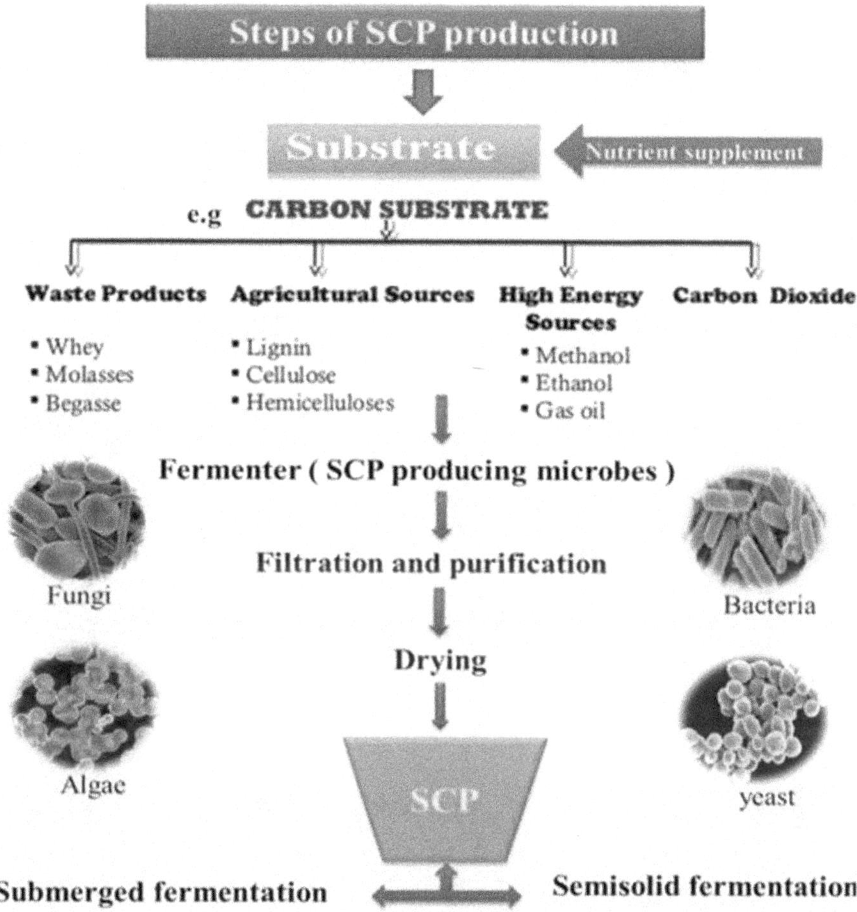

FIGURE 6.3 General steps during industrial production of SCP. Reprinted with permission from Aquaculture (Sharif et al., 2021), copyright 2021 Elsevier.

P. monodon and *L. vannamei* showed a significant increase in THC, DHC, Propo, and RBA (Shi et al., 2020; Traifalgar et al., 2012; Sivagnanavelmurugan et al., 2012; Arizo et al., 2015). Similarly, the carrageenan mixed with shrimp diet showed weight gain, SGR, total hemocyte count, and differential hemocyte count on *L. vannamei* and *P. monodon* (Chen et al., 2014; Jumah et al., 2020). In addition, the diet containing agar exhibited increased LYZ, PHA, RBA, and ACH50 activity on *P. bocourti* (Van Doan et al., 2014a, 2014b).

6.7 ALGAL-DERIVED PIGMENTS AS IMMUNE-STIMULANT IN AQUACULTURE

In aquaculture, Salmonid and crustacean coloring is a key quality characteristic by consumers. The appearance of reddish orange color is a characteristic feature of carotenoid richness and found to deposit on skin, muscle, exoskeleton, and gonads on original or modified state based on the species (Meyers and Chen, 1982). The predominant carotenoid astaxanthin was found to be present in most of the crustaceans and salmonids (Gentles and Haard, 1991). In the aquatic environment, the biosynthesis of astaxanthin was due to the presence of *Haematococcus* species, and it reaches the fish via the food chain, and it acts as natural coloration in most of the fishes (Lorenz, 1998). In farmed aquatic animals, there is no chance for astaxanthin production in the environment. Hence,

TABLE 6.2
Algal-derived Polysaccharides as Growth Promoter and Immunostimulant in Fish and Shrimp Aquaculture

Algal-derived Polysaccharides	Test Species	Concentration	Duration	Growth Performance	Immunological Responses	Disease Resistance	References
Fucoidans	*Oreochromis niloticus*	0.04–0.06 gkg^{-1}	15 days	n.a.	PHA, TP, LEU, PHI (↔)	n.a.	Isnansetyo et al., 2016
	O. niloticus	5–15 gkg^{-1}	30 days	FW, RC, WG, AFC, SUR, SL, HL, BH (↑)	n.a.	n.a.	Fabrini et al., 2017
	O. niloticus	0–1%	30 days	n.a.	GSH, GPx, SOD, CAT (↑) ALT, AST, ALP, CHO, Urea, Cr (↓)	n.a.	Abdel-Daim et al., 2020
	O. niloticus	n.d.	40 days	FCR, SUR (↑)	Hb, PCV, RBCs, WBCs, total protein, ALU, SOD, CAT, GPx (↑) MDA (↓)		Abdel-Warith et al., 2021
	O. niloticus	0.1–0.8%	214 days	SRG, Villi length and width (↑)	Total gut bacteria count (↓)	n.a.	Mahgoub et al., 2020
	Clariasgariepinus	4 and 6 gkg^{-1}	21 days	SUR (↑)	RBA, PHA, LYM, LYZ, NO, BCA (↑)	n.a.	Ei-Boshy et al., 2014
	Pelteobagrusfulvidraco	0.5–2 gkg^{-1}	84 days	SUR (↑)	SOD, CAT, LYZ, PHI, RBA (↑) MDA, TG, TC, HDL-C, LDL-C, GLU (↓)	*A. hydrophila*	Yang et al., 2014
	Lates calcarifer	5–10 gkg^{-1}	52 days	FW, FL, SGR, FI, FCR, SUR (↑)	MFA (↑) HSI (↔)	n.a.	Tuller et al., 2014
	Cyprinus carpio	0.1–10 gkg^{-1}	14 days	MOR (↓)	LYZ, SPO, IL-1β (↑)	*A. hydrophila* *E. tarda*	Rajendran et al., 2016
	Labeorohita	10–20 gkg^{-1}	60 days	WG, SGR, FCR, PER (↑)	RBA, MPO, LYZ, Ig, PHA, TLC (↑) A/G, GLU (↓)	*A. hydrophila*	Mir et al., 2017

Pagrus major	0.4%	56 days	WG, SGR (↔)	LYZ, NBT (↑)	*n.a.*	Sony et al., 2020
Fenneropenaeus chinensis	0.5–20 g kg^{-1}	14 days	MOR (↓)	THC, Propo, LYZ (↑) SOD (↔)	*V. harveyi*	Huang et al., 2006
Litopenaeusvannamei	3.5 g kg^{-1}	1 day	SUR (↑)	LAC, TP (↓) GLU (↔)	*V. campbellii*	Sánchez Campos et al., 2010
L. vannamei	0.0001–0.0004gkg^{-1}	21 days	SUR (↑)	THC, Propo, RBA (↑)	*V. alginolyticus*	Kitikiew et al., 2013
L. vannamei	1–2 g kg^{-1}	14 days	n.a.	LYZ, Propo, SOD, ACP (↑)	n.a.	Ying, 2008
L. vannamei	3 g kg^{-1}	15 days	SUR (↑) MOR (↓)	THC, Propo, SOD, TGs (↑)	White spot syndrome virus	Sinurat et al., 2016
L. vannamei	1 mg ml^{-1}	n.d.	n.d.	Propo, RBA (↑)	n.d.	Shi et al., 2020
Marsupenaeus japonicus	0.0001–0.001 g kg^{-1}	56 days	WG, PR, SGR (↑) FCR (↓)	THC, Propo, SBA (↑)	*V. alginolyticus*	Traifalgar et al., 2010
Macrobrachiumrosenbergii	0.0001–0.0005gkg^{-1}	28 days	SUR, WG, FCR (↑)	THC, Propo (↑)	White spot syndrome virus	Arizo et al., 2015
Penaeus japonicus	0.0001–0.001 g kg^{-1}	n.a.	SUR (↑)	n.a.	*V. harveyi*	Traifalgar et al., 2012
Penaeus monodon	0.0001–0.0004gkg^{-1}	10 days	SUR (↑)	ABA, PHA (↑)	White spot syndrome virus	Chotigeat et al., 2004
P. monodon	0.0005–0.010 g kg^{-1}	30 days	SUR, WG, SGR, PER (↑) FCR (↓)	n.a.	*V. harveyi*	Traifalgar et al., 2009
P. monodon	2.0 g kg^{-1}	10 days	SUR (↓)	PHA, ABA (↓)	*V. harveyi*	Traifalgar et al., 2013
P. monodon	0.0001–0.0004gkg^{-1}	20 days	MOR (↓)	n.a.	White spot syndrome virus	Sivagnanavelmurugan et al., 2012
P. monodon	0.0001–0.0004gkg^{-1}	1 day	WG, SGR (↑) MOR (↓)	n.a.	*V. parahaemolyticus*	Sivagnanavelmurugan et al., 2015
P. monodon	1–3 g kg^{-1}	45 days	MOR (↓)	THC, Propo, RBA, SOD, PHA (↑)	White spot syndrome virus	Immanuel et al., 2012

(Continued)

TABLE 6.2 (*Continued*)

Algal-derived Polysaccharides as Growth Promoter and Immunostimulant in Fish and Shrimp Aquaculture

Algal-derived Polysaccharides	Test Species	Concentration	Duration	Growth Performance	Immunological Responses	Disease Resistance	References
	P. monodon	1–3 g kg^{-1}	60 days	WG (↑) MOR (↓)	THC, Propo, RBA, SOD, PHA, BCA (↑)	*V. parahaemolyticus*	Sivagnanavelmurugan et al., 2014
Carrageenans							
	A. transmontanus	n.a.	16 days	SUR, BW (↑)	ALP, AP (↓)	n.a.	Gawlicka et al., 1996
	C. carpio	0.01–0.015 gkg^{-1}	6 days	SUR (↑)	n.a.	n.a.	Fujiki et al., 1997b
	C. carpio	0.00025 gkg^{-1}	21 days	SUR (↑)	ACH_{50}, PHA, RBA (↑)	n.a.	Fujiki et al., 1997a
	O. niloticus	5–20 gkg^{-1}	28 days	n.a.	n.a.	n.a.	Saboya et al., 2012
	L. vannamei	6 µg g^{-1}	5 days	SUR (↑)	THC, Propo, RBA, PHA (↑)	*V. alginolyticus*	Yeh and Chen, 2008
	L. vannamei	0.5–2.0 g kg^{-1}	21 days	SUR (↑)	HPTs, THC, PHA (↑)	*V. alginolyticus*	Chen et al., 2014
	Penaeus monodon	0.15–0.60 g kg^{-1}	14 days	WG, SGR, FCE (↑), PER (↔) MOR (↓)	n.a.	n.a.	Jumah et al., 2020
Agar							
	P. bocourti	1–3 gkg^{-1}	75 days	SGR, SUR (↑) FCR (↓)	LYZ (↑)	*A. hydrophila*	Van Doan et al., 2014a
	P. bocourti	2 gkg^{-1}	28 days	SUR, FCR, SGR (↑)	LYZ, PHA, RBA, ACH_{50} (↑)	*A. hydrophila*	Van Doan et al., 2014b

A/G: albumin/globulin ratio, ACH50: Alternative complement pathway, ALP: alkaline phosphate, ALT: Alanine aminotransferase, ALU: albumin, BCA: bactericidal activity, CAT: catalase, GLU: glucose, GSH: glutathione, GST: glutathione s-transferase, Hb: haemoglobin, HCT: haematocrit, HDL-C: higher density lipoprotein-cholesterol, HIS: Haemato somatic index, Ig: immunoglobulin, LDL-C: low density lipoprotein-cholesterol, LEU: Leucocytes, LYM: lymphocytes, LYZ: lysozyme activity, MDA: malondialdehyde activity, MFA: muscle fibre area, MPO: myeloperoxidase, NBT: nitro blue tetrazolium, NO: nitric oxide activity, PHA: phagocytic activity, PHI: phagocytic index, Propo: prophenoloxidase activity, RBA: respiratory burst activity, RBC: red blood cells, SBA: serum bactericidal activity, SOD: superoxide dismutase, SPO: serum peroxide, IL-1β: interleukin 1 beta, TC: total cholesterol, TG: triglyceride, THC: Total haemocyte count, TLC: total leucocyte count, TP: total protein, WBC: white blood cells

NA—Not available

it can be supplemented only through the feeds. The usage of astaxanthin in aquaculture is shown in Table 6.3. There are numerous reports available for the usage of astaxanthin in aquaculture practices and reported for the past two decades (Wang et al., 2015; Lim et al., 2019; Wu et al., 2021; Zhang et al., 2021).

Perhaps the largest potential use of natural astaxanthin produced by *Haematocoocus* is for salmonids and trout culture feeds. The flesh color of salmonids is the result of the absorption and deposition of dietary astaxanthin. Salmonids are unable to synthesize astaxanthin de nova, therefore,

TABLE 6.3
Algal-Derived Pigments (Astaxanthin) as Growth Promoter and Immunostimulant in Fish and Shrimp Aquaculture

Species	Inclusion level	Duration	Observations	References
Oncorhynchus mykiss	32 mg kg^{-1}	6 weeks	1. No significant effects on final weight, specific growth rate, and feed conversion ratio 2. Higher muscle astaxanthin retention	Choubert et al., 2006
O. mykiss	80 mg/kg	6 weeks	Enhanced gene expression and antioxidant enzymes	Wu et al., 2021
Pseudosciaenacrocea	0.03, 0.05, and 0.10 g 100 g^{-1} diet	66 days	1. Higher weight gain, condition factor, and whole-body lipid 2. Enhanced white blood cell, glucose, triglyceride, cholesterol, superoxide dismutase, catalase, glutathione peroxidase activities, lysozyme activities, and complement contents	Li et al., 2014
Lates calcarifer	50, 100, 150 mg kg^{-1}	90 days	1. Significant enhancements in hematological indices 2. Decreased alanine aminotransferase, aspartate aminotransferase, glucose, and cortisol 3. significantly stimulated immunological parameters	Lim et al., 2019
Cichlasomacitrinellum × *Cichlasomasynspilum*	400 mg/kg	50 days	Effectively enhance growth, skin coloration and the antioxidant capacity	Li et al., 2018
Litopenaeusvannamei	0, 50, or 100 ppm	4 weeks	1. Improved specific growth rate and weight gain 2. Promoting expression of immune related genes	Wang et al., 2020
L. vannamei	80 mg kg^{-1}	4 weeks	Improved immunological parameters, enhanced antioxidant status, promoted resistance to WSSV	Wang et al., 2015

(Continued)

TABLE 6.3 (*Continued*)
Algal-Derived Pigments (Astaxanthin) as Growth Promoter and Immunostimulant in Fish and Shrimp Aquaculture

Species	Inclusion level	Duration	Observations	References
L. vannamei	0, 50, 70, 90, 140 ppm	35 days	1. Higher growth performance 2. Lower cumulative mortality 3. Enhanced antioxidant enzymes	Liu et al., 2018
Marsupenaeus japonicus	50 and 100 mg kg^{-1}	9 weeks	1. Lower survival rate 2. Higher oxygen consumption rate	Chien and Shiau, 2005
Exopalaemoncarinicauda	400 mg kg^{-1}	56 days	1. Improved the survival rate and total weight gain 2. Enhanced the body pigmentation 3. Decreasing the activity of catalase, superoxide dismutase, glutathione peroxidase, and malonadehyde	Zhang et al., 2021

astaxanthin pigments must be supplied in the artificial aquaculture feeds (Choubert et al., 2006; Lu et al., 2021). Choubert et al. (2006) reported the possible effects of astaxanthin on the modulation of fish health and higher muscle astaxanthin retention of rainbow trout, *O. mykiss*. Trials were also conducted by Li et al. (2014) to investigate growth enhancement in *P. crocea* by feeding diets supplemented with astaxanthin (100 mg kg^{-1} diet) for 66 days. Previous studies demonstrated the ability of dietary astaxanthin to improve growth and survival in a variety of shrimp species. Administration of supplementary astaxanthin at 80 mg kg^{-1} diet was reported to enhance the growth performance of *L. vannamei* compared to those fed with a control diet by the end of 35 days (Liu et al., 2018). Zhang et al. (2021) observed that feeding astaxanthin incorporated diet at 400 mg kg^{-1} over 56 days, enhanced the body pigmentation and growth rate of *E. carinicauda*.

6.8 ALGAL-DERIVED PROTEIN (SINGLE-CELL PROTEIN) AS IMMUNOSTIMULANT IN AQUACULTURE

The alternative sources for fish and soybean meal are algal sources rich in proteins, including *Chlorella* sp., *Chondrus* sp., *Scenedesmus* sp., *Spirulina* sp., and *Porphyrium* sp. (Table 6.4). Algae consist of protein (40–60%), mineral salts (7%), nucleic acid (4–6%), chlorophyll, and bile pigments. The production of algal biomass for animal feeds, including aquafeeds is about 30% of the current world algal production. For SCP production, *Euglena gracilis* is the most preferred species due to the presence of high protein contents and increased digestion (Baker and Gunther, 2004). The diets consist of 5% *Spirulina platensis*, which can be replaced with *Artemia nauplii* meal in *Litopenaeus schmitti* larvae culture (Jaime-Ceballos et al., 2005). In salmonids, both whole-cell and processed microalgae have been used as feed ingredients (Kousoulaki et al., 2016; Katerina et al., 2020). The biomass of *Schizochytrium* was examined in the *S. salar* at the concentration up to 110 g/kg (Sprague et al., 2015). In another study, examination of *Nannochloropsis*, *Phaeodactylum*, and *Isochrysis* SCPs fed to *S. salar* has been found to have positive effects with average feed intake. However, *Nannochloropsis* and *Isochrysis* showed a clear effect on protein digestibility and less in *Phaeodactylum*. No growth results were observed in the digestive

TABLE 6.4
Algal-Derived Protein (Single-Cell Protein, SCP) as Growth Promoter and Immunostimulant in Fish and Shrimp Aquaculture

Species	Algal species used	Inclusion level	Duration	Observations	References
Oreochromis niloticus	*Nannochloropsisgaditana*	30%	42 days	Increased in apparent digestibility coefficient of protein and fat	Teuling et al., 2019
Oncorhynchus mykiss	*Scenedesmus almeriensis*	0, 5, 10, 20, and 40%	82 days	Evaluated on growth performance, protein utilization, fish health, biosecurity	Tomás-Almenar et al., 2018
Litopenaeusvannamei	*Arthrospira platensis*	50%	22 days	Fast digestion and assimilation rate	Gamboa-Delgado et al., 2019
Salmo salar	*Desmodesmus* sp.	10 or 20%	70 days	1. Lower lipid content 2. No significant effect on the growth indices (condition factor, specific growth rate) and survival	Kiron et al., 2016
S. salar	*Schizochytrium* sp.	50 g kg^{-1}	n.a.	1. No significant effects on fish blood plasma chemistry 2. Enhance lipid retention efficiency	Kousoulaki et al., 2016
S. salar	*Schizochytrium Limacinum*			Improved growth and filet quality	
S. salar	*Schizochytrium* sp.	11% and 5.5%	19 week	Increased level of fatty acids	Sprague et al., 2015
S. salar	*Entomoneis* spp.	2.5% and 5%	84 days	1. Increase in the omega-3 long-chain polyunsaturated fatty acid 2. Did not cause any major positive effects growth and feed efficiency	Norambuena et al., 2015
Litopenaeusschmitti	*Spirulina platensis*	0, 2.5, and 5%	120 h	Superior development index	Jaime-Ceballos et al., 2005
Litopenaeusvannamei	*Haematococcuspluvialis*	3, 6, 9, and 12%	8 week	Higher growth rate and lower feed conversion ratio	Ju et al., 2012
Marsupenaeus japonicus	*Nannochloropsis* sp.	10, 40, 70 g/kg	14 days	Improved growth rate, highest final body weight	Adissin et al., 2020
Litopenaeusvannamei	Green water meal	10, 20, 30, and 40%	44 days	No significant effect on survival, FCR, and without significant negative effect on survival, FCR, and SGR	Basri et al., 2015

aspects. The usage of diatom *Entomoneis* (19% protein, 0.4% lipid) as a feed additive on diets of *S. salar* showed no negative impacts on growth, feed intake, or FCR throughout the experimental period (Norambuena et al., 2015).

Ju et al. (2012) reported the use of a defatted *Haematococcus pluvialis* SCP in diets fed to pacific white shrimp. Furthermore, the addition of *Haematococcus* SCP serves as an excellent source of protein and carotenoids. The usage of maximum concentration up to 120 g/kg showed no growth loss, and changes in the protein quality were also assessed. There was an increase in astaxanthin content in the shrimp with every increment of the *Haematococcus* SCP, which was also noted as an increased redness of the shrimp. The effects and performance of *Nannochloropsis* derived SCP were evaluated in post-larval kuruma shrimp (*Marsupenaeus japonicus*) and tested up to a maximum of 70g/kg (Adissin et al., 2020). An oil extract from the *Nannochloropsis* was also evaluated at inclusion levels of 35 and 14 g/kg. The increased levels of SCP resulted in improved growth followed by feed utilization and no decline in feed uptake was noticed. In another study, Basri et al. (2015) reported that *Chlorella* sp. SCP was tested on *Litopenaeus vannamei* included in the diets at a maximum range of 340 g/kg. Moreover, with each inclusion level, the decline in the performance of white leg shrimp was noticed, and it suggested that this microalgal SCP was not suitable nutrition for shrimps.

6.9 CONCLUSION AND FUTURE PROSPECTS

Algal-derived products such as polysaccharides, pigments, and single-cell protein are essential high-value by-products that could be used as a natural immuno-stimulant for replacing antibiotics in the aquaculture industry. Few nutrients such as fucoidan, carrageenan, and agar are prebiotic substances that improve the gut microflora, immune response, and disease resistance of aquatic animals. These indigestible dietary substances utilized by gut microbiota improved the health of aquatic animals. Astaxanthin is one of the most essential carotenoid pigments in algae increases the antioxidant potential in the aquaculture feed industry. This pigment is responsible for the attractive appearance of most aquatic animal products, especially color, which contributes an important character quality criterion for marketing, price, and consumer demands of aquaculture products. Single-cell protein has a high nutritional value, supporting the survival of cultured organisms by enhancing the immune response and disease resistance capacity. Single-cell protein can also act as a source of β-carotene, enhancing the color of ornamental fish. The size and color of fish are regulating factors in the success of ornamental fish aquaculture. Both can be manipulated through the application of SCP derived from algae that contain large amounts of carotenoid pigments. Algal carotenoids can be used as a feed additive for the growth and coloration of ornamental fishes. Therefore, the conclusion can be made that it can easily replace conventional animal and protein sources in both humans and animal diets without any negative impact. Further studies are necessary to confirm the suitability of these ingredients in practical feed formulations.

ACKNOWLEDGMENTS

The first author would like to acknowledge Sri Vasavi College for providing support and essential facilities in writing this chapter.

REFERENCES

Abdel-Daim, M. M., Dawood, M. A., Aleya, L., and Alkahtani, S. (2020). Effects of fucoidan on the hematic indicators and antioxidative responses of Nile tilapia (*Oreochromis niloticus*) fed diets contaminated with aflatoxin B 1. Environmental Science and Pollution Research, 27(11), 12579–12586.

Abdel-Warith, A. W. A., Younis, E. M., Al-Asgah, N. A., Gewaily, M. S., El-Tonoby, S. M., and Dawood, M. A. (2021). Role of fucoidan on the growth behavior and blood metabolites and toxic effects of atrazine in nile tilapia *Oreochromis niloticus* (Linnaeus, 1758). Animals, 11(5), 1448.

Adissin, T. O., Manabu, I., Shunsuke, K., Saichiro, Y., Moss, A. S., and Dossou, S. (2020). Effects of dietary *Nannochloropsis* sp. powder and lipids on the growth performance and fatty acid composition of larval and postlarvalkuruma shrimp, *Marsupenaeus japonicus*. Aquaculture Nutrition, 26(1), 186–200.

Ahmed, A. B. A., Adel, M., Karimi, P., and Peidayesh, M. (2014). Pharmaceutical, cosmeceutical, and traditional applications of marine carbohydrates. Advances in Food and Nutrition Research, 73, 197–220.

Ambati, R. R., Phang, S. M., Ravi, S., and Aswathanarayana, R. G. (2014). Astaxanthin: Sources, extraction, stability, biological activities and its commercial applications—a review. Marine Drugs, 12(1), 128–152.

Arizo, M. A., Simeon, E. C., Layosa, M. J., Mortel, R. M., Pineda, C. M., Lim, J. J., and Maningas, M. B. (2015). Crude fucoidan from *Sargassum polycystum* stimulates growth and immune response of *Macrobrachium rosenbergii* against white spot syndrome virus (WSSV). Aquaculture, Aquarium, Conservation & Legislation, 8(4), 535–543.

Armisen, R., and Gaiatas, F. (2009). Agar. In: Phillips, G. O., and Williams, P. A. (eds.), Handbook of Hydrocolloids (pp. 82–107). Woodhead Publishing.

Baker, R., and Günther, C. (2004). The role of carotenoids in consumer choice and the likely benefits from their inclusion into products for human consumption. Trends in Food Science & Technology, 15(10), 484–488.

Basri, N. A., Shaleh, S. R. M., Matanjun, P., Noor, N. M., and Shapawi, R. (2015). The potential of microalgae meal as an ingredient in the diets of early juvenile Pacific white shrimp, *Litopenaeus vannamei*. Journal of Applied Phycology, 27(2), 857–863.

Batista, S., Pintado, M., Marques, A., Abreu, H., Silva, J. L., Jessen, F., and Valente, L. M. (2020). Use of technological processing of seaweed and microalgae as strategy to improve their apparent digestibility coefficients in European seabass (*Dicentrarchus labrax*) juveniles. Journal of Applied Phycology, 32(5), 3429–3446.

Becker, E. W. (2013). Microalgae for aquaculture: Nutritional aspects. Handbook of Microalgal Culture: Applied Phycology and Biotechnology, 2, 671–691.

Berteau, O., and Mulloy, B. (2003). Sulfated fucans, fresh perspectives: Structures, functions, and biological properties of sulfated fucans and an overview of enzymes active toward this class of polysaccharide. Glycobiology, 13(6), 29R–40R.

Bixler, H. J., and Porse, H. (2011). A decade of change in the seaweed hydrocolloids industry. Journal of Applied Phycology, 23(3), 321–335.

Boonsaner, M., and Hawker, D. W. (2013). Evaluation of food chain transfer of the antibiotic oxytetracycline and human risk assessment. Chemosphere, 93(6), 1009–1014.

Chen, Y. Y., Chen, J. C., Lin, Y. C., Putra, D. F., Kitikiew, S., Li, C. C., and Yeh, S. T. (2014). Shrimp that have received carrageenan via immersion and diet exhibit immunocompetence in phagocytosis despite a post-plateau in immune parameters. Fish & Shellfish Immunology, 36(2), 352–366.

Chien, Y. H., and Shiau, W. C. (2005). The effects of dietary supplementation of algae and synthetic astaxanthin on body astaxanthin, survival, growth, and low dissolved oxygen stress resistance of kuruma prawn, *Marsupenaeus japonicus* Bate. Journal of Experimental Marine Biology and Ecology, 318(2), 201–211.

Cho, D. H., Ramanan, R., Heo, J., Lee, J., Kim, B. H., Oh, H. M., and Kim, H. S. (2015). Enhancing microalgal biomass productivity by engineering a microalgal–bacterial community. Bioresource Technology, 175, 578–585.

Chotigeat, W., Tongsupa, S., Supamataya, K., and Phongdara, A. (2004). Effect of fucoidan on disease resistance of black tiger shrimp. Aquaculture, 233(1–4), 23–30.

Choubert, G., Mendes-Pinto, M. M., and Morais, R. (2006). Pigmenting efficacy of astaxanthin fed to rainbow trout *Oncorhynchus mykiss*: Effect of dietary astaxanthin and lipid sources. Aquaculture, 257(1–4), 429–436.

Corral-Rosales, C., Ricque-Marie, D., Cruz-Suárez, LE, Arjona, O., and Palacios, E. (2019). Fatty acids, sterols, phenolic compounds, and carotenoid changes in response to dietary inclusion of *Ulva clathrata* in shrimp *Litopenaeus vannamei* broodstock. Journal of Applied Phycology, 31(6), 4009–4020.

Costa, L. S., Fidelis, G. P., Telles, C. B. S., Dantas-Santos, N., Camara, R. B. G., Cordeiro, S. L., and Rocha, H. A. O. (2011). Antioxidant and antiproliferative activities of heterofucans from the seaweed *Sargassum filipendula*. Marine Drugs, 9(6), 952–966.

Del Campo, J. A., García-González, M., and Guerrero, M. G. (2007). Outdoor cultivation of microalgae for carotenoid production: Current state and perspectives. Applied Microbiology and Biotechnology, 74(6), 1163–1174.

Deprá, M. C., Mérida, L. G., de Menezes, C. R., Zepka, L. Q., and Jacob-Lopes, E. (2019). A new hybrid photobioreactor design for microalgae culture. Chemical Engineering Research and Design, 144, 1–10.

Dewi, A. K., Nursyam, H., and Hariati, A. M. (2014). Response of fermented Cladophora containing diet on growth performances and feed efficiency of Tilapia (Oreochromis sp.). International Journal of Agronomy and Agricultural Research, 5(6), 78–85.

El-Boshy, M., El-Ashram, A., Risha, E., Abdelhamid, F., Zahran, E., and Gab-Alla, A. (2014). Dietary fucoidan enhance the non-specific immune response and disease resistance in African catfish, *Clarias gariepinus*, immunosuppressed by cadmium chloride. Veterinary Immunology and Immunopathology, 162(3–4), 168–173.

Fabrini, B. C., Braga, W., Andrade, E., Paula, D., and Paulino, R. (2017). Sulfated polysaccharides in diets for Nile tilapia (*Oreochromis niloticus*) in the initial growth phase. Journal of Aquaculture and Research Development, 8(477), 2. https://doi.org/10.4172/2155-9546.1000477.

Fitton, J. H., Stringer, D. N., and Karpiniec, S. S. (2015). Therapies from fucoidan: An update. Marine Drugs, 13(9), 5920–5946.

Fujiki, K., Shin, D. H., Nakao, M., and Yanot, T. (1997a). Effects of κ-carrageenan on the non-specific defense system of carp *Cyprinus carpio*. Fisheries Science, 63(6), 934–938.

Fujiki, K., Shin, D. H., Nakao, M., and Yano, T. (1997b). Protective effect of κ-carrageenan against bacterial infections in carp *Cyprinus carpio*. Journal of the Faculty of Agriculture, Kyushu University, 42(1–2), 113–119.

Gamboa-Delgado, J., Morales-Navarro, Y. I., Nieto-López, M. G., Villarreal-Cavazos, D. A., and Cruz-Suárez, L. E. (2019). Assimilation of dietary nitrogen supplied by fish meal and microalgal biomass from Spirulina (*Arthrospira platensis*) and *Nannochloropsis oculata* in shrimp *Litopenaeus vannamei* fed compound diets. Journal of Applied Phycology, 31(4), 2379–2389.

Gautam, S., and Mannan, M. A. U. (2020). The role of algae in nutraceutical and pharmaceutical production. In: Bioactive Natural Products in Drug Discovery (pp. 665–685). Springer.

Gawlicka, A., McLaughlin, L., Hung, S. S., and de la Noüe, J. (1996). Limitations of carrageenan microbound diets for feeding white sturgeon, *Acipenser transmontanus*, larvae. Aquaculture, 141(3–4), 245–265.

Gentles, A., and Haard, N. F. (1991). Pigmentation of rainbow trout with enzyme-treated and spray-dried *Phaffia rhodozyma*. The Progressive Fish-Culturist, 53(1), 1–6.

Goda, A. M. S. (2008). Effect of dietary Ginseng herb (Ginsana® G115) supplementation on growth, feed utilization, and hematological indices of Nile Tilapia, *Oreochromis niloticus* (L.), fingerlings. Journal of the World Aquaculture Society, 39(2), 205–214.

Halfhide, T., Åkerstrøm, A., Lekang, O. I., Gislerød, H. R., and Ergas, S. J. (2014). Production of algal biomass, chlorophyll, starch and lipids using aquaculture wastewater under axenic and non-axenic conditions. Algal Research, 6, 152–159.

Hamed, I. (2016). The evolution and versatility of microalgal biotechnology: A review. Comprehensive Reviews in Food Science and Food Safety, 15(6), 1104–1123.

Han, D., Li, Y., and Hu, Q. (2013). Biology and commercial aspects of *Haematococcus pluvialis*. Handbook of microalgal culture: Applied Phycology and Biotechnology, 2, 388–405.

Huang, X., Zhou, H., and Zhang, H. (2006). The effect of *Sargassum fusiforme* polysaccharide extracts on vibriosis resistance and immune activity of the shrimp, *Fenneropenaeus chinensis*. Fish & Shellfish Immunology, 20(5), 750–757.

Immanuel, G., Sivagnanavelmurugan, M., Marudhupandi, T., Radhakrishnan, S., and Palavesam, A. (2012). The effect of fucoidan from brown seaweed *Sargassum wightii* on WSSV resistance and immune activity in shrimp *Penaeus monodon* (Fab). Fish & Shellfish Immunology, 32(4), 551–564.

Isnansetyo, A., Fikriyah, A., and Kasanah, N. (2016). Non-specific immune potentiating activity of fucoidan from a tropical brown algae (Phaeophyceae), *Sargassum cristaefolium* in tilapia (*Oreochromis niloticus*). Aquaculture International, 24(2), 465–477.

Jaime Ceballos, B., Villareal, H., Garcia, T., Pérez Jar, L., and Alfonso, E. (2005). Effect of Spirulina platensis meal as feed additive on growth, survival and development in *Litopenaeus schmitti* shimp larvae. Revista de Investigaciones Marina, 26(3), 235–241.

Ju, Z. Y., Deng, D. F., and Dominy, W. (2012). A defatted microalgae (Haematococcuspluvialis) meal as a protein ingredient to partially replace fishmeal in diets of Pacific white shrimp (*Litopenaeus vannamei*, Boone, 1931). Aquaculture, 354, 50–55.

Jumah, Y. U., Tumbokon, B. L., and Serrano Jr, A. E. (2020). Effects of dietary κ-carrageenan on growth and resistance to acute salinity stress in the black tiger shrimp *Penaeus monodon* post larvae. The Israeli Journal of Aquaculture-Bamidgeh, 72.

Kalla, A., Yoshimatsu, T., Araki, T., Zhang, D. M., Yamamoto, T., and Sakamoto, S. (2008). Use of Porphyra spheroplasts as feed additive for red sea bream. Fisheries Science, 74(1), 104–108.

Kamunde, C., Sappal, R., and Melegy, T. M. (2019). Brown seaweed (AquaArom) supplementation increases food intake and improves growth, antioxidant status and resistance to temperature stress in Atlantic salmon, Salmo salar. PLoS One, 14(7), e0219792.

Kang, C. D., Lee, J. S., Park, T. H., and Sim, S. J. (2005). Comparison of heterotrophic and photoautotrophic induction on astaxanthin production by *Haematococcus pluvialis*. Applied Microbiology and Biotechnology, 68(2), 237–241.

Katerina, K., Berge, G. M., Turid, M., Aleksei, K., Grete, B., Trine, Y., and Bente, R. (2020). Microalgal *Schizochytrium limacinum* biomass improves growth and filet quality when used long-term as a replacement for fish oil, in modern salmon diets. Frontiers in Marine Science, 7, 57.

Kiron, V., Sørensen, M., Huntley, M., Vasanth, G. K., Gong, Y., Dahle, D., and Palihawadana, A. M. (2016). Defatted biomass of the microalga, *Desmodesmus* sp., can replace fishmeal in the feeds for Atlantic salmon. Frontiers in Marine Science, 3, 67.

Kitikiew, S., Chen, J. C., Putra, D. F., Lin, Y. C., Yeh, S. T., and Liou, C. H. (2013). Fucoidan effectively provokes the innate immunity of white shrimp *Litopenaeus vannamei* and its resistance against experimental *Vibrio alginolyticus* infection. Fish & Shellfish Immunology, 34(1), 280–290.

Kousoulaki, K., Mørkøre, T., Nengas, I., Berge, R. K., and Sweetman, J. (2016). Microalgae and organic minerals enhance lipid retention efficiency and fillet quality in Atlantic salmon (*Salmo salar* L.). Aquaculture, 451, 47–57.

Li, F., Huang, S., Lu, X., Wang, J., Lin, M., An, Y., and Cai, M. (2018). Effects of dietary supplementation with algal astaxanthin on growth, pigmentation, and antioxidant capacity of the blood parrot (*Cichlasomacitrinellum*× *Cichlasomasynspilum*). Journal of Oceanology and Limnology, 36(5), 1851–1859.

Li, M., Wu, W., Zhou, P., Xie, F., Zhou, Q., and Mai, K. (2014). Comparison effect of dietary astaxanthin and Haematococcuspluvialis on growth performance, antioxidant status and immune response of large yellow croaker *Pseudosciaena crocea*. Aquaculture, 434, 227–232.

Lim, K. C., Yusoff, F. M., Shariff, M., Kamarudin, M. S., and Nagao, N. (2019). Dietary supplementation of astaxanthin enhances hemato-biochemistry and innate immunity of Asian seabass, *Lates calcarifer* (Bloch, 1790). Aquaculture, 512, 734339.

Liu, J., and Huang, Q. (2016). Screening of astaxanthin-hyperproducing *Haematococcus pluvialis* using Fourier transform infrared (FT-IR) and Raman microspectroscopy. Applied Spectroscopy, 70(10), 1639–1648.

Liu, X. H., Steele, J. C., and Meng, X. Z. (2017). Usage, residue, and human health risk of antibiotics in Chinese aquaculture: A review. Environmental Pollution, 223, 161–169.

Liu, X. H., Wang, B. J., Li, Y. F., Wang, L., and Liu, J. G. (2018). Effects of dietary botanical and synthetic astaxanthin on E/Z and R/S isomer composition, growth performance, and antioxidant capacity of white shrimp, *Litopenaeus vannamei*, in the nursery phase [For this article an Erratum has been published]. Invertebrate Survival Journal, 15(1), 131–140.

Liu, X. H., Zhang, H., Li, L., Fu, C., Tu, C., Huang, Y., and Christie, P. (2016). Levels, distributions and sources of veterinary antibiotics in the sediments of the Bohai Sea in China and surrounding estuaries. Marine Pollution Bulletin, 109(1), 597–602.

Lorenz, R. T. (1998). A review of the carotenoid, astaxanthin, as a pigment source and vitamin for cultured Penaeus prawn. www.cyanotech.com/pdfs/axbul51.pdf.

Lu, Q., Li, H., Zou, Y., Liu, H., and Yang, L. (2021). Astaxanthin as a microalgal metabolite for aquaculture: A review on the synthetic mechanisms, production techniques, and practical application. Algal Research, 54, 102178.

Lu, Q., Li, J., Wang, J., Li, K., Li, J., Han, P., and Zhou, W. (2017). Exploration of a mechanism for the production of highly unsaturated fatty acids in *Scenedesmus* sp. at low temperature grown on oil crop residue based medium. Bioresource Technology, 244, 542–551.

Mahgoub, H. A., El-Adl, M. A., Ghanem, H. M., and Martyniuk, C. J. (2020). The effect of fucoidan or potassium permanganate on growth performance, intestinal pathology, and antioxidant status in Nile tilapia (*Oreochromis niloticus*). Fish Physiology and Biochemistry, 46(6), 2109–2131.

Marino, F., Di Caro, G., Gugliandolo, C., Spano, A., Faggio, C., Genovese, G., . . . Santulli, A. (2016). Preliminary study on the in vitro and in vivo effects of Asparagopsistaxiformis bioactive phycoderivates on teleosts. Frontiers in Physiology, 7, 459.

McClure, D. D., Nightingale, J. K., Luiz, A., Black, S., Zhu, J., and Kavanagh, J. M. (2019). Pilot-scale production of lutein using *Chlorella vulgaris*. Algal Research, 44, 101707.

McHugh, D. J. (2003). A guide to the seaweed industry FAO Fisheries Technical Paper 441. Food and Agriculture Organization of the United Nations, Rome.

Mensi, F., Jamel, K., and Amor, E. A. (2005). Potential use of seaweeds in Nile tilapia (*Oreochromis niloticus*) diets. Mediterranean Fish Nutrition, 151–154.

Meyers, S. P., and Chen, H. M. (1982). Astaxanthin and its role in fish culture. In: Stickney, R. R., and Meyers, P. S. (eds.), Proceeding of the Warmwater Fish Culture (pp. 153–165). Louisiana State University.

Mir, I. N., Sahu, N. P., Pal, A. K., and Makesh, M. (2017). Synergistic effect of l-methionine and fucoidan rich extract in eliciting growth and non-specific immune response of *Labeorohita* fingerlings against *Aeromonas hydrophila*. Aquaculture, 479, 396–403.

Mobin, S., and Alam, F. (2017). Some promising microalgal species for commercial applications: A review. Energy Procedia, 110, 510–517.

Morioka, K., Naeshiro, K., Fujiwara, T., and Itoh, Y. (2008). Estimation of meat quality of cultured yellowtail *Seriola quinqueradiata* fed Porphyra supplemented diet. World Aquaculture, 20, 19–23.

Moutinho, S., Linares, F., Rodríguez, J. L., Sousa, V., and Valente, L. M. P. (2018). Inclusion of 10% seaweed meal in diets for juvenile and on-growing life stages of Senegalese sole (*Solea senegalensis*). Journal of Applied Phycology, 30 (6), 3589–3601.

Mustafa, G., Wakamatsu, S., Takeda, T. A., Umino, T., and Nakagawa, H. (1995). Effects of algae meal as feed additive on growth, feed efficiency, and body composition in red sea bream. Fisheries Science, 61(1), 25–28.

Mustafa, M. G. (1994). Effects of Ascophyllum and Spirulina meal as feed additives on growth performance and feed utilization of red sea bream, *Pagrus major*. Journal of Applied Biological Sciences, 33, 125–132.

Nakagawa, H. (1993). Optinum level of ulva meal diet supplement to minimize weight loss during wintering in black sea bream *Acanthopagrus schlegeli* (Bleeker). Asian Fisheries Science, 6, 139–148.

Nakagawa, H., Kasahara, S., Sugiyama, T., and Wada, I. (1984). Usefulness of Ulva-meal as feed supplementary in cultured black sea bream. Aquaculture Science, 32(1), 20–27.

Nakagawa, H., Kumai, H., Nakamura, M., Nanba, K., and Kasahara, S. (1986). Preventive effect of kelp meal supplement on nutritional disease due to sardine feeding in cultured yellow tail (*Seriola quinqueradiata*) (Pisces). In Proceedings of 3rd Symposium of Trace Nutrient Research (Vol. 3, pp. 31–37).

Nakazoe, J. I. (1986). Effect of supplementation of alga or lipids to the diets on the growth and body composition of nibbler *Girella punctata* Gray. Bulletin of Tokai Regional Fisheries Research Laboratory, 120, 43–51.

Nasseri, A. T., Rasoul-Amini, S., Morowvat, M. H., and Ghasemi, Y. (2011). Single cell protein: Production and process. American Journal of Food Technology, 6(2), 103–116.

Norambuena, F., Hermon, K., Skrzypczyk, V., Emery, J. A., Sharon, Y., Beard, A., and Turchini, G. M. (2015). Algae in fish feed: Performances and fatty acid metabolism in juvenile Atlantic salmon. PLoS One, 10(4), e0124042.

Onorato, C., and Rösch, C. (2020). Comparative life cycle assessment of astaxanthin production with *Haematococcus pluvialis* in different photobioreactor technologies. Algal Research, 50, 102005.

Peixoto, M. J., Magnoni, L., Gonçalves, J. F., Twijnstra, R. H., Kijjoa, A., Pereira, R., and Ozório, R. O. (2019). Effects of dietary supplementation of Gracilaria sp. extracts on fillet quality, oxidative stress, and immune responses in European seabass (*Dicentrarchus labrax*). Journal of Applied Phycology, 31(1), 761–770.

Pulz, O., and Gross, W. (2004). Valuable products from biotechnology of microalgae. Applied Microbiology and Biotechnology, 65(6), 635–648.

Radmer, R. J. (1996). Algal diversity and commercial algal products. Bioscience, 46(4), 263–270.

Rajendran, P., Subramani, P. A., and Michael, D. (2016). Polysaccharides from marine macroalga, *Padina gymnospora* improve the nonspecific and specific immune responses of *Cyprinus carpio* and protect it from different pathogens. Fish & Shellfish Immunology, 58, 220–228.

Ravindra, P. (2000). Value-added food: Single cell protein. Biotechnology Advances, 18(6), 459–479.

Rhein-Knudsen, N., Ale, M. T., and Meyer, A. S. (2015). Seaweed hydrocolloid production: An update on enzyme assisted extraction and modification technologies. Marine Drugs, 13(6), 3340–3359.

Saboya, J. P. D. S., Araujo, G. S., Silva, J. W. A. D., Sousa Junior, J. D., Maciel, R. L., and Farias, W. R. L. (2012). Effect of sulfated polysaccharides from the Rodophyta *Kappaphycus alvarezii* in post larva of Nile tilapia (*Oreochromis niloticus*) submitted to a stress situation. Acta Scientiarum. Animal Sciences, 34(3), 215–221.

Sánchez Campos, L. N., Díaz, F., Licea, A., Re, A. D., Lizárraga, M. L., Flores, M., and Tordoya Romero, C. (2010). Effect of hydrosoluble polysaccharides of *Macrocystis pyrifera* on physiological and metabolic responses of *Litopenaeus vannamei* infected with *Vibrio campbellii*. Hidrobiológica, 20(3), 246–255.

Sharif, M., Zafar, M. H., Aqib, A. I., Saeed, M., Farag, M. R., and Alagawany, M. (2021). Single cell protein: Sources, mechanism of production, nutritional value and its uses in aquaculture nutrition. Aquaculture, 531, 735885.

Shi, Y. Z., Suwaree, K., Chen, Y. Y., Hsu, C. H., and Chen, J. C. (2020). White shrimp *Litopenaeus vannamei* hemocytes receiving fucoidan release endogenous molecules that activate and synergize innate immunity in the presence of fucoidan. Aquaculture, 519, 734720.

Singh, R., Parihar, P., Singh, M., Bajguz, A., Kumar, J., Singh, S., and Prasad, S. M. (2017). Uncovering potential applications of cyanobacteria and algal metabolites in biology, agriculture and medicine: Current status and future prospects. Frontiers in Microbiology, 8, 515.

Sinurat, E., Saepudin, E., Peranginangin, R., and Hudiyono, S. (2016). Immunostimulatory activity of brown seaweed-derived fucoidans at different molecular weights and purity levels towards white spot syndrome virus (WSSV) in shrimp *Litopenaeus vannamei*. Journal of Applied Pharmaceutical Science, 6(10), 82–91.

Sivagnanavelmurugan, M., Karthik Ramnath, G., Jude Thaddaeus, B., Palavesam, A., and Immanuel, G. (2015). Effect of *Sargassum wightii* fucoidan on growth and disease resistance to *Vibrio parahaemolyticus* in *Penaeus monodon* post-larvae. Aquaculture Nutrition, 21(6), 960–969.

Sivagnanavelmurugan, M., Marudhupandi, T., Palavesam, A., and Immanuel, G. (2012). Antiviral effect of fucoidan extracted from the brown seaweed, *Sargassum wightii*, on shrimp *Penaeus monodon* post larvae against white spot syndrome virus. Journal of the World Aquaculture Society, 43(5), 697–706.

Sivagnanavelmurugan, M., Thaddaeus, B. J., Palavesam, A., and Immanuel, G. (2014). Dietary effect of *Sargassum wightii* fucoidan to enhance growth, prophenoloxidase gene expression of *Penaeus monodon* and immune resistance to *Vibrio parahaemolyticus*. Fish & Shellfish Immunology, 39(2), 439–449.

Sony, N. M., Hossain, M. S., Ishikawa, M., Koshio, S., and Yokoyama, S. (2020). Efficacy of mozuku fucoidan in alternative protein-based diet to improve growth, health performance, and stress resistance of juvenile red sea bream, *Pagrus major*. Fish Physiology and Biochemistry, 46(6), 2437–2455.

Sprague, M., Walton, J., Campbell, P. J., Strachan, F., Dick, J. R., and Bell, J. G. (2015). Replacement of fish oil with a DHA-rich algal meal derived from *Schizochytrium* sp. on the fatty acid and persistent organic pollutant levels in diets and flesh of Atlantic salmon (*Salmo salar*, L.) post-smolts. Food Chemistry, 185, 413–421.

Teuling, E., Wierenga, P. A., Agboola, J. O., Gruppen, H., and Schrama, J. W. (2019). Cell wall disruption increases bioavailability of *Nannochloropsis gaditana* nutrients for juvenile Nile tilapia (*Oreochromis niloticus*). Aquaculture, 499, 269–282.

Tomás-Almenar, C., Larrán, A. M., de Mercado, E., Sanz-Calvo, M. A., Hernández, D., Riaño, B., and García-González, M. C. (2018). *Scenedesmus almeriensis* from an integrated system waste-nutrient, as sustainable protein source for feed to rainbow trout (*Oncorhynchus mykiss*). Aquaculture, 497, 422–430.

Traifalgar, R. F. M., Corre, V. L., and Serrano, A. E. (2013). Efficacy of dietary immunostimulants to enhance the immunological responses and Vibriosis resistance of juvenile *Penaeus monodon*. Journal of Fisheries and Aquatic Science, 8(2), 340–354.

Traifalgar, R. F. M., Kira, H., Thanh Tung, H. A., Raafat Michael, F., Laining, A., Yokoyama, S., and Corre, V. (2010). Influence of dietary fucoidan supplementation on growth and immunological response of juvenile *Marsupenaeus japonicus*. Journal of the World Aquaculture Society, 41, 235–244.

Traifalgar, R. F. M., Koshio, S., Ishikawa, M., Serrano, A. E., and Corre, V. L. (2012). Fucoidan supplementation improves metamorphic survival and enhances vibriosis resistance of *Penaeus japonicus* larvae. Journal of Fisheries and Aquaculture, 3(1), 33.

Traifalgar, R. F. M., Serrano, A. E., Corre, V., Kira, H., Tung, H. T., Michael, F. R., and Koshio, S. (2009). Evaluation of dietary fucoidan supplementation effects on growth performance and vibriosis resistance of *Penaeus monodon* postlarvae. Aquaculture Science, 57(2), 167–174.

Tuller, J., De Santis, C., and Jerry, D. R. (2014). Dietary influence of Fucoidan supplementation on growth of *Lates calcarifer* (Bloch). Aquaculture Research, 45(4), 749–754.

Valente, L. M. P, Gouveia, A., Rema, P., Matos, J., Gomes, E. F., and Pinto, I. S. (2006). Evaluation of three seaweeds *Gracilaria bursa-pastoris*, *Ulva rigida* and *Gracilaria cornea* as dietary ingredients in European sea bass (*Dicentrarchus labrax*) juveniles. Aquaculture, 252(1), 85–91.

Van Doan, H., Doolgindachbaporn, S., and Suksri, A. (2014a). Effects of low molecular weight agar from seaweed on growth performance, immunity, and disease resistance of basa fish (*Pangasius bocourti*, Sauvage 1880). Pensee, 76(4).

Van Doan, H., Doolgindachbaporn, S., and Suksri, A. (2014b). Effects of low molecular weight agar and *Lactobacillus plantarum* on growth performance, immunity, and disease resistance of basa fish (*Pangasius bocourti*, Sauvage 1880). Fish & Shellfish Immunology, 41(2), 340–345.

Wang, H., Dai, A., Liu, F., and Guan, Y. (2015). Effects of dietary astaxanthin on the immune response, resistance to white spot syndrome virus and transcription of antioxidant enzyme genes in Pacific white shrimp *Litopenaeus vannamei*. Iranian Journal of Fisheries Sciences, 14(3), 699–718.

Wang, Y., Wang, B., Liu, M., Jiang, K., Wang, M., and Wang, L. (2020). Comparative transcriptome analysis reveals the potential influencing mechanism of dietary astaxanthin on growth and metabolism in *Litopenaeus vannamei*. Aquaculture Reports, 16, 100259.

Wijffels, R. H. (2008). Potential of sponges and microalgae for marine biotechnology. Trends in Biotechnology, 26(1), 26–31.

Wu, K., Cleveland, B. M., Portman, M., Sealey, W. M., and Lei, X. G. (2021). Supplemental microalgal DHA and astaxanthin affect astaxanthin metabolism and redox status of juvenile rainbow trout. Antioxidants, 10(1), 16.

Yaakob, Z., Ali, E., Zainal, A., Mohamad, M., and Takriff, M. S. (2014). An overview: Biomolecules from microalgae for animal feed and aquaculture. Journal of Biological Research-Thessaloniki, 21(1), 1–10.

Yang, Q., Yang, R., Li, M., Zhou, Q., Liang, X., and Elmada, Z. C. (2014). Effects of dietary fucoidan on the blood constituents, anti-oxidation and innate immunity of juvenile yellow catfish (*Pelteobagrus fulvidraco*). Fish & Shellfish Immunology, 41(2), 264–270.

Yang, W. N., Chen, P. W., and Huang, C. Y. (2017). Compositional characteristics and in vitro evaluations of antioxidant and neuroprotective properties of crude extracts of fucoidan prepared from compressional puffing-pretreated *Sargassum crassifolium*. Marine Drugs, 15(6), 183.

Yasuo, Y., Furuichi, M., and Urano, T. (1986). Effects of wakame Undaria pinnatifida and Ascophyllum nodosum on absorption of dietary nutrients, and blood sugar and plasma free amino-N levels of red sea bream. ???????, 52(10), 1817–1819.

Yeh, S. T., and Chen, J. C. (2008). Immunomodulation by carrageenans in the white shrimp *Litopenaeus vannamei* and its resistance against *Vibrio alginolyticus*. Aquaculture, 276(1–4), 22–28.

Yi, Y. H., and Chang, Y. J. (1994). Physiological effects of seamustard supplement diet on the growth and body composition of young rockfish, *Sebastes schlegeli*. Korean Journal of Fisheries and Aquatic Sciences, 27(1), 69–82.

Ying, H. E. (2008). Immunoregulation effect of *Sargassum* Polysaccharides on White-leg Shrimp [J]. Journal of Anhui Agricultural Sciences, 31.

Yone, Y., Furuichi, M., and Uranus, K. (1986). Effects of dietary wakame *Undaria pinnatifida* and *Ascophyllum nodosum* supplements on growth, feed efficiency, and proximate compositions of liver and muscle of red sea bream. Nippon Suisan Gakkaishi, 52(8), 1465–1468.

Zayed, A., and Ulber, R. (2020). Fucoidans: Downstream processes and recent applications. Marine Drugs, 18(3), 170.

Zhang, C., Jin, Y., Yu, Y., Xiang, J., and Li, F. (2021). Effects of natural astaxanthin from microalgae and chemically synthetic astaxanthin supplementation on two different varieties of the ridgetail white prawn (*Exopalaemon carinicauda*). Algal Research, 57, 102347.

Zhang, R., Tang, J., Li, J., Zheng, Q., Liu, D., Chen, Y., and Zhang, G. (2013). Antibiotics in the offshore waters of the Bohai Sea and the Yellow Sea in China: Occurrence, distribution and ecological risks. Environmental Pollution, 174, 71–77.

Theme II

Bioremediation Using Algae

7 Biological Fuel Cells with Oxygenic Photosynthetic Microorganisms for Wastewater Treatment

State of the Art and Perspectives in Biocathode

Serban N. Stamatin, Matei-Tom Iacob, Cristina Moisescu, Ana-Valentina Ardelean, Adrian Ghinea, Sanda Voinea, Ioan Stamatin, and Ioan I. Ardelean

7.1 INTRODUCTION

Biological fuel cells (BFCs) are bioelectrochemical devices that are using liquid and solid anthropogenic wastes to generate electricity or value-added chemicals. Wastewater originating from different industries can be treated in either single-chambered or in double-chambered BFCs with an ion exchange membrane that separates the anode from the cathode (Logan and Regan, 2006; Xiao et al., 2015; Saratale et al., 2017). The low electron yield is a key-drawback of these bioelectrochemical devices and hinders the production of larger amounts of bioelectricity.

Bioelectrochemical oxidation reactions are used to neutralize anthropogenic waste. Oxidation reactions in BFCs take place at the anode compartment. Electrons, generated at the anode, travel to the cathode where they are used in (bio) electrochemical reduction reactions, desalination cells, or to generate biomass (Figure 7.1). The use of oxygenic phototrophic microorganisms (OPhMs) in the anode compartment has been, thoroughly, discussed in the previous chapter (i.e., Chapter 5 of the book *Algal Biotechnology: Significant Applications for Industrial Development and Human Welfare*). The cathodic potential of OPhM will be explored in the chapter at hand. The reader should bear in mind that there are some reviews available in the literature that discuss BFCs (Enamala et al., 2020; Jaiswal et al., 2020; Kondaveeti et al., 2020). We will focus on the limitations imposed by (bio) cathodes and how such limitations can be overcome by OPhM.

Cyanobacteria and microalgae are the most studied OPhMs for wastewater treatment, bioelectricity generation, and biomass growth. OPhMs enriched biocathode was initially studied by Powell et al. (2009), however, without an emphasis on wastewater. Zhang et al. (2012) was the first report to discuss OPhMs in relation to wastewater treatment. They reported that electrochemically active (non-photosynthetic) bacteria in the anode compartment can oxidize organic matter, thus releasing electrons, protons, and carbon dioxide (CO_2) while microalgae in the cathode will assimilate the CO_2 taken from anode.

The electrons generated at the bioanode are not used by the biocathode in the setup mentioned previously. For high electron uptake efficiency, there is the need for a direct electron

DOI: 10.1201/9781003219156-10

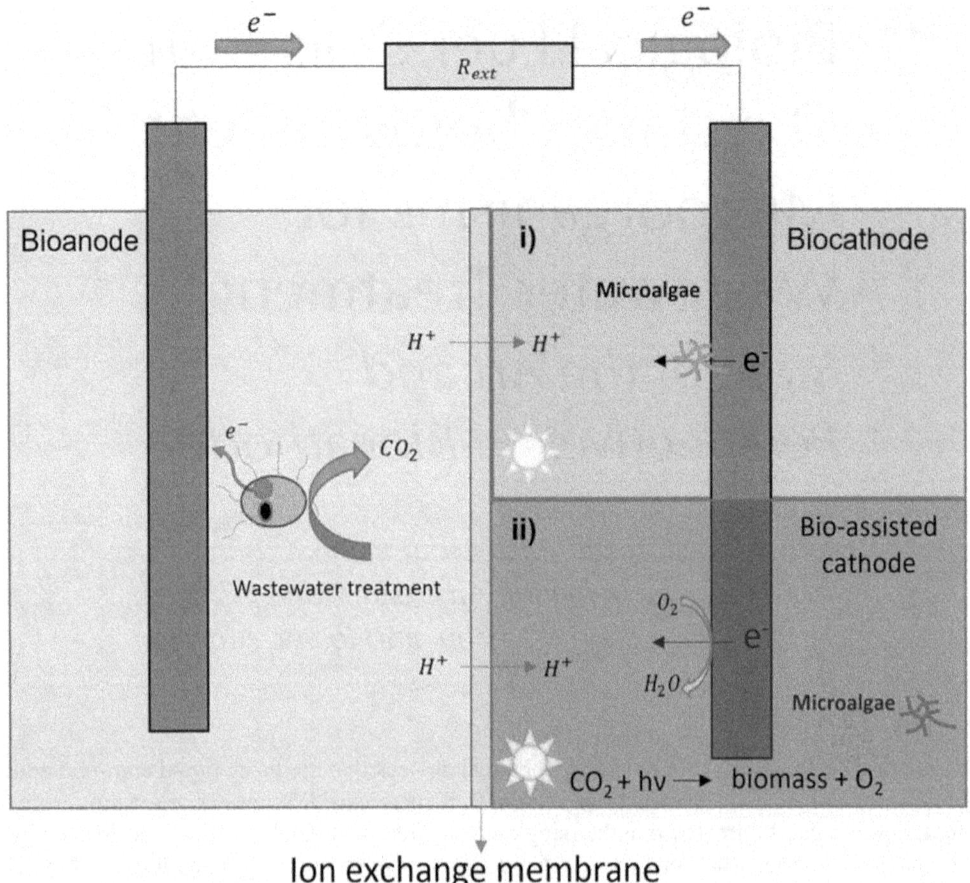

FIGURE 7.1 Different configurations of BFCs that use OPhM at the cathode for the following electrochemical reduction reactions: (i) bio-assisted cathode and (ii) direct electron transfer to the cathode (direct biocathode).

transfer from the abiotic cathode to OPhMs. The electron pathway at the biocathode has rarely been studied. Therefore, cathode limitations, BFCs setups using OPhMs, and electron pathways are discussed in this chapter.

7.2 BIO-ASSISTED CATHODES BASED ON OPHM

Anodes are created to take electrons generated by exoelectrogenic microorganisms with minimum resistance. The chief aim of the bioanode is to generate electricity by using the electrons from metabolic reactions. A cathode is needed to balance the overall reaction. Biocathodes are designed to facilitate the use of electrons generated at the (bio) anode via electrochemical reduction reactions. Enhancing the cathode reaction leads to faster electrogenesis at the anode, which accelerates the wastewater treatment (Venkata et al., 2014). Bioelectrochemical systems (BES) are very versatile devices, which led to a plethora of configurations (Kadier et al., 2020). Traditional BES were used to generate minor amounts of bioelectricity while treating wastewater, which morphed into microbial desalination cells, microbial carbon capture, microbial electrolysis cells, and many more (Kadier et al., 2020).

7.2.1 Cathode Limitations

The most common cathodic reaction in microbial fuel cells is the oxygen reduction reaction (ORR):

$$O_2 + 4H^+ + 4e^- \rightarrow 2H_2O$$

which has a standard potential equal to 1.229 V at atmospheric pressure, pH = 0 and room temperature. It is well-known that the ORR is an uphill reaction that generates an overpotential of approx. 0.4 V on all metal surfaces (Man et al., 2011). Carbons are the material of choice for the cathode compartment in BFCs, which are not suitable ORR catalysts. In this respect, the use of OPhM can enhance the BFC overall performance by increasing and maintaining a high oxygen concentration. Indeed, it has been shown that there is an oxygen supersaturation effect in microbial mats (Revsbech et al., 1983).

Non-ideal ORR catalysts such as carbon require mechanical aeration in order to sustain the electron transfer from the anode.

OPhM can aid wastewater treatment and generate bioelectricity by being incorporated in the cathode in two distinct ways: (1) making use of the oxygen liberated during the metabolism to drive the oxygen reduction reaction, which in turn accelerates the electron transfer from the anode and therefore enhances the wastewater treatment while removing the cost of mechanical aeration; (2) bioelectrosynthesis of useful chemicals by accepting electrons directly from the cathode.

Other configurations in which BFCs are used for wastewater treatment which is then fed to an alga photobioreactor (Jiang et al., 2013) fall beyond the general scope of this work.

7.2.2 Bio-Assisted Cathode

One of the research goals in microbial fuel cells field is to find how to increase the oxygen concentration in the cathode compartment by using the oxygen resulted from the algae metabolism:

$$6CO_2 + 12H^+ + 12e^- \rightarrow C_6H_{12}O_6 \text{ (biomass)} + 3O_2$$

Algae photobioreactors can connect algae to the cathode: (1) directly, in which the cathode comes into contact with the microalgae or (2) indirectly, in which the oxygen generated by the microalgae is collected at the top of the bioreactor and transferred to the cathode (Figure 7.1). Table 7.1 summarizes the power densities obtained for different configurations and algae. Usual power densities are in the mW/cm^2 range.

The power generated by the bioelectrochemical system is hindered by the cathode reaction, which is bottlenecked by the low concentration of oxygen in oxygen saturated aqueous solutions, that is, 1.2 μmol/ml. Mechanical aeration is needed in order to provide a continuous oxygen supply to the cathode which is energy intensive and cost sensitive. Microalgae produce oxygen during their metabolism, which is directly used at the cathode. Indeed, the oxygen generated by an algal biofilm in the cathode compartment can lead to a 20% increase in the power density (Yang et al., 2017). It was shown that the concentration of dissolved oxygen was at least larger at a cathode with microalgae than a abiotic cathode with mechanical oxygen supply (Xiao et al., 2012). The thin biofilm that microalgae produce on the surface of cathodes was found to limit oxygen diffusion, which enhances the power output (Gajda et al., 2013).

Interestingly, light affects the cathode potential and the power density (Del Campo et al., 2013; Wu et al., 2014). The dissolved oxygen increases linearly with the light intensity up to 3,500 lx. Similarly, the cathode potential increases linearly with the illumination power up to 3,000 lx (Wu et al., 2014). However, it was shown that a fluorescent light with a power density above 125 mW/cm^2 leads to a lower power output due to a less effective algal photosynthesis (Juang et al., 2012). An

TABLE 7.1
Summary of the Power Density by Using a Biocathode

Cathode Type	Power Density [mW/cm²]	OPhM	Additional Information	Reference
Bio-assisted cathode	5700	*Chlorella vulgaris*	Lipid production	Hou et al., 2016
Bio-assisted cathode	650	*Scenedesmus quadri*		Yang et al., 2017
Bio-assisted cathode	630	Mixed strains		Kakarla et al., 2015
Bio-assisted cathode	0.0045 0.03	*Scenedesmus obliquus*	Carbon paper Carbon brush	Kakarla and Min, 2014
Bio-assisted cathode	187	*C. vulgaris*		Liu et al., 2015
Bio-assisted cathode	100	*Desmodesmus*		Wu et al., 2014
Bio-assisted cathode	94.5 63.9	*Chlorella vulgaris* *Anabaena*		Venkata et al., 2014
Bio-assisted cathode	57	Mixed strains		Venkata et al., 2014
Bio-assisted cathode	50	Mixed strains		Nguyen et al., 2017
Bio-assisted cathode	15	*Chlorella vulgaris*		Del Campo et al., 2013
Bio-assisted cathode	0.03	*Scenedesmus acutus*	Lipids, C_{16-18}	Angioni et al., 2018
Bio-assisted cathode	0.98	*Chlorella vulgaris*		Velasquez-Orta et al., 2009
Bio-assisted cathode	0.76	*Ulva lactuca*	synthetic wastewater	Velasquez-Orta et al., 2009
Biocathode	41.5	*Synechoccus*		Naina Mohamed et al., 2020
Biocathode	2.7	*Chlorella vulgaris*	without wastewater treatment	Powell et al., 2009
Biocathode	0.063	*Scenedesmus quadricauda*	96% reduced TN 91% reduced TP	Yang et al., 2017
Biocathode	3720	*Chlorella vulgaris*	86% reduced NH_4^+-N 70% reduced TN 94% reduced TOC	Zhang et al., 2019
Biocathode	2485	*Chlorella vulgaris*		Zhou et al., 2012

algal biocathode coupled to a microbial anode has been tested under simulated day-night profiles for lagooning purposes (Lobato et al., 2013).

Another important aspect is the evolution of the pH in time. The ORR can undergo the direct 4 electrons–4 protons reaction or an indirect reaction in which the oxygen molecule is reduced to the peroxide by 2 electrons followed by another 2 electrons reduction to water. The reader should bear in mind that microorganisms are sensitive to the pH which can turn alkaline due to the 2 electron ORR. Liu et al. (2015) showed that the pH value can reach values of 8.5 in less than 24 h. The optimal pH for the cathode was determined to be 7.5 by Varanasi et al. (2020).

The cathode morphology was studied by Kakarla and Min (2014), who showed a seven-fold improvement of the carbon brush over the regular carbon paper. Angioni et al. (2018) compared the bare carbon cloth to a platinized electrode which led to an eight-fold improvement in the current density, albeit the setup was used to produce C_{16-18} neutral lipids. It was speculated that the polybenzimidazole membrane is more robust, which led to an increase in the current density compared to Nafion. The obvious limitations of ion exchange membranes created the need to find other electrochemical separators which are also biocompatible. In this respect, terracotta and earthenware can effectively separate the anolyte from the catholyte while providing ion conductivity (Park and Zeikus, 2002; Winfield et al., 2013; Jadhav et al., 2017; Yadav, 2020).

An issue that needs to be addressed is the CO_2 crossover through the membrane from the anode to the cathode, which leads to a decrease in the power density (Liu et al., 2015). The well-known Nafion proton exchange membrane has been the membrane of choice in algal-based BFC. Nafion can be permeable to gasses upon degradation that can arise from the high concentration of OH- species.

7.2.3 COD Reduction, Nutrient and Pollutant Removal

Chlorella sp. is the alga of choice for biocathodes for bioelectrosynthesis and electrogenesis. *Chlorella vulgaris* at the cathode was shown to increase the abundance of electrogenic bacteria at the anode (*Geobacter* and *Desulfobulbaceae*) (Song et al., 2020). CO_2 sequestration and biodiesel synthesis were found to have different performances. For example, *Chlorella vulgaris* showed a 44% increase in CO_2 fixation than *Chlorella* sp. as shown by Hu et al. (2016). The anode was treating the wastewater, while the algae assisted the cathode by maintaining the oxygen and aeration. Lipid production in both strains was found similar between light intensities of 9 and 12 W/m^2 (Hu et al., 2016). The optical density of the inoculum was studied for *C. vulgaris*, which showed the highest COD removal (i.e., 44%) at 150 mg/L. Similarly, the highest voltage (0.17 V) and power density (19.15 W/m^3) was achieved at 150 mg/L (Hou et al., 2016). Varanasi et al. (2020) showed that a 12h inoculum age is the most suitable for power generation. *C. vulgaris* bio-assisted the cathode in treating chocolate factory wastewater with 79% COD removal (Huarachi-Olivera et al., 2018). The anode effluent was transferred to the cathode in a BFC, where *C. vulgaris* was used to assist the cathode by providing oxygen via photosynthesis. (Commault et al., 2017). The maximum power density doubled when alga was used in the cathode and increased the COD removal by 15% (Commault et al., 2017).

A mixed microalgae culture consisting of *Chlorella* sp., *Desmodesmus* sp., and *Scenedismus* sp. was grown in a cathode together with a solution of known ammonium concentration (Kakarla and Min, 2019). Wastewater was treated at the anode, while ammonium removal rate was found to be affected by light cycles and photon flux density (Kakarla and Min, 2019). Abazarian et al. (2020) showed that the power density is directly linked to the illumination cycles. The maximum power density, that is 19.6 mW/m^2, was achieved during a daily 12 h illumination period. Increasing daily illumination to 16 h resulted in a decreased power density to 12.3 mW/m^2. Varanasi et al. (2020) confirmed that the 12:12 (light:dark) cycle enhances the power density. Colombo et al. (2017) studied the photosynthesis oxygen generation of *Spirulina* to assist the cathode in running the oxygen reduction reaction. *Spirulina* grew as fast as in the control experiment while nutrients were recovered, which showed that the *Spirulina* is not affected by the power generation mechanism in fuel cells.

Yang et al. (2017) proposed an algal biofilm that works in the cathode compartment together with a microbial cathode in a sediment BFC. Betaproteobacteria dominated the microbial community in the BFC without the algae biofilm, which was reduced considerably when the algae were added. It was speculated that betaproteobacteria was competing for nutrients with the microalgae. Deltaproteobacteria were found to be the most suitable candidates to work together with the algae biofilm.

Golenkinia sp. was studied in the cathode to remove total nitrogen (TN) and total phosphorous (TP) from the anaerobically digested kitchen waste, diluted 21 times (Hou et al., 2016). Nevertheless, the results were promising in terms of wastewater treatment with more than 90% of TN and TP.

The degradation of antibiotics is morphing into a remarkable research avenue. For example, Sun et al. (2020) placed anaerobic sludge in the anode and cathode. Different concentrations of the oxytetracycline antibiotic (OTC), and *C. vulgaris* were placed at the cathode in addition to the anaerobic sludge. Interestingly, the degradation of OTC supplied plenty of mediators to accelerate the transfer of electrons from the cathode as attested by the decreased charge transfer resistance. In consequence, a 5-fold increase in power density from 19 mW/m^2 at 0 mg/L OTC to 106 mW/m^2 at 5 mg/L OTC. However, the power density decreased linearly with increasing OTC concentration, reaching 56 mW/m^2 at 50 mg/L OTC. The same group reported the same system coupled with nitrogen removal (Sun et al., 2019). Interestingly, the electron transfer at the cathode was enhanced below 20 mg/L and inhibited above 50 mg/L. It was speculated that the alga and bacteria were poisoned by the OTC at large concentrations (Sun et al., 2019).

7.2.4 Microbial Desalination Cell

Kokabian and Gude (2013) put forward a photosynthetic microbial desalination cell. The anode and cathode compartment were separated by a third compartment which has an anion and cation exchange membrane facing the anode and cathode, respectively. Natural seawater or artificially salted water is placed in the third compartment where the chloride and sodium ion are travelling to the anode and cathode, respectively. *C. vulgaris* was placed in the cathode while a microbial consortium and anaerobic sludge was placed in the anode. The desalination compartment contained 200 ml of 10 g/L NaCl aqueous solution. The maximum cell potential was approx. 0.24 V for the setup with algal cathode, which was reached in less than three days. The abiotic cathode had a maximum potential of 0.22 V achieved in almost five days. Desalination rates increase linear with cell potential due to the stimulation of ion transfer in the desalination chamber. In consequence the algal biocathode realized a 40% desalination rate in less than one month, which is double the desalination rate of the abiotic cathode (Kokabian and Gude, 2013).

Elakkiya and Niju (2020) used ghee industry wastewater as anolyte and three different cathodes: ferricyanide (abiotic), mechanical aeration (abiotic), and a microalgae consortium collected from a lily pond. The treated water at the anode was used by the microalgae at the cathode. A 96% COD removal efficiency was determined only when microalgae were used. The power density was 13.7 mW/m^2 and 3.33 mW/m^2 for the abiotic cathode and microalgae cathode, respectively, which shows the negative effect on the power generation when the carbon cathode is covered by microalgae.

Bejjanki et al. (2021) used wastewater from a dairy farm and artificially salted water in a triple chambered microbial desalination cell. The dairy farm wastewater was placed at the anode with bacteria, the cathode was filled with BG11 medium and *Oscillatoria* sp., and the desalination compartment was filled with salted water with an increasing concentration. A 65% desalination efficiency and 80% COD removal was obtained using the system, which makes *Oscillatoria* sp. a potential candidate for enhanced desalination.

Ashwaniy et al. (2020) used *Scenedesmus abundans* in 50% petroleum refinery effluent in two different microbial desalination cells: (1) at the anode as support for bacterial growth and (2) at the cathode as a terminal electron acceptor. The results showed 70% COD, 81% BOD, 67% phosphorous, 61% sulphide, 67% total dissolved solids, and 62% total suspended solids. A 90% desalination performance was obtained only when the alga was used in the cathode compartment, which emphasizes the versatility of combining wastewater treatment with desalination technology.

7.2.5 Microbial Carbon Capture

CO_2 is produced at the anode of microbial fuel cells by microorganisms during metabolism. OPhM use solar illumination and CO_2 to generate carbohydrates. Wang et al. (2010) was the first to introduce the microbial carbon capture (MCC) in which CO_2 was transferred from the anode (fed with glucose) to the cathode compartment where *C. vulgaris* formed biomass. CO_2 presence in the cathode maintained catholyte pH and conductivity (Wang et al., 2010). MCCs were further advanced by coupling wastewater treatment, CO_2 sequestration, and bioelectricity (Pandit et al., 2012).

A mixture of CO_2/air was used at the cathode of MCC to achieve a better understanding on the effect of CO_2. *Anabaena* was placed in the cathode and sparged with air (control experiment) and a CO_2/air mixture. CO_2/air sparging in the cathode increased the power density from 29.7 mW/m^2 (air sparging) to 57.8 mW/m^2, stabilized the catholyte pH, and increased the cell potential (Pandit et al., 2012). The anodic CO_2 was not sufficient for the growth of *Chlorella sorokiniana* at the cathode of a MCC (Varanasi et al., 2020). Naina Mohamed et al. (2020) compared the cathode performance of *Synechococcus* sp. and *Chlorococcum* sp. Power densities were found to be light dependent, similar to the study by Wu et al. (2014) and Juang et al. (2012). When the electrolyte in the cathode compartment was enriched with a mediator, the power densities were as high as 41.5 mW/cm^{-2} and 30.2 mW/cm^{-2} for *Synechococcus* sp. and *Chlorococcuum* sp., respectively. Preliminary results showed that MCCs are a promising technology for using the CO_2 rich industrial effluents to increase biomass production and generate bioelectricity.

7.3 FUTURE PERSPECTIVES

7.3.1 OPhM as Direct Electron Acceptors from the Cathode

There are many possible anode-cathode combinations (Figure 7.1) especially when OPhM is used at the cathode (Fischer, 2018). Microbial fuel cells centered around the cathode reactions are called Microbial Electrosynthesis Cells (MEC). The understanding of the coupling between cell physiology and bioelectricity has been revised recently (Schofield et al., 2020), which pushes the biocathode research into a new age, that of microorganism electrosynthesis. Optimal biocathodes that are aimed at the synthesis of products should use DET from the cathode to microorganism without mediators in order to reduce the cost of further purification of the final product (Thrash and Coates, 2008; Lovley, 2011). Direct electron-microorganism transfer is rather new with the first report published in 2004 although on *G. metallireducens* (Gregory et al., 2004). Further on, Powell et al. (2011) has been among the first to make use of the direct electron transfer abilities of *C. vulgaris* albeit not connected to a wastewater anode. Powell et al. (2011) showed that biomass growing generates a higher electrochemical cell potential than *C. vulgaris* resting cells. The reactive oxygen species generated by cyanobacteria can act as a terminal electron acceptor (Cai et al., 2013). Interspecies electron transfer has been recently explored (Rotaru et al., 2012; Rotaru et al., 2014), which can be exploited as mixed cultures with OPhM. An interesting concept to extract electrons from OPhM is the confinement of microalgae in a Fabry-Perot optical microcavity, which can be further coupled to a bioelectrochemical system (Roxby et al., 2020).

7.3.2 Bioelectrosynthesis of Chemicals Coupled with Wastewater Treatment

There are plenty of microorganisms that use CO_2 in their metabolism to generate carbohydrates (Kondaveeti et al., 2020). Such mechanisms must be exploited as part of an international effort to integrate CO_2 into a circular economy approach. Nowadays, carbon can be captured, transformed, or sequestrated by several technologies that have a technology readiness level above 9 (Kondaveeti et al., 2020). However, a common drawback is the large size needed to make such technologies cost-effective, which limits decentralization. In this respect, (photo)electrocatalytic CO_2 reduction has been viewed as the solution. The limitations imposed on the reaction by what is now called *scaling relationship* (Man et al., 2011) are preventing electrocatalysis to further develop the knowledge and provide a solution. In this respect, microbial electrosynthesis has the potential to deliver an integrated system that can convert CO_2 to useful products. Several products are accessible through microbial electrosynthesis, such as acetic acid, ethanol, or even C6 products (i.e., caproic acid; Vassilev et al., 2018). In this respect, OPhM participating directly at the cathodes of microbial electrosynthesis cells should be further explored, taking also into account their massive contribution to CO_2 reduction on our planet.

ACKNOWLEDGMENT

This work has received funding from the Romanian Ministry of Research and Innovation under the PN-III-1.2-PCCDI-2017–0185 Project ECOTECH-GMP-Nr. 76 PCCDI/2018.

REFERENCES

Abazarian E., Gheshlaghi R., Mahdavi M. A. (2020). Impact of light/dark cycle on electrical and electrochemical characteristics of algal cathode sediment microbial fuel cells. *J. Power Sources.* **475**: 228686.

Angioni S., Millia L., Mustarelli P., Doria E., Temporiti M. E., Mannucci B., Corana F., Quartarone E. (2018). Photosynthetic microbial fuel cell with polybenzimidazole membrane: Synergy between bacteria and algae for wastewater removal and biorefinery. *Heliyon* **4**(3): e00560.

Ashwaniy V. R. V., Perumalsamy M., Pandian S. (2020). Enhancing the synergistic interaction of microalgae and bacteria for the reduction of organic compounds in petroleum refinery effluent. *Environ. Technol. Innov.* **19**: 100926.

Bejjanki D., Muthukumar K., Radhakrishnan T. K., Alagarsamy A., Pugazhendhi A., Mohamed S. N. (2021). Simultaneous bioelectricity generation and water desalination using Oscillatoria sp. as biocatalyst in photosynthetic microbial desalination cell. *Sci. Total Environ.* **754**: 142215.

Cai P.-J., Xiao X., He Y.-R., Li W.-W., Zang G.-L., Sheng G.-P., Lam M. H.-W., Yu L., Yu H.-Q. (2013). Reactive oxygen species (ROS) generated by cyanobacteria act as an electron acceptor in the biocathode of a bio-electrochemical system. *Biosensors and Bioelectronics.* **39**(1): 306–310.

Colombo A., Marzorati S., Lucchini G., Cristiani P., Pant D., Schievano A. (2017). Assisting cultivation of photosynthetic microorganisms by microbial fuel cells to enhance nutrients recovery from wastewater. *Bioresour. Technol.* **237**: 240–248.

Commault A. S., Laczka O., Siboni N., Tamburic B., Crosswell J. R., Seymour J. R., Ralph P. J. (2017). Electricity and biomass production in a bacteria-chlorella based microbial fuel cell treating wastewater. *J. Power Sources.* **356**: 299–309.

Del Campo G. A., Cañizares P., Rodrigo M. A., Fernández F. J., Lobato J. (2013). Microbial fuel cell with an algae-assisted cathode: A preliminary assessment. *J. Power Sources.* **242**: 638–645.

Elakkiya E., Niju S. (2020). Simultaneous treatment of lipid rich ghee industry wastewater and power production in algal biocathode based microbial fuel cell. Energy Sources, Part A: Recovery, Utilization, and Environmental Effects, DOI: 10.1080/15567036.2020.1823529.

Enamala M. K., Dixit R., Tangellapally A., Singh M., Dinakarrao S. M. P., Chavali M., Pamanji S. R., Ashokkumar V., Kadier A., Chandrasekhar K. (2020). Photosynthetic microorganisms (Algae) mediated bioelectricity generation in microbial fuel cell: Concise review. *Environ. Technol. Innovation.* **19**: 100959.

Fischer F. (2018). Photoelectrode, photovoltaic and photosynthetic microbial fuel cells. *Renew. Sustain. Energy Rev.* **90**: 16–27.

Gajda I., Greenman J., Melhuish, C., Ieropoulos, I. (2013). Photosynthetic cathodes for Microbial Fuel Cells. *Int. J. Hydrog. Energy.* **38**(26): 11559–11564.

Gregory K. B., Bond D. R., Lovley D. R. (2004). Graphite electrodes as electron donors for anaerobic respiration. *Environ Microbiol.* **6**(6): 596–604.

Hou Q., Pei H., Hu W., Jiang L., Yu Z. (2016). Mutual facilitations of food waste treatment, microbial fuel cell bioelectricity generation and *Chlorella vulgaris* lipid production. *Bioresour. Technol.* **203**: 50–55.

Hu X., Zhou J., Liu B. (2016). Effect of algal species and light intensity on the performance of an air-lift-type microbial carbon capture cell with an algae-assisted cathode. *RSC Adv.* **6**: 25094.

Huarachi-Olivera R., Dueñas-Gonza A., Yapo-Pari U., Vega P., Romero-Ugarte M., Tapia J., Molina L., Lazarte-Rivera A., Pacheco-Salazar D. G., Esparza-Mantilla M. (2018). Bioelectrogenesis with microbial fuel cells (MFCs) using the microalga *Chlorella vulgaris* and bacterial communities. *Electron. J. Biotechnol.* **31**: 34–43.

Jadhav D. A., Jain S. C., Ghangrekar M. M. (2017). Simultaneous wastewater treatment, algal biomass production and electricity generation in clayware microbial carbon capture cells. *Appl. Biochem. Biotechnol.* **183**(3): 1076–1092.

Jaiswal K. K, Kumar V., Arora N., Singh A., Vlaskin M. S., Sharma N., Rautela I., Nanda M., Chauhan P. K. (2020). Microalgae fuel cell for wastewater treatment: Recent advances and challenges. *Water Process. Engi.* **38**: 101549.

Jiang H., Luo S., Shi X., Dai M., Guo R. (2013). A system combining microbial fuel cell with photobioreactor for continuous domestic wastewater treatment and bioelectricity generation. *J. Central South University.* **20**(2): 488–494.

Juang D. F., Lee C. H., Hsueh S. C. (2012). Comparison of electrogenic capabilities of microbial fuel cell with different light power on algae grown cathode. *Bioresour. Technol.* **123**: 23–29.

Kadier A., Pratiksha J., Bin L., Mohd Sahaid K., Sanath K., Khulood Fahad Saud A., Ibrahim M. A-R., Gunda M. (2020). Biorefinery perspectives of microbial electrolysis cells (MECs) for hydrogen and valuable chemicals production through wastewater treatment. *Biofuel Res. J.* **25**: 1128–1142.

Kakarla R., Kim J. R., Jeon B-H., Min B (2015). Enhanced performance of an air-cathode microbial fuel cell with oxygen supply from an externally connected algal bioreactor. *Bioresour. Technol.* **195**: 210–216.

Kakarla R., Min B. (2014). Evaluation of microbial fuel cell operation using algae as an oxygen supplier: Carbon paper cathode vs. carbon brush cathode. *Bioprocess Biosyst. Eng.* **37**:2453–2461.

Kakarla R., Min B. (2019). Sustainable electricity generation and ammonium removal by microbial fuel cell with a microalgae assisted cathode at various environmental conditions. *Bioresour. Technol.* **284**: 161–167.

Kokabian B., Gude V. G. (2013). Photosynthetic microbial desalination cells (PMDCs) for clean energy, water and biomass production. *Environ. Sci.: Processes Impacts.* **15**: 2178.

Kondaveeti S., Abu-Reesh I. M., Mohanakrishna G., Bulut M., Pant D. (2020). Advanced routes of biological and bio-electrocatalytic Carbon Dioxide (CO2) mitigation toward carbon neutrality. *Front. Energy Res.* **8**: 94.

Liu T., Rao L., Yuan Y., Zhuang L. (2015). Bioelectricity generation in a microbial fuel cell with a self-sustainable photocathode. *The Scientific World Journal.* **2015**: 864568.

Lobato J., del Campo A. G., Fernández F. J., Cañizares P., Rodrigo M. A. (2013) Lagooning microbial fuel cells: A first approach by coupling electricity-producing microorganisms and algae. *Appl. Energy.* **110**: 220–226.

Logan B. E., Regan J. M. (2006). Electricity-producing bacterial communities in microbial fuel cells. *Trends in Microbiology.* **14**(12): P512–518.

Lovley D. R. (2011). Powering microbes with electricity: Direct electron transfer from electrodes to microbes. *Environ Microbiol Rep.* **3**(1): 27–35.

Man I. C., Su H.-Y., Calle-Vallejo F., Hansen H. A., Martínez J. I., Inoglu, N. G., Kitchin J., Jaramillo T. F., Nørskov J. K., Rossmeisl J. (2011). Universality in oxygen evolution electrocatalysis on oxide surfaces. *Chem. Cat. Chem.* **3**(7): 1159–1165.

Naina Mohamed S., Ajit Hiraman P., Muthukuma, K., Jayabalan T. (2020). Bioelectricity production from kitchen wastewater using microbial fuel cell with photosynthetic algal cathode. *Bioresour. Technol.* **295**: 122226.

Nguyen H. T. H., Kakarla R., Min B. (2017). Algae cathode microbial fuel cells for electricity generation and nutrient removal from landfill leachate wastewater. *Int. J. Hydrog. Energy.* **42(49)**: 29433–29442.

Pandit S., Nayak B. K., Das D. (2012). Microbial carbon capture cell using cyanobacteria for simultaneous power generation, carbon dioxide sequestration and wastewater treatment. *Bioresour. Technol.* **107**: 97–102.

Park D. H., Zeikus J. G. (2002). Improved fuel cell and electrode designs for producing electricity from microbial degradation. *Biotehnol. Bioner.* **81**(3): 348–355. DOI: 10.1002/bit.10501.

Powell E. E., Evitts R. W., Hill G. A., Bolster J. C. (2011). A microbial fuel cell with a photosynthetic microalgae cathodic half cell coupled to a yeast anodic half cell. *Energy Sources, Part A: Recovery, Utilization, and Environmental Effects* **33**(5): 440–448.

Powell E. E., Mapiour M. L., Evitts R. W., Hill G. A. (2009). Growth kinetics of Chlorella vulgaris and its use as a cathodic half cell. *Bioresour. Technol.* **100**: 269–274.

Revsbech N. P., Jorgensen B. B., Blackburn T. H., Cohen Y. (1983). Microelectrode studies of the photosynthesis and 02, H&5, and pH profiles of a microbial mat. *Limnol. Oceanogr.* **28**(6): 1062–1074.

Rotaru A. E., Shrestha P. M., Liu F., Markovaite B., Chen S., Nevin K. P., Lovley D. R. (2014). Direct interspecies electron transfer between *Geobacter metallireducens* and *Methanosarcina barkeri. App. Environ. Microbiol.* **80**(15): 4599–4605.

Rotaru A. E., Shrestha P. M., Liu F., Ueki T., Nevin K., Summers Z. M., Lovley D. R. (2012). Interspecies electron transfer via hydrogen and formate rather than direct electrical connections in cocultures of *Pelobacter carbinolicus* and *Geobacter sulfurreducens. App. Environ. Microbiol.* **78**(21): 7645–7651.

Roxby D. N., Yuan Z., Krishnamoorthy S., Wu P., Tu W.-C., Chang G.-E., Lau R., Chen Y.-C. (2020). Enhanced biophotocurrent generation in living photosynthetic optical resonator. *Adv. Sci.* **7**: 1903707.

Saratale R. G., Kuppam C., Mudhoo A., Saratale G. D., Periyasamy S., Zhen G., Koók L., Bakonyi P., Nemestóthy N., Kumar G. (2017). Bioelectrochemical systems using microalgae—a concise research update. *Chemosphere* **177**: 35–43.

Schofield Z., Meloni G. N., Tran T., Zerfass C., Sena G., Hayashi Y., Grant M., Contera S. A., Minteer S. D., Kim M., Prindle A., Rocha P., Djamgoz M. B. A., Pilizota T., Unwin P. R., Asally M., Soyer O. S. (2020). Bioelectrical understanding and engineering of cell biology. *J. R. Soc. Interface.* **17**: 20200013.

Song X., Wang W., Cao X., Wang Y., Zou L., Ge X., Zhao Y., Si Z., Wang Y. (2020). *Chlorella vulgaris* on the cathode promoted the performance of sediment microbial fuel cells for electrogenesis and pollutant removal. *Sci. Total Environ.* **728**: 138011.

Sun J., Li N., Yang P., Zhang Y., Yuan Y., Lu X., Zhang H. (2020). Simultaneous antibiotic degradation, nitrogen removal and power generation in a microalgae-bacteria powered biofuel cell designed for aquaculture wastewater treatment and energy recovery. *Int. J. Hydrog. Energy*. **45**(18): 10871–10881.

Sun J., Xu W., Yuan Y., Lu X., Kjellerup B. V., Xu X., Zhang H., Zhang Y. (2019). Bioelectrical power generation coupled with high-strength nitrogen removal using a photo-bioelectrochemical fuel cell under oxytetracycline stress. *Electrochim. Acta* **299**: 500–508.

Thrash J. C., Coates J. D. (2008). Review: Direct and indirect electrical stimulation of microbial metabolism. *Environ. Sci. Technol.* **42**(11): 3921–3931.

Varanasi J. L., Prasad S., Singh H., Das D. (2020). Improvement of bioelectricity generation and microalgal productivity with concomitant wastewater treatment in flat-plate microbial carbon capture cell. *Fuel* **263**: 116696.

Vassilev I., Hernandez P. A., Batlle-Vilanova P., Freguia S., Krömer J. O., Keller J., Ledezma P., Virdis, B. (2018). Microbial electrosynthesis of isobutyric, butyric, caproic acids, and corresponding alcohols from carbon dioxide. *ACS Sustain. Chem. Eng*. **6**(7): 8485–8493.

Venkata M. S., Srikanth S., Chiranjeevi P., Arora S., Chandra R. (2014). Algal biocathode for in situ terminal electron acceptor (TEA) production: Synergetic association of bacteria-microalgae metabolism for the functioning of biofuel cell. *Bioresour. Technol.* **166**: 566–574.

Velasquez–Orta S. B., Curtis T. P., Logan B. E. (2009). Energy from algae using microbial fuel cells. *Biotechnol. Bioeng*. **103**: 1068–1076.

Wang X., Feng Y., Liu J., Lee H., Li C., Li N., Ren N. (2010). Sequestration of CO2 discharged from anode by algal cathode in microbial carbon capture cells (MCCs). *Biosens. Bioelectron.* **25**: 2639–2643.

Winfield J., Greenman J., Huson D., Ieropoulos I. (2013). Comparing terracotta and earthenware for multiple functionalities in microbial fuel cells. *Bioprocess Biosyst. Eng*. **36**: 1913–1921.

Wu Y., Wang Z., Zheng Y., Xiao Y., Yang Z., Zhao F. (2014). Light intensity affects the performance of photo microbial fuel cells with Desmodesmus sp. A8 as cathodic microorganism. *Appl. Energy*. **116**: 86–90.

Xiao L., Young E. B., Berges J. A., He Z. (2012). Integrated Photo-Bioelectrochemical System for Contaminants Removal and Bioenergy Production. *Environ. Sci. Technol.* **46**: 11459–11466.

Xiao L., Young E. B., Grothjan J. J., Lyon S., Zhang H., He Z. (2015). Wastewater treatment and microbial communities in an integrated photo-bioelectrochemical system affected by different wastewater algal inocula. *Algal Res.* **12**: 446–454.

Yadav G., Sharma I., Ghangrekar M., Sena R. (2020). A live bio-cathode to enhance power output steered by bacteria-microalgae synergistic metabolism in microbial fuel cell. *J. Power Sources*. **449**: 227560.

Yang Z., Pei H., Hou Q., Jiang L., Zhang L., Nie C. (2017). Algal biofilm-assisted microbial fuel cell to enhance domestic wastewater treatment: Nutrient, organics removal and bioenergy production. *Chem. Eng. J.* **332**: 277–285.

Zhang G., Zhao Q., Jiao Y., Wang K., Lee D.-J., Ren N. (2012). Biocathode microbial fuel cell for efficient electricity recovery from dairy manure. *Biosens. Bioelectron.* **31**(1): 537–543.

Zhang Y., Zhao Y., Zhou M. (2019). A photosynthetic algal microbial fuel cell for treating swine wastewater. *Environ. Sci. Pollution Res.* doi:10.1007/s11356-018-3960-4.

Zhou M., He H., Jin T., Wang H. (2012). Power generation enhancement in novel microbial carbon capture cells with immobilized *Chlorella vulgaris*. *J. Power Sources*. **214**: 216–219.

8 Phycoremediation
A Prospective Technique for Human Welfare

B. Sankaran

8.1 INTRODUCTION

Eco-deterioration occurs worldwide due to continuous human interventions. Mankind is wondering whether their offspring would be able to survive up to the next two to three centuries. Human technologies and its applications are somehow or other responsible for environmental pollution. The developed nations are the major contributors for the maximum of it, and they are the one seeking for early solutions. Population explosion in the developing world also accounts for environmental pollution. In the act of catching up with the developed nations, the developing nations do contribute to an extent for ecological damage. The magnitude of resource consumption and the pollutants generated due to human activities are causing havoc to the biosphere. It is very difficult for humans to reduce the use of natural resource. In the conquest for the so called development, humans fail to cease the continuous impairment they cause to this heaven called Earth. Any further damage to Earth is going to affect every individual irrespective of their age, gender, religion, status, etc. and their surroundings.

Humans have sensed the alarming damage caused to the biosphere at least in the middle of the 20th century. International conventions and conferences were continuously organized in the later part of the 20th century for the reduction in environmental pollution. Microbe-driven technologies will be a solution for this long-term misery. Microbes have been an answer for pollution abatement for a long time now. In fact, they can fulfil the dream of Sustainable development.

The term phycoremediation was introduced by **John** (**2000**). The term phycoremediation is used to denote the remediation (removal, degradation, assimilation, etc.) from various types of algae and cyanobacteria (**Olguín and Sánchez-Galván 2012**). Like any major discoveries, phycoremediation is a replica of nature. Algal blooms occur in water bodies with high nutrient content. The intrinsic property of algae to effectively remove nutrients, metals, and organic compounds from wastewater is making them very suitable for bioremediation (**Laurens *et al.*, 2017**). Humans have utilized this inherent potential of microalgae to grow in nutrient-rich water for the purpose of bioremediation. High nutrient content in wastewater is very essential for phycoremediation as it will augment the growth of microalgae and aiding in removal of wastes.

Microalgae have been extensively used for various wastewater treatments since 1950s (**Oswald *et al.*, 1953**; **Oswald and Gotaas, 1957**; **Oswald and Golueke, 1960**). Phycoremediation is effectively used to remediate the acidic and metal wastewaters and industrial effluents by removing the heavy metals, organics, inorganics, xenobiotics, etc. Heavy metals can be removed from water by microalgae as they accumulate dissolved metals (**Ting, 1989**). Microalgae are very effective in sequestering of heavy metals as they have a large surface-to-volume ratio. They adsorb or absorb heavy metals and sediments in water bodies and in turn help in reduction of its toxicity in the water bodies. Globally treated wastewater is reused for various purposes, and it contributes to 5–10% of water supply (**Wurochekke *et al.*, 2016**).

DOI: 10.1201/9781003219156-11

8.2 WATER POLLUTION—A MAJOR THREAT TO MAN AND ENVIRONMENT

India is a water rich nation, which is endowed with river networks and Himalayan snow cover which suffice the water requirement of it. About 70% of its surface water resources and ground water are contaminated with disease-causing microbes and toxic pollutants (**Kumar and Murty, 2011**). The story of the world is the same with high water pollution rates. Water has become unsafe for common usage like consumption, irrigation, etc. Water gets polluted due to domestic and industrial usage in urban areas and agriculture in rural areas. Millions of liters of wastewater enter rivers and water bodies in India and impacts the economic growth of the concerned area. Directly it affects the agriculture yield of every nation. Environmental degradation and health issues will add up to the disaster.

Pollutants in water can be categorized into organic and inorganic pollutants. Organic pollutants include hydrocarbons, polychlorinated biphenyls, insecticides, detergents, etc. Inorganic pollutants include toxic heavy metals, nonmetallic nitrates and phosphates, radioactive isotopes, etc.

We need to preserve the quality of water as it gives us many benefits like fisheries, reduction of water-borne diseases, pure water supply for domestic, industries, and agriculture. In turn the biodiversity of water is preserved. Water is a natural resource with a regenerative capacity, and it can tolerate the pollutant load to an extent without compromising its quality. If the load goes beyond its regenerative capacity, it gets polluted.

8.2.1 Wastewater Treatment

Wastewater generated in both domestic and industrial waste contains organic and inorganic matter (dissolved and suspended), microbes (pathogenic and nonpathogenic). Conventional wastewater treatment method is performed in a stepwise process. Primary treatment is carried out using both physical and chemical methods to remove nonbiodegradable solid wastes and hazardous chemicals. Gravitation aids in settlement of solid wastes and formation of primary sludge in settlement tanks. The supernatant is less dense and has toxic chemicals which are neutralized using chemicals. Primary treatment is the most essential and non-avoidable step in wastewater treatment. Secondary treatment uses bacteria and fungi to remove both suspended and dissolved organic matter. Microalgae are presently thought of as a replacement, and it can be used either separately or as a biological consortium with other microbes. Being eco-friendly and cost-effective, they are the most preferred method as compared to usage of costly eco-toxic chemicals.

8.2.2 Phycoremediation vs Conventional Methods

The conventional methods involve both physical and chemical. The soluble heavy metal wastes are removed by expensive methods like electrochemical treatment, evaporative recovery, ion-exchange and precipitation. This is performed when the heavy metal contamination is between 10 and 100 ppm. Microbes including algae are capable of concentrating heavy metals many times than their surroundings. Nutrients like phosphates and nitrates are removed using crystallization and polymer hydrogels. Electrochemical denitrification is used to remove nitrates (**Zhang *et al.*, 2005**).

Nutrient removal with the aid of microalgae compares very favorably to other conventional technologies (**Levoie A and De la Noue, 1985**). It is capable of treating wastewaters both efficiently and economically without utilizing expensive energy-intensive processes as compared with conventional technologies (**Mayuri Chabukdhara, 2017**). In conventional bacterial treatment, both aerobic and anaerobic microbes used will not be harvested for further use. They generate very high amounts of activated sludge. Phycoremediation uses a technique which is both cost-effective and needs lesser energy inputs. The skilled labor requirement is less in phycoremediation.

Phycoremediation has several advantages over currently available chemical technologies due to its flexibility for a wide range of applications, ability to reduce heavy metals, and remove harmful pollutants and pathogens from wastewater, economical and easily removed from after remediation.

Phycoremediation not only improves the wastewater properties, it does not require high energy inputs as in methods already available for bioremediation. Algal cultivation at a large scale to remove wastes can lead to increased production of its biomass, which can be used as biofuel alternatives or biofertilizers (**Milano *et al.*, 2016**).

8.2.3 How Algae Thrive in Wastewater?

The presence of algae in stagnant wastewater infers that they can survive in wastewater. The presence of microalgae has been studied for many years now. **John** (**2000**) found green algae (*Mougeotia*) in highly acidic water pools in coal mines of Coolie (Australia). The green alga present there were a hyper-accumulator of heavy metals and capable of sequestering iron by biosorption. Microalgae grow well in water bodies with a wide range of organic and inorganic pollutants.

8.2.3.1 Mechanism of Heavy Metal Uptake

They have a high tolerance to heavy metals. Large surface area to volume ratio, phototactic movement, phytochelation and genetic manipulation of algal cells helps it in heavy metal utilization (**Cai *et al.*, 1995**). Heavy metal accumulation is comparable or higher than chemical sorbents, activated carbon, ion-exchange resin, zeolite, etc. (**Volesky, 1997**; **Mehta and Gaur, 2005**). Heavy metal is the most prevalent pollutant present in soil, water bodies, and its sediments. Microbial uptake of metal ions occurs by two different mechanisms, namely, biosorption and bioaccumulation.

8.2.3.1.1 Biosorption

Biosorption is passive physical attachment (physical adsorption) of heavy metals to biomolecules by chelation, complexation, or ion exchange. The cell walls of microalgae have functional groups to bind with heavy metals. It is a fast and reversible reaction. It is followed by a slower process called chemisorption. Dead cells accumulate heavy metals like the living cells and sometimes to a greater extent than living cells (**Leusch *et al.*, 1995**).

8.2.3.1.2 Bioaccumulation

It is a slower process in which the metal ions are taken within the living cell by active transport. Metal ion transporter protein helps the transport of heavy metals into microalgae through cell membrane. Bioaccumulation is dependent on the growth of the organism. Phytoplanktons affect trace metal chemistry of water bodies by both metal uptake and forming metal complexes with its organic exudates like polyhydroxamate siderophores and keeping a check on their concentration (**McKnight, 1980**; **Trick et al., 1983**). Microalgae hyper accumulate metals by a combination of various mechanisms like adsorption, active uptake, photosynthetic oxidation (**John, 2003**).

8.2.3.2 Mechanism of Nutrient Removal

Nitrogen or phosphorous are the major nutrients needed for growth and reproduction of microalgae. Nitrogen is more efficiently removed from gray water by microalgae than the phosphorous using anabolic pathway. Nitrogen diffuses faster through cell membrane than phosphorous (**Wurochekke *et al.*, 2019**).

Inorganic nutrients like nitrogen and phosphorous are a good nutrient source for luxuriant growth of microalgae. When algae grows in wastewater, it increases the pH, and it strips ammonia and precipitates phosphate and uses inorganic nitrogen and phosphorous from it for their assimilation (**P. S. Lau *et al.*, 1995**). Phosphorous can be extracted from domestic sewage using algae. *Chlorella* and *Scenedesmus* can remove 80% of ammonia, nitrite, and phosphorous from wastewater (**Ruiz-Marin *et al.*, 2010**).

8.3 ALGAE USED IN PHYCOREMEDIATION

Algae have been suitable for treating wastewater arising from different industries. Micro and macroalgae have the capability to bioremediate different kinds of pollutants. Microalgae have more advantages as compared with macroalgae. Microalgae usage is more convenient for phycoremediation due to its growth pace and ease in product recovery. Application of macroalgae is in situ as compared to microalgae, which is both in situ and ex situ. The species count of microalgae outlasts the macroalgae. The options provided by microalgae are plenty.

Microalgae can be used individually or in combination with bacteria as a consortium. The removal of heavy metals, pathogens, nutrients, organic matter from domestic wastewater is enhanced by microalgae, and it is used for production of high-value products. Microalgae produce oxygen by photosynthesis, which aids in aerobic respiration (**Raul Muñoz and Benoit Guieysse, 2006**). Few examples of microalgae utilized in phycoremdiation and removal of heavy metals are shown in Tables 8.1 and 8.2.

TABLE 8.1
Few Microalgae Used in Industrial Wastewater Treatment

S. No.	Algae	Industry	Reference
1	*Chroococcus*	Alginate	Sivasubramanian *et al.* (2009)
2	*Chlorella*	Beverage (Alcohol)	Alexei Solovchenko *et al.* (2014)
3	*Acutodesmus*, *Diplosphaera*	Dairy	Liu *et al.* (2016) Kaumeel Choksi *et al.* (2016)
4	*Chlorella*	Crude oil	Das and Deka (2019)
5	*Chlorococcum*	Detergent	V Sivasubramanian *et al.* (2011)
6	*Chlorella*, *Auxanochlorella*, *Monoraphidium*, Cyanobacteria	Domestic and municipal wastewater	Wang *et al.* (2010), Zhou *et al.* (2012), Hage *et al.* (2018), Badr, Omnia *et al.* (2019)
7	*Chlorella*, *Scenedesmus*	Leather tannery	Hanumantha Rao *et al.* (2011), Ballén-Segura, M *et al.* (2017)
8	*Scenedesmus*	Olive oil	Hodaifa *et al.* (2013)
9	*Chlorella*	Paper and Pulp	Dilek Feliz *et al.* (1999), Yel E *et al.* (2002)
10	Green algae Cyanobacteria	Pesticides	Cáceres *et al.* (2008) Hussein *et al.* (2016)
11	*Scenedesmus*, *Nannochloris*	Piggery waste	Jimenez-Perez *et al.* (2004)
12	*Anabaena*, *Chlorella*, *Nostoc*	Textile	Chu *et al.* (2009) Elsadany A (2018)

TABLE 8.2
Microalgae Utilized in Phycoremediation of Heavy Metals

S. No.	Microalgae	Heavy metal	Reference
1	*Botryococcus*	Arsenic	Podder and Majumder (2017)
2	*Merismopedia*	Cadmium	Imani *et al.* (2011) Dixit and Singh (2013) Mustafa A. Fawzy (2016)
3	*Chlorella*	Chromium	P. Rau *et al.* (2013)

TABLE 8.2 (*Continued*)
Microalgae Utilized in Phycoremediation of Heavy Metals

S. No.	Microalgae	Heavy metal	Reference
4	*Merismopedia*	Copper	Mustafa A. Fawzy (2016)
5	*Nostoc*	Lead	Imani *et al.* (2011) Dixit and Singh (2013)
6	*Oocystis*	Lithium	El Naggar *et al.* (2019)
7	*Amphora* *Dunaliella*	Nickel	Dahmen Ben Mousa *et al.* (2017)
8	*Chlorella*, *Dunaliella*	Titanium	Marchello *et al.* (2018) Ghazaei *et al.* (2020)
9	*Richterella* *Spirogyra*	Zinc	Abioye, Peter *et al.* (2015)

8.4 PHYCOREMEDIATION TECHNOLOGY

Secondary biological treatment of wastewater is performed using the following methods, like activated sludge systems, aerobic biofilters, anaerobic bioreactors, algal stabilization ponds, and land disposal systems (**Jais *et al.*, 2017**).

8.4.1 Activated Sludge Systems

The process deals with biological treatment of sewage and industrial wastewaters. It consists of three main components—aeration tank (bioreactor), settling tank (separating solids and treated wastewater), and activated sludge equipment (return settled activated sludge to aeration tank) (**Ardern and Lockett, 1914**).

8.4.2 Aerobic Biofilters

Aerobic bioreactor is a filtration system which uses living cells to capture and degrade pollutants. It is commonly used in dairy and domestic wastewater treatment and gray water recycling. Aerobic biofilter has a biofilm with polymeric substances, sand, rock, etc. Algae or microbes are attached to a solid bed of media. Wastewater is passed on through the biofilm for treatment. Aerobic biofilters are used in closed aquaculture systems to treat animal wastes.

8.4.3 Anaerobic Bioreactors

An anaerobic bioreactor possesses filtration membrane which filters out biomass, suspended and inert solids. It converts the waste BOD to biogas and reserves nitrogen and phosphorous for reuse (**Chang, 2014**). They can be used in treating effluents of agricultural, food, and beverage industries. It has its application in slaughterhouse, breweries, leather tanneries, dairies, sugar mills and starch, yeast and soft drink production units, processing units of potato, coffee, fruit, fish, and vegetables.

8.4.4 Algal Stabilization Ponds

Stabilization ponds are manmade ponds built for reducing organic content and removing pathogens from wastewater. The waste enters through inlet and after many days passes out through an outlet. It is used to treat sewage, stormwater, and biodegradable wastes. The pond consists of algae to treat the wastes. The tank is provided with mechanical aeration.

8.4.5 Algal Immobilization

Immobilization technique can be done in two ways, namely, passive and active. Passive technique is exploitation of the surface adhering capability of microalgae. Plastic, glass, and wood can be used to immobilize algae. Active technique is entrapping live or dead cells in synthetic or natural polymers. Polyurethane, polysulfone, polyvinyl alcohol, and epoxy resin are synthetic polymers used to entrap microalgae. Natural polymers used for encapsulating algae are alginate, agar, carrageenan, and proteins. Silica gel is a non-polymeric synthetic material used for algal immobilization (**Moreno-Garrido, 2013**). The removal of algae after treatment and recycling of some inoculum can be avoided when immobilization technique is used (**de la Noüe *et al.*, 1992**).

8.4.6 Land Disposal Systems

After primary and secondary treatment, the effluent is released to a land area. Microbes or algae in the land filters treated effluent as they pass through the soil. The treated effluent is used for the purpose of irrigation, and they are not potable.

8.5 PRODUCTS OBTAINED FROM ALGAE USED IN PHYCOREMEDIATION

Humans have been exploring nature for products of utilization from prehistoric times. His search was from both nonliving and living matter. Initially his search was for the food requirement. Later on he started for looking products for application in various allied areas. Bioprospecting is exploring the options for availability of products in living organisms. Algae are continuously explored for products. Already there are innumerable algae-based products available in the market. Phycoremediation helps us to take the value-added products we have been using from residual algal biomass.

8.5.1 Food and Feed

Microalgae have been consumed by human beings, livestock, and aquaculture. Microalgae can be readily used as feedstock for animals. They are rich in primary metabolites like proteins, carbohydrates, and fats. Biomass obtained through phycoremediation is inexpensive and can be used as a feed to cattle, fish, poultry, pig, etc. The benefit of using microalgae as feed is obtaining a better animal for human consumption both quantitatively and qualitatively (**Pulz and Gross, 2004**). Microalgae obtained from phycoremediation will have the residue of the contaminant in which it is grown. It is difficult to directly use it for consumption.

8.5.2 Renewable Energy

Algal biomass has three major components like carbohydrates, lipids, and proteins. Biomethane potential is known based on the biochemical composition. Based on the wastewater in which they are cultivated, the biochemical composition varies. When algae are grown in nitrogen and phosphorous starvation, they will accumulate lipid (**Menon *et al.*, 2013**) and carbohydrates (**Markou *et al.*, 2013**) respectively. A balanced biochemical composition of carbohydrates, lipids, and proteins is essential for biomethane production. *Chroococcus* biomass is an excellent microalgae with balance biochemical composition and suitable for biomethane production (**Prajapati *et al.*, 2014**).

The process of hydrogen metabolism by green algae was discovered by **Gaffron, 1944**. Biohydrogen production is the most achievable as they are the by-product of carbon fixation. The use of hydrogen as a fuel helps to reduce air pollution to a great extent. Hydrogen gas gets accumulated through systematic steps using *Chlamydomonas* (**Melis and Happe, 2001**). Usage of hydrogen indirectly reduces the dependence on fossil fuels, and that will reduce global warming.

Recently there has been an explosion in research possibilities of biodiesel production using microalgae. Microalgae through transesterification process produced high amount of lipids, and this potential is tapped for biodiesel production. The cost of biodiesel production using microalgae is expensive. Socio-political factors and public policies play a big role in feasibility of biodiesel production using microalgae (**Vincent Amanor-Boadu *et al.*, 2014**).

8.5.3 PUFA

Polyunsaturated fats are healthy essential fatty acids with omega-3 and omega-6 fats. Humans cannot synthesize these fatty acids and get it from their nutrition. PUFA requires for the function of human heart (**Ander *et al.*, 2003**; **Abdelhamid *et al.*, 2018**) and brain (**Bentsen, 2017**). The PUFA contents of *Isochrysis*, *Phaeodactylum*, and *Porphyridium* were to the standards of cod liver oil (**Handayani *et al.*, 2011**). The PUFA content of *Spirulina*, *Chlorella*, *Haematococcus*, and *Chlamydomonas* was analyzed by **Abdo Sayeda *et al.* (2015**).

8.5.4 Astaxanthin

Astaxanthin is an antioxidant used as a food supplement. It prevents many physiological disorders like diabetes (**Landon *et al.*, 2020**), cardiac (**Fassett and Coombes, 2012**), and neurodegenerative diseases like Alzheimer's and Parkinson's (**Galasso *et al.*, 2018**; **Syed Obaidur Rahman *et al.*, 2019**). It increases the immunity in both humans and animals. Astaxanthin is obtained from *Haematococcus pluvialis*, *Chlorella zofingiensis*, *Chlorococcum*, etc. (**Ambati *et al.*, 2014**). The efficiency of *Chlorococcum*, *Chlorella*, etc., in phycoremediation has been studied. It is already available for oral medications in different forms, like capsules, powders, syrups, tablets, etc. Topical creams and gels are also available for use. It is one of the most important carotenoid available for commercial purposes. It is used in skin care and has its application in cosmetic industry.

8.6 ENVIRONMENTAL SIGNIFICANCE OF PHYCOREMEDIATION

Phycoremediation qualifies for clean development mechanism originated in Kyoto protocol. It aids in reduction of greenhouse gas in the atmosphere by carbon sequestration while growing in industrial wastewater and flue gas released from power generators. It is a green technology available for the effective utilization in industries. Industrialized nations can earn certified emission reduction saleable credits by using CDM (**IPCC, 2007**) in developing countries and meet their emission reduction targets.

8.6.1 Carbon Footprint

The major concern of humans today is to reduce both greenhouse gas emission and freshwater consumption. As they are going to directly affect the survival of the future human generation. It is a right time to reduce the carbon and water footprint left by humans in the name of human development. Every human activity is directly or indirectly involved in release of greenhouse gases into the atmosphere. The measure of amount of greenhouse, particularly carbon dioxide, released due to human activity as an individual or an organization is known as carbon footprint.

Carbon credit is an attempt made by international traders to diminish the concentrations of greenhouse gases in the atmosphere. One carbon credit is equal to one ton of carbon dioxide or its equivalent gases. Microalgae used in phycoremediation can pave the way to generate carbon credits. In fact, it will help the establishments involving in phycoremediation to sell their carbon credits to commercial companies who want to lower their carbon footprint. Carbon markets help in reduction of greenhouse gases by trading carbon credits.

8.6.2 Carbon Sequestration

Carbon dioxide is the foremost greenhouse gas responsible for global warming. An effective technique to remove it from atmosphere is the need of the hour. At present carbon dioxide is injected into the geological formations and deep oceans. The most effective way to mitigate the global atmospheric carbon dioxide level is to utilize the producers of the biosphere. Both green plants and algae have the capability to fix it from atmosphere during its photosynthesis and in turn help in reducing the carbon released by humans. Microalgae are the best alternative for carbon fixation due to its high productivity, fast growth rate and adaptability to the surroundings, low cost in maintenance, and ease in handling. It can be used for both capture and storage of carbon dioxide (**Razzak *et al.*, 2013**; Zhao and Su, 2014). In fact, algae are good carbon sinks. About 1.8 tons of carbon dioxide are needed for production of one ton of algal biomass (**Yusuf, 2007**; Li *et al.*, 2008). Phycoremediation will not only reduce the atmospheric load of carbon dioxide but also increase the algal biomass for recovering many products. When compared with other microbes, microalgae are efficient in carbon dioxide sequestration (**Olguín, 2010**).

8.6.3 Water Footprint

Water footprint is the measure of water consumed, evaporated, and polluted by humans to produce goods and services. Water footprint is split into three categories as blue, green, and gray water footprint. Blue water footprint is the amount of surface water or ground water required to produce an item. Green water footprint is the amount of rainwater required to make an item. Gray water footprint is the amount of freshwater required to dilute the wastewater generated in manufacturing in order to maintain the water quality.

Phycoremediation will certainly not directly act on water footprint. In fact, it will not help in reduction of freshwater consumption by humans. But it is going to increase the quantity of freshwater available for human activities. It will reduce the load of water footprint created by mankind. When wastewater is treated with microalgae, they utilize the nutrients available in it for their growth. They convert the wastewater into usable water.

8.6.4 Pollution Control

Phycoremediation can control air and water pollution simultaneously. When microalgae are used for bioremediation of industrial wastes, they utilize the carbon dioxide from atmosphere and reduce greenhouse gas. Industrial flue gas is utilized as carbon source for the microalgae growing in industrial wastewater is an integrated approach to mitigate both air and water pollution (**Panga Kiran kumar *et al.*, 2018**). Microalgae remove water pollutants from the effluents and purify water into usable form. Microalgae remove the pollutants from wastewater and industrial effluent. Pollutants like xenobiotics, heavy metals, nutrients are removed efficiently by them. It neutralizes toxic wastes or detoxifies or transforms them into nontoxic forms.

Blue-green algae have been fixing atmospheric nitrogen in nature. They have been used as biofertilizers for a very long time (*Chlorella*–manure).

8.7 ADVANTAGES OF PHYCOREMEDIATION

Microalgae have a dual advantage of both reducing the toxicity of the effluent and its biomass yield used in production of many products. The technique is eco-friendly and removes impurities alone form the effluent and do not disturb the environment. Highly tolerant organisms to the variations caused due to addition of effluents with different quality. The technique is very flexible so that different bioreactors (batch, semicontinuous, or continuous) can be used as per the need and quality of the effluent. The technique requires a proper land, water, and nutrients for its application and do not interfere in other operations in industry. Phycoremediation is a simple technique and do not require any skilled labor for operations. It reduces global warming due to the carbon sequestering ability of algae for photosynthesis, and during which the oxygen release purifies the air. It is a cost-effective technique avoiding use of harmful costly chemicals. Microalgae uptakes heavy metals through biosorption and bioaccumulation. Heavy metals can be recovered and used for other purposes. The products like biofuels, biofertilizers, biochemicals obtained from microalgae can be used as food, fodder in nutraceuticals, and pharmaceutical industry after proper utilization. Algae have a very high primary productivity potential, and it will help us to get more useful wastes from the biomass used for treatment.

There is a need for more scientific insight to tap many more possibilities which can be unearthed.

8.8 MAJOR CHALLENGES

Phycoremediation is not an exception even it has its own pitfalls. Challenges can be classified into two like before, during, and after phycoremediation.

Challenges before phycoremediation

1. The availability of space for algal culture
2. Finding suitable microalgae for the purpose of phycoremediation
3. Determining the effluent characteristics
4. Genetically modified forms cannot be used as they are capable of adapting to conditions faster

Challenges during phycoremediation

5. When compared with other microbes used in bioremediation, the speed of growth of algae is relatively slower
6. Usage of antibiotics to maintain sterilized condition
7. Continuous supply of carbon dioxide in a bioreactor needs extra efforts and high cost
8. Maintaining temperature and availability of high quality light in photobioreactors

Challenges after phycoremediation

9. Disposal of microalgae which is adsorbed and absorbed heavy metals and pollutants
10. Usage of microalgae for production of food grade products as it will have adsorbed or absorbed pollutants
11. As they grow in contaminated effluent, there will be a low biomass production, and recovery of products in high quantity will be difficult
12. The products to be obtained will be lesser inside the algal cells
13. The downstream processing of harvesting algal cells and recovering the products will be expensive
14. If treated water has a fast-growing microalga, it will affect the aquatic ecosystem in which it is let off

8.9 PHYCOREMEDIATION—THE FUTURE OF HUMAN SURVIVAL

Phycoremediation should be the most sought-out technique for effluent treatment in all the industries with a few exceptions. It is a powerful technique used to reduce BOD, COD, sludge from industrial effluent. It reduces the operational cost of effluent treatment and corrects the pH. This technique helps to remove color, odor, and excess nutrients present in the effluent. The ability to fix carbon dioxide and nutrient uptake with the advantage of less energy and cost utilization makes phycoremediation a long-term sustainable method (**Hirata *et al.*, 1996; Murakami and Ikenouchi, 1997**).

Microalgae biomass after phycoremediation contains the nutrients and pollutants obtained from effluent which can be recycled for useful purposes (**Pizarro *et al.*, 2006**).

8.10 DIRECT AND INDIRECT BENEFITS OF PHYCOREMEDIATION

Water pollution

(i) It removes nutrients from effluents rich in organic matter
(ii) Removal and neutralizing of xenobiotic compounds
(iii) Acidic and metal wastewater treatment
(iv) Algal indicators can be used in biosensors to detect toxic compounds
(v) pH correction

Air pollution

(vi) Reduction in atmospheric CO_2 concentration
(vii) Usage of algae in biochimneys

Economy

(viii) Biofuel
(ix) Biofertilizers
(x) Pharmaceuticals
(xi) Nutraceuticals

Ecological services

(xii) Ecosystem restoration
(xiii) Soil conditioning agents
(xiv) Soil reclamation
(xv) Atmospheric oxygen restoration

REFERENCES

Abdelhamid, A. S., Martin, N., Bridges, C., Brainard, J. S., Wang, X., Brown, T. J., Hanson, S., Jimoh, O. F., Ajabnoor, S. M., Deane, K. H., Song, F., and Hooper, L. (2018). Polyunsaturated fatty acids for the primary and secondary prevention of cardiovascular disease. The Cochrane Database of Systematic Reviews, 7(7), CD012345.

Abdo, S., Ali, G., and El-Baz, F. (2015). Potential production of omega fatty acids from microalgae. International Journal of Pharmaceutical Sciences Review and Research, 34, 210–215.

Abioye, P., Adeoye, B. S., Abiodun, A., and Oyewole, O. (2015). Phycoremediation of zinc by Spirogyra and Richterella sp. isolated from pond. International Journal of Biochemistry and Molecular Biology. https://www.researchgate.net/publication/292607905

Ambati, R. R., Phang, S. M., Ravi, S., and Aswathanarayana, R. G. (2014). Astaxanthin: Sources, extraction, stability, biological activities and its commercial applications—a review. Marine Drugs, 12(1), 128–152.

Ander, B. P., Dupasquier, C. M., Prociuk, M. A., and Pierce, G. N. (2003). Polyunsaturated fatty acids and their effects on cardiovascular disease. Experimental and Clinical Cardiology, 8(4), 164–172.

Ardern, E., and Lockett, W. T. (1914). Experiments on the oxidation of sewage without the aid of filters. Journal of Society Chemical Indian, 33(10), 523–539.

Badr, O., Elshawaf, I., El-Garhy, H., Moustafa, M., and Farid, O. (2019). Antioxidant activity and phycoremediation ability of four cyanobacterial isolates obtained from a stressed aquatic system. Molecular Phylogenetics and Evolution, 134. doi:10.1016/j.ympev. 2019.01.018.

Ballén-Segura, M., Hernández, L., Parra, D. Vega, A., and Pérez, K. (2017). Using Scenedesmus sp. For the phycoremediation of tannery wastewater. Tecciencia, 12(21), 69–75.

Bentsen, H. (2017). Dietary polyunsaturated fatty acids, brain function and mental health. Microbial Ecology in Health and Disease, 28(Supl), 1281916.

Cáceres, T. P., Megharaj, M., and Naidu, R. (2008). Biodegradation of the pesticide fenamiphos by ten different species of green algae and cyanobacteria. Current Microbiology, 57(6), 643–646.

Cai, X. H., Logan, T., Gustafson, T., Traina, S., and Sayre, R. T. (1995). Applications of eukaryotic algae for the removal of heavy metals from water. Molecular Marine Biology and Biotechnology, 4, 338–344.

Chabukdhara, M., Kumar Gupta, S., and Gogoi, M. (2017). Phycoremediation of heavy metals coupled with generation of bioenergy. In: Gupta, S. K. et al. (eds), Algal biofuels. doi:10.1007/978-3-319-51010-1_9 © Springer International Publishing AG.

Chang, S. (2014). Anaerobic Membrane Bioreactors (AnMBR) for wastewater treatment. Advances in Chemical Engineering and Science, 4, 56–61.

Chu, W. L., See, Y. C., and Phang, S. M. (2009). Use of immobilised *Chlorella vulgaris* for the removal of colour from textile dyes. Journal of Applied Phycology, 21, 641.

Dahmen-Ben Moussa, I., Athmouni, K., Chtourou, H. *et al.* (2017). Phycoremediation potential, physiological, and biochemical response of *Amphora subtropica* and *Dunaliella* sp. to nickel pollution. Journal of Applied Phycology, 30, 931–941.

Das, B., and Deka, S. (2019). A cost-effective and environmentally sustainable process for phycoremediation of oil field formation water for its safe disposal and reuse. Scientific Reports, 9, 15232. https://doi.org/10.1038/s41598-019-51806-5.

de la Noüe., J, Laliberte, G., and Proulx., D. (1992). Algae and waste water. Journal of Applied Phycology, 4(3), 247–254.

Dilek, F., Taplamacioglu, H., and Tarlan, E. (1999). Colour and AOX removal from pulping effluents by algae. Applied Microbiology and Biotechnology, 52, 585–591.

Dixit, S., and Singh, D. P. (2013). Phycoremediation of lead and cadmium by employing Nostoc muscorum as biosorbent and optimization of its biosorption potential. International Journal of Phytoremediation, 15, 801–813.

El-Naggar, N. E. A., Hamouda, R. A., Rabei, N. H. *et al.* (2019). Phycoremediation of lithium ions from aqueous solutions using free and immobilized freshwater green alga *Oocystis solitaria*: Mathematical modeling for bioprocess optimization. Environmental Science and Pollution Research, 26, 19335–19351.

Elsadany, A. (2018). The use of microalgae in bioremediation of the textile wastewater effluent. Nature and Science, 16. doi:10.7537/marsnsj160318.11.

Fassett, R. G., and Coombes, J. S. (2012). Astaxanthin in cardiovascular health and disease. Molecules (Basel, Switzerland), 17(2), 2030–2048.

Gaffron, H. (1944). Photosynthesis, photoreduction and dark reduction of carbon dioxide in certain algae. Biological reviews of the Cambridge Philosophical Society, 19, 1–20.

Galasso, C., Orefice, I., Pellone, P., Cirino, P., Miele, R., Ianora, A., Brunet, C., and Sansone, C. (2018). On the neuroprotective role of astaxanthin: New perspectives? Marine Drugs, 16(8), 247.

Ghazaei, F., and Shariati, M. (2020). Effects of titanium nanoparticles on the photosynthesis, respiration, and physiological parameters in *Dunaliella salina and Dunaliella tertiolecta*. Protoplasma, 257(1) January, 75–88. doi:10.1007/s00709-019-01420-z. Epub 2019 Aug 1. PMID 31372761.

Hage, A., Luckett, N., and Holbrook, G. P. (2018). Phycoremediation of municipal wastewater by the cold-adapted microalga *Monoraphidium* sp. Dek19. Water Environment Research, 90(11), 1938–1946.

Handayani, N., Ariyanti, D., and Hadiyanto, H. (2011). Potential production of polyunsaturated fatty acids from microalgae. International Journal of Science and Engineering, 2. doi:10.12777/ijse.2.1.13–16.

Hanumantha Rao, P., Ranjith Kumar, R., Raghavan, B. G., Subramanian, V. V., and Sivasubramanian, V. (2011). Application of phycoremediation technology in the treatment of wastewater from a leather-processing chemical manufacturing facility. Water South Africa, 37, 7–14.

Hirata, S., Hayashitani, M., Taya, M., and Tone, S. (1996). Carbon dioxide fixation in batch culture of *Chlorella* sp. using a photobioreactor with a sunlight-collection device. Journal of Fermentation Bioengineering, 81, 470–472.

Hodaifa, G., Sanchez, S., Eugenia Martinez, M. P., and Orpez, R. (2013). Biomass production of *Scendesmus obliquus* from mixtures of urban and olive mill wastewaters used as culture medium. Applied Energy, 104, 345–352.

Hussein, M., Abdullah, A., Eltanahy, E., and Din, N. (2016). Phycoremediation of some pesticides by microchlorophyte alga, Chlorella sp. Journal of Fertilizers & Pesticides, 7. doi:10.4172/2471–2728.1000173.

Imani, S., Rezaei-Zarchi, S., Hashemi, M., Borna, H., Javid, A., Zand, A. M., and Abarghouei, H. B. (2011). Hg, Cd and Pb heavy metal bioremediation by *Dunaliella* alga. Journal of Medicinal Plants Research, 5, 2775–2780.

IPCC. (2007). Glossary J-P. In (book section): Annex I. In: Metz, B. *et al.* (eds), Climate Change 2007: Report of the intergovernmental panel on climate change. Cambridge University Press, Cambridge, UK and New York, NY.

Jais, N. M., Mohamed, R. M. S. R., and Al-Gheethi, A. A. (2017). The dual roles of phycoremediation of wet market wastewater for nutrients and heavy metals removal and microalgae biomass production. Clean Technologies and Environmental Policy, 19, 37–52.

Jimenez-Perez, M. V., Sánchez-Castillo, P., Romera, O., Fernandez-Moreno, D., & Pérez-Martınez, C. (2004). Growth and nutrient removal in free and immobilized planktonic green algae isolated from pig manure. Enzyme and Microbial Technology, 34(5), 392–398.

John, J. (2000). A self-sustainable remediation system for acidic mine voids In: Proceedings of the 4th international conference on diffuse pollution. International Association of Water Quality, Bangkok, 506–511.

John, J. (2003). Phycoremediation. In: Ambasht, R. S., and Ambasht, N. K. (eds), Modern trends in applied aquatic ecology. Springer, Boston, MA. https://doi.org/10.1007/978-1-4615-0221-0_6.

Kaumeel, C., Pancha, I., and Mishra, S. (2016). Microalgal biomass generation by phycoremediation of dairy industry wastewater: An integrated approach towards sustainable biofuel production. Bioresource Technology, 221, 455–460.

Kumar, S., and Murty, M. N. (2011). Water pollution in India: An economic appraisal. Environmental Science, Semantic Scholar. https://www.semanticscholar.org/paper/Water-Pollution-in-India%3A-An-Economic-Appraisal-Kumar-Murty/e33d4c0d0bc0e559345eb2bb1dba012a33e6d074#citing-papers

Landon, R., Gueguen, V., Petite, H., Letourneur, D., Pavon-Djavid, G., and Anagnostou, F. (2020). Impact of astaxanthin on diabetes pathogenesis and chronic complications. Marine Drugs, 18(7), 357.

Lau, P. S., Tam, N. F. Y., and Wong, Y. S. (1995). Effect of algal density on nutrient removal from primary settled waste water. Environmental Pollution, 89, 59–66.

Laurens, L. M., Chen-Glasser, M., and McMillan, J. D. (2017). A perspective on renewable bioenergy from photosynthetic algae as feedstock for biofuels and bioproducts. Algal Research, 24, 261–264.

Leusch, A., Holan, Z. R., and Volesky, B. (1995). Biosorption of heavy metals (Cd, Cu, Ni, Pb, Zn) by chemically-reinforced biomass of marine algae. Journal of Chemical Technology & Biotechnology, 62, 279–288.

Levoie, A., and De la Noue. (1985). Hyper concentrated cultures of Scenedesmus obliquus: A new approach for wastewater biological tertiary treatment. *Water Research*, 19, 1437–1442.

Li, Y., Horsman, M., Wu, N., Lan, C. Q., and Dubois-Calero, N. (2008). Biofuels from microalgae. Biotechnology Progress, 24(4), 815–820.

Liu, C., Subashchandrabose, S., Mallavarapu, M., Hu, Z., and Xiao, B. (2016). Diplosphaera sp. MM1—A microalga with phycoremediation and biomethane potential. Bioresource Technology, 218. doi:10.1016/j.biortech.2016.07.077.

Marchello, A. E., Barreto, D. M., and Lombardi, A. T. (2018). Effects of titanium dioxide nanoparticles in different metabolic pathways in the freshwater microalga *Chlorella sorokiniana* (Trebouxiophyceae). Water Air Soil Pollution, 229, 48.

Markou, G., Angelidaki, I., and Georgakakis, D. (2013). Carbohydrate-enriched cyanobacterial biomass as feedstock for bio-methane production through anaerobic digestion. Fuel, 111, 872–879.

McKnight, D. M., and Morel, F. M. M. (1980). Copper complexation by siderophores from filamentous blue green algae. Limnology and Oceanography, 25, 62–71.

Mehta, S. K., and Gaur, J. P. (2005). Use of algae for removing heavy metal ions from wastewater: Progress and prospects. Critical Reviews in Biotechnology, 25, 113–152.

Melis, A., and Happe, T. (2001). Hydrogen production. Green algae as a source of energy. Plant Physiology, 127, 740–748.

Menon, K. R., Balan, R., and Suraishkumar, G. K. (2013). Stress induced lipid production in *Chlorella vulgaris*: Relationship with specific intracellular reactive species levels. Biotechnology and Bioengineering, 110, 1627e1636.

Milano, J., Ong, H. C., Masjuki, H., Chong, W., Lam, M. K., et al. (2016). Microalgae biofuels as an alternative to fossil fuel for power generation. Renewable and Sustainable Energy Reviews, 58, 180–197.

Moreno-Garrido, I. (2013). Microalgal Immobilization Methods. Methods in molecular biology (Clifton, N.J.), 1051, 327–347. doi:10.1007/978-1-62703-550-7_22.

Muñoz, R., and Guieysse, B. (2006). Algal–bacterial processes for the treatment of hazardous contaminants: A review. Water Research, 40, 2799–2815.

Murakami, M., and Ikenouchi, M. (1997). The biological CO2 fixation and utilization project by RITE (2). Energy Conversion and Management, 38, S493–S497.

Mustafa, A. F. (2016). Phycoremediation and adsorption isotherms of cadmium and copper ions by Merismopedia tenuissima and their effect on growth and metabolism. Environmental Toxicology and Pharmacology, 46, 116–121.

Olguín, E. J. (2010). Phycoremediation and phytoremediation: Powerful tools for the mitigation of global change. Journal of Biotechnology, 150, 50–51. doi:10.1016/j.jbiotec.2010.08.134.

Olguín, E. J., and Sánchez-Galván, G. (2012). Heavy metal removal in phytofiltration and phycoremediation: The need to differentiate between bioadsorption and bioaccumulation. *New Biotechnol*ogy, 30, 3–8.

Oswald, W. J., and Golueke, C. G. (1960). Biological transformation of solar energy. Advances in Applied Microbiology, 2, 223–262.

Oswald, W. J., and Gotaas, H. B. (1957). Photosynthesis in sewage treatment. Transactions of the American Society for Civil Engineering, 122, 73–105.

Oswald, W. J., Gotaas, H. B., Ludwig, H. F., and Lynch, V. (1953). Algal symbiosis in oxidation ponds. Sewage and Industrial Waste, 25, 692–704.

Panga Kiran Kumar, S., Krishna, V., Verma, K., Pooja, K., Bhagawan, D., and Himabindu, V. (2018). Phycoremediation of sewage wastewater and industrial flue gases for biomass generation from microalgae. South African Journal of Chemical Engineering, 25, 133–146.

Pizarro, J., Santander, E., and Herrera, L. (2006). Nutrients measured on water bottle samples at station Bm_1997–08–30_1, Bajo Molle, Chile. Universidad Arturo Prat, Iquique, doi:10.1594/PANGAEA.547826.

Podder, M. S., and Majumder, C. B. (2017) Toxicity and bioremediation of As(III) and As(V) in the green microalgae Botryococcus braunii: A laboratory study. International Journal of Phytoremediation, 19(2), 157–173.

Prajapati, S. K., Kumar, P., Malik, A., and Vijay, V. K. (2014). Bioconversion of algae to methane and subsequent utilization of digestate for algae cultivation: A closed loop bioenergy generation process. Bioresource Technology, 158, 174–180.

Pulz, O., and Gross, W. (2004). Valuable products from biotechnology of microalgae. Applied Microbiology and Biotechnology, 57, 287–293.

Rajamani, S., Siripornadulsil, S., Falcao, V., Torres, M., Colepicolo, P., Sayre, R. (2007). Phycoremediation of Heavy Metals Using Transgenic Microalgae. In: León, R., Galván, A., Fernández, E. (eds) Transgenic Microalgae as Green Cell Factories. Advances in Experimental Medicine and Biology, vol 616. Springer, New York, NY. https://doi.org/10.1007/978-0-387-75532-8_9

Rao, P., Saisha, V., and Bhavikatti, S. S. (2013). Removal of chromium (VI) from synthetic waste water using immobilized algae. International Journal of Current Engineering and Technology. Special Issue 1 Sept. 2013. Proceedings of National Conference on, "Women in Science & Engineering" (NCWSE 2013), SDMCET Dharwad.

Razzak, S. A., Hossain, M. M., Lucky, R. A., and Bassi, A. S. (2013). Integrated CO2 capture, wastewater treatment and biofuel production by microalgae culturing—a review. Renewable Sustainable Energy Review, 27, 622–653.

Ruiz-Martin, A., Mendoza-Espinosa, L. G., and Stephenson, T. (2010). Growth and nutrient removal in free and immobilized green algae in batch and semi-continuous cultures treating real wastewater. Bioresource Technology, 101, 58–64.

Sivasubramanian, V., V.V. Subramanian, B.G. Raghavan and R. Ranjithkumar (2009). Large scale phycoremediation of acidic effluent from an alginate industry. ScienceAsia, 35, 220–226.

Sivasubramanian, V., Sankaran, B., Murali, R., and Subramanian, V. V. (2011). Micro algal technology to correct pH and reduce sludge in an acidic effluent from a detergent industry. Journal of Algal Biomass Utilization, 2, 95–102.

Solovchenko, A., Pogosyan, S., Chivkunova, O., Selyakh, I., Semenova, L., Voronova, E., Scherbakov, P., Konyukhov, I., Chekanov, K., Kirpichnikov, M., and Lobakova, E. (2014). Phycoremediation of alcohol distillery wastewater with a novel Chlorella sorokiniana strain cultivated in a photobioreactor monitored on-line via chlorophyll fluorescence. Algal Research, 6, 234–241.

Syed Obaidur, R., Prasad Panda, B., Parvez, S., Kaundal, M., Hussain, S., Akhtar, M., and Najmi, A. B. (2019). Neuroprotective role of astaxanthin in hippocampal insulin resistance induced by Aβ peptides in animal model of Alzheimer's disease. Biomedicine & Pharmacotherapy, 110, 47–58.

Ting, Y. P., Lawson, F., & Prince, I. G. (1989). Uptake of cadmium and zinc by the alga Chlorella vulgaris: Part 1. Individual ion species. Biotechnology and Bioengineering, 34(7), 990–999.

Trick CG, Anderen RJ, Gillam A. and Harrison PJ (1983). Prorocentrin—an extracellular siderophore produced by the marine dinoflagellate Prorocentrum minimum. Science. 219, 306–308.

Vincent Amanor-Boadu, Pfromm, P. H., and Nelson, R. (2014). Economic feasibility of algal biodiesel under alternative public policies. Renewable Energy, 67, 136–142.

Volesky, B. (1997). Removal and recovery of heavy metals by biosorption. In: Volesky, B. (ed), Biosorption of heavy metals. CRC Press, Boca Raton, 629–635.

Wang, L., Min, M., Li, Y., Chen, P., Chen, Y., Liu, Y., et al. (2010). Cultivation of green algae Chlorella sp. in different wastewaters from municipal wastewater treatment plant. Applied Biochemistry and Biotechnology, 162, 1174e1186.

Wurochekke, A. A., Mohamed, R. M. S., Al-Gheethi, A. A., Atiku, H., Amir, H. M., and Matias-Peralta, H. M (2016) Household grey water treatment methods using natural materials and their hybrid system. Journal of Water and Health, 14(6), 914–928.

Wurochekke, A. A., Radin Mohamed, R. M. S., Al-Gheethi, A. A. S., Noman, E. A., and Mohd Kassim, A. H. (2019). Phycoremediation: A green technology for nutrient removal from greywater. In: Radin Mohamed, R., Al-Gheethi, A., and Mohd Kassim, A. (eds), Management of greywater in developing countries. Water Science and Technology Library. Springer, Cham, Vol. 87.

Yel, E., Dilek, F., and Yetis, U. (2002). Effectiveness of algae in treatment of a wood-based pulp and paper industry wastewater. Bioresource Technology, 84, 1–5.

Yusuf, C. (2007). Biodiesel from microalgae. Biotechnology Advances, 25, 294–206.

Zhang, L., Jia, J., Zhu, Y., Zhu, N., Wang, Y., and Yang, J. (2005). Electro-chemically improved bio-degradation of municipal sewage. Biochemical Engineering Journal, 22(3), 239–244.

Zhao, B., and Su, Y. (2014). Process effect of microalgal carbon dioxide fixation and biomass production: A review. Renewable Sustainable Energy Review, 31, 121–132.

Zhou, W., Min, M., Li, Y., Hu, B., Ma, X., Cheng, Y., *et al.* (2012). A hetero-photoautotrophic two-stage cultivation process to improve wastewater nutrient removal and enhance algal lipid accumulation. Bioresource Technology, 110, 448–455.

9 Biosorption of Heavy Metal by Algae to Meet Clean Environment

The Need of the Hour for a Sustainable Future

Biswajita Pradhan, Sairendri Maharana, Srimanta Patra, Rabindra Nayak, Chhandashree Behera, Prajna Paramita Bhuyan, Soumya Ranjan Dash, and Mrutyunjay Jena

9.1 INTRODUCTION

Our ecosystem is facing serious threat due to increase of pollution day by day. Some of the major sources of pollutants include chemical by-products, pesticides, herbicides, weedicides, pharmaceuticals, plastic, textile, leather industry wastes, pigments, electroplating, storage, mining, smelting, batteries, and metallurgical processes. Unfortunately, due to rapid industrialization across the globe over the past few decades, many environmental issues have arisen, which are harmful to the ecosystem. The degradation of the ecosystem's quality is the most important environmental issue due to harmful heavy metals (Gautam et al. 2014; Naushad 2014).

The environmental contaminants such as heavy metals chiefly come from improper release of industrial effluents (Wu et al. 2015; Zhang et al. 2018). From UNESCO global statistics data in 2003 have shown that the distribution and use of water in different sector is 70% in agriculture, 22% in industry, and 8% in domestic (Al-Weshah 2003). A significant quantity of water has been yearly used in industrial activities, but an enormous fraction of this water is released as wastewater into environment that contributes a major portion for the environmental pollution. It is very well-known that the industrial wastewater has several toxic as well as heavy metals like Pb, Cd, Hg, Cr, and other metals and their ions released from various industrial processes (Ghasemi et al. 2014; Iyer et al. 2005; Qaiser et al. 2007). The heavy metals can't be degraded and have a very high accumulation propensity that contributes towards their high concentration in the ecosystem with several ill effects on human and other organism health (Qaiser et al. 2007; Volesky and Holan 1995). Several methods have been used for heavy metals removal from industrial wastes such as membrane filtration, chemical precipitation, electrochemical techniques, coagulation or flocculation, and ion exchange (Abdolali et al. 2016; Syukor et al. 2016). However, these conventional methods are not effective for the removal of heavy metals which are at low concentrations (Satapathy and Natarajan 2006; Wang et al. 2009).

In this context, biosorption method is often considered as a cost-effective method that offers flexibility and easy to operate and can successfully remove both high and low concentration of several heavy metals from the wastewater (Gupta et al. 2015; Vijayaraghavan and Balasubramanian 2015). The biosorption also minimizes the consumption of chemicals, reagent, and production of secondary sludge that permits the recovery of retained metal ions (Davis et al. 2003; Zhao et al. 2010). In addition to this, biosorption method is economically and ecologically significant and contributes sustainable development (Vijayaraghavan and Balasubramanian 2015).

DOI: 10.1201/9781003219156-12

However, these advantages mainly depend on what biomass type is used as biosorbents for heavy metals removal. This is why finding materials that are available in huge amounts, ease of collection, cost-effectiveness, and preparation that requires only a few stages are also highly desired (Satapathy and Natarajan 2006). The capabilities of algae to bruise in heavy metal contaminated environments is stimulating much interest of several biotechnological industries. The microalgae appear to be the appropriate agents for bioremediation as they require only inorganic nutrients and sunlight for their growth. The fast-growing algae are able to compartmentalize heavy metals within specific organelles (Kumar et al. 2007). Few strategies have been utilized for effective elimination of heavy metals from polluted water bodies. Although considerable progresses have been made for the large-scale applications of microalgal-based absorbents for remediation of heavy metals, still, the on-field application is in infancy. The main metal-removal processes (passive sorption onto algal biomass) are currently under investigation. Algae meet all the conditions to be an effective biosorbent and can be considered to be a favorable biological resource as they are existing in every region, do not need any special growth conditions, high productivity rate, require only a little easy ground work-stage, and have extraordinary efficiency to maintain heavy metal ions in aqueous solution (Davis et al. 2003; Hannachi et al. 2015). In addition to this, the biosorption process involves metabolism-independent mechanism in which metal ions from aqueous media are bound to the cellular surface walls. Non-viable algae as biosorbents are crucial and effective because they are easy to obtain and involve low-cost preparation. Under these circumstances, the algal biomass acts like a chemical substrate with various functional groups homogeneously distributed on the biomass surface that represent binding sites for heavy metal ions present in the aqueous solution. In this context, we have emphasized the critical role of algal biosorbents in heavy metal removal and the factors affecting this process with sustainable development for production of future generation cost-effective biosorbents (Figure 9.1).

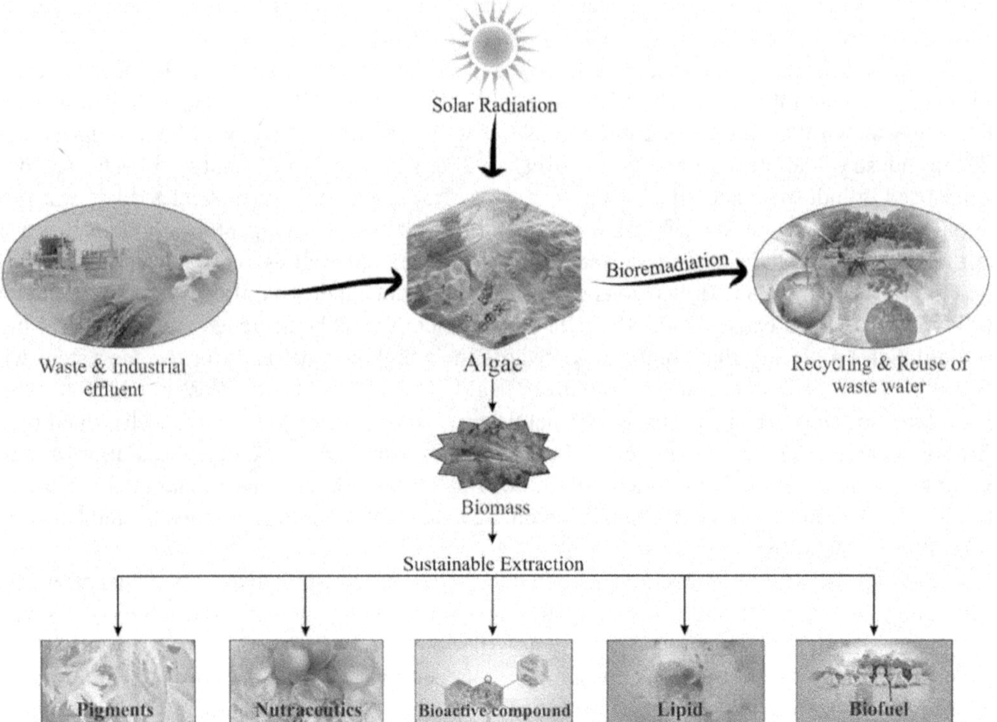

FIGURE 9.1 Potentiality of algae in biosorption of heavy metals and its sustainable extraction for humankind.

9.2 SOURCES OF TOXIC HEAVY METAL POLLUTION

The wastewater from industry contains major concentration of heavy metals which are released in huge quantities from several industrial activities. Hence, the release of wastes into the environment from industries without suitable treatment remains as the main source of heavy metal contamination and enhanced toxicity (Dixit et al. 2015; Gautam et al. 2014). Industrial activities like electroplating, conversion-coating, milling, and anodizing cleaning generate huge quantity of wastewater with effective presence of heavy metals, mainly lead, cadmium, nickel, chromium, zinc, copper, platinum, vanadium, silver, and titanium (Barakat 2011). In addition to this, the circuit board manufacturing residues is remaining also the additional source of industrial heavy metal containing wastewater with significant concentrations of copper, tin, nickel, and lead. Inorganic paint manufacturing also produces the wastewater that contains chromium and cadmium derivatives. Apart from these, the most common industrial activities such as aquaculture, intensive livestock manufacture, and energy production are chiefly accountable for release of wastewater and contains higher concentrations of hazardous heavy metals such as cadmium, copper, nickel, lead, chromium, and zinc. Therefore, to prevent heavy metal related pollution for sustainable environmental development, detoxification methods must be followed in the near future. Hence, appropriate biosorbent like marine algal biomass will allow such detoxification with effective heavy metals removal from industrial wastewater (Ayangbenro and Babalola 2017).

9.3 INTERACTIONS BETWEEN MICROALGAE AND HEAVY METALS

Heavy metals are defined as the elements with an atomic number greater than 20 and a density greater than 5 g/cm (Ali and Khan 2018). The metals like zinc, iron, nickel, manganese, cobalt, copper, and molybdenum are the essential metals for cellular metabolism and are the part of metalloproteins which play a pivotal role in numerous cellular functions that include electron transport and protects against reactive oxygen species (Chadd et al. 1996; Twining and Baines 2013). In addition to that, cadmium is also considered as an essential metal for some microalgae that can restore zinc as a catalyst of carbon anhydrase as reported in *Thalassiosira weissflogii* (Lane and Morel 2000; Xu et al. 2008). The intracellular concentration of heavy metal range from nano molar to femto molar and interspecies difference in metal stoichiometry (Twining et al. 2012). Heavy metals uptake into living cells usually occurs in two steps such as metal adsorption and transport across cell membrane. The transport across the cell membrane is generally considered to be rate limiting (Levy et al. 2008). The metal adsorption is metabolism independent in which metal ions are adsorbed into cell wall via interaction with the functional groups of major structural metabolites such as polysaccharides and proteins (Das et al. 2008). After adsorption into the cell wall, metal ions can enter into the cell membrane by binding either to the ion carriers or low molecular weight thiols such as cysteine via active transport process (Narula et al. 2015).

9.4 ALGAE AS BIOSORBENTS FOR REMOVAL OF HARMFUL HEAVY METALS

Algal biomass has different organic compounds like proteins, carbohydrates, and lipids that hold a number of various functional groups. The functional groups represents the binding sites for heavy metal ions in biosorption process (Javanbakht et al. 2014). Under these circumstances, the competence of biosorption process mainly depends on binding efficacy of functional groups with heavy metal ions (Table 9.1). The interaction will be affected by some chemical and physiological events such as the degree of dissociation of functional groups, speciation form of heavy metal ions, binding sites number form bio mass surface, the interaction time between the marine algal biomass (solid) and heavy metal ions (aqueous solution), pH and temperature (Marella et al. 2020).

TABLE 9.1
Biosorption of Different Heavy Metal Ion Using Different Algae under Different Optimal Condition

Metal	Algae	Max Sorption mg g^{-1}	Optimal pH	Initial Metal mg l^{-1}	Biomass conc. g l^{-1}	Temp. (°C)	Time Hour	References
Al	*Laminaria japonica*	75.27	4.5	–	1	–	30	(Lee et al. 2004)
Cd	*Ascophyllum nodusum*	87.7	6	50	0.5	–	2	(Romera et al. 2007)
	Chlorella vulgaris	85.3	4	200	0.75	20	2	(Aksu 2001)
	Chlamydomonas reinhardtii	42.6	6	–	–	23	1	(Tüzün et al. 2005)
	Cladophora fracta	4.08	5	8	–	25	1	(Lamaia et al. 2005)
	Laminaria japonica	136.1	4.5	50	0.5	–	30	(Lee et al. 2004)
	Ulva lactuta	29.2	5	10	1	20	1	(Sarı and Tuzen 2009)
Cr	*Laminaria japonica*	94.103	4.5	–	–	–	30	(Lee et al. 2004)
	Rhizoclonium	11.81	4	100	1	–	2	(Onyancha et al. 2008)
	Spirogyra condensate	14.82	5	–	–	–	2	(Onyancha et al. 2008)
	Chlorella vulgaris	140	1.5	250	1	25	2	(Gokhale et al. 2008)
	Chlmaydomonas reinhardtii	18.2	2		0.6	25	2	(Bayramoğlu et al. 2006)
	Dunaliella sp 1	58.3	2	100	1	25	72	(Dönmez and Aksu 2002)
	Dunaliellasp 2	45	2	100	1	25	72	(Dönmez and Aksu 2002)
Cu	*Ascophyllum nodusum*	58.8	4	50	0.5	–	2	(Romera et al. 2007)
	Aspaaragopsis armata	21.3	5	50	0.5	–	2	(Romera et al. 2007)
	Chlorella vulgaris	89.19	3.5	–	0.005	25	0.5	(Mehta and Gaur 2001)
	Chondrus crispus	40.5	4	50	0.5	25	2	(Romera et al. 2007)
	Cladophora fascicularis	102.30	5	–	2	25	–	(Deng et al. 2006)
	Fucus spiralis	70.9	4	50	0.25	23	2	(Romera et al. 2007)
	Laminaria japonica	85.15	5	–	0.25	–	2	(Ahmad et al. 2020)
	Saragassum sp.	72.5	5.5	–	1	22	3	(Karthikeyan et al. 2007)
	Sphaeroplea sp.	140.43	4	–	1	33	1.5	(Srinivasa Rao et al. 2005)
	Spirogyra insignis	19.3	4	–	1	33	1.25	(Srinivasa Rao et al. 2005)
	Spirogyra neglecta	115.3	4.5	100	0.1	25	0.16	(Singh et al. 2007)

TABLE 9.1 (*Continued*)
Biosorption of Different Heavy Metal Ion Using Different Algae under Different Optimal Condition

Metal	Algae	Max Sorption mg g^{-1}	Optimal pH	Initial Metal mg l^{-1}	Biomass conc. g l^{-1}	Temp. (°C)	Time Hour	References
	Ulothrix zonata	38.2	5	100	–	25	1	(Nuhoglu et al. 2002)
	Ulva fasciata	73.5	5.5	–	1	22	3	(Karthikeyan et al. 2007)
Hg	*Chlamydomonas*	72.2	6	50	–	25	1	(Tüzün et al. 2005)
Ni	*Ascophyllum nodusum*	43.3	6	50	0.5	–	2	(Romera et al. 2007)
	Chlorella miniate	1.367	7.4	–	–	–	24	(Wong et al. 2000)
	Chlorella vulgaris	0.641	7.4	200	–	25	0.33	(Akhtar et al. 2004)
	Chlorella vulgaris	59.29	4.5	5	–	–	1	(Mehta and Gaur 2001)
	Codium vermilara	13.2	6	50	0.5	–	2	(Romera et al. 2007)
	Fucus spiralis	50	6	50	0.5	–	2	(Romera et al. 2007)
	Spirogyra insigns	17.5	6	50	1	–	2	(Romera et al. 2007)
Pb	*Ascophyllum nodusum*	178.6	6	50	1	–	2	(Romera et al. 2007)
	Asparagopsis armata	63.6	4	50	1	–	2	(Romera et al. 2007)
	Chlamydomonas reinhardtii	96.3	5	–	–	25	1	(Tüzün et al. 2005)
	Codium Vermilaria	63.3	4	50	0.5	–	2	(Romera et al. 2007)
Se	*Cladophora*	74.9	5	10	8	20	1	(Sarı and Tuzen 2009)
U	*Chlorella vulgaris*	14.3	4.4	23.8	0.76	–	0.08	(Vogel et al. 2010)
Zn	*Ascophyllum nodusum*	42	6	50	0.5	–	–	(Romera et al. 2007)
	Condrus crispus	45.7	6	50	0.5	–	2	(Romera et al. 2007)
	Fucus spiralis	53.2	6	50	0.5	–	2	(Romera et al. 2007)
	Spirogyra insignis	21.1	6	50	1	–	2	(Romera et al. 2007)

9.5 PREPARATION AND CHARACTERIZATION OF ALGAE BIOSORBENTS

Algae are prevalent organisms distributed everywhere in the world. From a biological view, marine algae are placed in "plants" group and ranges from few micrometers (unicellular microscopic algae) to a few meters (multicellular macroscopic algae). Mostly, algae are classified according to their color and composition of pigments (Davis et al. 2003). Based upon these, it can be divided into three groups such as green algae (Chlorophyta), red algae (Rhodophyta), and brown algae (Phaeophyta). In addition to pigments, algae also have proteins, carbohydrates, and lipids, and its concentration and diversity vary from species to species. The green algal cell walls are mainly composed of cellulose, polysaccharides, and proteins (Romera et al. 2007; Wang and Chen 2009). The red algae have cellulose, polysaccharide, agar, and carraghenates as structural support (Romera et al. 2006). The brown algae usually contain alginic acid, cellulose, and sulfated polysaccharides like fucoidan for the structural support (Lodeiro et al. 2005; Romera et al. 2007). Differences in structure and composition of algal cell walls influence the ability of heavy metal ions uptake from aqueous medium. The cell walls constituents have diverse functional groups like carboxyl, carbonyl, hydroxyl, sulfate, and amino, which can interact with heavy metal ions in aqueous media via ion exchange and play a pivotal role in biosorption process (Fomina and Gadd 2014). Moreover, in biosorption process, certain organic compounds from algal source such as pigments and alginate can be released into the aqueous media and generate secondary pollution which decrease the algae biosorption capacity. Unfortunately, use of red and brown algae in biosorption processes of heavy metal ions has a significant disadvantage which radically limits the industrial applications. The leaching of organic compounds is trivial in green algae (Hamdy 2000). Therefore, the secondary pollution effect is minimal in green algae, and their biosorption efficacy is lower as compared to red and brown marine algae (Apiratikul and Pavasant 2008; Tobin et al. 1988; Yang and Chen 2008).

The cultivated and harvested algae are immediately washed with distilled water for removal of dissolved salts and other materials. Without proper washing, the dissolved materials will significantly change the ionic bond strength of aqueous solutions, and their presence will affect the efficacy of biosorption process. After proper washing, the drying of algae is done in air, humidity, and temperature. The last important stage in preparation of algal biomass is sieving and grinding that will provide approximately the equal size of biomass particles. Under these conditions, algae are considered as "low-cost" biosorbents, which require a few and simple stages of preparation. The presence of numerous and varied functional groups in the algal biosorbents make them effective for metal removal from aqueous effluents (Bashir et al. 2019).

9.6 OPTIMAL CONDITION FOR ALGAL BIOSORPTION PROCESS

The biosorption of metal ions from the aqueous media via using the algae as biosorbent takes place with maximum efficacy in certain physiological condition (Dönmez et al. 1999; Febrianto et al. 2009; Robalds et al. 2016). Such are as follows.

9.6.1 pH

The pH is the main fact, or which influences the rate of adsorption process. The studies of surface charge have shown that the ease of the use of free sites depend on pH. With risingin pH, surface of calcium alginate charged sites become more negative, and metal ions uptake increased with increasing in pH. Crist et al. (1994) found that the number of binding sites is reduced with decreasing in ph. The increasing of pH was evident during uptake of metal ion. Chen and Johns (1991) reported that under increased pH, additional negative sites were available for sorption of copper ions. The copper removal was increased at higher pH approximately about 0.1–1 unit from the initial adsorption. The results have demonstrated that pH change is due to mass balance resulted from ion absorption of copper. Schiewer and Volesky 1995 studied that additional sites were available for sorption of metal ion at higher pH values, and the consequence is the exclusion efficacy increased with pH. They used *Sargassum fluitans* in their study with pH values raised from 3.5 to 6.0. They also found that these values depend on speciation of harmful toxic heavy metals ions in aqueous solution and the dissociation degree of function groups from the algal biomass surface. At low pH, the heavy metal ions are mainly as a freely positive charged cation in the aqueous solution, while functional groups in algal surface area are mainly non-dissociated or positively charged, and that makes the chemical interactions. Only the solution of metal ions exist as negatively charged species and can be retained by marine algal biomass at very low pH in the initial solution (Murphy et al. 2008).

9.6.2 Influence of Ionic Strength

Earlier it has been noticed that ionic bond strength plays a significant role in uptake of metal ion. It is usually considered that the removal efficacy increasing with decreasing of ionic bond strength. Chen et al. (2006) studied the influence of ionic bond strength on adsorption of metal ion, and sodium perchlorate is used to adjust ionic bond strength at 0.005, 0.05, 0.5 mol/L resulted that when ionic bond strength decreased (0.5 mol/L to 0.005 mol/L), the removal efficacy was increased (80% to 95%). The result suggested that during the adsorption process, the struggle for functional groups between the metal ions and other ions played a pivotal role. At a fixed pH, the number of functional groups is fixed, so the sites available for uptake of metal ion get decreased with increasing of ionic bond strength. As a result of which, the removal of ion was less at higher ionic bond strength and vice versa.

9.6.3 Biomass Dosage

The amount of marine algal biomass is used to remove the harmful toxic heavy metals ions from a given volume of aqueous solution that determine the number of active sites be involved in the biosorption process, and the value of this parameter is important both from economic and technological consideration.

9.6.4 Temperature

Temperature is important for the biosorption process. Temperature plays an important role for metal ligand complex formation. Some of the previous studies claim that increased temperature in algal culture could possibly increase metal ion absorption capacity with minimal or no contemplation of structural change (Deng et al. 2006, 2007; Gupta and Rastogi 2008). The possibility reasons for increasing temperature to result in increase in biosorption of metal ion includes an increased number of active sites involved in uptake of metal ion, an increased tendency of active sites to metal ions absorption, mass transfer resistance reduction in diffusion layer by a reduction of thickness of the diffusion boundary layer around adsorbent groups and change of complex formation constant with temperature (Mehta and Gaur 2005). Conversely, few studies have proposed that the uptake of metal ion by some algae is exothermic (Tuzen et al. 2009). The capacity of metal ions uptake increase with decrease in temperature. Moreover, it has also been observed that temperature has no substantial influence on uptake of metal ion by algal cells (Lodeiro et al. 2006). These mismatched results apparently determined that optimum temperature is required for active biological reactions in the living cells. The variations in temperature causes different biosorption behaviors in several algal strains with different metal ions (Mehta and Gaur 2005).

9.6.5 Biosorption in Contact Time

The biosorption of heavy metal ion is mainly based on the contact time. However, based on the published articles, the kinetics of heavy metal ion biosorption on algal cell surface and the biosorption mechanism is algal species specific (Lee et al. 2016; Yaqub et al. 2012; Zhang et al. 2016). Biosorption can occur in two stages such as for algae biomass (ions passively adsorb to the cell membrane, and biosorption of metal ions occurs quickly within the first minutes) and for living algae (active absorption where heavy metal ions are gradually uptaken into algal cell) (Vogel et al. 2010). In addition, it was also reported that the uptake of uranium by *C. vulgaris* is more than 90% of dissolved uranium, which is adsorbed during the first five minutes (Vogel et al. (2010). Further, one more study (Tüzün et al. 2005) found that the biomass of microalgae *Chlamydomonas reinhardtii* rapidly adsorbed free ions of Cd, Pb, and Hg with the biosorption equilibrium achieved in 60 minutes. In conclusion, contact time has a better effect on biosorption capacity in living algae. For instance, Lamaia et al. (2005) calculated the uptake of Cd and Pb ions by *Cladophora fracta* and separately harvested after two, four, six, and eight days and found that the algal growth rate was decreased over time, and a greater biosorption capacity was observed in the old cultures. These results proposed that passive biosorption of heavy metals commence rapidly in first moments of contact, while a greater level of heavy metal bio-removal was achieved with longer contact times in living algal cell.

9.7 BIOREMEDIATION IN ALGAE

Industrialization has led to efficient discharge of heavy metals into the ecosystems. The metal pollutants can easily persist into the food chain that leads to ecosystem imbalance (Gosavi et al. 2004). Accumulation of heavy metals such as Cu, Cr, Zn, Cd, and Hg have numerous consequences, such as growth and developmental abnormalities, carcinogenesis, mental retardation, neuromuscular control defects, and renal failure with other illnesses in human beings (Gosavi et al. 2004). Metal ions elevated levels are usually cytotoxic and causes major damages to the cellular macromolecules. Unadventurous technologies such as ion exchange or lime precipitation are often ineffective and expensive for heavy metal ions removal at low concentrations, such as below 50 mg/L. In addition, these techniques are basically based on chemical replacement and physical displacement that generate toxic sludge possessing yet another problem. The disposal of these slugs adds further burden to technoeconomic feasibility in the treatment process. Algae can be provided solution to twin challenges of environmental pollution and energy security. They also have a great potential for removal of excess phosphorus and nitrogen from wastewater of the farm runoff. Algae can capture carbon dioxide in the flue gas from coal-fired power plants and reduce greenhouse gas and also promote enhanced algal biomass production that can be converted into biofuel latter. The algal species such as *Chlorella*, *Scenedesmus*, and *Spirulina* are extensively used for this proposal (Shuba and Kifle 2018).

The heavy metal absorption ability of algae has been recognized from last few decades (Olguín and Sánchez-Galván 2012; Rai et al. 1981). In the natural environments, algae play a pivotal role in controlling of metal concentration in water bodies (Dwivedi 2012). Algae has the ability to take up harmful toxic heavy metals from the environment. Metals penetrate into the algae through the process of adsorption (Demirbas 2008). The metal ions are adsorbed over the cell surface very quickly in few seconds or minutes through the process called as physical adsorption. Then, the ions are transported into the cytoplasm slowly through chemisorption. Polyphosphate bodies of the fresh water unicellular algae enable the storage of nutrients. Numerous studies have reported that metals such as Mg, Ti, Pb, Cd, Zn, Co, Sr, Ni, Cu, and Hg are sequestered in the polyphosphate bodies in case of green algae. These bodies perform two different functions in algae, such as to provide a "storage pool" for metals and can act as mediator for "detoxification mechanism". Microalgae are competent to produce peptides called metallothioneins, which are mostly posttranscriptionally synthesized class III metallothioneins or phytochelatins that have higher efficiency to bind the heavy metals (Kumar et al. 2015; Wang and Chen 2009).

Bioaccumulation studies have revealed the accumulation of the contaminants in the organisms. The algal species like *Chlorella* sp., *Westiellopsis prolifica*, *Anabaena inaequalis*, *Synechococcus* sp., *Stigeoclonium tenue* tolerate heavy metals. However, several species of *Anabaena* and *Chlorella* in addition to marine algae have been used for heavy metals removal (Pradhan, Patra, Dash et al. 2020; Pradhan, Patra, Maharana et al. 2020). But the operating conditions limit the practical application of these organisms in field (Pradhan, Patra, Maharana et al. 2020). The alga *Scenedesmus obliquus* was found to sequester some metals like phosphorus present in growth media. It was also able to accumulate higher concentration of Zn and Cd with high phosphorus concentration in the growth media (Zhang et al. 2016). A reduced selenium (Se) accumulation was also found in the similar condition (Zhang et al. 2016). Shehata and Badr (1980) cultured *Scenedesmus* sp. in different concentrations of Cu, Cd, Ni, Zn, and Pb to assess their effect on the growth of algae. The Ni solution was found to be less toxic as compared to the Cu for the *Scenedesmus* sp. growth. The algae tolerated high Pb concentrations (30 mg/L). Rai et al. (1998) studied both absorption and adsorption of Cd by a cyanobacterium, *Microcystis* in both on laboratory and field conditions. The naturally occurring cells displayed higher absorption efficacy for Ni and Cd as compared to laboratory cells. Several studies have been carried out to demonstrate the role of algae in the heavy metal's bioremediation. Some of the heavy metals such as Pb, Cu, Co, and Cd are removed by *Oedogonium rivulare* and *Cladophora glomerata* as short-term process. Other metals such as Ni, Cr, Mn, and Fe are continuously taken up by biosorption from aqueous solution by the freshwater filamentous algae like *Spirogyra hatillensis*.

Removal of Hg, Cd, and Pb from the aqueous solution by the use of low-cost adsorbents like marine macroalgae has gained much attention in the recent times. In a study, treated distillery wastewater from an anaerobic fixed-bed reactor in a pond microalga and concluded the removal of 84.1%, 90.2%, and 85.5% of ammonia, organic nitrogen, and total phosphorus respectively (Travieso et al. 1996). Kim et al. (1998) reported a removal of nitrogen and phosphorus is about 95.3% and 96% respectively by *Chlorella vulgaris* in secondarily 25% treated swine wastewater after the incubation of four days. Hodaifa et al. (2008) used *Scenedesmus obliquus* for the removal of potassium salts and other minerals present in the industrial wastewater obtained from olive oil extraction. The other wastewater treatment mechanism is immobilization of algal cells. The immobilization removes the harvesting stage, which is very difficult in treatment process. A gel matrix prevents the cells from freely moving in its environment. The reaction rates have increased in the immobilized cells because of its high cell density. Further, they have shown no cell washout. As a result, they are preferable to their free-living counterparts. Travieso et al. (1996) reported the higher nutrient removal from raw sewage treatment via internal immobilization of *Chlorella vulgaris* in the sodium alginate beads.

9.8 WASTEWATER TREATMENT AND BIOFUEL PRODUCTION USING MICROALGAE

Bio-removal of heavy metal ions by use of microalgae has been considered as an economically and environmentally sustainable approach for toxic metals removal from wastewaters as well as from the aquatic ecosystems (Mata et al. 2009). Being cost-effective, it also decreases the requirements of chemical remediation of wastewaters, less usage of freshwater consumption that boost up the suitability of introduction of algal components in the treatment process of wastewater (Kesaano and Sims 2014; Lee 2012; Lyon et al. 2015; Zeng et al. 2015). In addition to this, the production of an ample amount of valuable by-products such as nutrients and bioactive compounds also enhance their suitability, with higher production of algal biomass which leads to biodiesel, and bioethanol production is the sustainable approach (Park et al. 2011). Algal wastewater treatment and subsequent biofuel production can not only decrease the costs of algal biomass production but also efficiently remove the potential hazardous contamination such as toxic metal pollutants, residual nutrients, and even transgenic algae from wastewaters bodies (Ruiz-Marin et al. 2010). The coupled system is a helpful approach where nutrient as well as the removal of heavy metal ion is required prior to wastewater release. Furthermore, production of biofuels could also reduce the CO_2 sequestration from the industrial areas or power plants (Raeesossadati et al. 2014).

9.9 CONCLUSION AND FUTURE PROSPECTIVE

Low-cost cultivation, enhanced biomass production, higher harvesting, uptake of metal ion, and metal selectivity properties for huge scale metal removal makes algae a suitable candidate for bioremediation of wastewater. An inclusive description of biochemistry of microalgal substrates and their environmental reimbursement will be necessary to plausibly emphasize the advantage of algal biosorption over the conventional ion-exchange resins and routine chemical treatments. Further, more research at both fundamental and laboratory as well as on field trial will be needed to assist for optimizing the final biosorption efficacy to improve the economic sustainability and practical view for implementation of algal cell at large scale for bioremediation of heavy metal. Moreover, further studies should be focused on the adsorption mechanism and relationship between the binding capacity and alginate components. In conclusion, this piece of information will lead to decryption of effective use of algae for bioremediation of heavy metals and developing novel bio-absorbents for future use.

9.10 CONFLICT OF INTERESTS

The authors declare that there is no conflict of interest.

ACKNOWLEDGMENT

The authors are thankful to MoEF and CC, government of India, to carry out this piece of work. The authors are also thankful to Berhampur University for providing the necessary facilities to carry out the study.

REFERENCES

Abdolali A, Ngo HH, Guo W, Lu S, Chen S-S, Nguyen NC, Zhang X, Wang J, Wu Y. 2016. A breakthrough biosorbent in removing heavy metals: Equilibrium, kinetic, thermodynamic and mechanism analyses in a lab-scale study. Science of the Total Environment. 542:603–611.

Ahmad S, Pandey A, Pathak VV, Tyagi VV, Kothari R. 2020. Phycoremediation: Algae as eco-friendly tools for the removal of heavy metals from wastewaters. Bioremediation of industrial waste for environmental safety. Springer. p. 53–76.

Akhtar N, Iqbal J, Iqbal M. 2004. Removal and recovery of nickel (ii) from aqueous solution by loofa sponge-immobilized biomass of chlorella sorokiniana: Characterization studies. Journal of Hazardous Materials. 108(1–2):85–94.

Aksu Z. 2001. Equilibrium and kinetic modelling of cadmium (ii) biosorption by c. Vulgaris in a batch system: Effect of temperature. Separation and Purification Technology. 21(3):285–294.

Ali H, Khan E. 2018. What are heavy metals? Long-standing controversy over the scientific use of the term 'heavy metals'–proposal of a comprehensive definition. Toxicological & Environmental Chemistry. 100(1):6–19.

Al-Weshah R. 2003. The role of unesco in sustainable water resources management in the Arab world. Desalination. 152(1–3):1–13.

Apiratikul R, Pavasant P. 2008. Batch and column studies of biosorption of heavy metals by Caulerpa Lentillifera. Bioresource Technology. 99(8):2766–2777.

Ayangbenro AS, Babalola OO. 2017. A new strategy for heavy metal polluted environments: A review of microbial biosorbents. International Journal of Environmental Research and Public Health. 14(1):94.

Barakat M. 2011. New trends in removing heavy metals from industrial wastewater. Arabian Journal of Chemistry. 4(4):361–377.

Bashir A, Malik LA, Ahad S, Manzoor T, Bhat MA, Dar G, Pandith AH. 2019. Removal of heavy metal ions from aqueous system by ion-exchange and biosorption methods. Environmental Chemistry Letters. 17(2):729–754.

Bayramoğlu G, Tuzun I, Celik G, Yilmaz M, Arica MY. 2006. Biosorption of mercury (ii), cadmium (ii) and lead (ii) ions from aqueous system by microalgae chlamydomonas reinhardtii immobilized in alginate beads. International Journal of Mineral Processing. 81(1):35–43.

Chadd HE, Newman J, Mann NH, Carr NG. 1996. Identification of iron superoxide dismutase and a copper/zinc superoxide dismutase enzyme activity within the marine cyanobacterium synechococcus sp. Wh 7803. FEMS Microbiology Letters. 138(2–3):161–165.

Chen C-C, Coleman ML, Katz LE. 2006. Bridging the gap between macroscopic and spectroscopic studies of metal ion sorption at the oxide/water interface: Sr (ii), co (ii), and pb (ii) sorption to quartz. Environmental Science & Technology. 40(1):142–148.

Chen F, Johns MR. 1991. Effect of c/n ratio and aeration on the fatty acid composition of heterotrophicchlorella sorokiniana. Journal of Applied Phycology. 3(3):203–209.

Crist RH, Martin JR, Carr D, Watson J, Clarke HJ, Carr D. 1994. Interaction of metals and protons with algae. 4. Ion exchange vs adsorption models and a reassessment of scatchard plots; ion-exchange rates and equilibria compared with calcium alginate. Environmental Science & Technology. 28(11):1859–1866.

Das N, Vimala R, Karthika P. 2008. Biosorption of heavy metals–an overview. Indian Journal of Biotechnology. 7: 159–169.

Davis TA, Volesky B, Mucci A. 2003. A review of the biochemistry of heavy metal biosorption by brown algae. Water Research. 37(18):4311–4330.

Demirbas A. 2008. Heavy metal adsorption onto Agro-based waste materials: A review. Journal of Hazardous Materials. 157(2–3):220–229.

Deng L, Su Y, Su H, Wang X, Zhu X. 2006. Biosorption of copper (ii) and lead (ii) from aqueous solutions by nonliving green algae cladophora fascicularis: Equilibrium, kinetics and environmental effects. Adsorption. 12(4):267–277.

Deng L, Su Y, Su H, Wang X, Zhu X. 2007. Sorption and desorption of lead (ii) from wastewater by green algae Cladophora Fascicularis. Journal of Hazardous Materials. 143(1–2):220–225.

Dixit R, Malaviya D, Pandiyan K, Singh UB, Sahu A, Shukla R, Singh BP, Rai JP, Sharma PK, Lade H. 2015. Bioremediation of heavy metals from soil and aquatic environment: An overview of principles and criteria of fundamental processes. Sustainability. 7(2):2189–2212.

Dönmez G, Aksu Z. 2002. Removal of chromium (vi) from saline wastewaters by dunaliella species. Process Biochemistry. 38(5):751–762.

Dönmez GÇ, Aksu Z, Öztürk A, Kutsal T. 1999. A comparative study on heavy metal biosorption characteristics of some algae. Process Biochemistry. 34(9):885–892.

Dwivedi S. 2012. Bioremediation of heavy metal by algae: Current and future perspective. Journal of Advanced Laboratory Research in Biology. 3(3):195–199.

Febrianto J, Kosasih AN, Sunarso J, Ju Y-H, Indraswati N, Ismadji S. 2009. Equilibrium and kinetic studies in adsorption of heavy metals using biosorbent: A summary of recent studies. Journal of Hazardous Materials. 162(2–3):616–645.

Fomina M, Gadd GM. 2014. Biosorption: Current perspectives on concept, definition and application. Bioresource Technology. 160:3–14.

Gautam RK, Sharma SK, Mahiya S, Chattopadhyaya MC. 2014. Contamination of heavy metals in aquatic media: Transport, toxicity and technologies for remediation. Royal Society of Chemistry. p. 1–24.

Ghasemi M, Naushad M, Ghasemi N, Khosravi-Fard Y. 2014. Adsorption of pb (ii) from aqueous solution using new adsorbents prepared from agricultural waste: Adsorption isotherm and kinetic studies. Journal of Industrial and Engineering Chemistry. 20(4):2193–2199.

Gokhale S, Jyoti K, Lele S. 2008. Kinetic and equilibrium modeling of chromium (vi) biosorption on fresh and spent spirulina platensis/chlorella vulgaris biomass. Bioresource Technology. 99(9):3600–3608.

Gosavi K, Sammut J, Gifford S, Jankowski J. 2004. Macroalgal biomonitors of trace metal contamination in acid sulfate soil aquaculture ponds. Science of the Total Environment. 324(1–3):25–39.

Gupta V, Rastogi A. 2008. Biosorption of lead from aqueous solutions by green algae spirogyra species: Kinetics and equilibrium studies. Journal of Hazardous Materials. 152(1):407–414.

Gupta VK, Nayak A, Agarwal S. 2015. Bioadsorbents for remediation of heavy metals: Current status and their future prospects. Environmental Engineering Research. 20(1):1–18.

Hamdy A. 2000. Biosorption of heavy metals by marine algae. Current Microbiology. 41(4):232–238.

Hannachi Y, Rezgui A, Dekhil AB, Boubaker T. 2015. Removal of cadmium (ii) from aqueous solutions by biosorption onto the brown macroalga (dictyota dichotoma). Desalination and Water Treatment. 54(6):1663–1673.

Hodaifa G, Martínez ME, Sánchez S. 2008. Use of industrial wastewater from olive-oil extraction for biomass production of scenedesmus obliquus. Bioresource Technology. 99(5):1111–1117.

Iyer A, Mody K, Jha B. 2005. Biosorption of heavy metals by a marine bacterium. Marine Pollution Bulletin. 50(3):340–343.

Javanbakht V, Alavi SA, Zilouei H. 2014. Mechanisms of heavy metal removal using microorganisms as biosorbent. Water Science and Technology. 69(9):1775–1787.

Karthikeyan S, Balasubramanian R, Iyer C. 2007. Evaluation of the marine algae ulva fasciata and sargassum sp. For the biosorption of cu (ii) from aqueous solutions. Bioresource Technology. 98(2):452–455.

Kesaano M, Sims RC. 2014. Algal biofilm based technology for wastewater treatment. Algal Research. 5:231–240.

Kim SB, Lee SJ, Kim CK, Kwon GS, Yoon BD, Hee-Mock O. 1998. Selection of microalgae for advanced treatment of swine wastewater and optimization of treatment condition. Korean Journal of Applied Microbiology & Biotechnology. 26(1):76–82.

Kumar KS, Dahms H-U, Won E-J, Lee J-S, Shin K-H. 2015. Microalgae–a promising tool for heavy metal remediation. Ecotoxicology and Environmental Safety. 113:329–352.

Kumar YP, King P, Prasad V. 2007. Adsorption of zinc from aqueous solution using marine green algae—ulva fasciata sp. Chemical Engineering Journal. 129(1–3):161–166.

Lamaia C, Kruatrachuea M, Pokethitiyooka P, Upathamb ES, Soonthornsarathoola V. 2005. Toxicity and accumulation of lead and cadmium in the filamentous green alga cladophora fracta (of muller ex vahl) kutzing: A laboratory study. Science Asia. 31(2):121–127.

Lane TW, Morel FM. 2000. A biological function for cadmium in marine diatoms. Proceedings of the National Academy of Sciences. 97(9):4627–4631.

Lee H, Shim E, Yun H-S, Park Y-T, Kim D, Ji M-K, Kim C-K, Shin W-S, Choi J. 2016. Biosorption of cu (ii) by immobilized microalgae using silica: Kinetic, equilibrium, and thermodynamic study. Environmental Science and Pollution Research. 23(2):1025–1034.

Lee H, Suh J, Kim I, Yoon T. 2004. Effect of aluminum in two-metal biosorption by an algal biosorbent. Minerals Engineering. 17(4):487–493.

Lee JW. 2012. Advanced biofuels and bioproducts. Springer Science & Business Media.

Levy JL, Angel BM, Stauber JL, Poon WL, Simpson SL, Cheng SH, Jolley DF. 2008. Uptake and internalisation of copper by three marine microalgae: Comparison of copper-sensitive and copper-tolerant species. Aquatic Toxicology. 89(2):82–93.

Lodeiro P, Barriada JL, Herrero R, De Vicente MS. 2006. The marine macroalga cystoseira baccata as biosorbent for cadmium (ii) and lead (ii) removal: Kinetic and equilibrium studies. Environmental Pollution. 142(2):264–273.

Lodeiro P, Cordero B, Barriada JL, Herrero R, De Vicente MS. 2005. Biosorption of cadmium by biomass of brown marine macroalgae. Bioresource Technology. 96(16):1796–1803.

Lyon SR, Ahmadzadeh H, Murry MA. 2015. Algae-based wastewater treatment for biofuel production: Processes, species, and extraction methods. Biomass and biofuels from microalgae. Springer. p. 95–115.

Marella TK, Saxena A, Tiwari A. 2020. Diatom mediated heavy metal remediation: A review. Bioresource Technology. 305:123068.

Mata Y, Torres E, Blazquez M, Ballester A, González F, Munoz J. 2009. Gold (iii) biosorption and bioreduction with the brown alga fucus vesiculosus. Journal of Hazardous Materials. 166(2–3):612–618.

Mehta SK, Gaur JP. 2001. Characterization and optimization of ni and cu sorption from aqueous solution by chlorella vulgaris. Ecological Engineering. 18(1):1–13.

Mehta SK, Gaur JP. 2005. Use of algae for removing heavy metal ions from wastewater: Progress and prospects. Critical Reviews in Biotechnology. 25(3):113–152.

Murphy V, Hughes H, McLoughlin P. 2008. Comparative study of chromium biosorption by red, green and brown seaweed biomass. Chemosphere. 70(6):1128–1134.

Narula P, Mahajan A, Gurnani C, Kumar V, Mukhija S. 2015. Microalgae as an indispensable tool against heavy metals toxicity to plants: A review. International Journal of Pharmaceutical Sciences Review Research. 31(1):180.

Naushad M. 2014. Surfactant assisted nano-composite cation exchanger: Development, characterization and applications for the removal of toxic pb2+ from aqueous medium. Chemical Engineering Journal. 235:100–108.

Nuhoglu Y, Malkoc E, Gürses A, Canpolat N. 2002. The removal of cu (ii) from aqueous solutions by ulothrix zonata. Bioresource Technology. 85(3):331–333.

Olguín EJ, Sánchez-Galván G. 2012. Heavy metal removal in phytofiltration and phycoremediation: The need to differentiate between bioadsorption and bioaccumulation. New Biotechnology. 30(1):3–8.

Onyancha D, Mavura W, Ngila JC, Ongoma P, Chacha J. 2008. Studies of chromium removal from tannery wastewaters by algae biosorbents, spirogyra condensata and rhizoclonium hieroglyphicum. Journal of Hazardous Materials. 158(2–3):605–614.

Park J, Craggs R, Shilton A. 2011. Wastewater treatment high rate algal ponds for biofuel production. Bioresource Technology. 102(1):35–42.

Pradhan B, Patra S, Dash SR, Maharana S, Behera C, Jena M. 2020. Antioxidant responses against aluminum metal stress in geitlerinema amphibium. SN Applied Sciences. 2(5):800.

Pradhan B, Patra S, Maharana S, Behera C, Dash SR, Jena M. 2020. Demarcating antioxidant response against aluminum induced oxidative stress in *Westiellopsis prolifica* Janet. International Journal of Phytoremediation. 1941:1–14.

Qaiser S, Saleemi AR, Mahmood Ahmad M. 2007. Heavy metal uptake by agro based waste materials. Electronic Journal of Biotechnology. 10(3):409–416.

Raeesossadati M, Ahmadzadeh H, McHenry M, Moheimani N. 2014. Co2 bioremediation by microalgae in photobioreactors: Impacts of biomass and co2 concentrations, light, and temperature. Algal Research. 6:78–85.

Rai L, Gaur J, Kumar H. 1981. Phycology and heavy-metal pollution. Biological Reviews. 56(2):99–151.

Rai L, Singh S, Pradhan S. 1998. Biotechnological potential of naturally occurring and laboratory-grown microcystis in biosorption of ni 2+ and cd 2+. Current Science:461–464.

Robalds A, Naja GM, Klavins M. 2016. Highlighting inconsistencies regarding metal biosorption. Journal of Hazardous Materials. 304:553–556.

Romera E, Gonzalez F, Ballester A, Blazquez M, Munoz J. 2006. Biosorption with algae: A statistical review. Critical Reviews in Biotechnology. 26(4):223–235.

Romera E, González F, Ballester A, Blázquez M, Munoz J. 2007. Comparative study of biosorption of heavy metals using different types of algae. Bioresource Technology. 98(17):3344–3353.

Ruiz-Marin A, Mendoza-Espinosa LG, Stephenson T. 2010. Growth and nutrient removal in free and immobilized green algae in batch and semi-continuous cultures treating real wastewater. Bioresource Technology. 101(1):58–64.

Sarı A, Tuzen M. 2009. Equilibrium, thermodynamic and kinetic studies on aluminum biosorption from aqueous solution by brown algae (padina pavonica) biomass. Journal of Hazardous Materials. 171(1–3):973–979.

Satapathy D, Natarajan G. 2006. Potassium bromate modification of the granular activated carbon and its effect on nickel adsorption. Adsorption. 12(2):147–154.

Schiewer S, Volesky B. 1995. Modeling of the proton-metal ion exchange in biosorption. Environmental Science & Technology. 29(12):3049–3058.

Shehata SA, Badr SA. 1980. Growth response of scenedesmus to different concentrations of copper, cadmium, nickel, zinc, and lead. Environment International. 4(5–6):431–434.

Shuba ES, Kifle D. 2018. Microalgae to biofuels: 'Promising' alternative and renewable energy, review. Renewable and Sustainable Energy Reviews. 81:743–755.

Singh A, Kumar D, Gaur J. 2007. Copper (ii) and lead (ii) sorption from aqueous solution by non-living spirogyra neglecta. Bioresource Technology. 98(18):3622–3629.

Srinivasa Rao P, Kalyani S, Suresh Reddy K, Krishnaiah A. 2005. Comparison of biosorption of nickel (ii) and copper (ii) ions from aqueous solution by *sphaeroplea algae* and acid treated *sphaeroplea algae*. Separation Science and Technology. 40(15):3149–3165.

Syukor AA, Sulaiman S, Siddique MNI, Zularisam A, Said M. 2016. Integration of phytogreen for heavy metal removal from wastewater. Journal of Cleaner Production. 112:3124–3131.

Tobin J, Cooper D, Neufeld R. 1988. The effects of cation competition on metal adsorption by Rhizopus Arrhizus biomass. Biotechnology and Bioengineering. 31(3):282–286.

Travieso L, Benitez F, Weiland P, Sanchez E, Dupeyron R, Dominguez A. 1996. Experiments on immobilization of microalgae for nutrient removal in wastewater treatments. Bioresource Technology. 55(3):181–186.

Tuzen M, Verep B, Ogretmen AO, Soylak M. 2009. Trace element content in marine algae species from the black sea, turkey. Environmental Monitoring and Assessment. 151(1–4):363–368.

Tüzün I, Bayramoğlu G, Yalçın E, Başaran G, Celik G, Arıca MY. 2005. Equilibrium and kinetic studies on biosorption of hg (ii), cd (ii) and pb (ii) ions onto microalgae *chlamydomonas reinhardtii*. Journal of Environmental Management. 77(2):85–92.

Twining BS, Baines SB. 2013. The trace metal composition of marine phytoplankton. Annual Review of Marine Science. 5:191–215.

Twining BS, Baines SB, Vogt S, Nelson DM. 2012. Role of diatoms in nickel biogeochemistry in the ocean. Global Biogeochemical Cycles. 26(4).

Vijayaraghavan K, Balasubramanian R. 2015. Is biosorption suitable for decontamination of metal-bearing wastewaters? A critical review on the state-of-the-art of biosorption processes and future directions. Journal of Environmental Management. 160:283–296.

Vogel M, Günther A, Rossberg A, Li B, Bernhard G, Raff J. 2010. Biosorption of u (vi) by the green algae chlorella vulgaris in dependence of pH value and cell activity. Science of the Total Environment. 409(2):384–395.

Volesky B, Holan Z. 1995. Biosorption of heavy metals. Biotechnology Progress. 11(3):235–250.

Wang J, Chen C. 2009. Biosorbents for heavy metals removal and their future. Biotechnology Advances. 27(2):195–226.

Wang XS, Li ZZ, Sun C. 2009. A comparative study of removal of cu (ii) from aqueous solutions by locally low-cost materials: Marine macroalgae and agricultural by-products. Desalination. 235(1–3):146–159.

Wong J, Wong Y, Tam N. 2000. Nickel biosorption by two chlorella species, c. Vulgaris (a commercial species) and c. Miniata (a local isolate). Bioresource Technology. 73(2):133–137.

Wu S, Wallace S, Brix H, Kuschk P, Kirui WK, Masi F, Dong R. 2015. Treatment of industrial effluents in constructed wetlands: Challenges, operational strategies and overall performance. Environmental Pollution. 201:107–120.

Xu Y, Feng L, Jeffrey PD, Shi Y, Morel FM. 2008. Structure and metal exchange in the cadmium carbonic anhydrase of marine diatoms. Nature. 452(7183):56–61.

Yang L, Chen JP. 2008. Biosorption of hexavalent chromium onto raw and chemically modified sargassum sp. Bioresource Technology. 99(2):297–307.

Yaqub A, Mughal M, Adnan A, Khan W, Anjum K. 2012. Biosorption of hexavalent chromium by spirogyra spp.: Equilibrium, kinetics and thermodynamics. The Journal of Animal and Plant Sciences. 22(2):408–415.

Zeng X, Guo X, Su G, Danquah MK, Zhang S, Lu Y, Sun Y, Lin L. 2015. Bioprocess considerations for microalgal-based wastewater treatment and biomass production. Renewable and Sustainable Energy Reviews. 42:1385–1392.

Zhang L, Zhao B, Xu G, Guan Y. 2018. Characterizing fluvial heavy metal pollutions under different rainfall conditions: Implication for aquatic environment protection. Science of the Total Environment. 635:1495–1506.

Zhang X, Zhao X, Wan C, Chen B, Bai F. 2016. Efficient biosorption of cadmium by the self-flocculating microalga scenedesmus obliquus as-6–1. Algal Research. 16:427–433.

Zhao G, Wu X, Tan X, Wang X. 2010. Sorption of heavy metal ions from aqueous solutions: A review. The Open Colloid Science Journal. 4(1).

10 Bioremediation of Hexavalent Chromium Using Microalgae

Pritikrishna Majhi, Bidhu Bhusan Mukut, and Saubhagya Manjari Samantaray

10.1 INTRODUCTION

Nowadays, hasty urbanization, progressive industrialization, as well as improved civilization are the main cause for freshwater consumption. Several anthropogenic practices, such as industrial, mining, and agricultural activities lead to the discharge of heavy metals into the fresh water, which ultimately affect the natural aquatic and terrestrial ecosystems. Moreover, the toxic and nonbiodegradable heavy metals are bioaccumulated and biomagnified through the food chain causing severe human health problems, which includes growth inhibition, cancer, organ damage, nervous system damage, etc. The metals which are toxic even at low concentrations are Cr, Mn, Fe, Cu, Zn, Hg, Pb, and Cd. Hence, improvement of water quality is a major problem in the 2030 Agenda and on Sustainable Development Goals, recognizing the importance of good-quality water for a sustainable development of society. However, the development and implementation of cost-effective processes for the removal and the detoxification of metals from wastewater is a major concern in the modern society.

There are several conventional methods used for the bioremediation of heavy metals from wastewater such as precipitation and sludge separation, chemical oxidation or reduction, ion exchange, reverse osmosis, cementation, coagulation, flocculation, electrochemical treatment, membrane processes, and evaporation (Lee et al., 2012; Goharshadi and Moghaddam, 2015). The major shortcomings associated with the conventional technologies are (i) use of expensive chemicals; (ii) incomplete, unsafe metal removal and the generation of toxic sludge which needs proper disposal mode of operations; and (iii) high energy consumption. Therefore, there is a need for searching low-cost and easily available adsorbents. Biological materials for the heavy metal removal have been recommended as a simple, cost-effective, efficient, and environmentally friendly technique (Zeraatkar et al., 2016). Bioremediation is the process where some specific microorganisms are used to convert hazardous pollutants present in both soil and water to nonhazardous components (Dwivedi, 2012). Whatever barriers we define, in fact, in nature the process of biological remediation involves both plants and microbes, while microbial interaction with soil has a very important role (Igiri et al., 2018). However, the direct use of microbes having some distinctive characteristics such as catabolism and the value-added products from them such as bio surfactants, enzymes, and proteins is an advanced approach to enhance and boost their remediation capacity (Schenk et al., 2012; Le et al., 2017). In order to degrade recalcitrant, different types of alternatives are also available, such as use of microbial fuel cell (MFC) and biofilm mediated bioremediation, which are applied to cleanup of heavy metal contaminated environment. In this prospect, microalgae-based bioremediation of heavy metals proves as a sustainable process by many researchers.

10.2 POTENTIAL ROLE OF MICROALGAE TOWARDS HEAVY METAL BIOREMEDIATION

Algae are the large and diverse assemblage of photosynthetic and microscopic microbes available abundantly in aquatic as well as terrestrial environments. Microalgae are extremely advantageous

DOI: 10.1201/9781003219156-13

in terms of economical and eco-friendly aspects due to their high metal recovery capability, growth rate, and surface area to volume ratio, short regeneration time (Pahlavanzadeh et al., 2010; Rawat et al., 2011; Abdel-Ghani and El-Chaghaby, 2014) and, hence, provide an effective solution for amelioration of industrial effluents. The phenomenon of microalgae based heavy metal remediation is broadly categorized into two categories: (1) bioaccumulation by living cells and (2) biosorption by nonliving, nongrowing biomass or biomass products. Bioaccumulation of heavy metals by living microalgae occurs in two steps: the first step (i.e., passive removal) takes place rapidly and is essentially independent of cell metabolism, occurring in both living and nonliving cells. Here, heavy metal ions are adsorbed to functional groups present on the cell surface by electrostatic interactions. This process includes physical adsorption, ion exchange, chemisorption, coordination, complexation, chelation, micro-precipitation, entrapment in the structural polysaccharide network, and diffusion through the cell wall and membrane.

10.3 UNDERLYING PRINCIPLES OF HEAVY METAL BIOREMEDIATION EXHIBITED BY MICROALGAE

10.3.1 Biosorption by Microalgal Biomass

Biosorption is a reversible, independent, and rapid surface phenomenon of the biomass (Bulgariu and Gavrilescu, 2015). In this process, the microorganisms are engaged to remove and recover heavy metal ions from waste sources. The uptake of heavy metal ions and radioactive compounds could be attributed to the physicochemical interaction of heavy metal ions with the cellular compounds of microorganism. The cell wall of microalgae contains sugars (such as starch, cellulose, hemicelluloses), proteins, and lipids which are associated with different functional groups such as phosphate ($-PO_3$), carboxyl (-COOH), hydroxyl (-OH), amino ($-NH_2$), and sulfhydryl ($-SH_2$) in the cell wall (Wang and Chen, 2009; Widjaja et al., 2009). These functional groups overall confer negative charge to the microalgal cell wall, which leads to proficient binding of heavy metal cations (Anastopoulos and Kyzas, 2015). Biosorption can be categorized into physical adsorption and chemical adsorption process (Tripathi et al., 2019). Physical adsorption involves the van der Waals force and electrostatic interaction, while chemical biosorption refers to chemical reaction such as complexation, ion exchange, and microprecipitation (Bulgariu and Gavrilescu, 2015). Biosorption process has several advantages over conventional methods of heavy metal treatment, for example (1) it relies on cheaper biomaterials as an adsorbent, (2) highly selective in recovery and removal of heavy metal cations, (3) minimal requirements of expensive reagents that issue disposal and space problems, (4) it can be operated over a wide range of physiochemical conditions such as temperature, pH, etc., and (5) low operation cost.

10.3.2 Bioaccumulation

Bioaccumulation is a toxico-kinetic process that balances the sensitivity of living organisms towards chemicals (Chojnacka, 2010), where sensitivity depends upon the types of organisms and chemicals involved (Mishra and Malik, 2013). The bioaccumulating organisms should have significant bio-transformational capabilities that include the conversion of toxic to nontoxic compound and, hence, reduce the toxicity of the heavy metal (Mishra and Malik, 2013). The bioaccumulation becomes beneficial only when they reduce a high number of metals (Tripathy et al., 2019). This process is normally slower than biosorption process and is inhibited by lower temperatures. The accumulation of heavy metal ions in microalgae is processed via two physiological processes, for example, initial rapid uptake of metal ion (passive process) followed by a slower uptake called active process.

10.3.3 Biotransformation

Transformations of heavy metals by microbes involve different enzymatic reactions such as reduction, oxidation, demethylation, and methylation. They mostly remove heavy metals from the environment (Kisielowska et al., 2010) by transforming metal ions to their metallic form. Those reactions take place either inside vacuoles or on the cell surface (Jaishankar et al., 2014).

10.3.4 Bioprecipitation and Biocrystallization

Microbial activity leads crystallization or precipitation of heavy metals which furthermore causes transformation and lowers their toxicity. Those microbial precipitation and crystallization processes take part in biogeochemical cycles by accumulating iron and manganese, generating microfossils and mineralizing silver and manganese. Moreover, precipitation of heavy metals on the inner or outer cell surface results in creating enzymatic activity and several secondary metabolites (Jaishankar et al., 2014).

10.3.5 Bioleaching of Metals

Bioleaching is generally based upon mobilization of metals from indigent ores and industrial wastes. Two major processes are involved in it, such as synthesis of various organic acids like gluconic acid, citric acid, oxalic acid, etc. and secretion of complex forming agents in the environment. Microbes that possess such metal leaching abilities have relatively high resistance to pH and temperature (Medfu Tarekegn et al., 2020). This biological method of leaching is mainly applied in biohydrometallurgy in order to leach from sulfide and oxide minerals. Metals such as arsine, antimony, bismuth, zinc, cobalt, gold, lead, copper, molybdenum, nickel, vanadium, and uranium can be recovered by bio hydrometallurgical process.

10.4 FACTORS INFLUENCING HEAVY METAL BIOREMEDIATION BY MICROALGAE

Metal ion recovery and remediation by microalgal cell is influenced by several biotic and abiotic factors. The biotic factors include **(i) types of species:** metal ion uptake and remediation capacity is varied from microalgal species to species. Even the alga belonging to the same group may have different adsorption capacity (Kumar et al., 2015). **(ii) Tolerance to metal ions:** most of the algae have acquired successful adaptation mechanisms after exposure to pollutants. Moreover, the tolerating capacity of each microalga towards heavy metal ion varies. The tolerating capacity of microalgae is highly dependent upon the defense response against oxidative damages. Several mechanisms have been adopted by the microalgae to reduce heavy metal toxicity such as production of heavy metal binding factors and proteins (metallothioneins, GSH, and phytochelatin conjugates), removal of toxic heavy metals from cells by ion-selective metal transporters. **(iii) Size and volume of microalgae:** size of the microalgae affects the sensitivity to metal toxicity. The smaller-sized microalgae can effectively sequester heavy metals as they have larger surface to volume ratio (Kumar et al., 2015). Abiotic factors include **(i) pH:** pH is the most important parameters which influence microalgal metal adsorption as it affects the solubility and toxicity of heavy metals in water stream. pH dependent metal uptake by microalgae cells closely related to the metal chemistry in solution as well as acid-base properties of various functional groups present on the microalgal cell surface. In low pH, the cell wall ligands are closely associated with the hydronium ions, thereby obstructing the interaction between heavy metal ions and cell wall ligands. Whereas, in increased pH, more ligands such as carboxyl, phosphate, imidazole, and amino groups (negatively charged) are exposed to the heavy metal cations and thereby facilitating the metal ion uptake. Hence, it is necessary to determine the optimal pH for algae–metal interactions. **(ii) Ionic strength of wastewater:** in case of microalgal heavy metal treatment, the presence of high concentration of monovalent cations increases the ionic strength of wastewater, thereby decreasing the metal absorption capacity. **(iii) Temperature:** stability of metal ions, ligands, as well as the solubility of metal ions are temperature dependent. In general, higher temperature favors greater solubility of metal ions and hence reduces the biosorption of metal ions.

10.5 CHALLENGES ASSOCIATED WITH MICROALGAL HEAVY METAL REMOVAL AND RECOVERY

Even though microalgae provide several striking benefits for bioremediation of heavy metals, it still has some challenges like nutritional changes, biomass harvesting, contamination by other microbes, high content of turbidity, and difficulty in downstream processing. In some cases, the microalgal-based heavy metal remediation can be a proficient and significant alternative to conventional methods only when a suitable pretreatment is followed or when it is combined with other technologies. Moreover, potent microalgal strains should be collected, isolated, and developed with better treatment methods in order to improve the capability of heavy metal reclamation and to minimize the operational charges.

10.6 FUTURE PROSPECTS

Microalgal remediation confers many advantages over other bioremediation processes as they have some unique properties like large adsorbing surface area, low-cost raw material, and do not generate toxic sludge. Extensive study on accumulation of heavy metals using algae has been done many years ago for biomonitoring or bioremediation motive. However, further studies should focus on the mechanism of adsorption such as the relationship between the binding capacity and alginate components, commercialization of microalgal technology, and biosorption properties of each alginate component. This work will lead to effective use of algae and, subsequently, accelerate the development of more effective biosorbents.

10.7 CONCLUSION

The present study has marked the bioremediation of emerging pollutants and the characterization of the microalgal biomass along with the production of innovative components. Alternative methods are imperative due to the environmental issues involving water pollution, water scarcity and the restoration of aquatic ecosystems over the past few decades. Furthermore, this scenario has increased the interest in such kind of research work within a limited period. Microalgae can remove nutrients, organic contaminants, and metals and also sequester carbon dioxide for which they are meant to be suitable bioremediators of waste effluents. However, several hypotheses and their application gaps are still remaining. Moreover, it is also important to standardize optimal cultivation parameters along with harvesting techniques as several biotic and abiotic parameters influence the microalgal growth and performance which must be thoroughly established and species-specific. Methods for wastewater treatment using microalgae are still in its infancy, hence improvement of photobioreactors and open systems require to reduce the energetic costs and to develop large-scale cultivation protocols for the bioremediation purpose. Nowadays, the application of microalgae to wastewater bioremediation research has evolved and is also focused on coupling bioremediation with the production of value-added products, contributing towards globular economy and multipurpose solution.

ACKNOWLEDGMENTS

The authors acknowledge their gratitude to the Department of Microbiology, Odisha University of Agriculture and Technology, for the technical support provided. We also thank Chittaranjan Sahoo for sharing very useful information and drafting of the chapter.

REFERENCES

Abdel-Ghani, N.T., and El-Chaghaby, G.A. (2014). Biosorption for metal ions removal from aqueous solutions: A review of recent studies. *International Journal of Latest Research in Science and Technology*, 3(1), 24–42. www.mnkjournals.com/ijlrst.htm

Anastopoulos, I., and Kyzas, G.Z. (2015). Progress in batch biosorption of heavy metals onto algae. *Journal of Molecular Liquids*, 209, 77–86. https://doi.org/10.1016/j.molliq.2015.05.023.

Bulgariu, L., and Gavrilescu, M. (2015). Bioremediation of heavy metals by microalgae. In *Handbook of marine microalgae* (pp. 457–469). Academic Press.

Chojnacka, K. (2010). Biosorption and bioaccumulation–the prospects for practical applications. *Environment International*, 36(3), 299–307.

Dwivedi, S. (2012). Bioremediation of heavy metal by algae: Current and future perspective. *Journal of Advanced Laboratory Research in Biology*, 3(3).

Goharshadi, E.K., and Moghaddam, M.B. (2015). Adsorption of hexavalent chromium ions from aqueous solution by graphene nanosheets: Kinetic and thermodynamic studies. *International Journal Environmental Science Technology*, 12(7), 2153–2160.

Igiri, B.E., Okoduwa, S.I.R., Idoko, G.O., Akabuogu, E.P., Abraham, O.A., and Ibe, K.E. (2018). Toxicity and bioremediation of heavy metals contaminated ecosystem from tannery wastewater: A review, *Journal of Toxicology*. https://doi.org/10.1155/2018/2568038

Jaishankar, M., Tseten, T., Anbalagan, N., Mathew, B.B., and Beeregowda, K.N. (2014). Toxicity, mechanism and health effects of some heavy metals. *Interdisciplinary Toxicology*, 7(2), 60.

Kisielowska, E., Hołda, A., and Niedoba, T. (2010). Removal of heavy metals from coal medium with application of biotechnological methods. *Górnictwo i Geoinżynieria*, 34, 93–104.

Kumar, K.S., Dahms, H.U., Won, E.J., Lee, J.S., and Shin, K.H. (2015). Microalgae– A promising tool for heavy metal remediation. *Ecotoxicology and Environmental Safety*, 113, 329–352.

Le, T.T., Son, M.H., Nam, I.H., Yoon, H., Kang, Y.G., and Chang, Y.S. (2017). Transformation of hexabromocyclododecane in contaminated soil in association with microbial diversity. *Journal of Hazardous Materials*, 325, 82–89.

Lee, S.M., Laldawngliana, C., and Tiwari, D. (2012). Iron oxide nano-particles immobilized-sand material in the treatment of Cu(II), Cd(II) and Pb(II) contaminated waste waters. *Journal of Chemical Engineering*, 103–111.

Medfu Tarekegn, M., Zewdu Salilih, F., and Ishetu, A.I. (2020). Microbes used as a tool for bioremediation of heavy metal from the environment. *Cogent Food & Agriculture*, 6(1), 1783174.

Mishra, A., and Malik, A. (2013). Recent advances in microbial metal bioaccumulation. *Critical Reviews in Environmental Science and Technology*, 43(11), 1162–1222.

Pahlavanzadeh, H., Keshtkara, R., Safdari, J., and Abadi, Z. (2010). Biosorption of nickel(II) from aqueous solution by brown algae: Equilibrium, dynamic and thermodynamic studies. *Journal of Hazardous Material*, 175, 304–310.

Rawat, I., Kumar, R.R., Mutanda, T., and Bux, F. (2011). Dual role of microalgae: Phycoremediation of domestic wastewater and biomass production for sustainable biofuels production. *Applied Energy*, 88, 3411–3424.

Schenk, P.M., Carvalhais, L. C., and Kazan, K. (2012). Unravelling plant microbe interactions: Canmultispecies transcriptomics help? *Trends in Biotechnology*, 30(3), 177–184.

Tripathi, S., Arora, N., Gupta, P., Pruthi, P.A., Poluri, K.M., and Pruthi, V. (2019). Microalgae: An emerging source for mitigation of heavy metals and their potential implications for biodiesel production. In *Advanced biofuels* (pp. 97–128). Woodhead Publishing.

Wang, J., and Chen, C. (2009). Biosorbents for heavy metals removal and their future. *Biotechnology Advances*, 27(2), 195–226.

Widjaja, A., Chien, C.C., and Ju, Y.H. (2009). Study of increasing lipid production from fresh water microalgae *Chlorella vulgaris*. *Journal of the Taiwan Institute of Chemical Engineers*, 40(1), 13–20.

Zeraatkar, A.K., Ahmadzadeh H., Talebi A.F., Moheimani, N.R., and McHenry, M.P. (2016). Potential use of algae for heavy metal bioremediation, a critical review. *Journal of Environmental Management*, 1–15.

11 Role of Microalgae in Value Addition and Bioremediation of Wastewater

Samadhan Yuvraj Bagul, Pandiyan K, Khadke GN, and Dolly Wattal Dhar

11.1 INTRODUCTION

Microalgae are photoautotrophic microorganisms (0.5–30 μm in size) which cannot be seen with the naked eye (Waterbury 2006). They are unicellular or multicellular microorganisms and categorized in eukaryotic (Rhodophyta, Chlorophyta, Pheaophyta) or prokaryotic (Cyanophyta) domain found in diverse environments (Bagul, Tripathi *et al.* 2018; Khan *et al.* 2018). Microalgae are part of food chain in aquatic environment such as freshwater as well as marine water with high rate of growth and play important role in CO_2 sequestration (Bagul, Chakdar *et al.* 2018). These have the ability to use CO_2 and light from atmosphere and convert it to carbohydrates, proteins, and lipids. They are fast-growing microorganisms which convert 9–10% solar energy into biomass (77g/biomass/m^2/day), that is, approximately 280 ton/ha/year. Amongst microalgae, cyanobacteria are capable of fixing atmospheric nitrogen and convert it into more available form for the plants. Microalgae have been suggested to be promising candidates for bioremediation due to their wide adaptive nature under diverse ecological conditions as well as their growing ability irrespective of seasons (Bagul, Prasanna *et al.* 2017). Some like green algae (*Chlorella*, *Chlorococcum*, *Scenedesmus*, *Chlamydomonas*, etc.) and cyanobacteria (*Nostoc*, *Anabaena*, *Leptolyngbya*, *Phormidium*, etc.) have been used by different researchers for wastewater treatment (Bagul, Prasanna *et al.* 2017; Markou and Georgakakis 2011; Sharma and Khan 2013). Apart from bioremediation, microalgae have wide applications in agriculture, pharmaceutical, food, and feed industries.

11.2 WASTEWATER GENERATION IN INDIA

Approximately 80% of sewage generated in India flows into water bodies without prior treatment resulting in polluted water and is unsuitable for consumption. The enhanced nitrate levels due to unregulated sewage flow into the aquatic environment are higher than that of prescribed concentrations in the entire country's groundwater (Source: Times of India). The total reported production of sewage from India is 61,754 million liters per day (MLD), and the current capacity for sewage treatment is only 22,963 MLD. This disparity in wastewater facility generation and treatment results in approximately 3,8791 MLD of untreated sewage, i.e., discharged directly into local water bodies (CPCB 2016). In addition to sewage wastewater, various industries also produce effluents about 501 MLD including nearly 201.4 MLD from the paper and pulp industry. Figure 11.1 shows effluent generation from different industries in India (Source: http:/www.indiaenvironmentportal.org.in). There are 234 sewage water treatment plants (STPs) in India, and most of these were built from 1978–1979 onwards under different river action plans. Just 5% of these are situated along the banks of major rivers in towns and cities (CPCB 2005). Wastewater facilities have been built in the urban areas, but 26,468 MLD sewage water requires a treatment facility, and the study does not account for sewage from rural communities (Choudhary *et al.* 2016).

 DOI: 10.1201/9781003219156-14

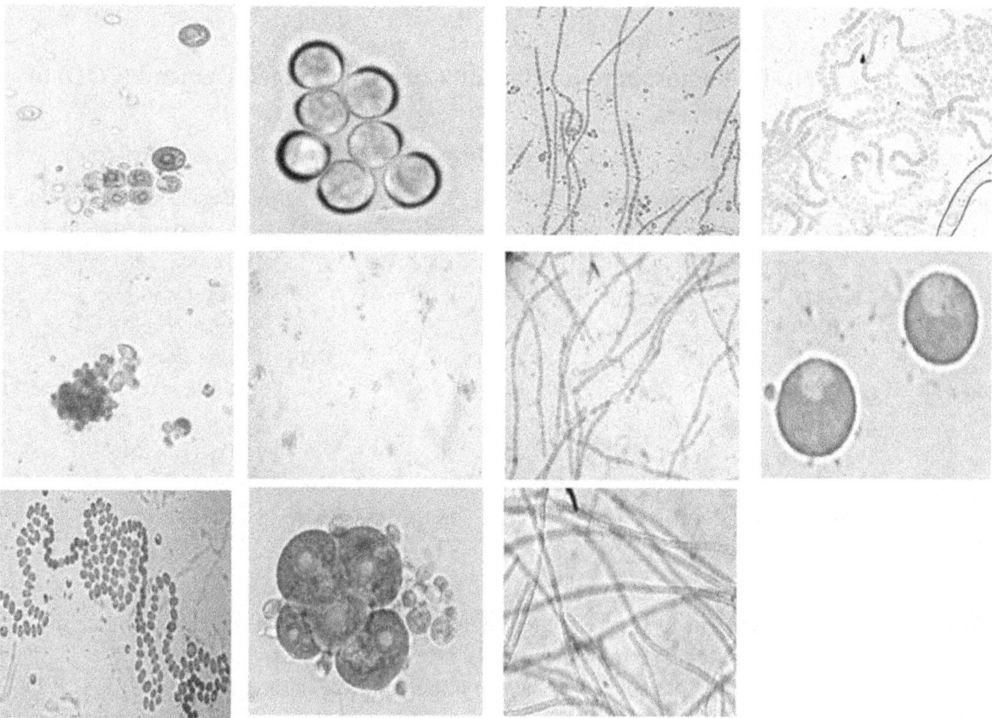

FIGURE 11.1 Wastewater generated from different industries in India.

11.3 COMPOSITION OF WASTEWATER

Wastewater is a complex waste comprising of different nutrients, heavy metals, and native microalgal flora (Figure 11.1). Effluent could be generated from different sources like dairy, sewage, piggery, tannery, paper, and pulp, while others may contain nutrients and contaminants as per the raw material used in processing. Generally, wastewaters are rich source of nutrients such as nitrogen and phosphorus, which may lead to eutrophication of lakes and upset the ecological balance (Cai *et al.* 2013). It is a tedious process to remove inorganic nitrogen and phosphorus from such waters. However, microalgae have the ability to scavenge nutrients from wastewater and utilize them for their growth. This may additionally have an advantage to remove heavy load of nutrients and reduce the concentration of these nutrients (Ahluwalia and Goyal 2007). The concentrations of total nitrogen and phosphorous reported are 15–90 mg/L (nitrogen) and 5–20 mg/L (phosphorous) in secondary treated municipal sewage wastewater, and in domestic wastewaters, the contents of these are lower. The concentration of total N and P varies and depends on the type of wastewater originating from the various sources, such as livestock breeding, and agriculture, such as anaerobic digested poultry litter effluent, swine or dairy manure, and wastewaters which can be sewage, industrial, and agricultural (Table 11.1). The concentrations of total N and total P in wastewater usually ranges from 185–3213 mg/L and 30–987 mg/L for (Chiu *et al.* 2015). Nitrogen in the form of ammonia and nitrates are the most common nitrogenous compounds found in wastewaters, and algae can use nitrate and ammonium ions directly from surrounding water. Nitrogen is utilized to synthesize amino acids, nucleotides, chlorophylls, and phycobilins, and the common nitrogen sources are nitrate, ammonia, and urea (Li *et al.* 2008; Xiong *et al.* 2008). Ammonium can be used more directly than nitrate to synthesize cellular nitrogenous compounds, and algae can convert nitrate to ammonium by using enzyme, nitrate reductase (Graham 2009). Some algae can take up nitrogen from organic sources such as yeast extract and glycine (Xiong *et al.* 2008). Most microalgal species

TABLE 11.1
Total Nitrogen (TN), Total Phosphorous (TP), and Chemical Oxygen Demand (COD) of Different Wastewater Reported

Wastewater Type	Source	TN (mg/L)	TP (mg/L)	COD mg O_2/l	Reference
Municipal	Secondary Treated Sewage	39	2.5	–	Bagul, Prasanna *et al.* 2017
	Sewage wastewater	41.88	3.68	149.75	Sharma and Khan 2013
	Sewage wastewater	15.3	9.8	–	Tripathi *et al.* 2019
Agricultural	Dairy	185–2636	30–727	–	Cai *et al.* 2013
	Poultry	1570	154	–	Singh *et al.* 2011
	Swine	510	76	5200	Chen *et al.* 2020
Industrial	Textile	0.1	1.51	51	El-Kassas and Mohamed 2014
	Winery	110	52	–	Cai *et al.* 2013
	Tannery	273	6	4000	Das *et al.* 2016
	Paper mill	9.9	30.25	3000	Usha *et al.* 2016
	Olive mill	58.9	43.1	5839	Malvis *et al.* 2018

utilize ammonium as chemical form of nitrogen than other forms, and it is a cheap source of nitrogen in wastewater or effluents that can be effectively used for their cultivation (Razzak *et al.* 2013). Phosphorus is also an essential nutrient that plays an important role in the metabolism of ATP, DNA, RNA, and phospholipids. This is commonly found in the form of H_2PO_{4-} or HPO_4^{2-}, and wastewater can provide the cheapest phosphorus source. Phosphorous availability has considerable impact on growth and photosynthesis (Razzak *et al.* 2013). Other than nitrogen and phosphorous, heavy metals, antibiotics, and hormones are also present in wastewater from poultry and piggery, which cause environmental problems as well as detrimental to human health. However, many algae show tolerance against cadmium, mercury, and other organic chemicals (Yin *et al.* 2011). Microalgal genera, such as *Chlorella*, *Ankistrodesmus*, and *Scenedesmus* species have been shown to grow in olive oil, mill, and paper industry wastewaters (Gupta *et al.* 2016). El-Kassas and Mohamed (2014) reported various groups of microalgae in the textile effluent. Among them, the members of Chlorophyta (*Chlorella vulgaris*, *Scenedesmus bijugatus*, *Ankistrodesmus falcatus*, *Kirchneriella contorta*, etc.), Bacillariophyta (*Amphora ovalis*, *Cyclotella meneghiniana*, *Navicula gracilis*, *Nitzschia apiculata*, etc.), Cyanophyta (*Aphanocapsa delicatissima*, *Chroococcus dispersus*, *Merismopedia punctate*, *Oscillatoria limnetica*, etc.), and Euglenophyta (*Euglena caudate*, *Phacus curvicauda*) have been reported. Bagul (2017) studied the native microalgal flora of secondary treated sewage wastewater from Delhi, India, and observed species of *Nostoc*, *Anabaena*, *Leptolyngbya*, *Scenedesmus*, *Chlamydomonas*, and *Chlorella* sp.

11.4 BIOREMEDIATION OF WASTEWATER BY MICROALGAE

The composition of municipal wastewater varies significantly from one location to another and typically contains human and other organic wastes, nutrients, microorganisms, and household discharges (Chiu *et al.* 2015). Selection of robust growing strains for wastewater bioremediation is of utmost importance because level of nutrients in different wastewaters varies, as mentioned in Table 11.1. A large number of microalgal genera belonging to Chlorophyceae have shown ability to grow on variety of wastewaters with potential to remove nutrients, and total nitrogen/phosphorous removal efficiency depends on microalgal species utilized (Figure 11.2). These can efficiently remove nitrogen, phosphorus, and heavy metals from wastewater (Cabanelas *et al.* 2013; Wang *et al.* 2010). *Chlorella* sp. showed total N removal efficiency in the range of 23–100%, whereas total P removal efficiency was in 20–100% range (Bohutskyi *et al.* 2015; Wu *et al.* 2014).

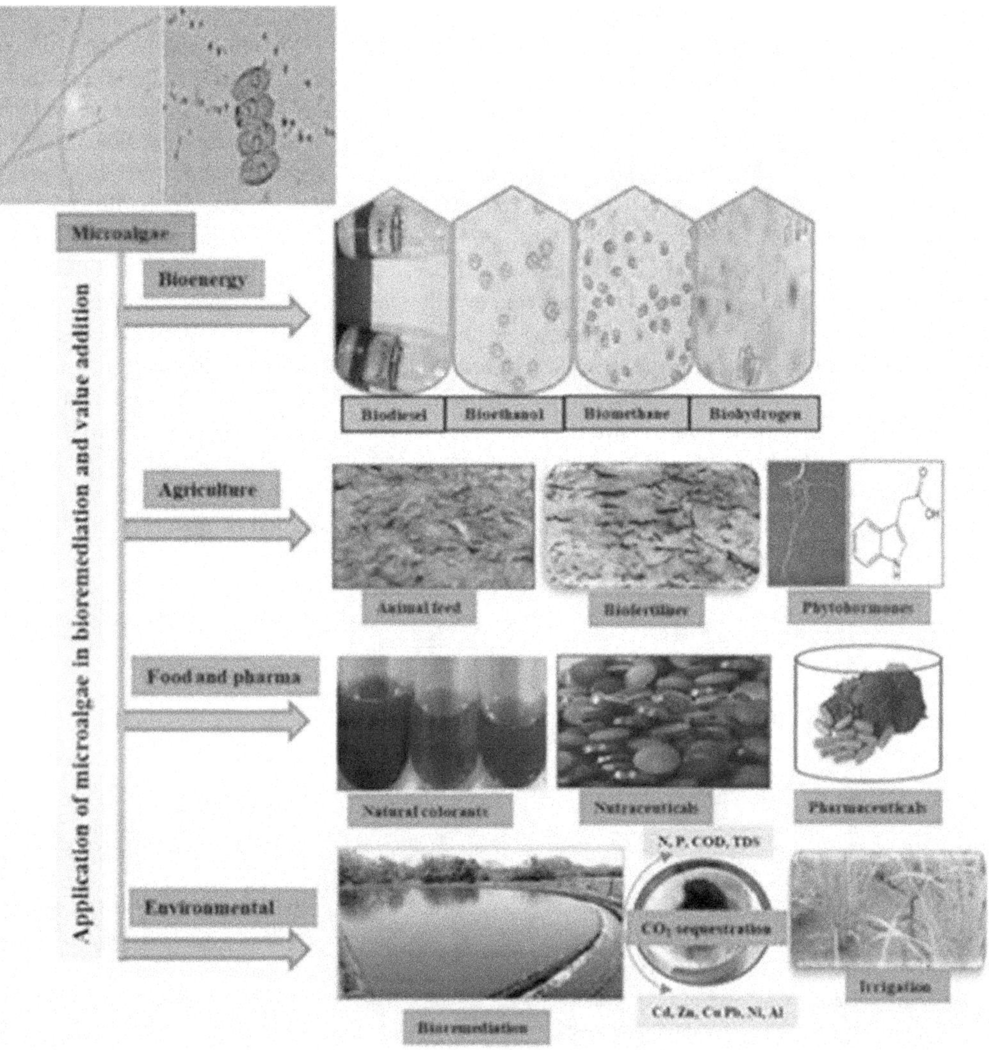

FIGURE 11.2 Applications of microalgae in bioremediation of wastewater and value-added products.

Besides nutrient scavenging potential, these can remove other pollutants as well (Caporgno *et al.* 2015; Ledda *et al.* 2015). *Chlorella vulgaris* was demonstrated to remove over 90% of N and 80% of P from primary treated sewage (Lau *et al.* 1995). *Actinastrum* sp., *Heynigia* sp., *Hindakia* sp., *Chlorella* sp., *Scenedesmus* sp., *Micractinium* sp., *Pediastrum* sp., *Chlamydomonas* sp., *Dictyosphaerium* sp., *Botryococcus* sp., and *Coelastrum* sp. proved to be potential N and P scavengers and could also remove other trace elements from wastewaters (Zhou *et al.* 2014). Lutzu *et al.* (2015) studied brewery wastewater treatment with *Scenedesmus dimorphus*, which could reduce more than 99% of total nitrogen and total phosphorous. Reduction in chemical oxygen demand (COD) up to 65% was also observed. *Desmodesmus* sp. CHX1 was used for treatment of piggery wastewater and recorded 78.46% and 91.66% removal of ammonical nitrogen and total phosphorous after seven days of incubation (Luo *et al.* 2018). Caporgno *et al.* (2015) showed N and P removal of 96% and 95% by *Chlorella kessleri* and 99% and 98% by *Chlorella vulgaris* from urban wastewater. Besides this, *Chlorella vulgaris* and *Coenochloris pyrenoidosa* removed large number of contaminants such as phenols, nitro- phenols, chlorophenols, and bisphenol A (Pham *et al.* 2013). In dairy effluents, microalgal growth reduced total dissolved solids from

1300 mg/L to 1040 mg/L after 20 days of incubation (Silambarasan *et al.* 2012). Three microalgal strains of *Chlorella vulgaris*, *Chlamydomonas debaryana* AT24, and *Chlamydomonas reinhardtii* effectively treated swine wastewater (Hasan *et al.* 2014). Zhang *et al.* (2014) conducted batch experiment for removal of contaminants from livestock wastewater and reported that *Scenedesmus dimorphus* could remove 85% each for 17α-estradiol and estrone, while 95% reduction was achieved for 17β-estradiol and estriol during eight days. Elumalai *et al.* (2013) showed an average 60% reduction in TDS by using different species of Cyanophyceae and Chlorophyceae. Similarly, Tripathi *et al.* (2019) reported 64.58%, 86.82%, and 88.2% reduction in TDS, BOD, and COD after treatment with *Scenedesmus* sp. ISTGA1. They also reported 100% reduction in heavy metals such as cadmium, nickel, lead, and cobalt. Das *et al.* (2016) used salt tolerant strain of *Chlorella* for treatment of tannery wastewater and showed 94.74% and 41% reduction in COD and TDS. The sulphate concentration was also reduced to 67%. Some cyanobacteria and algae may also remove xenobiotics from the environment by sorption, transformation, and degradation (Olguin 2003). Microalgae can assimilate phosphorous, and it can also get precipitated chemically induced by alkaline pH of the growth medium. (Sara *et al.* 2014). Nutrient composition of wastewaters can significantly affect growth and lipid production in microalgae (Cai *et al.* 2013). The nutrient removal efficiency of microalgae from wastewaters studied by different investigators has been presented in Table 11.2.

TABLE 11.2
Nutrient Removal Efficiency of Microalgae from Different Wastewater

Algal Strain	Growth Medium	Removal efficiency (%)		Reference
		Nitrogen	Phosphate	
Monoraphidium sp.	Sterile filtered domestic wastewater	51–95	74–95	Larsdotter *et al.* (2007)
Chlorella sp.	Sewage wastewater	89.9	91.9	Bagul, Prasanna *et al.* 2017
Chlorella vulgaris and *Scenedesmus dimorphus*	1:1 dilution of wastewater with fresh water	90	20–55	Gonzalez *et al.* (1997)
11 high-lipid content microalgae and *Scenedesmus* sp. LX1	Secondary effluent from domestic wastewater treatment plant in Beijing	98.5%	98%	Xin *et al.* (2010)
Scenedesmus sp. ISTGA1	Sewage wastewater	100%	100%	Tripathi *et al.* (2019)
Mixed consortia	Sewage wastewater	82.32	82.69	Sharma *et al.* 2020
Chlorella sorokiniana UTEX 2805, co-immobilized with bacterium *Azospirillum brasilense*	Carbon-free synthetic wastewater containing (mg/L): NaCl, 7; $CaCl_2$, 4; $MgSO_47H_2O$, 2; K_2HPO_4, 21.7; KH_2PO_4, 8.5; Na_2HPO_4, 33.4; and NH_4Cl, 10	100%	NA	de-Bashan *et al.* (2008)
Consortium of 15 native algal isolates	85%–90% carpet industry effluents with 10%–15% municipal sewage	96%		Chinnasamy *et al.* (2010)
Algae inoculums collected from local ponds treating municipal or winery wastewater, and from a creek	Primary clarifier effluent collected at municipal wastewater treatment facility	99	99	Woertz *et al.* (2009)

TABLE 11.2 (*Continued*)
Nutrient Removal Efficiency of Microalgae from Different Wastewater

Algal Strain	Growth Medium	Removal efficiency (%)		Reference
		Nitrogen	Phosphate	
Euglena sp.	Sewage treatment plant	93	66	Mahapatra *et al.* (2013)
Chlamydomonas Polypyrenoideum	Dairy industry wastewater	74–90	70	Kothari *et al.* (2013)
Chlorella Mexicana	Piggery wastewater	62	28	Abou-Shanab *et al.* (2013)
Chlamydomonas sp. TAI-2	Industrial wastewater	100	33	Wu *et al.* (2012)
Chlorella sorokiniana AK-1	Swine wastewater	78.3	97.7	Chen *et al.* 2020
Neochloris aquatica CL-M1	Swine wastewater	96.2	100	Wang *et al.* (2017)
Scenedesmus obliquus	Poultry wastewater	97.1	99.3	Oliveira *et al.* (2018)

11.5 VALUE-ADDED PRODUCTS FROM MICROALGAE

Microalgae are well-known for their potential in the area of value addition. Many microalgae such as *Chlorella, Isochrysis, Pavlova, Phaeodactylum, Tetraselmis, Skeletonema*, and *Thalassiosira* are are reported to have high protein content (Borowitzka 1997). *Spirulina platensis* is reported as one of the richest protein source of microbial origin having 460–630 g kg^{-1} on dry weight basis (Lupatini *et al.* 2017). Microalgae produce commercially important pigments such as phycocyanin, phycoerythrin, β-carotene, chlorophyll, and astaxanthin, which are used in therapies for tumorigenesis, neuronal disorders, and optical disease (Khan *et al.* 2018). Besides this, phycocyanin (a blue color) and phycoerythrin (a red color pigment) have also been used for dyes and colorants. These pigments have wide applications in cosmetics, food, and pharmaceutical industries due to their noncarcinogenic and nontoxic effects (Dikshit *et al.* 2018). *Dunaliella salina* and *Spirulina* (*Arthrospira* sp.) have been reported to produce β carotene and nutritional protein as food supplement (Liang *et al.* 2004; Metting 1996). Microalgae extracts are widely used in cosmetic industries for face, skin, and hair care and also in products which protect the skin from sunlight (Stolz and Obermayer 2005). Eicosapentaenoic acid from marine alga *Nannochloropsis* sp. has been used as a feed for rotifers (Zhang *et al.* 2001). Prasanna *et al.* (2010, 2013) and Manjunath *et al.* (2010) reported cyanobacteria for biocontrol of *Pythium aphanidermatum*, causing damping off disease in potato and *Fusarium* wilt in tomato plants. They also showed phytohormone production from cyanobacteria (Prasanna *et al.* 2010).

Microalgae are rich source of lipids and can accumulate up to 80% of lipids (Chisti 2007). Lipids are made up of triglycerides, which are converted into biodiesel by transesterification process (Bagul, Bharti *et al.* 2017). Microalgae that are rich in starch content and lack lignin in the cell wall have been reported for bioethanol production as well. Table 11.3 and Figure 11.2 depict different value-added products from microalgae reported by different researchers.

11.6 CONCLUSIONS AND FUTURE PROSPECTUS

Wastewater treatment is generally a costly affair requiring a large amount of money. Due to the wider adaptive nature in diverse ecological niches, microalgae are considered to be a potential, sustainable, and economically cheap alternative for the process of wastewater bioremediation. Apart from the bioremediation potential, the biomass can be harnessed for the beneficial value-added products. Nevertheless, highly effective strains, methodology, suitability of strains to different kinds of wastewaters, optimized nutrient conditions and competition with native microalgal flora as well as zooplanktons are the other important factors which need more attention in near future to make the whole process as economically feasible under the biorefinery concept.

TABLE 11.3
Value-Added Products from Microalgae

Microalgae	Value-Added Product	Application	Reference
Chlorella vulgaris	Biomethane	Fuel for cooking food	Mahdy *et al.* 2016
Chlorella sorokiniana and *Scenedesmus quadricauda*	Biogas	Fuel for cooking	Shchegolkova *et al.* 2018
Chlorella sp.	Biodiesel	Fuel	Bagul, Bharti *et al.* 2017
Anabaena variabilis (CCC421)	Phycoerythrin	Natural colorant	Chakdar and Pabbi 2012
Geitlerinema sp. H8DM	Phycocyanin	Natural colorant	Patel *et al.* 2018
Synechocystis PCC6803	Ethanol production		Dexter and Pengcheng 2009
Aphanothece sacrum	Sacran	Moisturizing agent	Okajima-Kaneko *et al.* 2009
Halomicronema sp. A27DM	Phycoerythrin	Natural colorant	Parmar *et al.* 2011
Aulosira fertilissima	Poly-β-hydroxybutyrate	Bioplastic	Samantaray and Mallick 2012
Anabaena variabilis RPAN59, *A. laxa* RPAN8	Defense related enzymes, Hydrolytic enzymes	Antifungal activity against plant pathogens	Prasanna *et al.* 2008, 2013
Haematococcus pluvialis	Astaxanthin	Antioxidant	Christian *et al.* 2018
Chromochloris zofingiensis	Astaxanthin	Antioxidant	Chen *et al.* 2017
Asterarcys quadricellulare PUMCC 5.1.1	Carotenoids	Natural colorants, antioxidants	Singh *et al.* 2019
Dunaliella salina CCAP 19/20	Β-carotene	natural colorants, antioxidants	Saha *et al.* 2018
Galdieria sp. strain USBA-GBX-832	PUFA	Health benefits	Lopez *et al.* 2019
Acutodesmus obliquus CN01	Α-linolenic acid	Health benefits	Othman *et al.* 2019
Oscilatoria sp. CMMA 1600	Mycosporine like amino acids (MAAs)	Cosmetics	Geraldes *et al.* 2019
Laminaria japonica	Fucoxanthin	Skin whitening	Thomas and Kim 2013
Chlorella sp. NKG 042401	γ-linolenic acid	Reducing blood pressure	Miura *et al.* 1993
Nostoc linckia	Borophycin	Antitumor activity against cancer	Hemscheidt *et al.* 1994

ACKNOWLEDGMENTS

Authors are grateful to the Indian Council of Agricultural Research-Directorate of Medicinal and Aromatic Plants, Anand, Gujarat, India, for technical support.

REFERENCES

Abou-Shanab RA, Ji MK, Kim HC, Paeng KJ and Jeon BH (2013). Microalgal species growing on piggery wastewater as a valuable candidate for nutrient removal and biodiesel production. Journal of Environmental Management, 115, 257–264.

Ahluwalia SS and Goyal D (2007). Microbial and plant derived biomass for removal of heavy metals from wastewater. Bioresource Technology, 98, 2243–2257.

Bagul SY (2017). Suitability of waste water grown microalgae for biodiesel production. Ph.D. Dissertation thesis. Centre for Conservation and Utilization of Blue Green Algae, Indian Agricultural Research Institute, New Delhi. p. 152.

Bagul SY, Bharti RK and Dhar DW (2017). Assessing biodiesel quality parameters for wastewater grown Chlorella sp. Water Science and Technology, 76(3), 719–727. https://doi.org/10.2166/wst.2017.223

Bagul SY, Chakdar H, Pandiyan K and Das K (2018). Conservation and application of microalgae for biofuel production in: Microbial resource conservation (Sharma and Verma eds.) Springer Nature. Soil Biology, 54, 335–352. https://doi.org/10.1007/978-3-319-96971-8_12

Bagul SY, Prasanna R and Dhar DW (2017). Biomass productivity and nutrient removal study with wastewater grown microalgae for biodiesel production. Green Farming, 8(3).

Bagul SY, Tripathi S, Chakdar H, Karthikeyan N, Pandiayan K, Singh A and Kumar M (2018). Exploration and characterization of cyanobacteria from different ecological niches of India for phycobilins production. International Journal of Current Microbiology and Applied Sciences, 7(12), 2822–2834. doi:10.20546/ijcmas.2018.712.321

Bohutskyi P, Liu K, Nasr LK, Byers N, Rosenberg JN and Oyler GA (2015). Bioprospecting of microalgae for integrated biomass production and phytoremediation of unsterilized wastewater and anaerobic digestion centrate. Applied Microbiology and Biotechnology. http://dx.doi.org/10.1007/s00253-015-6603-4.

Borowitzka MA (1997). Microalgae for aquaculture: Opportunities and constraints. Journal of Applied Phycology, 9, 393–401.

Cabanelas ITD, Ruiz J, Arbib Z, Chinalia FA, Garrido-Pérez C, Rogalla F, Nascimento IA and Perales JA (2013). Comparing the use of different domestic wastewaters for coupling microalgal production and nutrient removal. Bioresource Technology, 131, 429–436.

Cai T, Park SY and Li Y (2013). Nutrient recovery from wastewater streams by microalgae: status and prospects. Renew Sustain Energy Rev 19, 360–369.

Caporgno MP, Talebb A, Olkiewicz M, Font J, Pruvost J, Legrandb J and Bengoa C (2015). Microalgae cultivation in urban wastewater: Nutrient removal and biomass production for biodiesel and methane. Algal Research, 10, 232–239.

Central Pollution Control Board (2005). Parivesh Sewage Pollution - News Letter. Central Pollution Control Board, Ministry of Environment and Forests, Govt. of India, Parivesh Bhawan, East Arjun Nagar, Delhi 110 032 http://cpcbenvis.nic.in/newsletter/sewagepollution/contentsewagepoll-0205.htm

Central Pollution Control Board (2016) News Bulletin, 1. http://cpcb.nic.in/upload/Latest/Latest_123_SUMMARY_BOOK_FS.pdf

Chakdar H and Pabbi S (2012). Extraction and purification of Phycoerythrin from *Anabaena variabilis* (CCC421). Phykos, 42(1), 25–31.

Chen CY, Kuo EW, Nagarajan D, Ho SH, Dong CD, Lee DJ and Chang JS (2020). Cultivating *Chlorella sorokiniana* AK-1 with swine wastewater for simultaneous wastewater treatment and algal biomass production. Bioresource Technology, 122814. doi:10.1016/j.biortech.2020.122814.

Chen J, Liu L and Wei D (2017). Enhanced production of astaxanthin by *Chromo chloriszo fingiensis* in a microplate-based culture system under high light irradiation. Bioresource Technology, 245, 8–529.

Chinnasamy S, Bhatnagar A, Hunt RW and Das KC (2010). Microalgae cultivation in a waste water dominated by carpet mill effluents for biofuel applications. Bioresource Technololgy, 101:3097–3105

Chiu SY, Kao CY, Chen TY, Chang YB, Kuo CM and Lin CH (2015). Cultivation of microalgal *Chlorella* for biomass and lipid production using wastewater as nutrient resource. Bioresource Technology, 184, 179–189.

Chisti Y (2007). Biodiesel from microalgae. Biotechnology Advances, 25:294–306.

Choudhary P, Prajapati SK and Malik A (2016). Screening native microalgal consortia for biomass production and nutrient removal from rural waste waters for bioenergy applications. Ecological Engineering (91), 221–230.

Christian D, Zhang J, Sawdon AJ and Penga CA (2018). Enhanced astaxanthin accumulation in *Haematococcus pluvialis* using high carbon dioxide concentration and light illumination. Bioresource Technology, 256, 548–551. https://doi.org/10.1016/j.biortech.2018.02.074

Das C, Naseera K, Ram A, Meena RM and Ramaiah N (2016). Bioremediation of tannery wastewater by a salt-tolerant strain of *Chlorella vulgaris*. Journal of Applied Phycology, 29(1), 235–243. doi:10.1007/s10811-016-0910-8

De-Bashan LE, Trejo A, Huss VA, Hernandez JP and Bashan Y (2008). *Chlorella sorokiniana* UTEX 2805, a heat and intense, sunlight-tolerant microalga with potential for removing ammonium from wastewater. Bioresource Technology, 99(11), 4980–4989.

Dexter J and Pengcheng F (2009). Metabolic engineering of cyanobacteria for ethanol production. Energy & Environmental Science. 2(8), 857–864. doi:10.1039/b811937f.

Dikshit R and Tallapragada P (2018). Comparative study of natural and artificial flavoring agents and dyes. In Natural and Artificial Flavoring Agents and Food Dyes (pp. 83–111). https://doi.org/10.1111/tpj.13173

El-Kassas HY and Mohamed LA (2014). Bioremediation of the textile waste effluent by *Chlorella vulgaris*. Egyptian Journal of Aquatic Research, 40, 301–308

Elumalai S, Saravanan GK, Ramganesh S, Sakhtival R and Prakasam V (2013). Phycoremediation of textile dye industrial effluent from Tirupur district, Tamil Nadu, India. International Journal of Science Innovations and Discoveries, 3, 31–37.

Geraldes V, Jacinavicius FR, Genuário DB and Pinto E (2019). Identification and distribution of mycosporine-like amino acids in Brazilian cyanobacteria using ultrahigh-performance liquid chromatography with diode array detection coupled to quadrupole time-of-flight mass spectrometry. Rapid Communications in Mass Spectrometry, 34, 1–10.https://doi.org/10.1002/rcm.8634

González LE, Cañizares RO and Baena S (1997). Efficiency of ammonia and phosphorus removal from a Colombian agroindustrial wastewater by the microalgae Chlorella vulgaris and *Scenedesmus dimorphus*. Bioresource Technology, 60(3), 259–262.

Graham LE (2009). Algae (Wilbur B, editor). Pearson Education Inc. San Francisco.

Gupta SK, Ansari FA, Shriwastav A, Sahoo NK, Rawat I and Bux F (2016). Dual role of *Chlorella sorokiniana* and *Scenedesmus obliquus* for comprehensive wastewater treatment and biomass production for biofuels. Journal of Cleaner Production, 115, 255–264.

Hasan R, Zhang B, Wang L and Shahbazi A (2014). Bioremediation of Swine Wastewater and Biofuel Potential by using *Chlorella vulgaris*, *Chlamydomonas reinhardtii*, and *Chlamydomonas debaryana*. Journal of Petroleum & J Environmental Biotechnology, 5, 3 doi:10.4172/2157–7463.1000175.

Hemscheidt T, Puglisi MP, Larsen LK, Patterson GM, Moore RE, Rios JL and Clardy J. (1994). Structure and biosynthesis of borophycin, a new boeseken complex of boric acid from a marine strain of the blue-green alga *Nostoc linckia*. The Journal of Organic Chemistry, 59(12), 3467–3471. www.indiaenvironmentportal.org.in/files/file/Discharge%20of%20Sewage%20into%20Ganga%20River.pdf

Khan MI, Shin JH, and Kim JD (2018). The promising future of microalgae: Current status, challenges, and optimization of a sustainable and renewable industry for biofuels, feed, and other products. Microbial Cell Factories, 17(1), 36.

Kothari R, Prasad R, Kumar V and Singh DP (2013). Production of biodiesel from microalgae *Chlamydomonas polypyrenoideum* grown on dairy industry wastewater. Bioresource Technology, 144, 499–503.

Larsdotter K, La CJ and Dalhammar G (2007). Biologically mediated phosphorus precipitation in wastewater treatment with microalgae. Environmental Technology, 28(9), 953–960.

Lau PS, Tam NFY and Wong YS (1995). Effect of algal density on nutrient removal from primary settled wastewater. Environmental Pollution, 89(1), 59–66.

Ledda C, Idà A, Allemand D, Mariani P and Adani F (2015). Production of wild *Chlorella* sp. cultivated in digested and membrane-pretreated swine manure derived from a full-scale operation plant. Algal Research, 12, 68–73.

Li Y, Wang B, Wu N and Lan CQ (2008). Effects of nitrogen sources on cell growth and lipid production of *Neochloris oleoabundans*. Applied Microbiology and Biotechnology, 81(4), 629–636.

Liang S, Liu X, Chen F and Chen Z (2004) Current microalgal health food R & D activities in China. In Asian pacific phycology in the 21st century: Prospects and challenges (pp. 45–48). Springer, Dordrecht.

López G, Yate C, Ramos FA, Cala MP, Restrepo S and Baena S (2019). Production of polyunsaturated fatty acids and lipids from autotrophic, mixotrophic and heterotrophic cultivation of *Galdieria sp.* strain USBA-GBX-832. Scientific Reports, 9, Article number: 10791

Luo LZ, Shao Y, Luo S, Zeng FJ and Tian GM (2018). Nutrient removal from piggery wastewater by *Desmodesmus* sp. CHX1 and its cultivation conditions optimization. *Environmental Technology*. 40(21), 2739–2746.

Lupatini AL, Colla LM, Canan C and Colla E (2017). Potential application of microalga *Spirulina platensis* as a protein source. Journal of the Science of Food and Agriculture, 97(3) February, 724–732. doi:10.1002/jsfa.7987.

Lutzu GA, Zhang W and Liu T (2015). Feasibility of using brewery wastewater for biodiesel production and nutrient removal by *Scenedesmus dimorphus*. Environmental Technology. http://dx.doi.org/10.1080/09593330.2015.1121292

Mahapatra DM, Chanakya HN and Ramachandra TV (2013). *Euglena* sp. as a suitable source of lipids for potential use as biofuel and sustainable waste water treatment. Journal of Applied Phycology, 25, 855–865.

Mahdy A, Ballesteros M and González-Fernández C (2016). Enzymatic pretreatment of Chlorella vulgaris for biogas production: Influence of urban wastewater as a sole nutrient source on macromolecular profile and biocatalyst efficiency. *Bioresource Technology*, 199, 319–325.

Malvis A, Hodaifa G, Halioui M, Seyedsalehi M and Sánchez S (2018). Integrated process for olive oil mill wastewater treatment and its revalorization through the generation of high added value algal biomass. Water Research. 151, 332–342. doi:10.1016/j.watres.2018.12.026

Manjunath M, Prasanna R, Nain L, Dureja P, Singh R, Kumar A and Kaushik BD (2010). Biocontrol potential of cyanobacterial metabolites against damping off disease caused by Pythium aphanidermatum in solanaceous vegetables. Archives of Phytopathology and Plant Protection, 43(7), 666–677.

Markou G and Georgakakis D (2011). Cultivation of filamentous cyanobacteria (bluegreen algae) in agro-industrial wastes and wastewaters: A review. Applied Energy, 88(10), 3389–340.

Metting FB (1996). Biodiversity and application of microalgae. Journal of Industrial Microbiology, 17, 477–489.

Miura Y, Sode K, Nakamura N, Matsunaga N and Matsunaga T (1993). Production of γ-linolenic acid from the marine green alga *Chlorella* sp. NKG 042401. FEMS Microbiology Letters, 107(2–3), 163–167.

Okajima-Kaneko M, Miyazato S and Kaneko T (2009). Chemically cross-linking effects on the sorption of heavy metal ions to hydrogels of cyanobacterial mega molecules, sacran. Transactions of the Materials Research Society of Japan, 34(2), 359–362.

Olguín and EJ (2003). Phycoremediation key issues for cost-effective nutrient removal processes, Biotechnology advances, 22(1–2), 81–91.

Oliveira AC, Barata A, Batista AP and Gouveia L (2018). *Scenedesmus obliquus* in poultry wastewater bioremediation. Environmental Technology, 1–10. doi:10.1080/09593330.2018.1488003

Othman, JH, Ibrahim Z, Hara H, Yahya NA, Koji I and Mohamad SE (2019). Production of α-linolenic Acid by an oleaginous green algae *Acutodesmusobliquus* Isolated from Malaysia. Journal of Pure and Applied Microbiology, 13(3), 1297–1306.

Parmar A, Singh NK, Pandey A, Gnansounou E and Madamwar D (2011). Cyanobacteria and microalgae: A positive prospect for biofuels. Bioresource Technology, 102(22), 10163–10172.

Patel VK, Sundaram S, Patel AK and Kalra A (2018). Characterization of seven species of cyanobacteria for high-quality biomass production. Arabian Journal for Science and Engineering, 43(1), 109–121.

Pham M, Schideman L, Scott J, Rajagopalan N and Plewa MJ (2013). Chemical and biological characterization of waste water generated from hydrothermal liquefaction of *Spirulina*. Environmental Science & Technology, 47, 2131–2138.

Prasanna R, Chaudhary V, Gupta V, Babu S, Kumar A, Singh R, Singh SY and Nain L (2013). Cyanobacteria mediated plant growth promotion and bioprotection against Fusarium wilt in tomato. European Journal of Plant Pathology, 136, 337–353.

Prasanna R, Joshi M, Rana A and Nain A (2010). Modulation of IAA production in cyanobacteria by tryptophan and light. Polish Journal of Microbiology, 59(2), 99–105.

Prasanna R, Nain L, Tripathi R, Gupta V, Chaudhary V, Middha S, Joshi M, Brahma RA and Kaushik D (2008). Evaluation of fungicidal activity of extracellular filtrates of cyanobacteria—possible role of hydrolytic enzymes First published: 28 May https://doi.org/10.1002/jobm.200700199

Razzak SA, Hossain MM, Lucky RA, Bassi AS and Lasa HD (2013). Integrated CO2 capture, wastewater treatment and biofuel production by microalgae culturing—a review. Renewable and Sustainable Energy Reviews, 27, 622–653.

Saha SK, Kazipet N and Murray P (2018). The carotenogenic *Dunaliella salina* CCAP 19/20 produces enhanced levels of carotenoid under specific nutrients limitation. BioMed Research International. |Article ID 7532897 | https://doi.org/10.1155/2018/7532897

Samantaray S and Mallick N (2012). Production and characterization of poly-β-hydroxybutyrate (PHB) polymer from *Aulosira fertilissima*. Journal of Applied Phycology, 24(4), 803–814.

Sara RA, Nima MN, Saeedeh S, Azam S, Aboozar K, Pegah M, Mohammad AM and Younes G (2014). Removal of nitrogen and phosphorus from wastewater using microalgae free cells in batch culture system. Biocatalysis and Agricultural Biotechnology, 3, 126–131.

Sharma GK and Khan SA (2013). Bioremediation of sewage wastewater using selective algae for manure production. International Journal of Environmental Engineering and Management, 4(6), 573–580.

Sharma J, Kumar V, Kumar SS, Malyan SK, Mathimani T, Bishnoi NR and Pugazhendhi A (2020). Microalgal consortia for municipal wastewater treatment—Lipid augmentation and fatty acid profiling for biodiesel production. Journal of Photochemistry and Photobiology B: Biology, 202, January, 111638.

Shchegolkova N, Shurshin K, Pogosyan S, Voronova E, Matorin D and Karyakin D (2018). Microalgae cultivation for wastewater treatment and biogas production at Moscow wastewater treatment plant. Water Science and Technology. 78(1), 69–80.

Silambarasan T, Vikramathithan M, Dhandapani R, Mukesh Kumar DJ and Kalaichelvan PT. (2012). Biological treatment of dairy effluent by microalgae. World Journal of Science and Technology, 2(7), 132–134.

Singh DP, Khattar JS, Rajput A, Chaudhary R and Singh R (2019). High production of carotenoids by the green microalga *Asterarcys quadricellulare* PUMCC 5.1.1 under optimized culture conditions. PLoS One, 14(9), e0221930. https://doi.org/10.1371/journal.pone.0221930

Singh M, Reynolds DL and Das KC (2011). Microalgal system for treatment of effluent from poultry litter anaerobic digestion. Bioresource Technology, 102(23), 10841–10848. doi:10.1016/j.biortech.2011.09.037.

Source: http://timesofindia.indiatimes.com/home/environment/pollution/Around-80-of-sewage-in-Indian-cities-flows-into-water-systems/articleshow/18804660.cms

Source: www.indiaenvironmentportal.org.in/files/file/Discharge%20of%20Sewage%20into%20Ganga%20River.pdf

Stolz P and Obermayer B (2005). Manufacturing microalgae for skin care. Cosmetics and Toiletries, 120(3), 99–106.

Thomas NV and Kim SK (2013). Beneficial effects of marine algal compounds in cosmeceuticals. Marine Drugs, 11(1), January, 146–164.

Tripathi R, Gupta S and Thakur IS (2019). An integrated approach for phycoremediation of wastewater and sustainable biodiesel production by green microalgae, *Scenedesmus sp*. ISTGA1. Renewable Energy, 135, e617–e625.

Usha MT, Sarat Chandra T, Sarada R and Chauhan VS (2016). Removal of nutrients and organic pollution load from pulp and paper mill effluent by microalgae in outdoor open pond. Bioresource Technology, 214, 856–860. doi:10.1016/j.biortech.2016.04.060

Wang L, Li Y, Chen P, Min M, Chen Y, Zhu J and Ruan RR (2010). Anaerobic digested dairy manure as a nutrient supplement for cultivation of oil-rich green microalgae *Chlorella sp*. Bioresource Technology, 101, 2623–2628.

Wang Y, Ho SH, Cheng CL, Nagarajan D, Guo WQ, Lin C, Li S, Ren N and Chang JS (2017). Nutrients and COD removal of swine wastewater with an isolated microalgal strain *Neochloris aquatica* CL-M1 accumulating high carbohydrate content used for biobutanol production. Bioresource Technology, 242, 7–14.

Waterbury, J.B. (2006). The Cyanobacteria-Isolation, Purification and Identification. In: Dworkin, M., Falkow, S., Rosenberg, E., Schleifer, KH., Stackebrandt, E. (eds) The Prokaryotes. Springer, New York, NY. https://doi.org/10.1007/0-387-30744-3_38.

Woertz I, Feffer A, Lundquist T and Nelson Y (2009). Algae grown on dairy and municipal wastewater for simultaneous nutrient removal and lipid production for biofuel feedstock. Journal of Environmental Engineering, 135, 1115–1122.

Wu L, Chen P, Huang A and Lee C (2012). The feasibility of biodiesel production by microalgae using industrial waste water. Bioresource Technology, 113, 14–18.

Wu YH, Hu HY, Yu Y, Zhang TY, Zhu SF, Zhuang LL, Zhang X and Lu Y (2014). Microalgal species for sustainable biomass lipid production using wastewater as resource: A review. Renew Sustain Energy Review, 33, 675–688.

Xin L, Hong-ying H and Jia Y (2010). Lipid accumulation and nutrient removal properties of a newly isolated freshwater microalga, *Scenedesmus sp*. LX1, growing in secondary effluent. New Biotechnology, 27, 59–63.

Xiong W, Li X, Xiang J and Wu Q (2008). High-density fermentation of microalga *Chlorellaprotothecoides* in bioreactor for microbio-diesel production. Applied Microbiology Biotechnology, 78, 29–36.

Yin X, Wang L, Duan G, Sun G (2011). Characterization of arsenate transformation and identification of arsenate reductase in a green alga Chlamydomonas reinhardtii. Journal of Environmental Science, 23, 1186–1193.

Zhang CW, Zmora O, Kopel R and Richmond A (2001). An industrial-size flat plate glass reactor for mass production of *Nannochloropsis* sp. (Eustigmatophyceae). Aquaculture, 195(1–2), 35–49.

Zhang Y, Habteselassie MY, Resurreccion EP, Mantripragada V, Peng S, Bauer S and Colosi LM (2014). Evaluating removal of steroid estrogens by a model alga as a possible sustainability benefit of hypothetical integrated algae cultivation and wastewater treatment systems. ACS Sustainable Chemistry & Engineering, 2(11), 2544–2553.

Zhou W, Chen P, Min M, Ma X, Wang J, Griffith R, Hussain F, Peng P, Xie Q, Li Y, Shi J, Meng J and Ruan R (2014). Environment-enhancing algal biofuel production using wastewaters. Renewable and Sustainable Energy Reviews, 36, 256–269.

12 Microalgae for Plastic Biodegradation and Bioplastics Production

Jyothi K, Krishna Prasad M, Nivedita Sahu, and S. Sridhar

12.1 INTRODUCTION

Plastics of marine environment are often consumed by marine species which influence the food habits of fish, and their feed shrinks their multiplication, development, and population-level (Galloway et al., 2017). Ultimately, these adulterated marine species might be ingested by social groups as seafood, which indirectly upsets human health. The intake of toxic organic pollutants in the alignment of polymers (i.e., plasticizers, biocides, and flame retardants) imposes a danger to the communal clusters (Rochman et al., 2015). The fishing and aquaculture business is extremely friable as the yield, profitability, viability, and security are influenced by the plastic waste of the marine atmosphere (Rochman et al., 2015). Pollution due to plastic debris in aquatic environments causes vulnerable effects on marine life, including ocean faunae and coral reefs. This marine debris causes assimilation, entanglement, enervation, and choking to the marine types, causing condensed life quality, partial abilities to avoid predators, lessened reproductive capability, damage to feeding capacity, and demise (Stephanis et al., 2013). Since microplastics have been noticed in food and air samples, there are growing concerns about the possible issues of microplastics on social welfare. Human pressures such as the dumping of plastic waste into the marine system have adversely impacted the welfare of human and marine ecosystems (Galloway et al., 2017). The continuous entry of plastic waste into the world's oceans from terrestrial bases will rise extremely in the next era (Jambeck et al., 2015). These plastic wastes may degrade into minor fragments of microplastic (0.1 mm–5 mm) distributed in the ecosystem over geographical spans, leading to the necessity of time demanding, expensive, and laborious supervision in eliminating these aquatic plastics. Tourists stay for shorter time in seashore environments and evade some sites due to the existence of plastic debris on the coastline, which causes a decrease in the finances of tourism income (Hartley et al., 2013). Other concerns because of the occurrence of plastic wastes are unexpected injuries such as high-pitched debris, tangled in nets and contact to unhygienic things, harmful to physical and psychological welfare, and intensively inhabited by a variety of unscrupulous species (Kirstein et al., 2016).

Plastics are artificial carbon-based polymers which are aquaphobic, inactive, high-molecular-weight elongated chains of monomers linked by covalent bonds (Brigham, 2018). Their characteristics include flexible, malleable, durable, robust, frivolous, and low-priced, making them fit for the making of a range of products along with domestic objects, produce wrapping, and shop bags. Conventionally, artificial plastics are produced using refined petroleum products, whereas synthetic polymers made of carbon-carbon bonds are obtained in a contained environment (Shrivastava, 2018). These traditional plastics manufactured by heavy crude oil can cause fossil reserve exhaustion, climate variation, and greenhouse gases productions (Abdul et al., 2020). Nevertheless, the biotransformation of plastics depends on their chemical construction but not on the basis of the monomers (Siracusa and Blanco, 2020). For instance, generally used plastics, polyethene terephthalate (PET), polyvinyl chloride (PVC), high-density polyethylene (HDPE), low-density polyethylene

DOI: 10.1201/9781003219156-15

(LDPE), polystyrene (PS), polypropylene (PP), and varied plastics are non-biotransformable plastics although the initial monomers of polymers, including PE, PP, PVC, and PET might be resulting from biotic sources. These plastics synthesized by high molecular weight because of their repetitions of minor monomer units (Alshehrei, 2017). Transformable plastics could be separated into four groups: compostable plastics, phototransformable plastics, bio-based plastics, and biotransformable plastics. Biodegradable plastics generally break down while interrelating with liquid, enzymes, UV, and regular changes in pH. These can be formed by renewable resources comprising composites of animals, existing plants, algae, and microorganisms. Polyhydroxyalkanoate (PHA) is an illustration of biodegradable plastic which is biodegradable and has comparable properties to traditional plastics. This biodegradable bioplastic methodology is exceptionally resource-efficient with the advantages of saving energy and reducing carbon dioxide emissions (Thielen, 2014).

Plastics can be rapidly condensed to their basic constituents with the help of biological agents. Various microorganisms can be able to change some plastic polymers biologically into simpler products through aerobic and anaerobic mechanisms (Ahmed et al., 2018). A biological agent can exploit the biological polymer as a nutritious substrate for energy and progress, emerging in microbial biomass as the end produce of whole biotransformation (Kumar et al., 2017). In recent times, it was observed that several microalgae activate biotransformation of polymers, and the energy essential for transformation is condensed since the produced enzymes with modest or several toxin systems comprise a decrease in activation energy to decline the biochemical bonds in the polymer (Bhuyar, 2018).

Microalgae that can develop on waste supplies with more lipid gathering can be a possibly improved candidate for bioplastic manufacture as it would not contest with food bases (Khoo et al., 2020). Microalgal biomass offers the capability of mass production and greenhouse gas absorption because of its quick development and more carbon fixing efficiency when they propagate (Ho et al., 2020). Methods to yield microalgal bioplastics comprise of (1) compounds produced by merging microalgal biomass, petroleum, or bio-dependent polymers and extracts (2) biopolymers like polyhydroxybutyrates (PHBs) and starch cultured within microalgal cells (Onen et al., 2020). This review confers the present position and development of microalgae in plastic degradation by illuminating the research of plastics and the related marine pollution to highpoint the serious need for evolving environmentally friendly methods and products. This review also focuses on the mechanism of biodegradation with the identification of possible microalgae species and the outcome of microplastics on algae. Consecutively, the probability of algae as a foundation of bioplastics is systematically studied, along with merging microalgae with supplementary ingredients and genetic engineering for microalgae species that produce biopolymer. This evaluation too emphasized the challenges of research concerning microalgae and plastics.

12.2 MECHANISM OF PLASTIC BIOTRANSFORMATION BY ALGAE

Microalgae settle on artificial substrata like polythene tops in sewage water, and these are considered nontoxic (Sarmah and Rout, 2018) (Table 12.1). Attachment of algae superficially will produce ligninolytic and exopolysaccharide enzymes which start the biotransformation, and it is crucial for plastic biodegradation (Sharma et al., 2014). The algal enzymes so formed will connect with macromolecules of the plastic exterior and starts the biodegradation (Chinaglia et al., 2018). The polymer is exploited by algae as a carbon source as the species developing on the PE surface have greater cellular contents, such as protein and carbohydrates and a higher specific growth rate (Sarmah and Rout, 2018). Researchers noticed surface degradation on the cross-section of the algal-colonized PE sheets (Kumar et al., 2017). The methods of biodegradation comprising snarling, deterioration, hydrolysis and diffusion, deprivation of leakage composites, and pigment coloration through dissemination into the polymers were noticed (Kumar et al., 2017). Blue-green alga (cyanobacterium), *Anabaena spiroides*, exhibited the maximum proportion of LDPE degradation (8.2%), subsequently diatom *Naviculapupula* (5%) and green alga *Scenedesmus dimorphus* (4%). *Phormidium*

TABLE 12.1
Establishment of Algae on Plastic Source

Algae Species	Plastic source	References
Scenedesmus dimorphus, Anabaena spiroides, and *Naviculapupula*	Discarded waste polyethylene bags	Sarmah and Rout, 2018
Oscillatoria princeps, O. acuminate, O. vizagapatensis, O. limnetica, O. earlei, O. limosa, O. chalybea, and *O. salina*	Discarded waste polyethylene bags, Submerged polythene	Chinaglia et al., 2018
Phormidium calcicola, Lyngbya cinerascens, Nostoc carneum, Spirulina major, Hydrocoleum sp., *Chlorella* sp., *Pithophora* sp., and *Scenedesmus quadricauda*	Polythene carry bags	
Calothrix marchica, Anomoeoneis sp., *Oedogonium* sp., *Arthrospira platensis, Naviculaminuta, Nitzschia intermedia, Spirogyra* sp., and *Synedra tabulate*	Solid native sewage dumping sites	Otsuki et al., 2004
Phormidium lucidum, Oscillatoria subbrevis, Lyngbyadiguetii, and *Cylindrospermum muscicola*	Discarded waste polythene bags Native dirt water drainage system	Yurtsever et al., 2017
Phormidium tenue, Oscillatoria tenuis, Microcystis aeruginosa, Closterium constatum, and *Chlorella vulgaris*	Waste polythene Several ponds, lakes	Rochman et al., 2015
Coleochaete soluta, Chaetophora, Aphanochaete, Gloeotaenium, Oedogonium, Oocystis, Oscillatoria, Phormidium, Chroococcus, Fragilaria, Navicula, and *Cymbella*	Waste polythene materials Oligotrophic water bodies	Sarmah et al., 2014

lucidum and *Oscillatoria subbrevis* (freshwater nontoxic cyanobacteria) can initiate the PE surface and biodegrading LDPE efficiently without any pretreatment. These strains are readily accessible and fast-growing (Sarmah and Rout, 2018). Microalgae *Chlorella vulgaris* along with *Aeromonas hydrophilia* bacteria were observed to be efficient in the biotransformation of bisphenol A (BPA), which is a widely exploited polymer in the plastic industry (Gulnaz and Dincer, 2009). The outcomes specified that the BPA was simply deteriorated by algae, and its concentrations were underneath detection confines after 168 h without estrogenic activities (Gulnaz and Dincer, 2009). Hirooka et al. (2005) employed green alga *Chlorella fusca var. vacuolate* in an observation that BPA was degraded to composites without estrogenic action. Besides, microalgae could be genetically modified, that can produce and secrete plastic-degrading enzymes. Green microalgae *Chlamydomonas reinhardtii* was renovated to express PETase, and the cell lysate of the degradant was raised with PET, ensuing indents and pits on the PET film surface along with TPA, which is the completely degraded method of the PET (Kim et al., 2020).

12.3 INFLUENCE OF MICROPLASTIC ON ALGAE

When macroplastics, or specifically, factory-made drugs or individual care products, degraded abiotically, microplastics (<5 mm) were produced. Microplastics are further degraded into minor units of size <100 mm (nanoplastics) which might be spotted ubiquitously. The Antarctic region, an area with the minimum populace and maximum unreachability is too polluted with microplastics (Waller et al., 2017). Microplastic particles can captivate harmful complexes including biological contaminants and heavy metals. They comprise destructive extracts and arrive in the food chain at the stage of microorganisms or small creatures because of their insignificant size and physical characteristics (Law, 2017).

Taipale et al. (2019) identified mixotrophic algae (*Cryptomonas* sp.) nurturing on microbiome settled on PE microplastic, where it sequestered/cut off carbon of the polyethylene microplastic (PE-MP) to produce essential u-6 and u-3 polyunsaturated fatty acids (PUFA). It was observed that the microbes taking over the microplastic cause higher growth rates of the algae. Nevertheless, direct contact with the PE-MP or its discharging chemicals has a poisonous influence on mixotrophic algae (*Cryptomonas* sp.) as there was a non-appearance of microbes exploiting the chemicals sheathing the plastic surface. Khoironi and Anggoro (2019) conveyed that the more the dose of microplastics, the lesser the progress of microalgae, presenting the influence of microplastics on the development of *Spirulina* sp. The reason is the existence of microplastics in cultivation might produce shading effects which cause decreased light intensity and upset microalgal photosynthesis (Yurtsever et al., 2017). According to Zhang et al. (2017) investigation, the adverse influence of microplastic on microalgae was due to the collaboration amongst microalgae and microplastic, for example, accumulation and adsorption, but not because of the shading effect. This clarifies that the properties of microplastic on microalgae hang onto the element dimensions of microplastic. Liu et al. (2020) stated that microplastic with greater size produced severe effects by obstructing the passage of light and disturbing photosynthesis while microplastic with minor size damaged the microalgal cell wall by captivating on its exterior.

Canniff and Hoang (2018) reported opposite findings that *Raphidocelis subcapitata* showed advanced growth in media comprising plastic microbeads (63–75 mm). And also, the findings of Chae et al. (2019) presented that cell development and photosynthetic action of marine microalga *Dunaliella salina* were stimulated without influence on cell morphology when microplastics were bigger than the algal cells around 200 mm in diameter. Chae et al. (2019) reasoned the encouraged growth may be due to the little amounts of chemicals which might be leaked from microplastics, for instance, stabilizers, phthalates, and endocrine degraders. Additional research aiming at how microplastics influence microalgae which play a critical role as prime manufacturers of ecologies is essential. It is vital to explore the capability of microalgae biodegrading microplastic.

12.4 MICROALGAE AS A RESOURCE OF BIOPLASTICS

Bioplastics are plastics which are prepared from living or inexhaustible bases, including food crops, and possess the same function as petroleum-based plastics (Mekonnen et al., 2013). Bioplastics could be made of diverse materials which have diverse prospects. Usually, bioplastics fall into three groups: (i) bioderived but nonbiodegradable plastics (PE, PP, PET, polytrimethylene terephthalate (PTT) or polyester elastomers (TPC-ET)); (ii) bioderived and compostable plastics (polylactic acid (PLA), PHA, starch, cellulose); (iii) remnant resource-dependent plastics are biodegradable (polybutylene adipate terephthalate (PBAT)). Several sources are exploited to prepare bioplastics, chiefly agronomy-related crops, including wheat, soy proteins, corn, gelatin, and collagen. This elevates the debate on the viability of these feedstocks, for example, the competition among land and water supplies for social consumption (Bastos, 2018). Moreover, the process of taking out composites, particularly polymers of plants for the preparation of bioplastics, is difficult because of the existence of covered cell walls (Lodish et al., 2000). Besides, the "green" plastics prepared from food crops, including cassava and sago, face the concerns of reduced water struggle and mechanical possessions (Machmud et al., 2013). Consequently, algae have been evolving as an innovative and probable biomass for making bioplastics as algae could be cultured on non-cultivable lands with a short collecting time (Chew et al., 2017). Algae are resistant to unfavorable environmental conditions, could upgrade wastewater, and exploit carbon dioxide as a nutrient supply for biomass cultivation (Zhang et al., 2017). In the preparation of algae-based plastic, the encapsulation of nonbiodegradable polymers, such as polyolefin, in the thermoplastic algal blends could grab and accumulate carbon dioxide in biomass form forever. Accordingly, carbon dioxide do not release again into the atmosphere (Jyothi, 2018), thus relieving the greenhouse effect. In total, algal-based bioplastics help as a favorable and nontoxic substitute that can decrease the use of fossil fuels, improve the plastic value, and lessen undesirable ecological influences carried by the extreme use of petroleum-based plastics (Beckstrom et al., 2020).

Some methods are employed for the manufacture of microalgae-based bioplastics, including the direct practice of microalgae biomass, merging with supplementary materials, intermediate biorefinery processing, and genetic engineering to produce ideal polymer-synthesizing microalgae strains (Rahman and Miller, 2017).

12.4.1 Direct Usage of Microalgae Biomass

Protein and carbohydrate-based polymers of the algal biomass could be exploited as bioplastics composites. At present, starch, cellulose, PHA, PHB, PLA, PE, PVC, and protein-dependent polymers are some complexes of algae biomass that were conditioned to synthesize biodegradable plastics (Karan et al., 2019). Among these, PHA is the utmost recommended to yield bioplastics since it could be biotransformed by enzymatic act. Further PHB, a kind of PHA, newly appeared as a novel polymer to harvest bioplastics because of the good barricade for oxygen. The bacteria *Ralstoniaeutropha* and *Bacillus megaterium* are identified to yield PHB as a source of carbon intracellularly (Hempel et al., 2011).

There are some illustrations of factory-made bioplastics by designers (Rasul et al., 2017). For example, Eric Klarenbeek and Maartje Dros, the inventers of Dutch, shaped a bioplastic product finished from algae that might completely swap the conventional plastics. They also developed AlgaeLab to nurture algae to yield starch as a resource for the bioplastic. Besides, Austeja Platukyte established biotransformable materials from algae, comprising agar of algae and layered calcium carbonate, which can exchange petroleum-based plastics. While the byproducts are of less weight, they are water resistant, robust, and strong. Furthermore, bioplastics can be applied as compost to maintain soil moisture (Wen et al., 2020). Similarly, Ari Johnsson combined red algae powder with water to make a substitute bottle for the conventional plastic bottle. The bottle will retain its shape when it is filled with liquid, and it will decompose when unfilled. The water reserved in the bottle are harmless to drink (Wen et al., 2020). In a nutshell, algae act as a capable replacing feedstock for bioplastics manufacture to exchange the traditional petroleum plastics, thus resolving the ocean plastic contamination.

12.4.2 Merging with Other Materials

The microalgae biomass could be merged with additional ingredients during the manufacture of bioplastics to extend their duration, expand their properties, and improve their machine-driven presentation. Blending materials include petroleum plastics and natural products such as cellulose, or starch, or polymers (Rahman and Miller, 2017). The merging materials can be developed from algae biomass, for example, PLA, PHA, cellulose, starch, and protein. Shi et al. (2012) conveyed that the programmed characteristics of the microalgae-comprising plastic films were found similar to the plastic films (polyurethane or PE) in spite of the drawback of low tensile strength, mainly the extension at disruption. Otsuki et al. (2004) used maleic anhydride as a blending material to change PE for an improved interface among *Chlorella* sp. and PE resulting in the greater stretchy strength and thermal flexibility of the complex made of *Chlorella* sp. and modified PE. Thus, merging the algae-based plastic goods with other supplies can advance the physical and chemical possessions of the bioplastics. Chiellini et al. (2008) blended starch and polyvinyl alcohol (PVA) with algae, *Ulva armoricana* to manufacture bioplastics. The starch condensed the quantity required of PVA by nearly 40% with maintaining good cohesion. And similarly the degeneration frequency of the merged product displays the possibility of this source as a biodegradable and eco-compatible compound since it will ultimately be predisposed in the liquid and solid media. Zeller et al. (2013)

studied the thermo-mechanical polymerization of *Chlorella* sp. and *Spirulina* sp. for the production of algal bioplastics and their thermos-plastic mixtures. The result showed *Chlorella* sp. exhibited an improved bioplastic behavior but lesser blending presentation than *Spirulina* sp. biomass.

Mathiot et al. (2019) employed microalgae strain, *C. reinhardtii* 11032A for making starch-based bioplastics, and subsequently plasticization with glycerol. The starchy bioplastics were observed to have agreeable plasticization potential. Machmud et al. (2013) exploited the red algae, *Eucheuma cottonii*, as the resource for plastics manufacture via filtration technique. To yield bioplastics, red algae were combined distinctly with latex of *Artocarpus altilis* and *Calostropis gigantea* to swap the usage of glycerol as a plasticizer since the augmented concentration of glycerol condensed the depth and concentration of the bioplastics. Thus, the blend of bioplastics assisted by merging with the composites taken out from microalgae biomass can aid to expand the mechanical possessions of the end products. Throughout the assortment of merging materials, it is vital to see that the composites are required to be biotransformable to evade undesirable ecological influence.

12.4.3 Biorefinery Approach to Producing PHAs

A biorefinery approach for producing PHAs probably diminishes the total price of the system (Rizwan et al., 2015). Microalgae biorefineries systems produce multiple product streams similar to petroleum industries. Biorefineries systems exploiting lipids for biodiesel, hydrolyzing cell walls for solvent or plastic production, carbohydrates for food, and remaining biomass for anaerobic digestion could allow an integrated system to be cost inexpensive. By growing microalgae on domestic wastewater that could supply free sources of nutrients, costs could be even further reduced (Rahman and Miller, 2017).

One such method is using wastewater microalgae as a feedstock for a biorefinery model to yield multiple product streams (Anthony et al., 2013). In this process, domestic wastewater microalgae species such as *Scenedesmus obliquus* were grown in solar simulated bioreactors and harvested with several cationic starches (Anthony and Sims, 2013). Here the harvested microalgae biomass was processed using the wet lipid extraction procedure (WLEP) and produced three product streams. The first stage removed approximately 81% of all transesterifiable lipids from wet microalgae biomass through an acid/base hydrolysis, and lipids were converted to fatty acid methyl esters (FAME) or biodiesel. The first stage of the WLEP removed the widely held chlorophyll in the FAME [67, 68]. In the second stage, hydrolyzed microalgae biomass was exploited to make acetone, butanol, and ethanol through anaerobic fermentation by *Clostridium saccharoperbutylacetonicum* N1–4 (Ellis et al., 2012). In the third stage of the WLEP, the aqueous phase from stage two was used for recombinant *E. coli* cultivation and additional bioproduct generation (Figure 12.1). This research confirmed that biodiesel, solvents, and *E. coli* biomass could be synthesized from a single microalgae source.

The kind of harvesting method also affected the product streams produced by the WLEP. Using centrifugation, the solvents produced were 1.44 g/L butanol and 0.71 g/L acetone. By cationic corn and potato starches for microalgae harvesting produced 1.89 and1.9 g/L butanol and 0.92 and 0.85 g/L acetone. The highest yield of FAME achieved from the WLEP was in the centrifuged algae sample. In the case of recombinant *E. coli* growth, centrifuged microalgae media gave the highest levels of growth related to the other harvesting techniques (Anthony et al., 2013). A biorefinery that utilized the WLEP, centrifugation, and coagulation/flocculation was the harvesting method of choice as they produced the highest yields of bioproducts (Anthony et al., 2013).

12.4.4 Bioplastic Production by Genetically Engineered Microalgae

PHB is a thermoplastic, biodegradable polyester formed by bacteria. The algae strains can be modified to produce composites for bioplastics manufacture through genetic engineering (Jyothi, 2018). Employing genetic engineering by implanting the bacterial PHB making genes into microalgae could lessen the making charge as the bacterial fermentation structure for manufacturing PHB is expensive

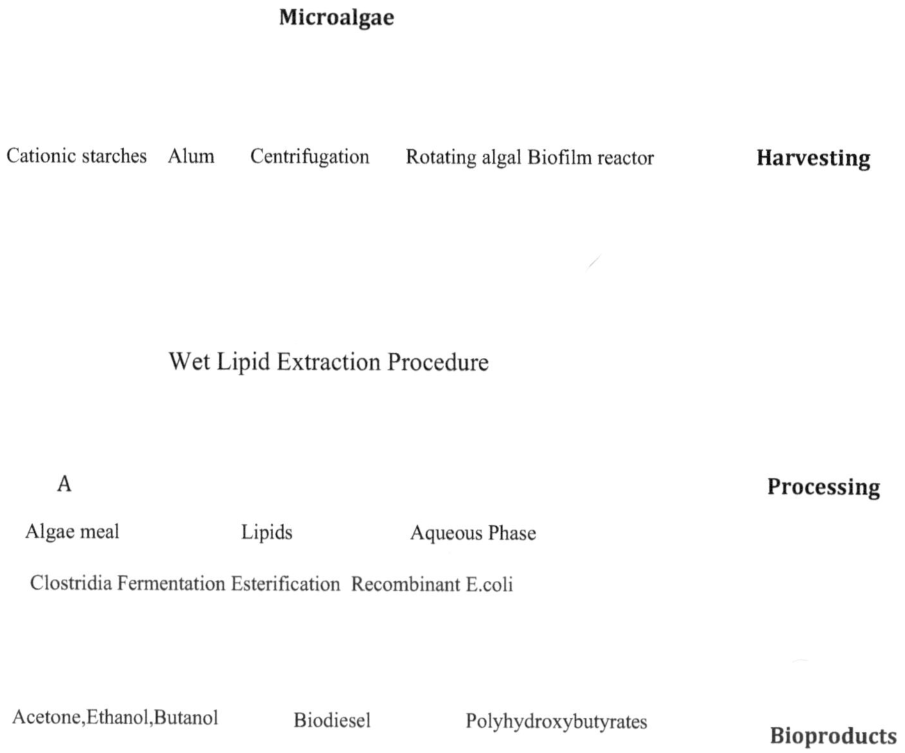

FIGURE 12.1 Wastewater microalgae harvesting and processing through wet lipid extraction procedure to produce multiple products in a biorefinery system.

for bioplastic construction (Hempel et al., 2011). Chaogang et al. (2010) altered the *Chlamydomonas reinhardtii* species with two expression vectors comprising phbB and phbC genes of *R. eutropha* for the making of PHB. This resulted in the existence of PHB particles in the cytol of transgenic algae. This encourages the study of improved creation of polyhydroxybutyrate in the genetically engineered algae and the buildup of polyhydroxybutyrate in the chloroplast. The genome of *C. reinhardtii* was fully sequenced in 2007 and observed to have similar phylogenetic properties as plants. *Chlamydomonas reinhardtii* can grow photoautotrophically or with acetate and has a doubling time of 10 h. Hempel et al. (2011) presented the biosynthesis of bacterial PHB path into the cytosolic section of diatom *Phaeodactylum tricornutum*. This method is inexpensive and eco-friendly and can produce PHB content of up to 11% of algal dry weightiness. In addition to *C. reinhardtii*, some other strains that are currently suitable for genetic manipulation are *Cyanidioschyzon merolae*, *Nannochloropsis* sp., *Ostreococcus tauri*, *Phaeodactylum tricornutum*, and *Thalassiosira pseudonana* (Chun et al., 2017). Henceforth, genetic engineering is the encouraging method to change the genes of specific algae strains to increase the manufacture of the composites of attention, such as starch or polymers, in lesser time to yield bioplastics (Chun et al., 2017). Algae are genetically modest as related to other complex organisms (Gimpel et al., 2015), which simplifies the genetic engineering practice. Genetic engineering is extremely probable by altering the DNA segments of the algae species to improve the making of polymers or other molecules for the production of bioplastics.

12.5 CHALLENGES OF ALGAL BIOPLASTICS

The development and production of algal plastics on a huge measure slowed down by a few disputes. These difficulties are required to be overwhelmed to avoid the waning of the plastic contamination and exhaustion of fossil fuels.

1. First, the identification of appropriate algae to yield polymers for bioplastics with diverse properties is crucial. The algae biomass conformation differs among the species. *Microcystis aeruginosa* had the maximum content of PHB (0.6 mg mL^{-1}) among the verified microalgae strains and could be exploited for making bioplastic with good plasticizing capacity (Abdo and Ali, 2019). Similarly, the abstraction of starch from *Scenedesmus almeriensis* was more successful than from *Neochloris oleoabundans* (Johnsson and Steuer, 2018).
2. During the planning of bioplastics, the assortment of suitable polymers mined from algae is a task to yield sustainable bioplastics with decent flexibility and strength as related to traditional plastics. Some aspects must be taken into account through the collection of suitable polymers, for instance, biodegradability, feedstock renewability, degradation rate, brittleness, consumer satisfactoriness, size of polymer, molecular weight, and humidity content (Rochman et al., 2015).
3. Throughout the synthesizing procedure of bioplastics, it is necessary to contemplate the influences on the atmosphere when the bioplastics are being degenerated, for instance, specific degradation situations, sluggish disintegration process, or accretion of the produced methane and other destructive gas. It is vital to guarantee that the bioplastics could be degraded in any ecological situations and do not produce injurious gases (Rasul et al., 2017).
4. The culture of algal system is one more task that wants to be overwhelmed prior to the marketing of algal plastics. Large-scale culturing is required to mass culture the algal biomass to harvest polymers or other amalgams. The algal biomass can be cultured in closed photobioreactor systems or the open pond, and these culture arrangements have their rewards and drawbacks (Shi et al., 2012).
5. Additionally, the treatment of bioplastics leftover is vital. There are limited approaches that can be practical to waste management of bioplastics such as chemical and mechanical reprocessing, incineration, anaerobic digestion, composting, and landfilling. Amongst these approaches, composting is the utmost appropriate technique for waste treatment of bioplastics since the biotransformable bioplastics destroy in less period as related to conservative plastics (Jankova, 2018).
6. Besides, the problem of unpleasant odors in bioplastics prepared from algae lipids has to be overwhelmed. Wang et al. (2016) discovered a foul-smelling odor when the plastic goods synthesized from microalgae *Nannochloropsis* sp., and planktonic algae (blend with PE or PP) were confined to greater portions of fatty acids, specifically polyunsaturated fatty acids. Absorbents would be needed to eliminate or lessen the odor as this will disturb the development of algal-based bioplastics.
7. Blesin et al. (2017) stated that consumers are unaware of the term "bioplastics or there is an absence of cognizance amongst the customers about the usage of bioplastics". Some customers take up that bioplastics are assessed to be costly due to the use of carbon-based materials.

12.6 CONCLUSIONS

Microalgae come up with the biotransformation of plastic debris by the enzymes that deteriorate the organic bonds of plastic polymers. The usage of microalgae to renovate plastics into metabolites, for example, water, carbon dioxide, and novel biomass, is of prime importance. As the diversity of plastics is explored, extra biodegradation studies utilizing algae must be performed, and the efficiency must be assessed. Widespread studies on how microplastics disturb microalgae will extremely assist in the considering of the effects of its elemental concentration on biodegradation research. An additional concern is that microalgae that contain proteins, hydrocarbon, lipids, and other high-value-added compounds are frequently exploited for food produce and should be free from contaminants, such as microplastics (Khoo et al., 2020). It is significant to examine the consequences of microplastics on the well-being of marine organisms and humans. Using microalgae for plastic biotransformation offers numerous benefits, and henceforth considered a probable solution. Bioplastics formed employing microalgae are cheap and ecologically harmless to substitute conservative plastics. Yet, the studies on algal bioplastics are in the childhood stage and unfeasible to be marketed on an industrial scale. With the additional exploration of the algal role in bioplastic degradations, a great number of bioplastics could be formed from algae biomass sustainably soon.

REFERENCES

Abdo, S.M., G.H. Ali, Analysis of polyhydroxybutrate and bioplastic productionfrom microalgae, Bull. Natl. Res. Cent. 43 (2019) 97.

Abdul-Latif, N.I.S., M.Y. Ong, S. Nomanbhay, B. Salman, P.L. Show, Estimation of carbon dioxide (CO2) reduction by utilization of algal biomass bioplastic in Malaysia using carbon emission pinch analysis (CEPA), Bioengineered. 11 (2020) 154–164.

Ahmed, T., M. Shahid, F. Azeem, I. Rasul, A. Shah, M. Noman, A. Hameed, N. Manzoor, D.I. Manzoor, S. Muhammad, Biodegradation of plastics: Current scenario and prospects for environmental safety, Environ. Sci. Pollut. Res. 25 (2018) 1–12.

Alshehrei, F., Biodegradation of synthetic and natural plastic by microorganisms, J. Appl. Environ. Microbiol. 5 (2017) 8–19.

Anthony, R.J., J.T. Ellis, A. Sathish, A. Rahman, C.D. Miller, R.C. Sims, Effect of coagulant/flocculants on bioproductsfrom microalgae, Bioresour. Technol. 149 (2013) 65–70.

Anthony, R.J., R. Sims, Cationic starch for microalgae and total phosphorus removal from wastewater, J. Appl. Polym. Sci. 130 (2013) 2572–2578.

Bastos Lima M., Toward multipurpose agriculture: Food, fuels, flex crops, and prospects for a bioeconomy, Global Environ. Polit. 18 (2018) 143–150.

Beckstrom, B.D., M.H. Wilson, M. Crocker, J.C. Quinn, Bioplastic feedstock production from microalgae with fuel co-products: A techno-economic and life cycle impact assessment, Algal Res. 46 (2020) 101769.

Bhuyar, P., Biodegradation of plastic waste by using microalgae and their toxins, in: Biopolymers & Bioplastics & Polymer Science and Engineering Conferences, Las Vegas, 2018.

Blesin, J.M., M. Jaspersen, W. Mohring, Boosting plastics' image? Communicative challenges of innovative bioplastics, E-plastory-J. Plast. History. 1 (2017) 2.

Brigham C., Chapter 3.22—biopolymers: Biodegradable alternatives to traditional plastics, in: B. Torok, T. Dransfield (Eds.), Green Chem, Elsevier, 2018, 753–770.

Canniff, P.M., T.C. Hoang, Microplastic ingestion by *Daphnia magna* and its enhancement on algal growth, Sci. Total Environ. 633 (2018) 500–507.

Chae, Y., D. Kim, Y.-J. An, Effects of micro-sized polyethylene spheres on the marine microalga *Dunaliella salina*: Focusing on the algal cell to plastic particle size ratio, Aquat. Toxicol. 216 (2019) 105296.

Chaogang, W., H. Zhangli, L. Anping, J. Baohui, Biosynthesis of poly-3-hydroxybutyrate (phb) in the transgenic green alga *Chlamydomonas Reinhardtii*1, J. Phycol. 46 (2010) 396–402.

Chew, K.W., J.Y. Yap, P.L. Show, N.H. Suan, J.C. Juan, T.C. Ling, D.-J. Lee, J.- S. Chang, Microalgae biorefinery: High value products perspectives, Bioresour. Technol. 229 (2017) 53–62.

Chiellini, E., P. Cinelli, V.I. Ilieva, M. Martera, Biodegradable thermoplastic composites based on polyvinyl alcohol and algae, Biomacromolecules. 9 (2008) 1007–1013.

Chinaglia, S., M. Tosin, F. Degli-Innocenti, Biodegradation rate of biodegradable plastics at molecular level, Polym. Degrad. Stabil. 147 (2018) 237–244.

Chun, Y., D. Mulcahy, L. Zou, I.S. Kim, A short review of membrane fouling in forward osmosis processes, Membranes. 7 (2017) 30.

Ellis, J.T., N.N. Hengge, R.C. Sims, C.D. Miller, Acetone, butanol, and ethanol production from wastewater algae, Bioresour. Technol. 111 (2012) 491–495.

Galloway, T.S., M. Cole, C. Lewis, Interactions of microplastic debris throughout the marine ecosystem, Nat. Ecol. Evol. 1 (2017) 1–8.

Gimpel, J.A., V. Henriquez, S.P. Mayfield, In metabolic engineering of eukaryotic microalgae: Potential and challenges come with great diversity, Front. Microbiol. 6 (2015) 1376.

Gulnaz, O., S. Dincer, Biodegradation of bisphenol a by *Chlorella vulgaris* and *Aeromonas hydrophilia,* J. Appl. Biol. Sci. 3 (2009) 79–84.

Hartley, B., S. Pahl, R. Thompson, Baseline evaluation of stakeholder perceptions and attitudes towards issues surrounding marine litter, DeliverableD2. 1 (2013).

Hempel, F., A.S. Bozarth, N. Lindenkamp, A. Klingl, S. Zauner, U. Linne, A. Steinbüchel, U.G. Maier, Microalgae as bioreactors for bioplastic production, Microb. Cell Factories. 10 (2011) 81.

Hirooka, T., H. Nagase, K. Uchida, Y. Hiroshige, Y. Ehara, J.-I. Nishikawa, T. Nishihara, K. Miyamoto, Z. Hirata, Biodegradation of bisphenol a and disappearance of its estrogenic activity by the green alga *Chlorella fusca* var. *vacuolata*, Environ. Toxicol. Chem. 24 (2005) 1896–1901.

Ho, S.-H., C. Zhang, F. Tao, C. Zhang, W.-H. Chen, Microalgal torrefaction for solid biofuel production, Trends Biotechnol. (2020).

Jambeck, J.R., R. Geyer, C. Wilcox, T.R. Siegler, M. Perryman, A. Andrady, R. Narayan, K.L. Law, Plastic waste inputs from land into the ocean, Science 347 (2015) 768771.

Jankova, V., Bioplastics: Opportunities and Challenges, Department of Environmental Sciences, Central European University, 2018.

Johnsson, N., F. Steuer, Bioplastic Material from Microalgae: Extraction of Starch and PHA from Microalgae to Create a Bioplastic Material, School of Industrial Engineering and Management, KTH Royal Institute of Technology, 2018.

Kaparapu, J., Polyhydroxyalkanoate (PHA) production by genetically engineered microalgae: A review, J. New Biol. Rep. 7(2) (2018) 68–73.

Karan, H., C. Funk, M. Grabert, M. Oey, B. Hankamer, Green bioplastics as part of a circular bioeconomy, Trends Plant Sci. 24 (2019) 237–249.

Khoironi, A., S. Anggoro, Evaluation of the interaction among microalgae Spirulina sp, plastics polyethylene terephthalate and polypropylene in freshwater environment, J. Ecol. Eng. 20 (2019).

Khoo, K.S., W.Y. Chia, D.Y.Y. Tang, P.L. Show, K.W. Chew, W.-H. Chen, Nanomaterials utilization in biomass for biofuel and bioenergy production, Energies 13 (2020) 892.

Kim, J.W., S.B. Park, Q.-G. Tran, D.H. Cho, D.Y. Choi, Y.J. Lee, H.S. Kim, Functional expression of polyethylene terephthalate-degrading enzyme (PETase) in green microalgae, Microb. Cell Factories. 19 (2020) 97.

Kirstein, I.V., S. Kirmizi, A. Wichels, A. Garin-Fernandez, R. Erler, M. Löder, G. Gerdts, Dangerous hitchhikers? Evidence for potentially pathogenic Vibrio spp. on microplastic particles, Mar. Environ. Res. 120 (2016) 1–8.

Kumar, R.V., G. Kanna, S. Elumalai, Biodegradation of polyethylene by green photosynthetic microalgae, J. Biorem. Biodegrad. 8 (2017) 2.

Law, K.L., Plastics in the marine environment, Ann. Rev. Mar. Sci. 9 (2017) 205–229.

Liu, G., R. Jiang, J. You, D.C.G. Muir, E.Y. Zeng, Microplastic impacts on microalgae growth: Effects of size and humic acid, Environ. Sci. Technol. 54 (2020) 1782–1789.

Lodish, H., A. Berk, S. Zipursky, P. Matsudaira, D. Baltimore, J. Darnell, The Dynamic Plant Cell Wall, WH Freeman, 2000.

Machmud, M.N., R. Fahmi, R. Abdullah, C. Kokarkin, Characteristics of red algae bioplastics/latex blends under tension, Int. J. Eng. Sci. 5 (2013) 81–88.

Mathiot, C., P. Ponge, B. Gallard, J.-F. Sassi, F. Delrue, N. Le Moigne, Microalgae starch-based bioplastics: Screening of ten strains and plasticization of unfractionated microalgae by extrusion, Carbohydr. Polym. 208 (2019) 142–151.

Mekonnen, T., P. Mussone, H. Khalil, D. Bressler, Progress in bio-based plastics and plasticizing modifications, J. Mater. Chem. A 1 (2013) 13379–13398.

Onen, S., Z.K. Chong, M.A. Kücüker, N. Wieczorek, U. Cengiz, K. Kuchta, Bioplastic production from microalgae: A review, Int. J. Environ. Res. Publ. Health. 17 (2020).

Otsuki, T., F. Zhang, H. Kabeya, T. Hirotsu, Synthesis and tensile properties of anovel composite of Chlorella and polyethylene, J. Appl. Polym. Sci. 92 (2004) 812–816.

Rahman, A., C.D. Miller, Chapter 6—microalgae as a source of bioplastics, in: R.P. Rastogi, D. Madamwar, A. Pandey (Eds.), Algal Green Chemistry, Elsevier, 2017, 121–138.

Rasul, I., F. Azeem, M.H. Siddique, S. Muzammil, A. Rasul, A. Munawar, M. Afzal, M.A. Ali, H. Nadeem, Chapter 8—algae biotechnology: A green light for engineered algae, in: K.M. Zia, M. Zuber, M. Ali (Eds.), Algae Based Polymers, Blends, and Composites, Elsevier, 2017, 301–334.

Rizwan, M., J.H. Lee, R. Gani, Optimal design of microalgae-based biorefinery: Economics, opportunities and challenges, Appl. Energy 150 (2015) 69–79.

Rochman, C.M., A. Tahir, S.L. Williams, D.V. Baxa, R. Lam, J.T. Miller, F.-C. Teh, S. Werorilangi, S.J. Teh, Anthropogenic debris in seafood: Plastic debris andfibers from textiles in fish and bivalves sold for human consumption, Sci. Rep. 5 (2015) 14340.

Sarmah, P., J. Rout, Algal colonization on polythene carry bags in a domestic solid waste dumping site of Silchar town in Assam, Phykos 48 (2018).

Sharma, M., A. Dubey, A. Pareek, Algal flora on degrading polythene waste, CIBTech J Microbiol. 3 (2014) 43–47, 67–77.

Shi, B., G. Wideman, J.H. Wang, A new approach of BioCO2 fixation by thermoplasticprocessing of microalgae, J. Polym. Environ. 20 (2012) 124–131.

Shrivastava, A., Introduction to Plastics Engineering, William Andrew, 2018.

Siracusa,V., and Blanco, I., Bio-Polyethylene (Bio-PE),Bio-Polypropylene (Bio-PP) and Bio-Poly (ethylene terephthalate) (Bio-PET):Recent Developments in Bio-Based Polymers Analogous to Petroleum-Derived Ones for Packaging and Engineering Applications. Polymers (2020), 12, 1641.

de Stephanis, R., J. Giménez, E. Carpinelli, C. Gutierrez-Exposito, A. Cañadas, As main meal for sperm whales: Plastics debris, Mar. Pollut. Bull. 69 (2013) 206e214.

Taipale, S.J., E. Peltomaa, J.V.K. Kukkonen, M.J. Kainz, P. Kautonen, M. Tiirola, Tracing the fate of microplastic carbon in the aquatic food web by compound-specific isotope analysis, Sci. Rep. 9 (2019) 19894.

Thielen, M., Bioplastics: Plants and Crops Raw Materials Products, Fachagentur Nachwachsende Rohstoffe eV (FNR) Agency for Renewable Resources, 2014.

Waller, C.L., H.J. Griffiths, C.M. Waluda, S.E. Thorpe, I. Loaiza, B. Moreno, C.O. Pacherres, K.A. Hughes, Microplastics in the Antarctic marine system: An emerging area of research, Sci. Total Environ. 598 (2017) 220–227.

Wang, K., A. Mandal, E. Ayton, R. Hunt, M.A. Zeller, S. Sharma, Chapter 6 -modification of protein rich algalbiomass to form bioplastics and odorremoval, in: G. Singh Dhillon (Ed.), Protein Byproducts, Academic Press, 2016, 107–117.

Wen Yi Chia, Doris Ying Ying Tang, KuanShiong Khoo, Andrew Ng Kay Lup, Kit Wayne Chew, Nature's fight against plastic pollution: Algae for plastic biodegradation and bioplastics production, Environ. Sci. Ecotechnol. 4 (2020) 100065.

Yurtsever, M., T. OngunSevindik, H. Tunca, The impact of PS microplastics on green algae Chlorella vulgaris growth, in: 15th International Conference on Environmental Science and Technology, 2017. Rhodes, Greece.

Zeller, M.A., R. Hunt, A. Jones, S. Sharma, Bioplastics and their thermoplastic blends from *Spirulina* and *Chlorella* microalgae, J. Appl. Polym. Sci. 130 (2013) 3263–3275.

Zhang, C., X. Chen, J. Wang, L. Tan, Toxic effects of microplastic on marine microalgae *Skeletonema costatum*: Interactions between microplastic and algae, Environ. Pollut. 220 (2017) 1282–1288.

Theme III

Algal-Omics and Applications

13 Nuclear Characteristics and Chromosome of Green Algae

A Review

Iccha Purak Singh and Nivedita Sahu

13.1 INTRODUCTION

13.1.1 Systematics and General Features of Green Algae

Fritsch (1935, 1944) placed all green algae under single-class Chlorophyceae with nine orders (Volvocales, Chlorococcales, Ulotrichales, Cladophorales, Chaetophorales, Conjugales, Oedogoniales, Siphonales, and Charales). In recent classifications charophytes are treated separately (Round, 1965; Stewart and Mattox, 1975; Mattox and Stewart, 1984). Most of the green algae are inhabitants of fresh water, a few are epiphytic, parasitic, or marine.

The thallus structure of green algae range from motile unicellular to nonmotile unicellular to colony/coenobium to unbranched filamentous to branched filamentous to heterotrichous or parenchymatous or siphonaceous or highly specialized forms (Charales) (Figure 13.1).

The green algae have eukaryotic cell structure having cell wall; nucleus and membrane bound cell organelles. The inner layer of cell wall is generally cellulosic and photosynthetic pigments, that is, chlorophyll a, b, carotenoids, and xanthophylls are present in the same proportion as that of higher plants. The pigments are located in definite chloroplasts. The photosynthate is generally starch. The flagella when present range from 2–4- ∞, are equal in length, and are all of whiplash type (Isokontae).

13.1.2 Nuclear Structure of Green Algae

In majority, green algal cells are uninucleate (e.g. In *Chlamydomomas, Oedogonium, Spirogyra, Zygnema*, etc.), but in some members of Cladophorales and Siphonales, the cells are multinucleate (e.g. *Cladophora, Rhizoclonium, Bryopsis, Caulerpa*, etc.). The nucleus of green algal cells possesses a definite two layered nuclear membrane. The outer nuclear membrane continues with endoplasmic reticulum. The nuclear membrane has numerous pores. The interphase nucleus of most of the green algal cells contains uncoiled and expanded chromatin structures along with one or more nucleoli and a few deeply stainable lumps (chromocentres). The nucleoli and chromatin material remain suspended in the granular matrix of the nucleus.

13.1.3 Cell Cycle and Process of Cell Division

Nuclear division is mitotic in vegetative cells and meiotic in reproductive cells. Both mitosis and meiosis are like higher plants. Cell division involves division of nucleus (Karyokinesis) followed by Cytokinesis. During interphase (resting phase): time gap between two divisions, the nucleus as well as cell prepares for next division and comprise G1, S, and G2 substages. During this stage, proteins, enzymes, and other necessary molecules are synthesized and replication of DNA takes place (Figure 13.2).

DOI: 10.1201/9781003219156-17

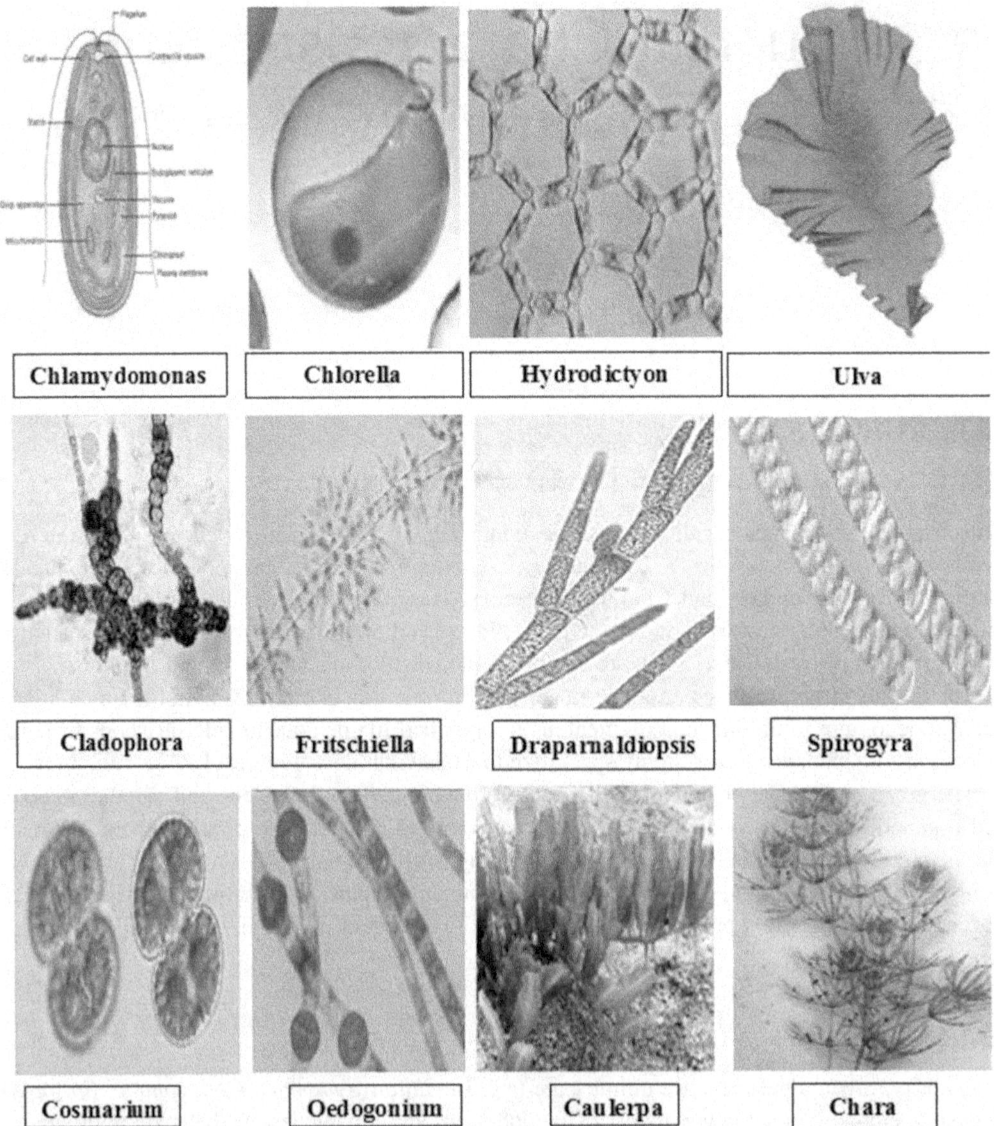

FIGURE 13.1 Thallus organization in green algae.

Mitosis involves prophase, metaphase, anaphase, and telophase as that of higher plants. The nuclear membrane and nucleoli degenerate at the end of prophase and are reorganized after the telophase. Significant contributions for chromosome number and karyotypic analysis of green algae were made by a number of researchers.

The process of mitosis in green algae follows the pattern as that of higher plants. The interphase nucleus enlarges considerably (2–3 times) with the onset of prophase. The granular structure in prophase nucleus later results in discontinuously stained long chromatin threads as a network. Late prophase is marked by disappearance of nuclear membrane and nucleolous. The chromatin threads start condensation to form rod like chromosomes. At metaphase, chromosomes are arranged on the equatorial plate. Chromosomes at this stage are thick and rod shaped with smooth outline. Chromosomes get longitudinally divided into two chromatids attached only at centromere.

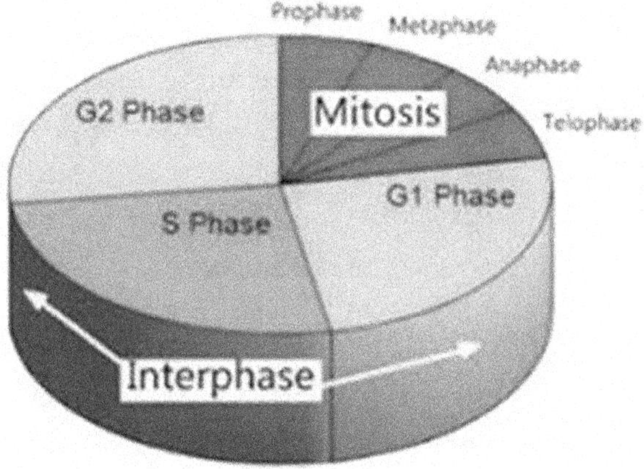

FIGURE 13.2 Cell cycle pattern in green algae.

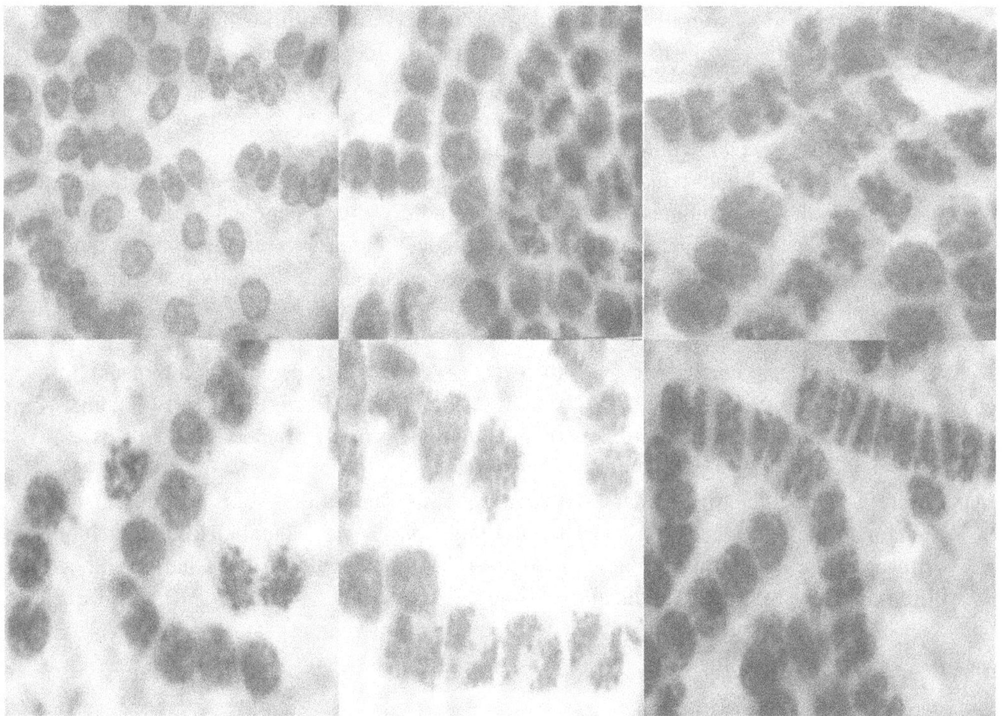

FIGURE 13.3 Different stages of mitosis in *Chara corallina*: (1) interphase, (2) prophase, (3 & 4) metaphase, (5) anaphase, (6) telophase.

Two chromatids of a chromosome start moving apart towards opposite poles indicating onset of anaphase (Figure 13.3).

The chromatids assume different shapes (V, J, and I) depending upon the position of centromere (median, submedian, subterminal, and terminal) (Figure 13.4).

The position of centromere can be traced comfortably in orders like Chaetophorales, Oedogoniales, and Charales. At telophase two sets of chromatids undergo de-condensation, and nuclear membrane and nucleolous reappear, organizing two identical daughter nuclei.

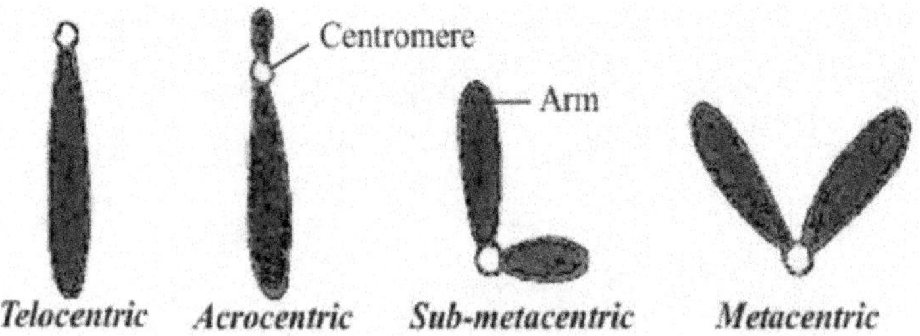

FIGURE 13.4 Types of chromosomes on the position of centromere.

Cell plate is laid down in between the two daughter nuclei at telophase. These daughter cells undergo a period of rest before undergoing next division.

Two modes of nuclear and cell division are reported in green algae.

i) The mitosis is intranuclear (with nuclear envelope not completely degraded at metaphase) and is interrupted by spindle microtubules. A characteristic additional set of microtubules are disposed transversely to long axis of spindle at the equator. It occurs at telophase and is termed as phycoplast. The phycoplast facilitates cytokinesis either by furrowing or cell plate formation. In this case, the two daughter nuclei tend to lie closer to each other. Example *Ulothrix, Oedogonium, Draparnaldia, Fritschiella*, etc.
ii) In other case, spindle and nuclear envelope are open at metaphase (completely degraded). At telophase the microtubules of intranuclear spindle are persistent and result in formation of phragmoplast. These fibers (vertical microtubules) increase in number to form a barrel-shaped region as a result of which the two daughter nuclei are distantly placed, example, charophytes, *Spirogyra, Coleochaete*, etc.

After conducting an extensive survey of the process of mitosis (especially cytokinesis) in several genera of green algae, Pickett-Heaps (1967, 1972), Marchant and Pickett Heaps (1973), and Mattox and Stewart (1984) have suggested that green algae may be broadly divided into two classes.

A) Chlorophyceae including orders Volvocales, Chlorococcales, Microsporales, Ulvales, Chaetophorales, and Oedogoniales showing Phycoplast (two nuclei lie close to each other)
B) Chlorophyceae including orders Coleochaetales, Conjugales, and Charales showing Phragmoplast (two nuclei lie distantly) (Figure 13.5)

13.1.4 Chromosome Types Recorded in Green Algae

About 250 taxa of green algae have been analyzed cytologically with the help of light microscopic study. On the basis of extensive studies, following chromosome types may be recognized in green algae (Sarma, 1982)

- Apparently long chromosomes
- Extremely small chromosomes
- Polycentric or diffused centric chromosomes
- Nucleolar organizing chromosomes
- Lampbrush chromosomes

Apparently long chromosomes with a recognizable centromere, as in members of Oedogoniales, Cladophorales, and Charales. Chromosomes range up to 12 μm and may have single distinct

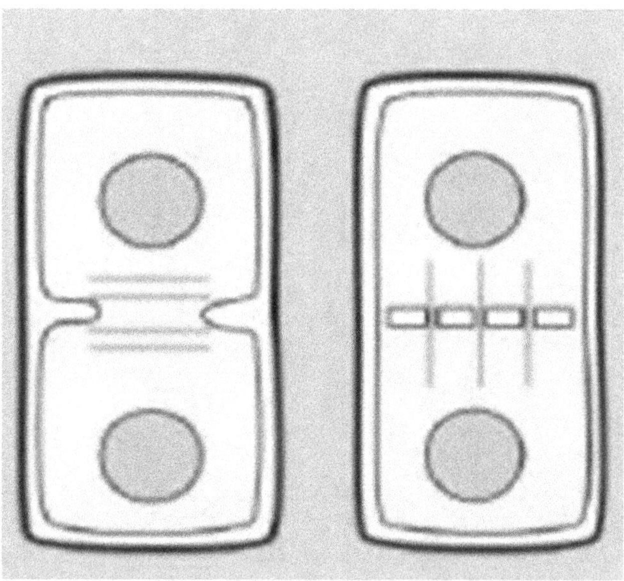

FIGURE 13.5 Different modes of cytokinesis: (1) phycoplast and (2) phragmoplast.

localized centromere (Singh and Choudhary, 1990). V, J, and I shaped configurations of anaphase chromosomes indicate metacentric, sub-metacentric, and telocentric nature of chromosomes.

Extremely small chromosomes (0.2 µm–1.0 µm) without any indication of the position of the centromere under light microscope as in most of the Volvocales, Chlorococcales, Ulotrichales, Chaetophorales, and some species of filamentous Conjugales and desmids. Later on ultrastructural studies provided clues of localized centromere even in very minute chromosomes and are able to move towards opposite poles during anaphase through well-organized spindle.

Polycentric or diffused centric chromosomes, centromere not localized, but is diffused throughout length, for example, several taxa of *Spirogyra* and desmids. In case of polycentric chromosomes of Conjugales, sister chromatids at metaphase and anaphase stages lie parallel to each other indicating their poly or diffused centric nature (Godward, 1954).

Nucleolar organizing chromosomes (NO chromosomes) have been recorded in several algae. Godward (1956) has critically analyzed the structure, behavior, and role of secondary constriction in chromosomes of Conjugales and mentioned about the presence of terminal satellite separated from main body of chromosome by sub terminal constriction. Further Electron micrographic studies have shown that during development of nucleolous, the organizer appear as fibrillar material (nucleolemma) and interpreted this as chromatin expanded for transcription.

Lampbrush chromosomes have been recorded in the primary nucleus of green alga *Acetabularia* of Siphonales (Spring et al., 1975; Koop, 1979; De and Berger, 1990), studied detailed morphology, meiotic pairing using fluorescence microscopy, and reported certain chromosomes having distinct loops (Figure 13.6).

Electron microscopy of spread preparation of isolated nuclear components of *Acetabularia* revealed that although the lampbrush chromosomes are small (4–30 µm long), they possess typical loops (up to 20 µm long). Chromomere-like nodules (1–2 µm in diameter) and large axial globules (2–4 µm in diameter) are comparable to those of animal lampbrush chromosomes. Association of some of these chromosomes with nucleolar structure and nuclear envelope were also recognized.

The discovery of lampbrush chromosomes in *Acetabularia* proved as milestone in history of elucidation of chromosome structure in algae and in understanding the possible evolutionary trends in genome organization during evolution of angiosperms.

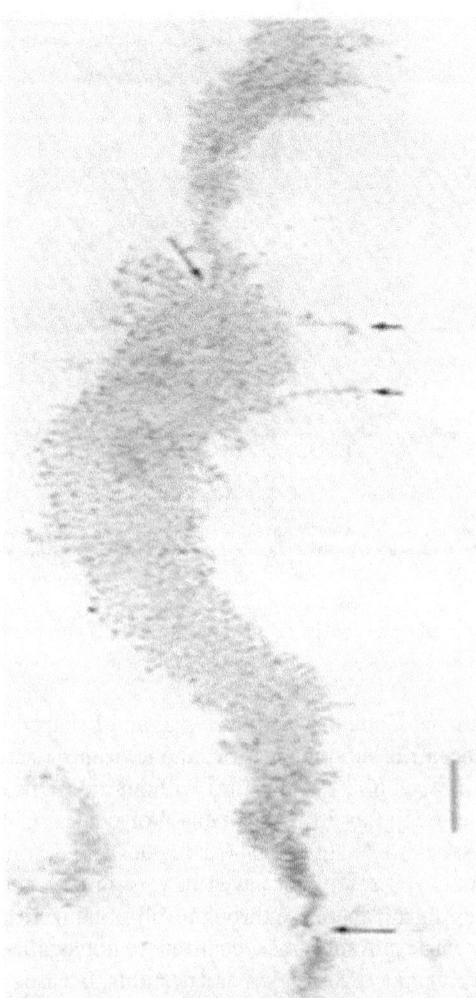

FIGURE 13.6 Lampbrush-type chromosomes in the primary nucleus of the green alga *Acetabularia mediterranea*.

Source: Spring et al. (1975)

13.1.5 Karyotypes in Green Algae

The karyotype of a species or taxon illustrates its chromosome complement, including number, length, centromeric position, breadth, nucleolar organizer chromosomes, etc. Its graphical representation is known as Ideogram/Karyogram (Figure 13.7).

Every individual species possesses its distinct karyotype. Karyotypic analysis have been made in green algae especially Charophyceae by many cytologists viz. Godward (1948–1950), Sunderlingam (1962), Sinha (1958, 1974), Sarma (1964), Guerlesquin (1963, 1967), Sarma and Khan (1965b), Sinha and Noor (1967), Noor (1967), Verma (1967), Sinha and Ahmad (1971), Chatterjee (1971), Noor and Mukherjee (1977), Ramjee, P and Bhatnagar, S K (1978a), Purak (1986).

Generally, differences in karyotypes result in some morphological differences also but sometimes species having similar karyotypes may be entirely different morphologically.

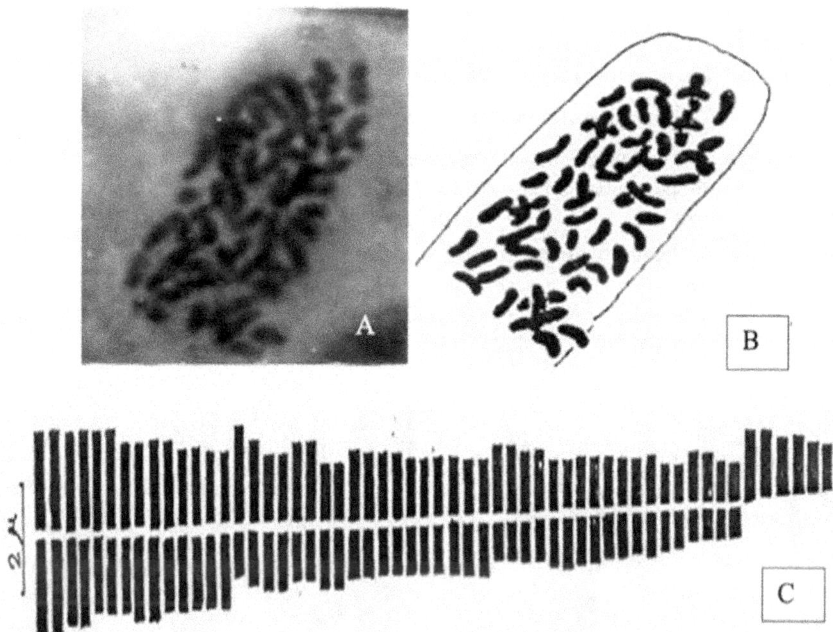

FIGURE 13.7 Seven karyological features of *Chara fibrosa* var. *fibrosa* f. *fibrosa* (Nordst.) R D W (A) photomicrograph of metaphase plate (n = 56); (B) outline drawing of metaphase; (C) karyogram showing 56 chromosomes.

Source: Purak (1986)

Karyological studies play an important role in understanding the evolutionary pattern as opined by Stebbins (1971) for higher plants and help in solving systematic position of certain controversial algal genera.

a) Sphaeroplea and Microspora are more clearly related to Ulotrichales than to Cladophorales or Siphonales (Sarma, 1962) based on their karyology.
b) Consideration of Schizomeris as an ecophene of *Stigeoclonium* (Campbell and Sarafis, 1972) has been opposed by Choudhary (1975) and Sarma and Choudhary (1975b) on cytological grounds, and according to them, *Schizomeris* deserves a generic status.

Karyotypes with low chromosome number, chromosomes of almost of same size, large size of chromosome, with median and submedian centromere are taken as primitive, whereas high number of chromosomes, small size of chromosomes, with appreciable difference in size of chromosomes (smallest and largest chromosome), terminal and sub-terminal position of centromere are considered as asymmetric and advanced.

Analysis of karyotype symmetry can be used as a tool to trace intervarietal, interspecific, or intergeneric karyotypic evolution. Karyotypic analysis of different taxa is helpful in tracing evolutionary sequence.

Karyotypes have been classified into 12 categories (1A- 4C) as per Stebbins (1958, 1971), taking into account both the degree of difference between the largest (LC) and smallest (SC) chromosome of complement and position of centromere on chromosome (Table 13.1).

The orders like Ulotrichales, Siphonales, Chlorococcales have asymmetric karyotypes and are advanced over orders like Cladophorales, Oedogoniales, and Charales having symmetric karyotypes.

TABLE 13.1
Karyotype Categories as per Stebbins (1958)

Karyotype Categories as per Stebbins (1958)

	P = No. of chromosomes with sm centromere/No. of total chromosomes			
LC/SC	Proportion of chromosome with arm ratio< 2:1			
	0.00	0.01–0.50	0.51–0.99	1.00
<2:1	1A (most primitive)	2A	3A	4A
2:1–4:1	1B	2B	3B	4B
>4:1	1C	2C	3C	4C (most advanced)

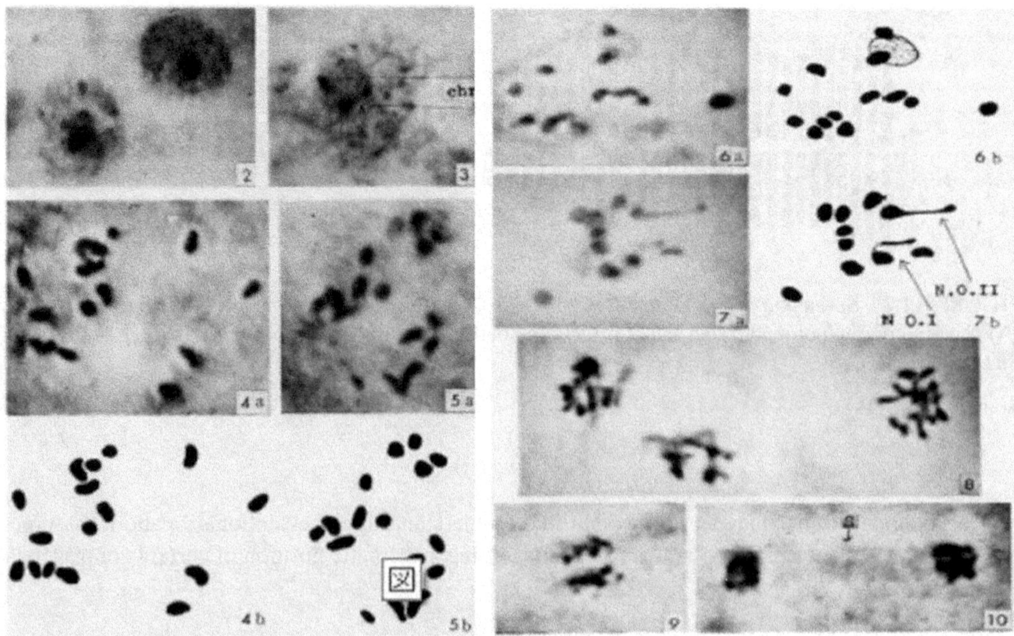

FIGURE 13.8 Courtesy: Sarma YSRK (1962). Nuclear cytology of *Sphaeroplea annulina* (Roth) Ag. *Cytologia* 27:71–78.

13.1.6 Chromosome Numbers Recorded in Green Algae

(Basic chromosome number, polyploid and aneuploid chromosome counts)

In green algae chromosome number ranges from n = 2 in some species of Spirogyra to n = 592 in a Desmid *Netrium digitus* (King, 1960).

The lowest chromosome number in Volvocales is n = 4 (*Chlamydomonas monoiae*) and highest is n = 38±4 in *Chlamydomonas eugametus*). The chromosomes are mostly minute dot like. Aneuploidy is prevalent in the group (Sarma, 1982).

Chromosome number in Chlorococcales varies from n = 4 (*Scenedesmus platydiscus*) to n = 80 (*Eraemosphaera viridis*).

Ulotrichales show a range of n = 3 (*Prasiola japonica*) to n = 48 (*Hormidium crenulatum*). The chromosomes are extremely small in *Ulothrix zonata*, *Ulva lactuca*, and *Enteromorpha*. Sarma (1962) reported n = 16 for *Sphaeroplea annulina* of Ulotrichales. The prophase nuclei show two nucleolar organizing chromosomes, and he also mentioned that *Sphaeroplea* should be included as a genus under Ulotrichales on cytological grounds (Figure 13.8).

Sinha (1963) reported 24 chromosomes at mitotic metaphase and 12 bivalents at meiotic metaphase in *Cladophora flexuosa*. Out of 12 bivalents, 7 bivalents show one chiasma each and the rest 5 bivalents show two chiasmata each. Sinha (1967) reported polyploid counts 48, 72, and 96 in four freshwater forms of *Cladophora glomerata*, where as its marine form exhibits diploid count of 24 chromosomes (Figure 13.9).

N = 4 is lowest chromosome count for Chaetophorales (*Draparnaldiopsis indica*) and *Fritschiella tuberosa* whereas n = 56 forms the highest count (*Trentepohlia treubiana* and *T. unarata*). Michetti et al. (2010) have reported three, five, and eight chromosomes in four species of *Stigeoclonium* from Argentina.

The different chromosome counts for different genera of Conjugales show that euploidy has played a significant role in speciation of *Spirogyra*, but on the whole, aneuploidy plays a role in species diversification. Abhayavardhani and Sarma (1983) recorded n = 6, 12, 14, and 24 chromosome numbers for *Spirogyra* links. Most common chromosome number in *Spirogyra* is n = 24.

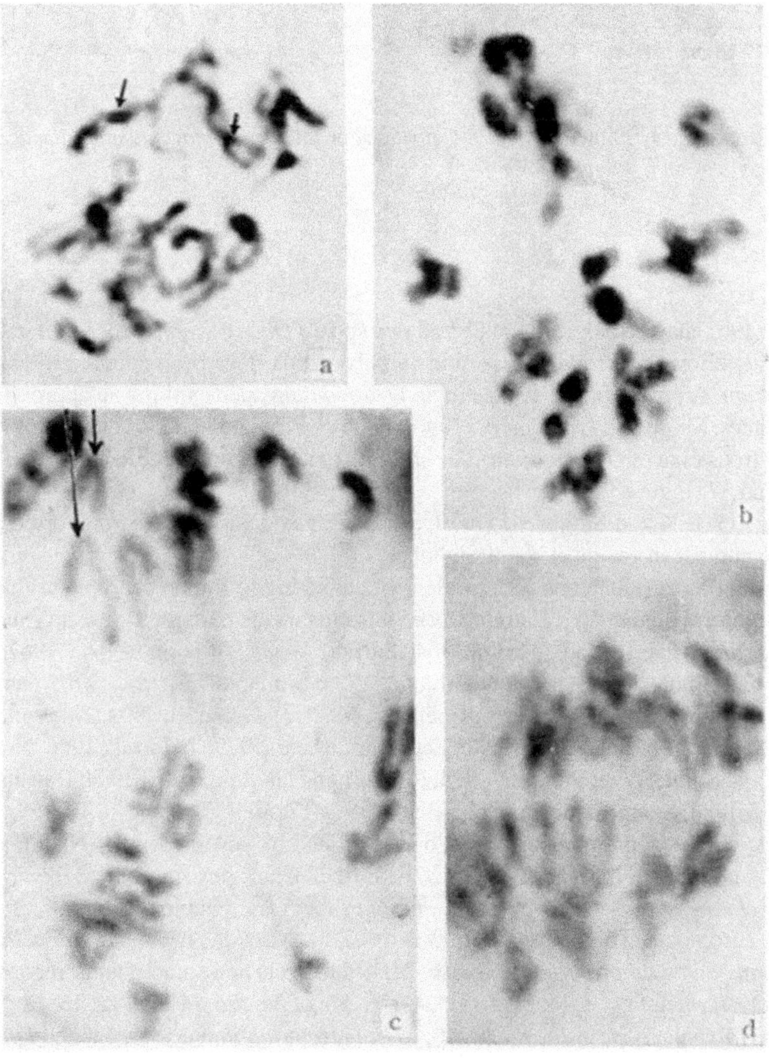

FIGURE 13.9 Courtesy: Sinha JP (1963). Cytotaxonomical Studies on *Cladophora flexuosa* (Griff.) Harv., a *Marine* sp. *Cytologia* 28:1–11. Twelve bivalents at meiotic metaphase.

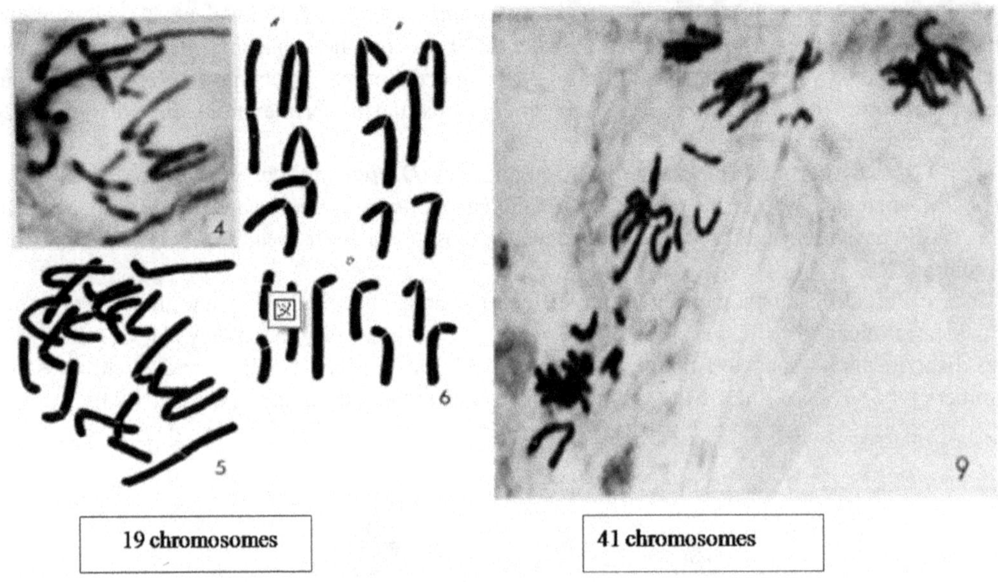

FIGURE 13.10 Courtesy: Sinha JP (1962). Cytological studies on *Oedogonium cardiacum* Wittrock and another *Oedogonium* sp. *Cytologia* 28:194–200.

Sinha J P (1962) has cytologically analyzed two spp. of *Oedogonium*. Process of nuclear division during mitosis follows a normal course, and 19 chromatids have been counted in anaphase plates of *O. cardiacum*. A pre-metaphase stage in an *Oedogonium* species show an approximate count of 41 chromosomes. Singh and Choudhary (1990) made cytological and karyological investigation on nine macrandrous taxa of *Oedogonium*. The chromosome numbers recorded were n = 13, 17, 19, 21, 22, 25, 32, and 37 (Figure 13.10).

Chowdhary, Y B K and Singh, S J (1970) and Verma and Kumar (1985) reported 6, 10, and 20 chromosome number in various taxa of Siphonales.

More than 150 taxa (Khan and Sarma,1967a,b) belonging to Indian charophytes have been subjected to cytological studies by different workers from various parts of the world (Sundaralingam, 1946; Sarma and Khan,1965a,b; Sarma,1964; Sarma, 1968; Sinha and Noor, 1967, 1971; Noor, 1965; Sinha and Sinha, 1989; Sinha and Ahmad, 1973; Sinha and Verma, 1970; Chatterjee, 1970, 1975, 1976; Chakrabarty and Ray, 2016; Chennaveeraiah and Bharati, 1974; Noor and Mukherjee, 1975, 1977; Ramjee and Bhatnagar, 1978b; Patel and Jawale, 1979a,b; Purak, 1986; Noor and Purak 1989; Purak and Sinha, 1998; Pundhir, 1994; Pundhir and Gautam, 1993, 1994; Ray and Chatterjee, 1986a,b, 1987, 1988; Subramanian and Ganesan, 1983) (Figures 13.11 and 13.12).

The lowest chromosome number of n = 6 is exhibited by several taxa of Nitella while highest number n = 70 is recorded in *Chara zeylanica*. Almost all Charophycean genera like *Chara, Nitella, Tolypella, Lychnothamnus*, and *L. amprothamnium* show varied euploid series.

The basic chromosome number for *Chara* is n = 7, for *Nitella* is n = 3, for *Tolypella* is n = 5, for *Lychnothamnus* and *Lamprothanium* is n = 7. Different chromosome counts recorded in varied Indian taxa of *Chara* are n = 7, 14, 21, 28, 35, 42, 49, 56; of *Nitella* n = 6, 9, 12, 15, 18, 21, 24, 27, 30, 36, 42, 48; of *Tolypella* n = 10, 11, 15, 20, 50; of *Lamprothamnium* n = 14, 42; and *Lychnothamnus* n = 14, 28, 35.

Some aneuploid counts (8, 11, 19, 26, 27, 29, 48, 60) have also been recorded in a few charophyte taxa by Noor and Mukherjee (1975, 1977), Sarma and Bala (1984), Bhatnagar and Johri

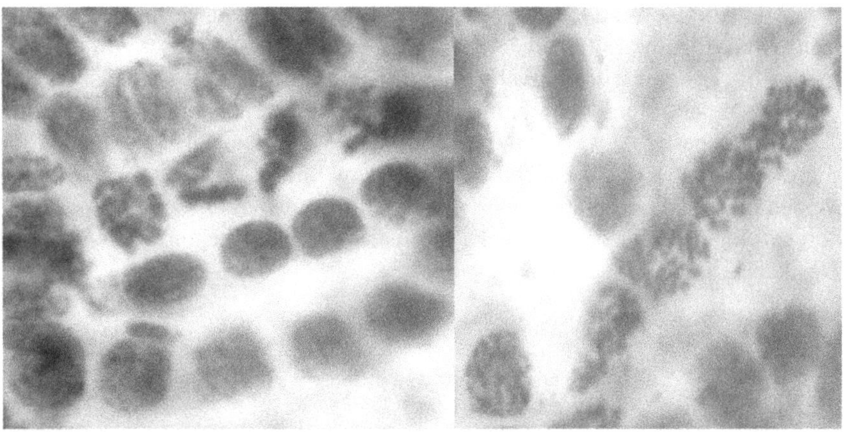

FIGURE 13.11 *Chara fibrosa*—prophase, metaphase, anaphase, and telophase of mitosis.

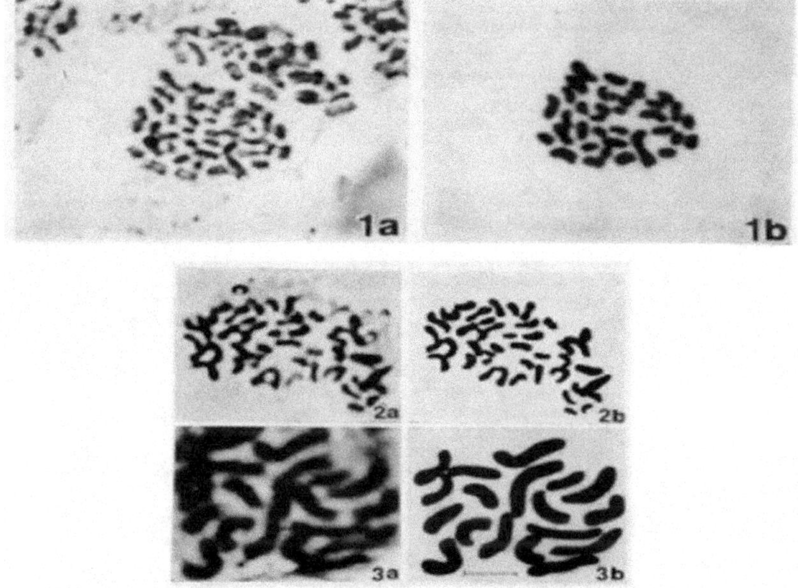

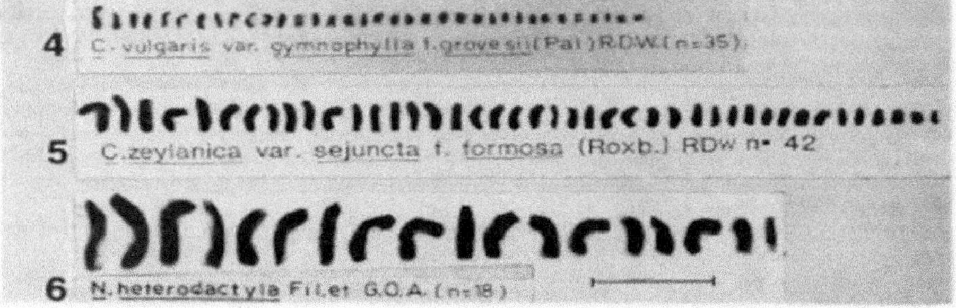

FIGURE 13.12 Courtesy of Subrahmanyam and Chowdary Y. B. K. (1992). Karyological Studies of Some Indian Charophyta. *Cytologia* 57:409–415: *Chara vulgaris* n = 35, *Chara zeylanica* n = 42, *Nitella heterodactyla* n = 18.

Purak (1986) recorded n = 18, 24, 36, and 42 for 9 taxa of *Nitella* and n = 14, 28, 42, and 56 for 9 taxa of *Chara* (Figure 13.13 a and b and Figure 13.14 a and b).

(1987). Aneuploid Chromosome Counts were studied by Prasad and Verma (1985), Subramanian and Ganesan (1983) (Figures 13.15 and 13.16).

The position of centromere is generally determined on the basis of shape and bending of the chromosome as the centromeres in charophytes are not clearly demarcated, resulting in large variations in karyotypes from worker to worker. Giemsastain C-banding technique for location of centromere was employed in some charophyte taxa by Bhatnagar et al. (1989). Bhatnagar et al. (1996) revealed that the centromere consists of constitutive heterochromatin.

Based on cytological evidence, the view of Chaetophoralean origin of charophytes put forward by Desikachary and Sunderlingam (1962).

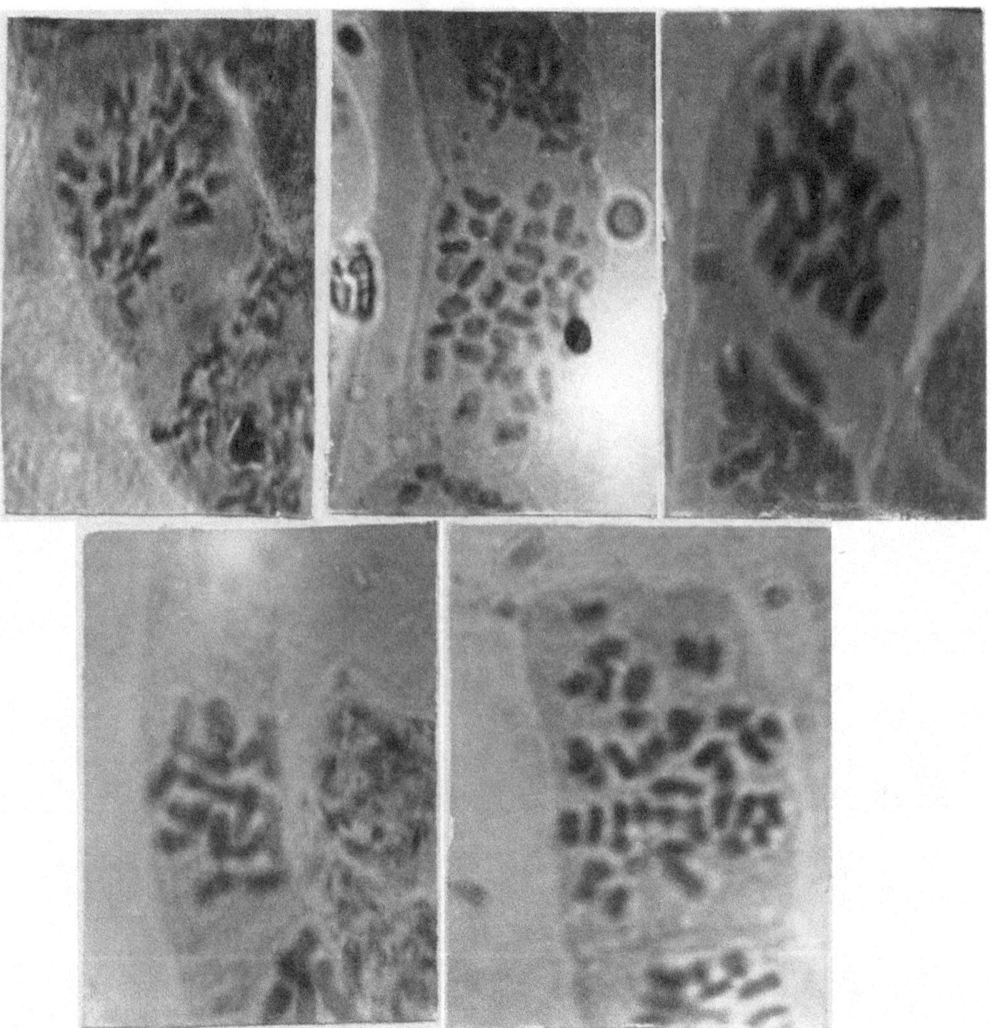

FIGURE 13.13A (i) *Nitella furcata* f. *microcarpa* n = 36; (ii) *Nitella furcata* f. *oligospira* n = 18; (iii) *Nitella furcata* f. *oligospira* n = 24; (iv) *N. pseudoflabellata* f. *pseudoflabellata* n = 36; (v) *Nitella gracilis* f. *confervacea* n = 36.

Source: (Purak, 1986)

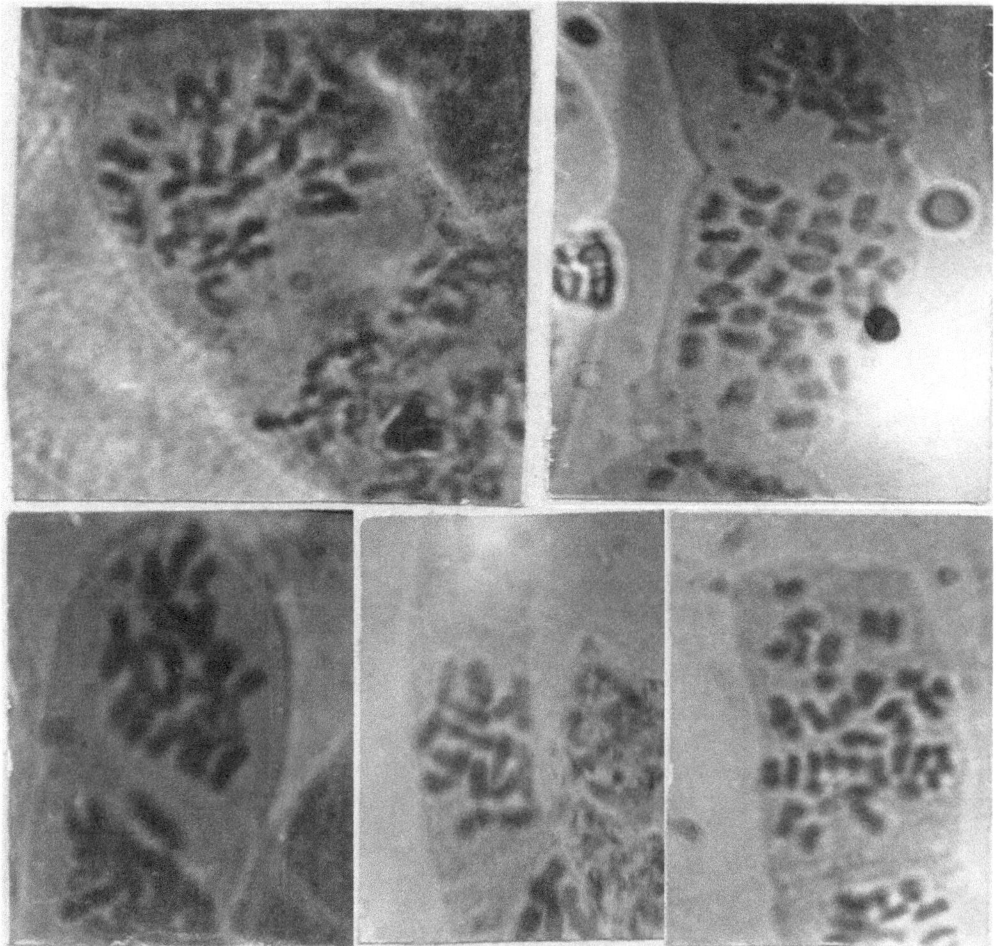

FIGURE 13.13B (i) *Nitella hyalina* f. *hyalina* n = 18; (ii) *Nitella translucens* f. *axillaris* n = 42; (iii) *Nitella hyaline* f. *formosa* n = 24; (iv) *Nitella hyalina* f. *maxima* n = 18; (v) *Nitella tricellularis* var. *tricellularis* n = 36.

Source: (Purak, 1986)

13.1.7 Work Done on DNA Isolation and Molecular Characterization by RAPD Markers of a Few Charophyte Taxa

C-TAB method was employed for isolation of Genomic DNA from four species of *Chara* and three species of *Nitella* (Bhatnagar et al., 2005). Abrol and Bhatnagar (2006) analyzed molecular characteristics such as band frequency, RAPD polymorphism, genetic identity index, band sharing frequency, and genetic distance for 12 charophyte taxa using random primers.

Sarma et al. (2014) employed CTAB protocol for genomic DNA isolation from three samples of *Nitella mirabilis* from Assam Purak et al. (2018) analyzed in molecular characterization using RAPD markers and constructed phylogenetic tree based on PHYLIP (Version 6.a3, UPGMA) of few charophyte taxa found in Ranchi.

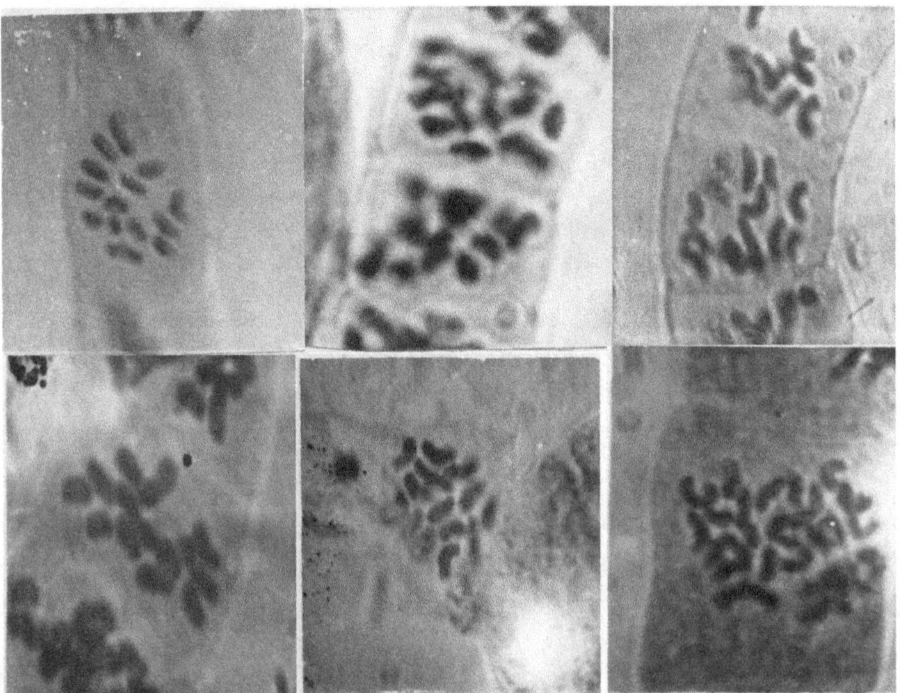

FIGURE 13.14A (i) *Chara braunii f. braunii* n = 14; (ii) *Chara braunii f. kurzii* n = 14; (iii) *C. brauni f. coromandelina* n = 14; (iv) *C. braunii f. novimexicana* n = 14; (v) *C. braunii f. oahuensis* n = 28; (vi) *C. braunii f. perrottetii* n = 14.

Source: (Purak, 1986)

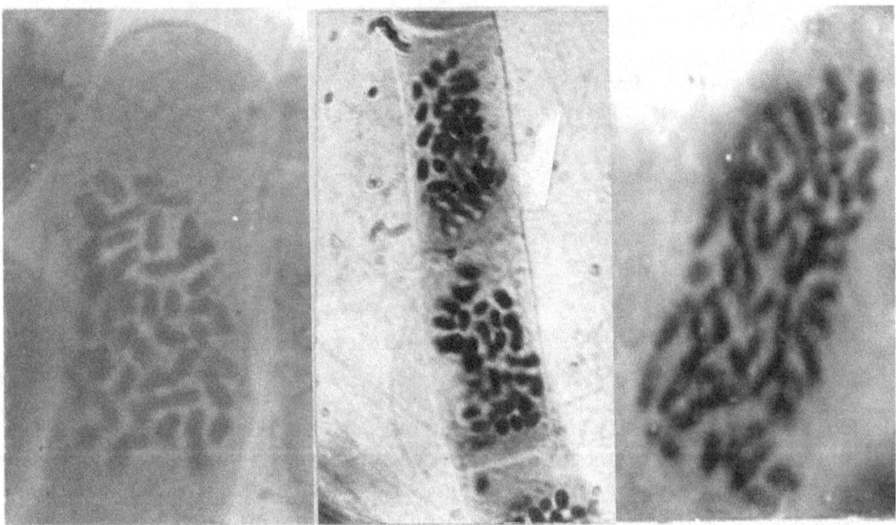

FIGURE 13.14B (i) *Chara corallina f. corallina* n = 42; (ii) *Chara corallina f. kyusyensis* n = 42; (iii) *Chara fibrosa f. tylacantha* n = 56.

Source: (Purak, 1986)

13.1.8 Resistance or Susceptibility of Nuclear Structure Towards Physical and Chemical Mutagens

In general, it appears green algae are generally more tolerant to chemical mutagens than the higher plants. High doses of chemicals can induce abnormalities as chromosome breaks, clumped metaphase, anaphasic bridges, laggards etc.

The effect of various chemical mutagens viz. EMS, colchicine, antibiotics, triacontanol, butachlor, herbicides, pesticides, fungicides (Bavistin), etc. have been worked on by members of green algae such as *Cladophora, Rhizoclonium, Oedogonium, Spirogyra Chara, Nitella,* and *Tolypella.* Significant contributions have been made by Sarma and Tripathi (1973, 1974, 1976a, 1976b), Srivastava and Sarma (1980a, 1980b), Vedajanani and Sarma (1978b), Abhayavardhani and Sarma (1983), Bhatnagar and Johri (1985, 1987), Singh et al. (1988), Bhatnagar et al. (1989), Noor and Purak (1989), Purak and Noor (1991, 2018), Purak (1986, 1990, 1991, 1993), Pal and Chatterjee (1987, 1989), Bhatnagar and Kirti (1988), Subrahmanyam et al. (1991, 1992) etc. (Table 13.2) (Figures 13.17–13.22).

Mutagenic chemicals usually act in two ways on the nuclear division of biological organisms. Some chemicals act on the chromatin matter resulting in different chromosomal alterations viz. stickiness, breakage, clumping of chromosomes, and pycnosis, while others make a thrust on spindle formation during cell division, disrupting the normal pattern of the proteins. The spindle disturbance results in cytological abnormalities like scattered and multipolar distribution of chromosomes, polyploidy, and C-mitosis by Aruna and Vidyavati (2010), Vidyavati (1992).

Colchicine is a remarkable alkaloid polyploidising agent and is well-known for pretreatment chemical in cytological investigations, which acts on spindle during nuclear division cycle resulting into well-scattered chromosome configurations. Polyploidy, as an effect of colchicine, has already been reported by Noor (1965, 1966) in *Rhizoclonium implexum* and Vedajanani and Sarma (1978a) in *Spirogyra azygospora.*

Sarma and Tripathi (1973) have investigated karyological effects of colchicine, maleic hydrazide, caffeine, and theobromine on *Oedogonium acmandrium* mentioning cytological aberrations. Almost similar chromosomal alterations have also been recorded using various antibiotics in different algal members by a number of workers viz. Vedajananin and Sarma (1978a) in *Spirogyra azygospora* using chloramphenicol, oxytetracycline, and gentamicin.

Srivastava and Sarma (1980a) in *Oedogonium gunnii* with penicillin, streptomycin, and tetracycline; Abhayavardhani and Sarma (1980) in *Spirogyra azygospora* with griseofulvin; Sarma and Abhayavardhani (1980) in *Oedogonium gunnii* with rifamycin, griseofulvin, and polymixin; Mogili and Vidyavati (1986) in *Cladophora crispata* with aureofungin, rifamycin, gentamicin, and mitomycin D; Purak (1992) in *Rhizoclonium stagnale* with penicillin; and Purak (1993) in *Rhizoclonium majus* with Kinetin.

Sarma and Tripathi (1976a, 1976b) have studied effects of chemicals like colchicine, maleic hydrazide, caffeine, theobromine, gibberellic acid, 2, 4-D, coumarin, and acenapthene on four taxa viz. *Chara globularis, Chara braunii, Nitella flagelliformis,* and *Nitella furcata* to analyze their mitotic patterns in antheridial filaments.

Mutagenicity of 1-triaconatanol has been analyzed by Bhatnagar and Johri (1985) on *Chara braunii*and by Bhatnagar et al. (1989) on *Tolypella prolifera.* They have recommended its application as polyploidizing agent like colchicine. Quite a number of abnormalities like broken chromatid threads, chromatin bridges, scattered metaphases, laggards, rings at metaphase and anaphase, and chromosome clumping have been encountered.

Effects of copper sulfate and diurone on mitotic processes of *Chara globularis, C. Setosa, C. zeylanica,* and *C. corallina* have been studied by Pal and Chatterjee (1987, 1989), recording inhibiting mitotic index with clumping of chromosomes, stickiness, deformed chromatin matter, budding of chromatin, micro-nuclei formation, chromatid bridges at anaphase, metaphase arrest, c-mitosis, and polyploidy.

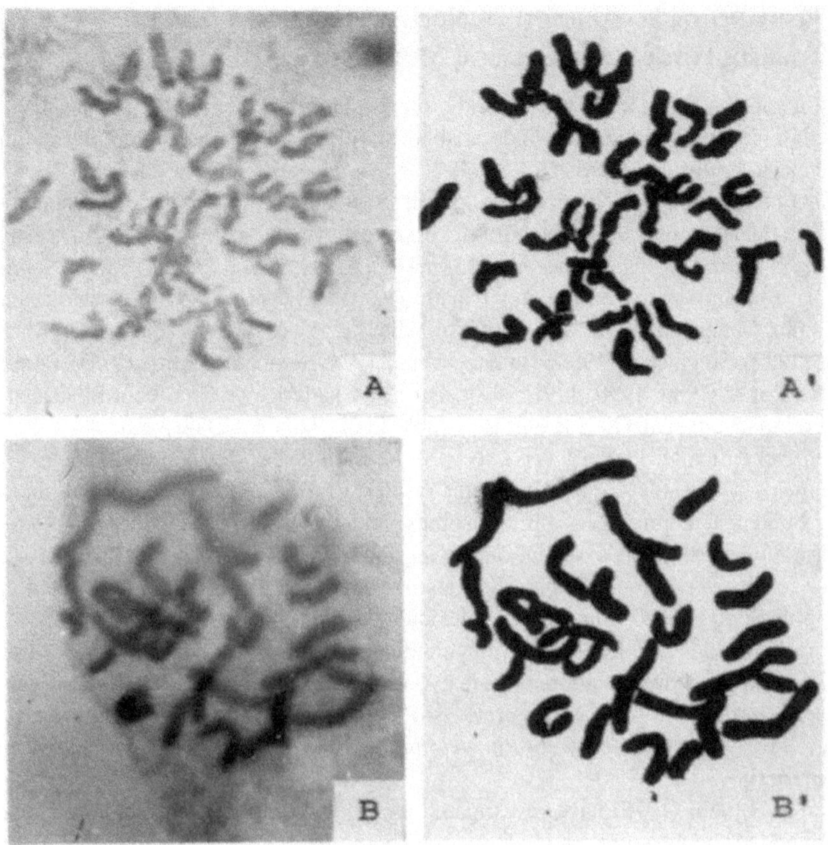

FIGURE 13.15 Courtesy: Bhatnagar S K and Johri M (1986a, 1986b): Some aneuploids in Indian Charophyta. Acta Bot. Neeth. 35 (4): 377–381.

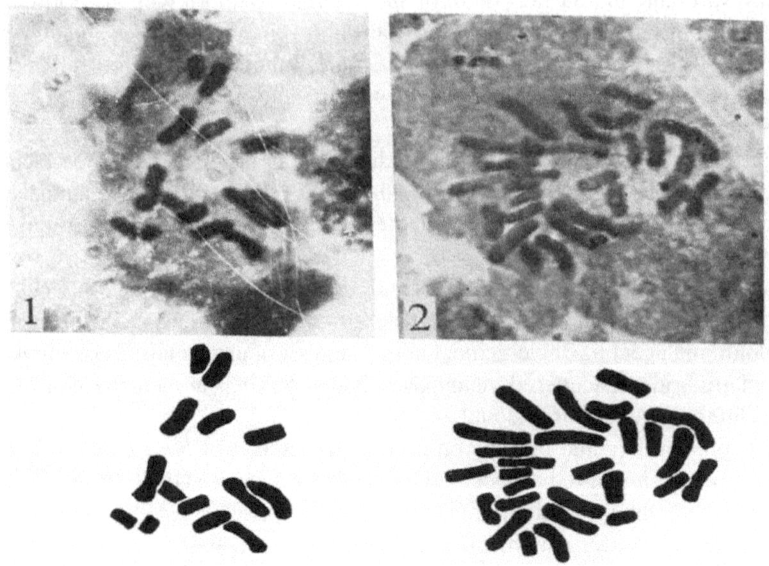

FIGURE 13.16 Courtesy: Prasad P K and Verma B N (1985): Aneuploid count for *Chara setosa* Klein ex. Willd. Cytologia 50:241–245.

TABLE 13.2
Effect of Chemicals on Chromosomes of Green Algae

S N	Chemical Used	Taxon	Author/s/year
1	Colchicine	*Chara vulgaris*	Campbell and Sarafis (1972)
2	Maleic hydrazide	*Chara vulgaris*	Marchant et al. (1973)
3	Colchicine	*Rhizoclonium implexum*	Noor (1965, 1966)
4	Colchicine	*Spirogyra azygospora*	Vedajanani and Sarma (1978b)
5	Hydroquinone	*Chara zeylanica*	Sarma (1982)
6	Copper sulphate Diurone	*Chara globularis* *Chara setosa* *Chara zeylanica* *Chara corallina*	Pal and Chatterjee (1987, 1989)
7	Colchicine MH Caffeine Theobromine	*Oedogonium acmandrium*	Sarma and Tripathi (1973)
8	Gibberellic Acid Coumarin 2,4-D Acenaphthene	*Oedogonium acmandrium*	Sarma and Tripathi (1974)
9	Colchicine Maleic hydrazide Caffeine Theobromine	*Chara braunii* *Chara globularis* *Nitella flagelliformis* *Nitella furcata*	Sarma Tripathi (1976a)
10	Giberellic Acid Coumarin Acenaphthene 2,4-D	*Chara braunii* *Chara globularis* *Nitella flagelliformis* *Nitella furcata*	Sarma Tripathi (1976b)
11	Chloramphenicol Oxytetracycline Gentamicin	*Spirogyra azygospora*	Vedajanani and Sarma (1978b)
12	NTG (N-Methyl-N-Nitro-N- Nitrosoguanidine)	*Spirogyra paradoxa Rao*	Vedajanani et al. (1980)
13	Penicillin Streptomycin trtracycline	*Oedogonium gunnii*	Srivastava and Sarma (1980a)
14	Griseofulvib	*Spirogyra azygospora*	Abhayavardhani and Sarma (1980b)
15	rifamycin, griseofulvin polymixin,	*Oedogonium gunnii*	Sarma and Abhayavardhani (1983)
16	Paracetamol Phencerin	*Cladophora crispata*	Mogili and Vidyavati (1985)
17	Aureofungin Rifamycin Gentamicin Mitomycin D Penicillin	*Cladophora crispata*	Mogili and Vidyavati (1986)
18	Paracetamol Aspirin	*Spirogyra verrucosa*	Devi and Nizam (1986)

(*Continued*)

TABLE 13.2 (*Continued*)
Effect of Chemicals on Chromosomes of Green Algae

S N	Chemical Used	Taxon	Author/s/year
19	Maleic hydrazide Colchicine	*Chara delicatula* *Nitella mirabilis*	Bhatnagar (1981)
20	Chlorflurenol Morphactin Coumarin	*Nitella hyaline* *Nitella mucronata*	Sarma (1984), Sarma (1961–62)
21	Streptomycin Chloramphenicol	*Chara braunii*	Noor and Purak (1991)
22	Kinetin	*Chara braunii*	Purak (1990)
23	Chloramphenicol	*Chara corallina*	Purak and Noor (1991)
24	Penicillin	*Rhizoclonium stagnale*	Purak (1992)
25	Kinetin	*Rhizoclonium majus*	Purak (1993)
26	Neem extract	*Nitella hyalina*	Noor et al. (2002)
27	2,4-D	*Chara corallina* *Chara fibrosa*	Purak (1991)
28	IAA NAA GA	*Chara braunii* *Chara corallina*	Purak (1991)
29	Bavistin	*Chara braunii*	Purak (1991)
30	Penicillin	*Chara wallichi*	Singh et al. (1988)
31	Streptomycin penicillin	*Chara braunii* *Chara fibrosa* *Nitella hyalina* *Nitella furcata*	Sinha and Sinha (1992)
32	1-triacontanol	*Chara braunii*	Bhatnagar and Johri (1985)
33	EMS	*Chara delicatula* *Nitella mirabilis* *N hyaline* *N muctonata*	Bhatnagar and Johri (1987)
34	Penicillin	*Chara wallichi*	Singh et al. (1988)
35	1- triacontanol	*Tolypella prolifera*	Bhatnagar et al. (1989)
36	Butachlor	*Chara wallichi* *Nitella mirabilis*	Bhatnagar and Kirti (1988, 1990)

Noor and Purak (1989) have investigated the effects of two antibiotics viz. streptomycin and chloramphenicol on the antheridial cells of *Chara braunii* Gm., where chloramphenicol is more inhibitory towards survival pattern than streptomycin. Cytological alterations viz. clumping of chromosomes, chromosomal fragments, micro-nuclei, anaphasic bridges, and degenerate nuclei have been recorded by them, and lower concentration of chloramphenicol causes duplication of chromosomes when compared to streptomycin doses.

Purak (1990) has analyzed effects of various concentrations viz. 10, 25, 50, and 100 ppm of kinetin on the nuclear division of the spermatocytes of *Chara braunii* Gm. This kinetin accelerates the rate of nuclear division up to 50 ppm and induces lagging of chromosomes, erosion and fragmentation of chromosomes, anaphasic bridges, and organization of chromosomes in two to

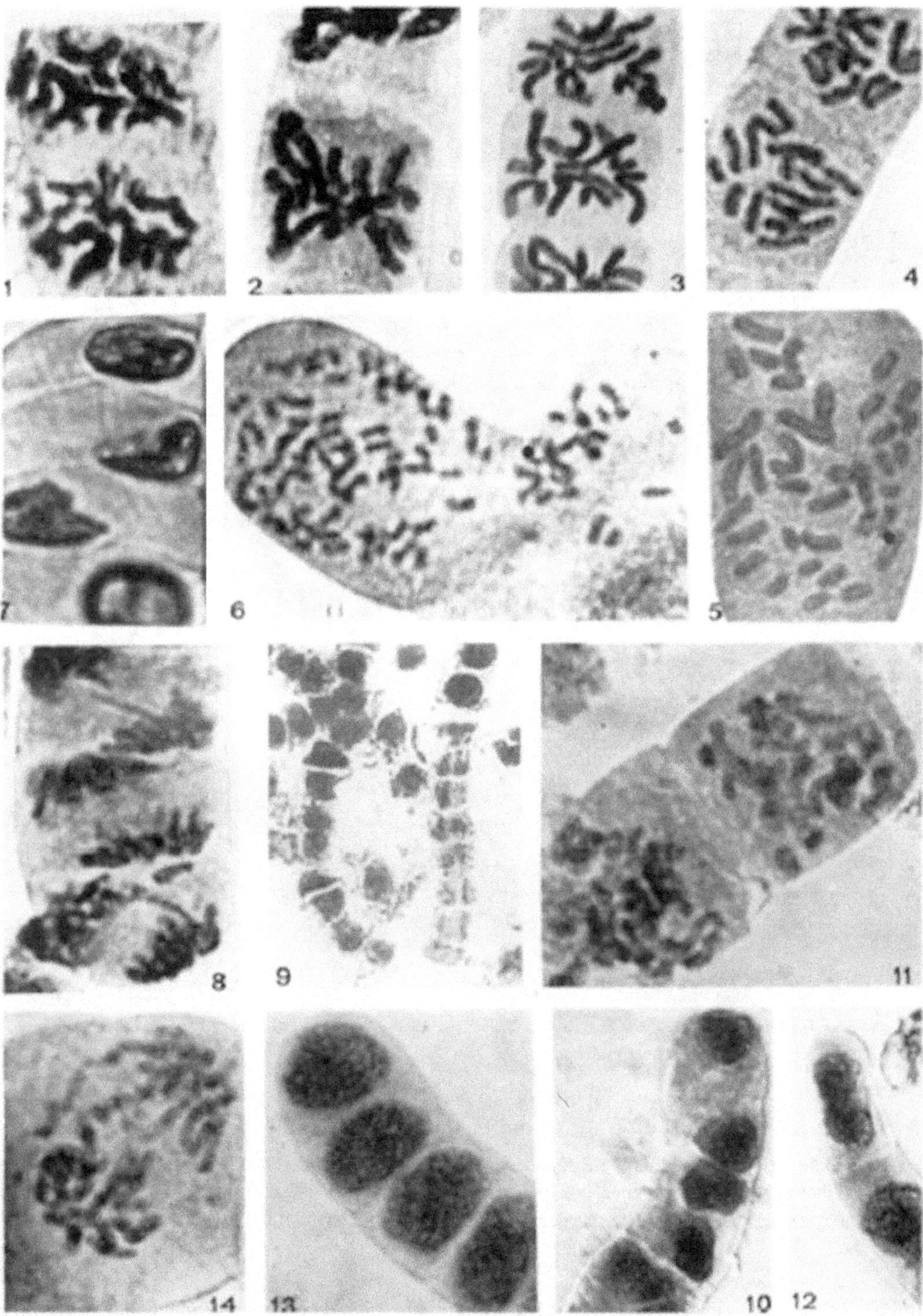

FIGURE 13.17 Courtesy: Sarma Y S R K and Tripathi, S. N (1976). Effects of Chemicals on Some Members of Indian Charophyta I Caryologia 29(3):247–262. Chemicals used: colchicine, MH, caffeine.

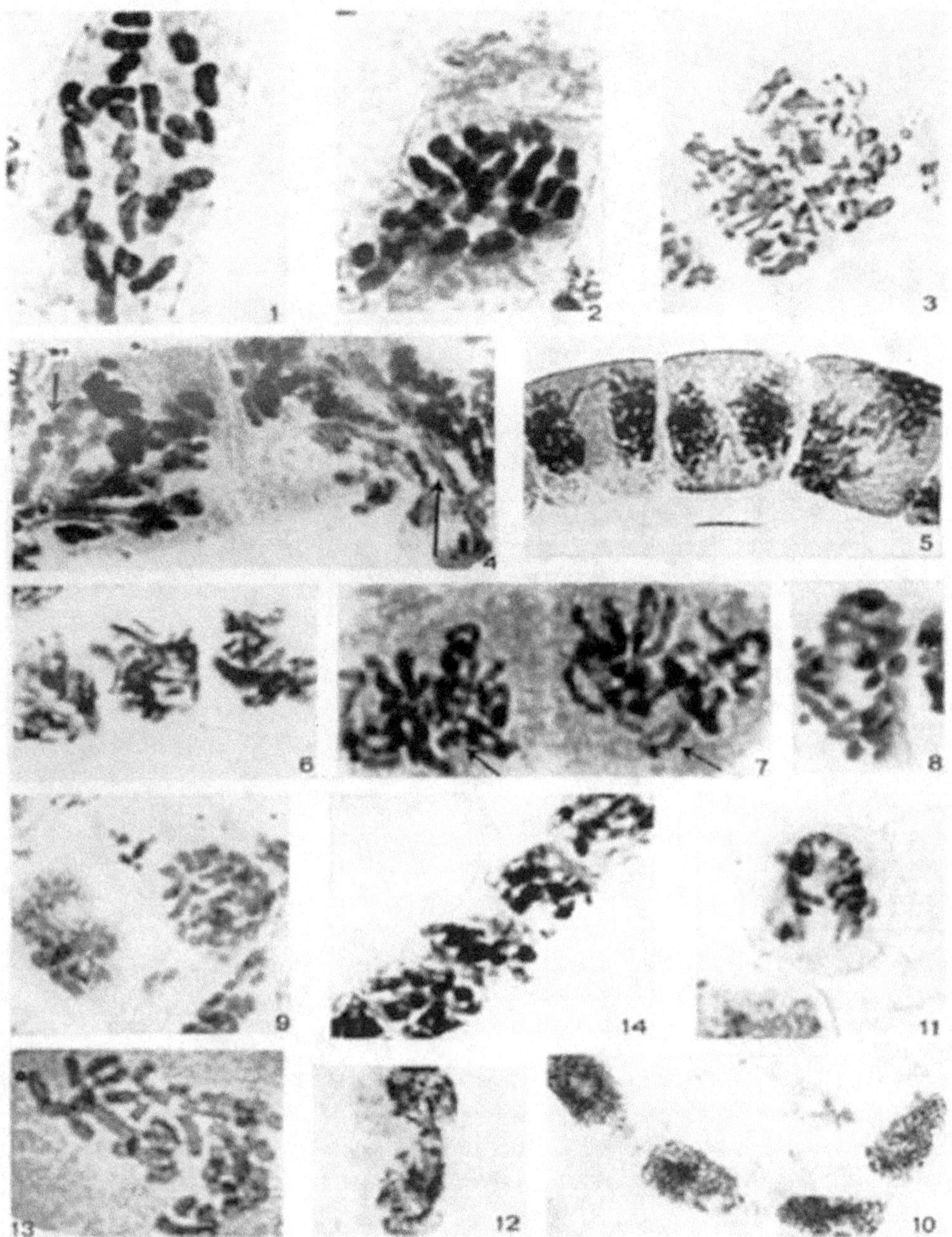

FIGURE 13.18 Courtesy: Sarma Y S R K and Tripathi, S. N (1976). Effects of Chemicals on Some Members of Indian Charophyta. II Caryologia 263–276. Chemicals used: GA, 2, 4-D, Coumarin, Acenaphthene.

three groups at metaphase. However, only a few cells in 25 and 50 ppm treatments have exhibited diploid cells.

Purak and Noor (1991) have analyzed the effects of varied concentrations of chloramphenicol on growth pattern and cytological behavior of *Chara corallina*, which proved to be mito-depressive and inhibitory to growth leading to death of algal specimen within six days. This drug induces

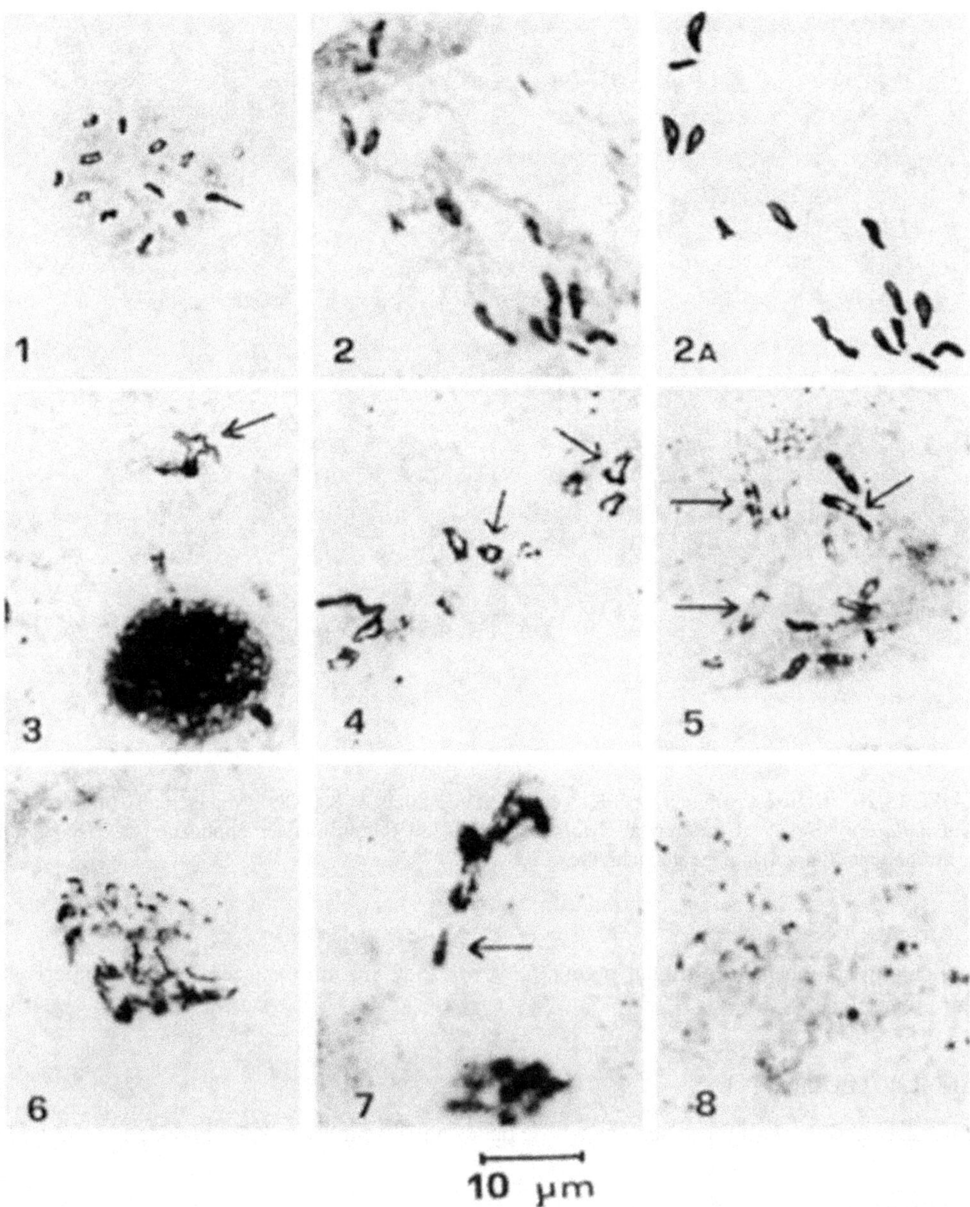

FIGURE 13.19 Courtesy: Vedajanani, K., Abhaya Vardhani P., and Sarma Y.S.R.K (1980). Cytological Effects of N-Methyl-N-Nitro-Nitrosoguanidine (NTG) on *Spirogyra paradoxa* Rao Caryologia, 33:2, 299–305.

cytological abnormalities like clumping of chromosomes, formation of micro-nuclei, bi- and tri-nucleate cells, anaphasic bridges, rotation of anaphase axes, and grouping of chromosomes at meta-phases. These findings show mutagenical potential of this antibiotic.

From the extensive work done regarding effects of chemicals on karyology of green algae, it can be easily concluded that green algae are more resistant in comparison to higher plants. Some of the chemicals in lower concentration are stimulatory to mitotic division, whereas higher concentrations are inhibitory to division. It can be inferred further charophytes are more resistant to

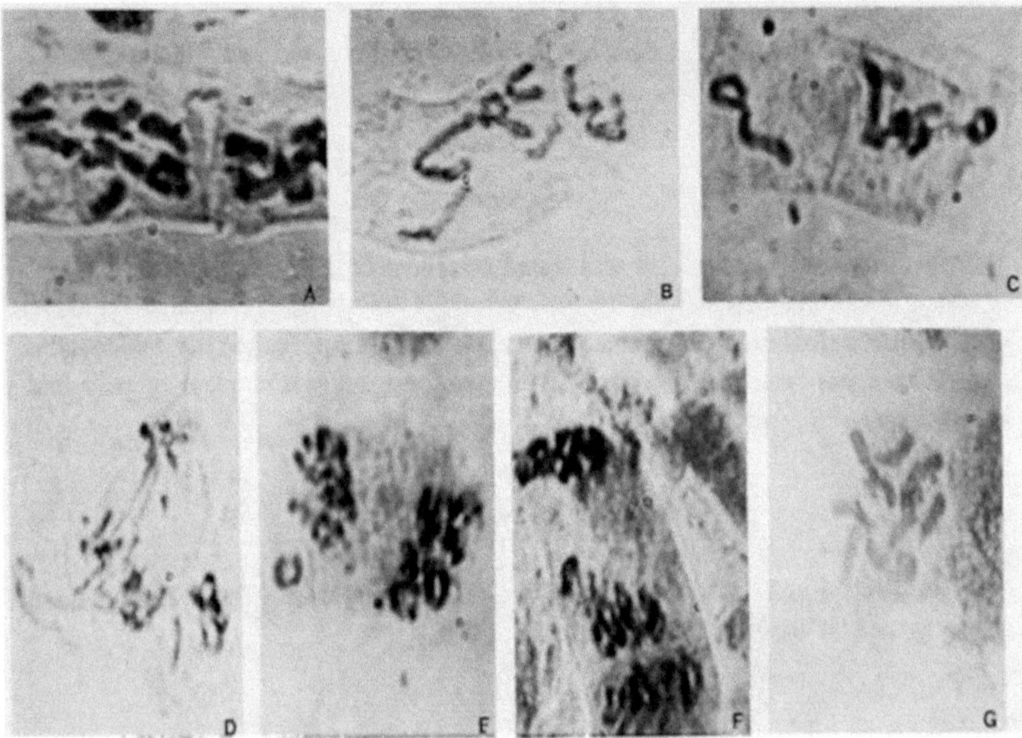

FIGURE 13.20 Courtesy: Bhatnagar S K, Verma, A, and Singh V K (1989). Mutagenicity of Triacontanol in *Tolypellaprolifera* (Div. Charophyta) *Cytologia* 54:183–189. Recommended application of triacontanol in place of polyploidy-inducing agent colchicine.

chemicals in comparison to higher plants; however, they are more susceptible in comparison to other green algae.

ACKNOWLEDGMENT

Author acknowledges all those persons whose plates and research work are included in this review.

REFERENCES

Abhayavardhani, P and Sarma, Y S R K (1980a). Effects of two insecticides (Kkitin and Anthio) on the karyology of Spirogyra paradoxa Rao. J. Cytol. Genet. 15; 26–31.

Abhayavardhani, P and Sarma, Y S R K (1980b). Effects of ethylmethane sulphoiute (EMS) on the karyology of Spirogyra paradoxa Rao (Chlorophyceae). Nucleus 23: 208–212.

Abhayavardhani, P and Sarma, Y S R K (1981). Effects of grisiofulvin on the karyology of Spirogyra paradoxa Rao. Curr. Sci. 60: 691–693.

Abhayavardhani, P and Sarma, Y S R K (1983). Cytological and cytotaxonomical studies on the genus Spirogyra link (Conjugates, Chtorophyceae) Cytologia 48: 467–482.

Abrol, D and Bhatnagar, S K (2006). Biodiversity of a few Indian charophyte taxa based on molecular characterization and construction of phylogenetic tree. African Journal of Biotechnology 5(17): 1511–1518.

Bhatnagar, S K (1981). Cytotaxonomy of charophytes of Rohilkh and division and different effect of different chemicals on selected taxa. Ph.D. thesis, Rohilkh and University, Bareilly.

Bhatnagar, S K, Abrol, D and Kumar, D (2005). First protocol for DNA isolation in Indian Charophyta. Journal of Biological Research 3: 109–111.

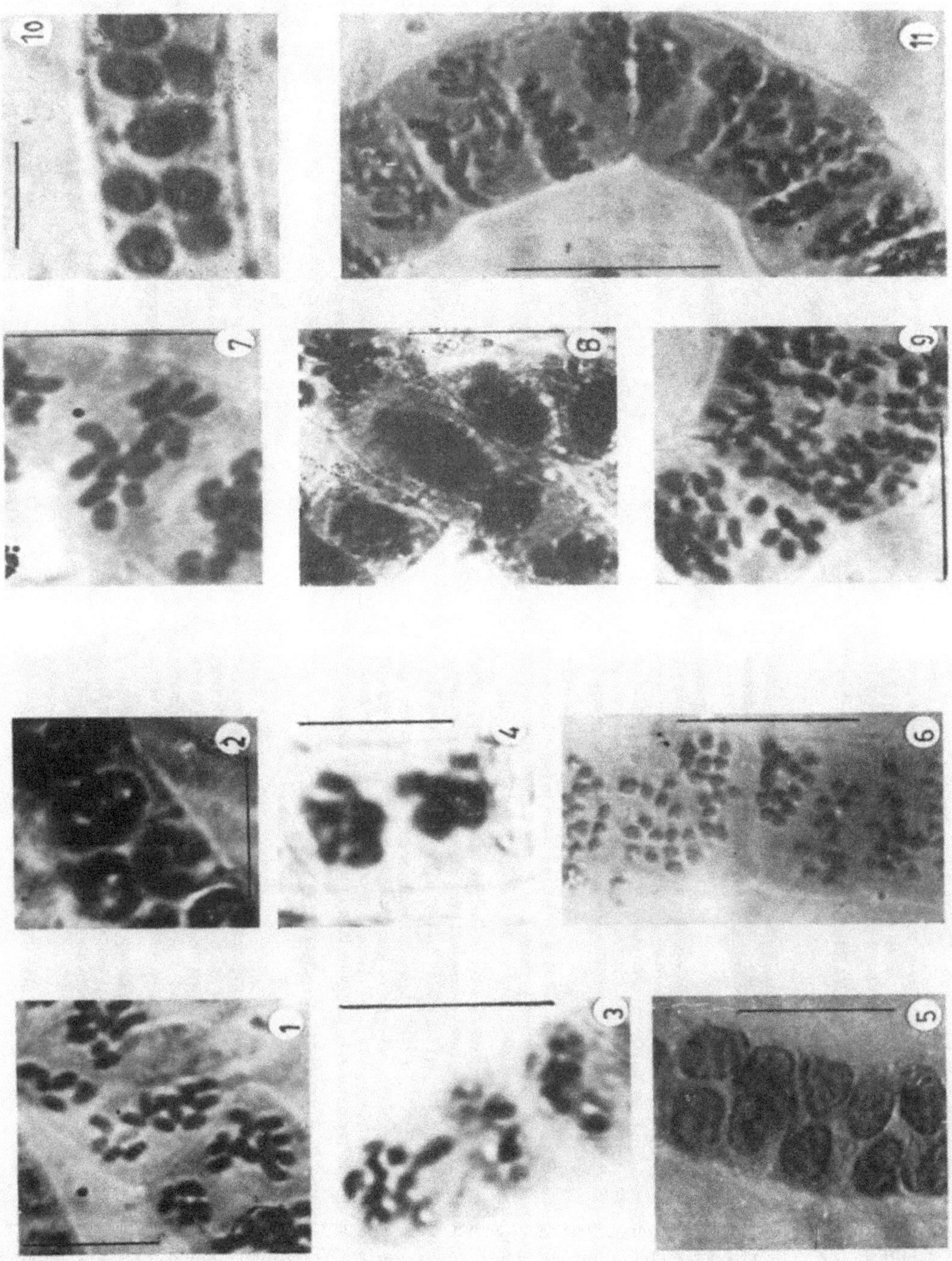

FIGURE 13.21 Courtesy: Noor M N and Purak I (1989). Effects of two antibiotics on nuclear division of *Chara braunii* Gm (Characeae) *Cryptogamie Algol* 10(2): 143–152

Streptomycin and Chloramphenicol.

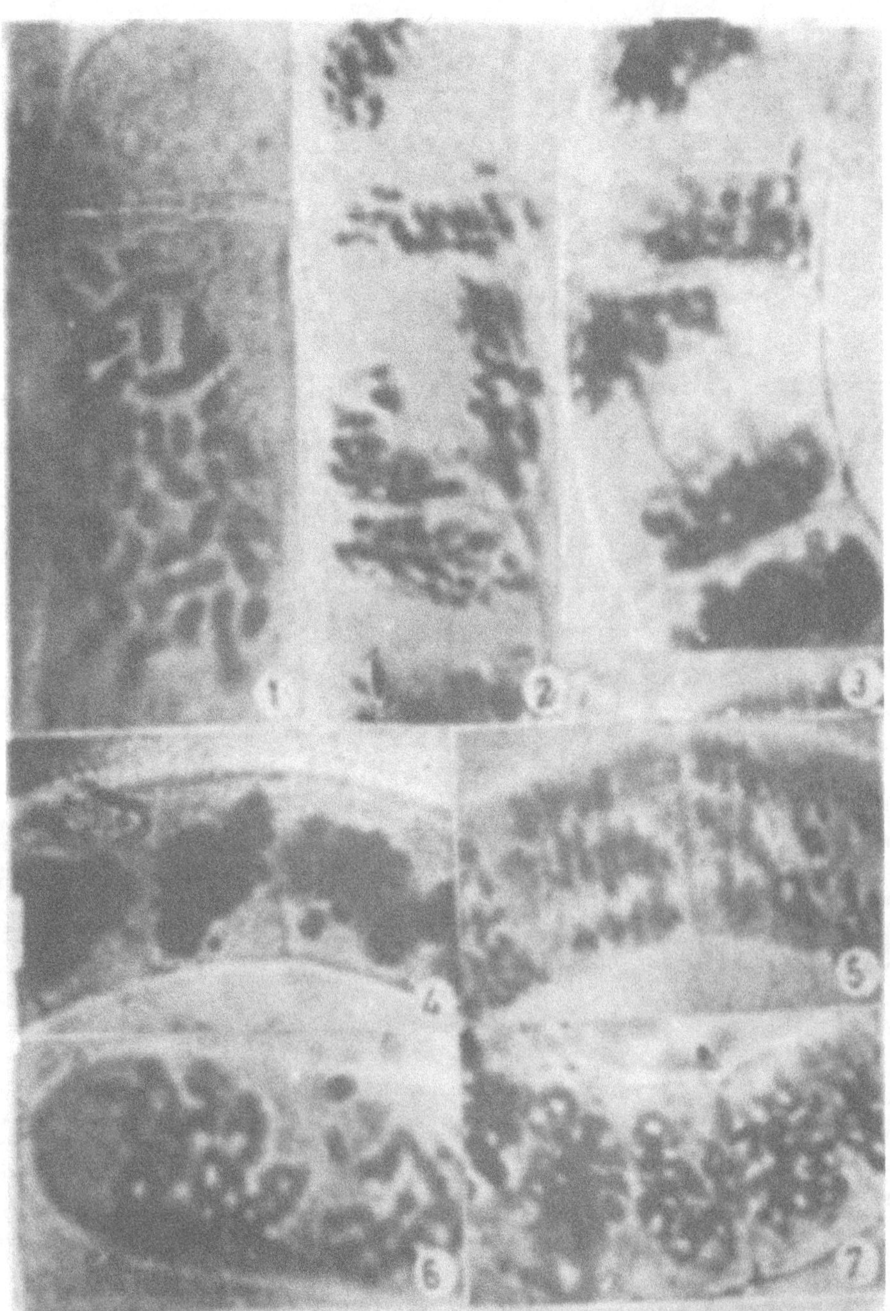

FIGURE 13.22 (1–7) Effect of chloramphenicol on *Chara corallina*: (1) normal count n = 42; (2) rotation of anaphase plates; (3) anaphasic bridge; (4) clumped metaphases; (5) laggards; (6) rings at metaphase; (7) erosion of chromosomes.

Courtesy: Purak I and Noor M N (1991). Effect of Chloramphenicol on growth pattern and cytological behaviour of *Chara corallina*. Phykos 30(1&2): 123–128.

Bhatnagar, S K and Johri, M (1985). Mutagenic efficacies of tri-aconatol in *Chara braunii* Gm. (Charopohyta) with reference to its application in chromosome analysis. Cytologia 6(4): 273–280.
Bhatnagar, S K and Johri, M (1987). Radiomimetic efficacies of synthetic bioregulants on chromosomes of Indian Charophyta 1. Morphactin: Chlorfl urenol. Cytologia 8(4): 301–317.
Bhatnagar SK and Kirti (1988). Mutagenicity of Butachlor in Chara wallichii (Div. Charophyta). Vegetos 1(1): 64–66.
Bhatnagar SK and Kirti (1990) Optimized RAPD-PCR protocol for genomic DNA amplification in Indian Charophyta. J Ind Bot Soc 87(3 & 4): 291–293.
Bhatnagar, S K, Verma, A and Singh, V K (1989). C-banding technique for kinetochore and heterochromatin differentiation in *Nitella mirabilis* (div. Charophyta). Current Science 58(1): 377–378.
Campbell, E O and Sarafis, V (1972). Shizomeris—a growth form of *Stigeoclonium tenue* (Chlorophyta: Chaetophoraceae) Journal of Phycology 8(3): 276–282.
Chakrabarty, R and Ray, S (2016). Chromosomal variations and cytotaxonomical considerations in two populations of *Nitella hyalina* (Charophyceae, Characeae) from West Bengal, India. Phykos 46(2): 14–19.
Chatterjee, P (1971). An analysis of the structure and behaviour of chromosomes as an aid in the taxonomy of certain higher and lower groups of plants and the cytological effects of certain chemical and physical agents in the latter. Ph.D. thesis, Calcutta University, Calcutta.
Chatterjee, P (1975). Some additions to the charophytes of West Bengal. Bulletin of the Botanical Society of Bengal 29: 105–109.
Chatterjee, P (1976). Cytotaxonomical studies of West Bengal Charophyta: Karyotype analysis in Chara braunii. Hydrobiologia 49: 171–174.
Chennaveeraiah, M S and Bharti, S G (1974). Morphological and cytological observations on *Chara gymnopitys* A Br. Cytologia 39(3): 443–451.
Chowdhary, Y B K and Singh, S J (1970). Cytological observations on two siphonaceous marine algae. Botanica Marina 13: 3–5.
De, D N and Berger, S (1990). Karyology of *Acetabularia mediterranea.* Protoplasma 155: 19–28.
Devi P and Nizam J (1986) Effect of certain pharmaceutical drugs on Spirogyra verrucosa (Rao) Kolkwitz & Krieger. J Cytolenet 20: 72–76.
Desikachary, T V and Sunderlingam, V S (1962). Affinities and Interrelationships of the Characeae. Phycologia 2(1): 9–16.
Fritsch, F E (1935). The structure and Reproduction of the algae, Vol. 1. Cambridge University Press, Cambridge.
Fritsch, F E (1944). Present day classification of algae. Botanical Review 10(4): 232–277.
Godward, M B E (1954). The diffuse centromere of polycentric chromosomes in Spirogyra. Annals of Botany 18: 143–156.
Godward, M B E (1956). Cytotaxonomy of Spirogyra 1: S. submargaritata, S. subechinata and S. Britannica spp. novae. Journal of the Linnean Society (Bot.) 55: 532–546.
Guerlesquin, M (1963). Contribution a l'etude chromosomiqua des Charophycaes d' Europe occidentals et d Afrique du Nord-2. Revue géńérale de botanique 70: 355.
Guerlesquin, M (1967). Recherches caryoptipiques et cytotaxonomiques ches les Charophycees d'Europe occidentale et d'Afrique du Nord. Bull Soc. Sei. Bretagne 41: 1–265.
King, G C (1960). The Cytology of the Desmids: The chromosomes. The New Phytologist 65–72.
Koop, H U (1979). Observations on the morphogenetic capacity of 'residual nuelei' in Acetabularia. Protoplasma 101: 363–371.
Labh, L and Verma, B N (1985). New records of chromosome numbers for the charophytes. Cytologia 50: 55–58.
Marchant, H J and Pickett Heaps, J D (1973). Mitosis and cytokinesis in *Coleochaete scutata.* Journal of Phycology 9: 461–471.
Mattox, K R and Stewart, K D (1984). Classification of the green algae: A concept based on comparative cytology. In Systematics of the green algae. Eds. D E G Irvine and D John, 41, 42, 57, 58. Academic Press, London.
Michetti, K M, Leonardi, P and Caceres, E J (2010). Morphology, cytology and taxonomic remarks of four species of Stigeoclonium (Chaetophorales, Chlorophyceae) from Argentina. Phycological Research 58(1): 35–43.
Noor, M N (1965). Cytotaxonomic and cultural studies of some Freshwater Chlorophyceae of Chota Nagpur Plateau. Ph.D. thesis, Ranchi University, Ranchi (India).
Noor MN (1966) Effects of Colchicine on Rhizoclonium implexum. J Ranchi Univ 3: 9.
Noor, M N and Mukherjee, S (1975). On the aneuploid chromosome number in *Chara hydropitys* Reich. from India. Cytologia 40: 803–807.

Noor, M N and Mukherjee, S (1977). Some new records of chromosome numbers in Indian Charophyta. Cytologia 42: 227–232.

Noor, M N and Purak, I (1989). Effects of two antibiotics on nuclear division of *Chara braunii* gm. Characeae. Cryptogamie Algologie 10(2): 143–152.

Noor MN, Purak I and Sarkar G (2002) Efficacious mutagenicity of neem (Azadirachta indica L.) extract on nuclear cytology and survival pattern in Nitella hyalina. In: Recent trends in Charophyte Research. Eds: MN Noor, SK Bhatnagar and AKM Nurul Islam. Pp 209–216.

Pal, R. and Chatterjee, P(1987) Algicidal action of Diurone in the control of *Chara*—a rice pest. *Proc. Indian Acad. Sci.* **97**, 359–363.

Pal, R. and Chatterjee, P (1989) Cytological Effects of Two Common Algicides on the Mitotic Cell Division in Antheridial Filaments of *Chara* Species, CYTOLOGIA, 54:1, 173–178.

Pickett-Heaps, J D (1967). Ultrastructure and differentiation in Chara. II. Mitosis. Australian Journal of Biological Sciences 20: 883–894.

Pickett-Heaps, J D (1972). Variation in mitosis and cytokinesis in plant cells; Its significance in phylogeny and evolution of ultrastructural systems. Cyiobios 5: 59–77.

Prasad, P K and Verma, B N (1985). Aneuploid count for *Chara setosa* Klein ex. Willd. Cytologia 50: 241–245.

Pundhir, H S (1994). Studies on the karyomorphology of *Chara hornemannii* f. *longifolia* (Rob.) RDW: A new record for India. J. Mendel 11(3–4): 157–160.

Pundhir, H S and Gautam, A (1993). Occurrence of *Nitella hyalina* var. *hyalina* f. *Brachyactis* (A.Br.) Feldm R.D.W.A new record from India. Phykos 32: 99–104.

Pundhir, H S and Gautam, A (1994). A new chromosome count for *Chara hornemannii* f.
longifoliar (Rob.) RDW: A new record for India. Cytologia 59: 305–307.

Purak, I (1986). Cytotaxonomic and mutagenic studies of some members of Charophyceae and other green algae. Ph.D thesis, Ranchi University, Ranchi (India).

Purak, I (1990). Effects of Kinetin on nuclear division of *Chara braunii* Gm. (Characeae) Vegetos 3(2): 190–192.

Purak, I (1991). Effects of Bavistin on nuclear division of *Chara braunii* Gm. (Characeae) Biojounal 3(1): 65–68.

Purak, I (1992). Effects of penicillin on nuclear division of *Rhizoclonium stagnale* wolle (Cladophoraceae). Vegetos 5(1&2): 52–59.

Purak, I (1993). Effects of Kinetin on nuclear division of *Rhizoclonium majus* wolle (Cladophoraceae). Journal of Applied Biology 3(1–2): 1–6.

Purak, I, Kumar, D, Guru, S D and Anand, R (2018). Molecular characterization and construction of phylogenetic tree of few charophyte taxa found in Ranchi. Pariplex, Indian Journal of Research 7(10): 406–408.

Purak, I and Noor, M N (1991). Effect of chloramphenicol on growth pattern and cytological behavior in *Chara coralline*. Phykos 30(1&2): 123–128.

Purak, I and Noor, M N (2018). Morphological and cytological analysis of *Chara fibrosa* var. *fibrosa* form *tylacantha* (Nordst.) R D W. In Bioprospecting of algae (Prof J P Sinha Memorial Volume). Eds. M N Noor, S K Bhatnagar, S K Sinha: 11–14. Society for Plant Research, New Delhi. Purak, I and Sinha, S (1998). Morphological and cytological studies of *Nitella translucens* var. *axillaris* form *axillaris* (A.Br.) R D W J. Applied Biology 8(1): 1–6.

Ramjee, P and Bhatnagar, S K (1978a). Significance of a new chromosome number in *Nitella mirabilis* (Nordst ex. Gr.) em. RDW. Hydrobiologia 57: 99–101.

Ray, S and Chatterjee, P (1987). Karyomorphological investigation of three populations of *Chara corallina* var. and f. *corallina* in West Bengal. Cytologia 52: 455–458.

Ray, S and Chatterjee, P (1988). Cytotaxonomical study of West Bengal Charophyta: A new chromosome number count of *Nitella stuartii* and its karyotype. Acta Botanica Neerlandica 37(4): 523–525.

Round, F E (1965). The biology of the algae. Arnold Publisher, London.

Sarma, T A and Bala, V (1984). New chromosome counts in four taxa of chara from Punjab. Journal of Cytology & Genetics 19: 27–32.

Sarma, Y S R K (1961–62). Effects of colchicine on two members of Chlorophyceae. B.H.U. J. Sci. Res. 112: 377–383.

Sarma, Y S R K (1962). Nuclear Cytology of Sphaeroplea annulina (Roth) Ag., and its Bearing on the Systematic Position of Sphaeroplea. Cytologia 27: 72–78.

Sarma, Y S R K (1982). Chromosome numbers in algae. The Nucleus 25: 66–108.

Sarma, Y S R K (1984). Contributions to the study of algae with particular reference to their karyology. J. Indian bot. Soc. 63: 215–225, 1984.

SarmaYSRK and Abhayavardhani P(1983) Effects of mitomycin-C, rifamycin SVand polymycin B on Spirogyra paradoxa Rao. In: Current approaches in Cytogenetics (Eds) RP Sinha and U Sinha. Spectrum publishing House, Patna. Pp 29–36.

Sarma, Y S R K and Tripathi, S N (1973). Effects of caffeine and theobromine on the karyology of a green alga *Oedogonium acmandrium* Elfving. The Nucleus 16: 167–172.

Sarma, Y S R K and Tripathi, S N (1974). Effects of gibberellic acid on the green alga, Oedogonium *acmandrium* Elfving. Indian Journal of Experimental Biology 12: 204–206.

Sarma, Y S R K and Tripathi, S N (1976a). Effects of chemicals on some members of Indian Charophyta. I Caryologia 29(3): 247–262.

Sarma, Y S R K and Tripathi, S N (1976b). Effects of chemicals on some members of Indian Charophyta. II Caryologia: 263–276.

Sinha, S and Purak, I (1995). Taxonomy of charophya-trends and perspectives–a review article. In Book recent trends in algal taxonomy, Vol I: 179–220. APC Publications.

Sinha, S and Sinha, J P (1989). Study on mitotic chromosomal counts in some members of Charales of Chhotanagpur (Bihar). Biojournal 1: 69–72.

Sinha, S and Sinha, J P (1992). Karyomorphological studies on three species of *Chara*. Journal of Applied Biology 2: 7–11.

Spring, H, Scheer, U, Franke, W W and Trendenburg M F (1975). Lampbrush-type chromosomes in the primary nucleus of the green alga *Acetabularia mediterranea*. Chromosoma 50(1): 25–43.

Stebbins G L (1958). Variation and evolution in plants. Columbia University Press, New York.

Stebbins, G L (1971). Chromosomal evolution in higher plants. Addison Wesley Publishing Co. Stewart, K D and Mattox, K R (1975). Comparative cytology, evolution and classifi cation of the green algae with some consideration of the origin of other organisms with chlorophylls a and b. Botan Review 41: 104–135.

Subrahmanyam, B V S and Chowdary, Y B K (1992). Karyological studies of Some Indian charophyta. Cytologia 57: 409–415.

Subrahmanyam, B V S, Chowdary, Y B K and Sarma, Y S R K (1991). Aneuploidy in two taxa of Nitella (division Charophyta) from Karnataka, India. The Nucleus 34: 174–176.

Subramanian, D and Ganesan, R (1983). Cytotaxonomical studies of Charophyta from Tamil Nadu: 184–191. Proceedings of the all India applied Phycological Congress, Kanpur, India.

Sundaralingam, V S (1946). The cytology and spermatogenesis in *Chara zeylanica* Willd. Journal of the Indian Botanical Society, M.O.P. Iyengar commemoration volume, Silver Jubilee Session, Allahabad 1946: 289–303.

Vedajanani, K, Abhaya V P and Sarma, Y S R K (1980). Cytological effects of N-Methyl-N-Nitro-NNitrosoguanidine (NTG) on *Spirogyra Paradoxa* Rao. Caryologia 33(2): 299–305.

Vedajanani, K and Sarma Y S R K (1978a). Effects of antibiotics on green alga *Spirogyra azygospora* Singh. Indian Journal of Experimental Biology 16(2): 845–848.

Vedajanani, K and Sarma, Y S R K (1978b). Karyological studies on Indian Conjugales 1. *Spirogyra* Link. Phykos 17: 1–16.

Verma, B N (1967). A new chromosome count of *Chara braunii* Gmelin. Proceedings Indian Science Congress: 305.

Verma, B N and Labh, L (1986). Karyomorphological studies in the genus *Chara* Linn. Cytologia 51: 501–506.

Vidyavati (1992). Cytogenetics of algae-1. Effect of chemicals. In Indian phycological review, Vol 1. Ed. M Khan: 131–152. M/S Bishen Singh Mahindra Pal Singh, Dehradun

14 Current Strategies of Stress Tolerance in Microalgae and Plants

D. Mubarak Ali, Y. Saurabh, R. Sathya, S. Hemalatha, and Nooruddin Thajuddin

14.1 INTRODUCTION

In the 21st century, the world suffered from many stresses, particularly water shortage, which led to pollution of different kinds. The salt stress which causes salinization is also correlated to water scarcity. Salinity of soil is a major abiotic stress and a detrimental environmental factor affecting the plant productivity (Yarra et al., 2012). It is caused when sodium and chloride salts are in excess amount and plant growth and development gets impaired because it hinders with the water absorption from saline soil (Ge et al., 2017). Due to these factors, plants have evolved adaptive mechanisms at all levels viz. physiological, biochemical, and molecular levels (Zhu et al., 2012). It has been projected that salinization of soils will reduce 30% of soil adaptability by 2050 (Song and Wang, 2015). To understand the salt stress phenomenon and molecular basis, researchers have utilized model plants and model algae. This has given insight to correlate the evolutionary basis involved in both plants and algae (Zhu, 2001). The omics approach has also given widespread help to the scientists working in the area of plant abiotic stress. This work has thrown light on some of the potential genes being upregulated. Another approach using plant biotechnology is the molecular characterization of genes involved in salt stress. Few existing reports include the molecular cloning and characterization of NAC family genes, rice 6-phosphogluconate dehydrogenase gene, carotene biosynthesis related, and so on (Yang et al., 2011; Huang et al., 2003; Chen et al., 2008). Overexpression of Caffeic Acid O Methyltransferase 1 (COMT1), DREB; ERF Transcription Factor, MdDMR6 genes in various plants have displayed salt tolerance (Sun et al., 2019, 2020; Zhao et al., 2018; Fan et al., 2016). Hence, in addition to the management practices of land and water, the modern tools of omics approaches will surely pave a way for providing solution of salt stress to increase plant productivity. Table 14.1 is a comprehensive report on the various stresses on the microalgae and their significance with special reference to omics.

14.2 BIOTIC STRESS TO PLANTS AND MICROALGAE

Plants and microalgae are environmentally and economically very valuable throughout life (Hussain and Usman, 2019). Plants are a major source of food for almost all life-forms. The yield and production of plants is being affected by various kinds of stress factors. Any reduction in the yield will have an impact in the economy of the country. Thus, it is important to study the various stress factors affecting the growth, yield, and production of plants and microalgae.

Stress is a condition which is caused due to a change internally or in the external environment and affects growth, development, and productivity. A plant or microalgae under stressed conditions may respond by altering their metabolism or gene expression to overcome it. The stress to plants and microalgae can be beneficial or non-beneficial in an industrial prospective. The stress can be caused either by living factors or nonliving factors (Borowitzka, 2018). In biotic stress, the stress is

 DOI: 10.1201/9781003219156-18

TABLE 14.1
Comprehensive Report on the Various Stresses on the Microalgae and Their Significance with Special Reference to Omics

S. No	Microalgae Used	Stress Induced	Significance	References
1	*Synechocystis* sp.	Salt	ABC transporters and hypothetical proteins	(Qiao et al., 2013)
2	*Anabaena doliolum*	Salt	Under high concentration PRK significantly accumulated	(Rai et al., 2013)
3	*Picochlorum atomus*	Salinity	To produce high yield of biomass	(Von Alvensleben et al., 2013)
4	*Picochlorum SE3*	Salt	Encodes six copies of the NHX8/ salt overly sensitive 1 (SOS1) gene, helpful for the expulsion of sodium ions from the cell	(Foflonker et al., 2015)
5	*Tetraselmis* sp.	Salinity	To study the effect of fatty acid under nutrition depletion, the genes are unregulated by activating the Omega-3 fatty acids	(Adarme-Vega et al., 2014)
6	*Nitzschia laevis*	Salt	Increases the degree of unsaturated fatty acid	(Kim et al., 2014)
7	*Gracilariopsis tenuifrons*	Environment	Induced the photosynthetic activity, pigment composition, and antioxidant activity	(Zubia et al., 2014)
8	*Desmodesmus armatus* *Mesotaenium* sp. *S. quadricauda* *Tetraedron* sp.	Salinity	Increases the total lipid and fatty acid content	(Von Alvensleben et al., 2016)
9	*Amphora subtropica* *Dunaliella* sp.	Salt	To induce the chlorophyll pigment, lipid content, and carotenoids	(BenMoussa-Dahmen et al., 2016)
10	*Chlamydomonas* sp.	Salt	Increases carbohydrate and lipid synthesis genes and also increase the pool size of ADP-glucose and malonyl-CoA of starch degradation genes to produce biodiesel	(Ho et al., 2017)
11	*Acutodesmus dimorphus*	Salinity and Oxidative	To increase the lipid accumulation and decreases growth rate, which causes improved biofuel production	(Chokshi et al., 2017)
12	*Tetraselmis suecica*	Oxidative	To produce carotenoids having antioxidant and expression of genes and proteins on protective activity of human cells	(Sansone et al., 2017)
13	*Chlamydomonas reinhardtii*	Salinity	Upregulated PRK	(Sithtisarn et al., 2017)
14	*Chlorella* sp. *Chlorella* sp. *AE10* *Chlorella* sp. *S30*	Environmental	Causes loss to photosynthesis, oxidative phosphorylation, fatty acid biosynthesis, and tyrosine metabolism	(Li et al., 2018)
15	*Schizochytrium* sp.	Salinity	To improve the antioxidant system and lipid accumulation	(Sun et al., 2018)

(*Continued*)

TABLE 14.1 (*Continued*)
Comprehensive Report on the Various Stresses on the Microalgae and Their Significance with Special Reference to Omics

S. No	Microalgae Used	Stress Induced	Significance	References
16	*Arthrospira platensis*	Temperature	Upregulation of glycosyl transferase in proteome analysis	(Tahara et al., 2018)
17	*Haematococcus pluvialis*	Abiotic stress	Increase nitric oxide and salicylic acid causes melatonin, which enhances the biosynthesis of astaxanthin	(Ding et al., 2019a)
18	*Dunaliella* sp. *Dunaliella salina* *Dunaliella tertiolecta*	Salt	To regulate detoxify ROS related genes, Ca^{2+} signal transduction, and lipid accumulation	(Panahi et al., 2019)
19	*Monoraphidium* sp.	Abiotic stress	To increased ROS and oxidase activity which in turn increased the lipid synthesis to many folds	(Li et al., 2019)
20	*Scenedesmus* sp. *Scenedesmus obliquus* *Scenedesmus* sp. *IITRIND2 regular letters*	Salinity	To maintain extra polysaccharide homeostasis and restrict the ions channels, which resolves the cellular damage	(Arora et al., 2019)
21	*Acutodesmus dimorphus*	Salinity	It showed that the reactive oxygen species known to inhibit the photosynthesis mechanism by blocking the photo system II	(Li et al., 2019)
22	*Cylindrotheca closterium*	Low Temperature	To enhances the production of polyunsaturated fatty acid (PUFA) and metabolites	(Almeyda et al., 2020)
23	*Anabaena laxa* *Nostoc muscorum*	Oxidative	Induce antioxidant defense and R-metalaxyl phycoremediation	(Hamed et al., 2020)
24	*Chromochloris zofingiensis*	Environment	Enhancement of astaxanthin production and biosynthesis of lipids	(Chen et al., 2020)
25	*Interfilum* *Klebsormidium* sp.	Abiotic stress	Synthesis and accumulation of mycosporine-like amino acids (MAA)	(Hartmann et al., 2020)

caused by organisms such as bacteria, fungi, viruses, parasites, insects, nematodes, weeds, native, or cultivated plants. In abiotic stress, the stress is caused by negative environmental factors such as drought, flood, salinity, temperature changes, pH changes, heavy metals (Gull et al., 2019). An overview of stress tolerance in microalgae and plants respective to the potential applications is shown in Figure 14.1.

Since the biotic stress is caused by other living organisms that invade the crop plants or microalgae for their survival. There are different types of biotic-stress-causing agents, which cause diseases and directly utilize and deprive the nutrients and causes reduction in the production or death of the host. Thus biotic stress is a major cause for pre- and postharvest loss. The natural defense mechanism against the stress is controlled by the genes in its genetic code. Thus, different methods and mechanisms have been developed to overcome various types of biotic stresses occurring in plants and microalgae through research. The biotic stress can be overcome by altering genes and studying the mechanism of invasion. Genetically modified crops developed have proven to be biotic stress resistant to a large extent.

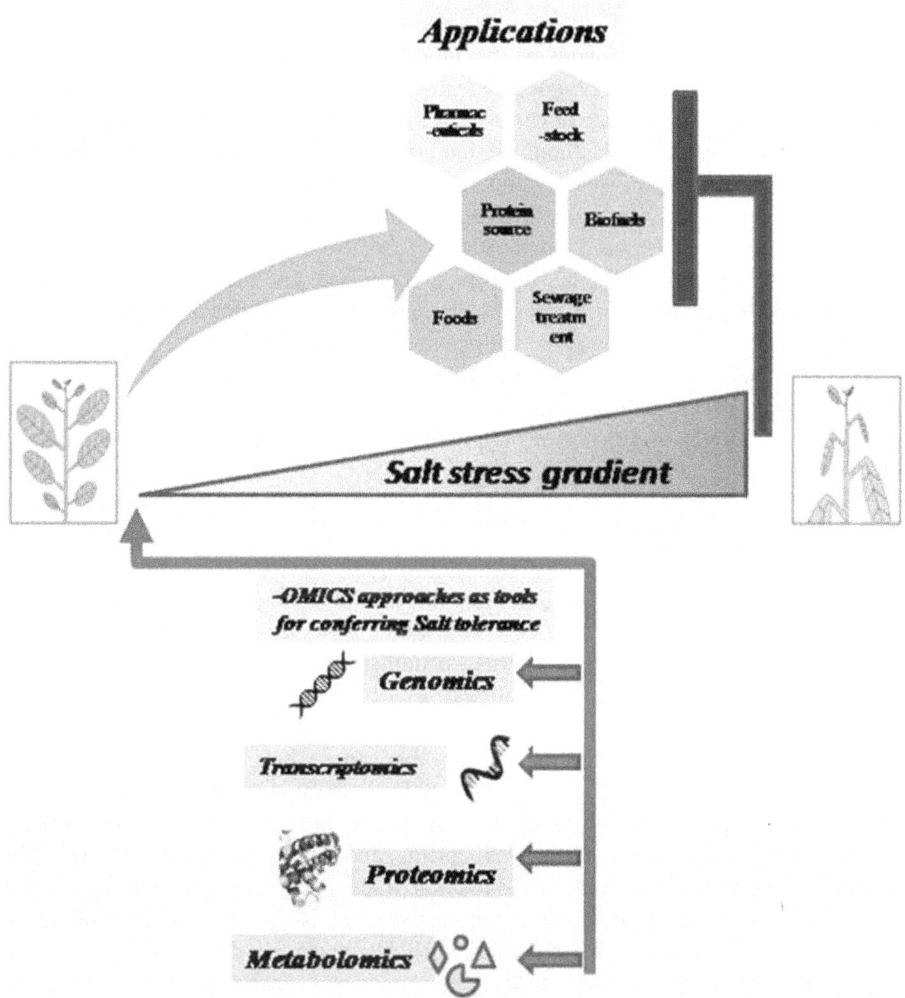

FIGURE 14.1 Overview of stress tolerance in microalgae and plants respective to the potential applications.

14.2.1 Fungal Biotic Stress

Fungi are a group of organisms which belong to eukaryotic microorganisms. These organisms are classified as kingdom fungi. In plants, fungi are more prone to cause stress or disease than any other pathogen. Several crop plants are affected by many fungal diseases. They affect almost all parts of the plants, like fruits, leaves, stem, flowers, seeds, and root. These can adversely damage the crop production and crop rotation process. The stress can be caused in preharvest or postharvest during storage. Preharvest stress is caused by affecting seeds, leaves, stem, root, and fruit. Fungal stress in seeds can cause poor germination, which is due to poor draining. In leaves, it can cause drying of leaves, rolling of leaves, spots and powdery substance in leaves, color changes in leaves. In fruits it may cause almost 50 to 100% losses by forming spots, lesions on the fruits, reduction in size, number, and even cause rotting of the fruit. In stems it causes discoloration, lesions, and rotting. In roots it causes discoloration and rotting (Hussain and Usman, 2019).

14.2.2 Bacterial Biotic Stress

Bacteria are tiny, single cellular organisms that bloom in various environments. Bacteria are both pathogenic and beneficial. The biotic stress is caused by the pathogenic bacteria that invade the host. Bacteria establish a living space in various parts of the crop plants, especially in the xylem and phloem and cause various diseases. Plants are more prone to bacterial biotic stress during rainfall that causes the outspread of spore due to humidity. No matter how the pathogens are, they require a wound or natural opening such as stomata to enter into the host. The pathogens enter directly into the plant system through air-exposed stomata. These can also be vector borne and are transmitted by sucking insects like planthoppers, leafhoppers, and psyllids. The stress can cause different types of symptoms, like hydrolysis (water soaked) or lesions that lend a wet, dark, and usually sunken and/or translucent appearance to the affected area, galls or overgrowth, wilts, leaf spots, specks and blights, soft rot, as well as scabs and cankers. Bacteria infect the host by releasing Ti or Ri plasmid that cause the mosaic structure (Barton et al., 2017). A few pathogenic bacteria derived from plants can generate harmful toxins or infuse special proteins, which usually leads to host death or breakdown of enzymes.

14.2.3 Viral Biotic Stress

Viruses are intracellular pathogenic particles that infect other living organisms. Most plant viruses are either rod shaped or isometric. Viruses also invade a wide variety of plants, and the common diseases are tobacco mosaic virus, potato virus, cucumber mosaic virus, rose mosaic virus, tomato spotted wilt virus, etc. Since plants have cell wall as a barrier, the viruses require a host to inter and infect the plant system through wounds or openings. The vectors may be herbivore, sap-feeding insects, aphids, leafhoppers, whiteflies, beetles, or other source that come in contact with the plant. Once the virus has entered the plant, it uses the phloem for its transport to the target site. The plant's defense mechanism is evaded to promote the viral replication. The infection causes chlorosis, necrosis, tissue proliferation, phyllody, leaf curling, and other physiological changes. The viral proteins affect the photosynthesis and carbon fixation process, carbon partitioning and metabolism, manipulate amino acid metabolism, manipulate stress response proteins by controlling the relative oxygen species, ROS-scavenging enzymes, such as superoxide dismutases (SOD), catalases, peroxidases, and thioredoxins; detoxify hydrogen peroxide and superoxide anions, chaperones, and related proteins; manipulating cell wall biogenesis and metabolism; and manipulating translation, protein processing, and protein degradation (Alexander and Cilia, 2016).

14.2.4 Nematode Biotic Stress

Nematodes are parasitic on plants, are active, slender unsegment roundworms (also called Nemas or eelworms). They usually live in soil and attack small roots, but some species inhabit and feed in bulbs, buds, stems, leaves, or flowers. Nematode infections are widely seen in moderate temperature and high humidity (Abdulkhair and Alghuthaymi, 2016). Plant parasitic nematodes are the key components which causes considerable yield losses in citrus-making sectors (Kumar and Das, 2019). These parasites suck the sap or juice of the plant, in which the enzymes are injected for digestion and then feed on it. Nematode feeding reduces yield and also makes the plant more prone to other infection caused by bacteria, fungi, and viruses. The symptoms such as loss of green color and yellowing, wilting on hot, bright days, stunting, dieback of twigs and shoots, and lack of response to water and fertilizer are commonly observed during nematode infection. Often, the root systems may be discolored, decayed, extremely branched, or decreased. The plants develop root injury because of secretion of toxins in the saliva of the scrounger, the tissues respond by degeneration or by producing enlargement of cells (Escobar et al., 2015). Common nematode diseases are root-knot, root-lesion, golden nematode of potato, citrus nematode.

14.4.5 Plant Parasitic Biotic Stress

Parasitic plants grow on the host plant to derive nutrition for its survival. It feeds through specialized organs called haustoria which penetrate host vascular bundles. Holoparasites lack the capability to photosynthesize and concert CO_2 to essential carbon source. Hemiparasites are partially photosynthetic but still utilize the host carbon (Těšitel, 2016). The parasitic plant can be parasitic to the stem or root competing for nutrition produced by the host plant. This causes stress in the host plant and limits the productivity. The invasion can cause stunned growth of branches or even die. Some plant parasites are witch weed, dodder, mistletoe (Abdulkhair and Alghuthaymi, 2016).

14.3 METABOLOMICS ON SEVEN STRESS TOLERANCE IN MICROALGAE UNDER SALINITY

Metabolomics is an emerging field of study which enables profiling of huge variety of small metabolites in a biological specimen (Clish, 2015). The profiling of vast variety of metabolites can be performed on single extract which represents the physiological state of a cell, its gene function and regulation. Advancements in metabolomics study have enabled the identification of known and unknown metabolites using the following techniques: nuclear magnetic resonance (NMR) spectroscopy and mass spectrometry (MS) (Kumar et al., 2017). Omics studies, specifically metabolomics studies over a nonspecific screening for all the metabolites or small intermediate compounds in a metabolic pathway helps 22 in determining the chemical inducers or inhibitors of particular biomolecules synthesis (Wase et al., 2019).

Microalgae are unicellular photosynthetic cells which serve as the richest source of carbon compounds that includes bioethanol, biopharmaceuticals, and nutraceuticals, and their growth is highly affected by the pH of the medium in which they are cultured. Maximum yield is observed at pH 9.0, and an increase in the pH increases the salinity (stress condition) of the culturing medium (Khan et al., 2018). Microalgal biomass are exploited not only to increase the yield of biofuels, pigments, and biopharmaceuticals but also to meet the nutrient needs of increasing population, as a single-cell protein (SCP) because of their ability to accumulate proteins (Lupatini et al., 2017). They are also found to be useful in removing the nitrogen, phosphorous, and BOD in the treatment of municipal wastewater (Wang et al., 2017). Microalgae survive in high saline condition by accumulation compatible solutes, excluding harmful ion and by synthesizing antioxidants (You et al., 2019). Salt tolerance in microalgae is responded with the biosynthesis of certain solute molecules, such as carotenoid and lipid synthesis are elaborated.

Carotenoids are the widespread group of natural pigments, and their biosynthesis pathways are used as taxonomic markers to differentiate among diverse microalgal phyla (Varela et al., 2015). Carotenoids synthesis involves the biosynthesis of beta-carotenes, an antioxidant molecule or the upregulation of carotenoids, genes such as canthaxanthin and astaxanthin (Paliwal et al., 2017). Microalga, *Dunaliella salina* can accumulate 10%–14% of beta-carotenes as dry cellular weight under high saline condition (Han et al., 2019). *Dunaliella tertiolecta* cultured under high saline (high NaCl) condition exhibited reduction in the accumulation of their pigment contents (Chen et al., 2019). When the saline condition is increased, the chlorophyll concentration also increased, specifically in the microalgae *Chlorella* sp., and some microalgae are used in the wastewater treatment (Vo et al., 2018). The carotenoids acts as the defense mechanism for algal cells besides superoxide dismutase (SOD) by capturing the free radicals produced during the stress condition, which makes them suitable for the purpose of high quality feed for organisms (Zhang et al., 2019a).

Halotolerant microalgae known to exhibit saline tolerance by three stages involving: alarm response, regulation, and acclimatization, which include influx of water and efflux of Na^+ ions (Arora et al., 2019). *Chlamydomonas* sp. JSC4 showed increase in synthesis of lipid and carbohydrate content when produced at 2% salt stresses condition than at 0% saline condition. This is due to enlarged pool size of ADP-glucose and malonyl-CoA, respectively (Ho et al., 2017). Saline stresses induced in microalgae culturing not only increases the lipid synthesis but also alters the type of lipid composition. Therefore, the salt concentration to be used can be altered depending on the type of lipid to be synthesized (Alishah Aratboni et al., 2019). High saline condition reduces the utilization of CO_2 and nutrients leading to the NADPH production. It also reduces the light penetration, reducing chlorophyll synthesis that causes oxidative stress and release of reactive oxygen species (ROS) (Srivastava et al., 2017). Chokshi et al. (2017) reported that the salinity stress studies conducted on *Acutodesmus dimorphus* showed that the ROS known to restrain the photosynthesis mechanism by blocking the photo system II (PS II) system. *Monoraphidium* sp. QLY-1 culture grown in fulvic acid (plant growth regulator) added medium along with saline stress showed that it increases the abiotic stress, that is, increased ROS and oxidase activity which in turn increased the lipid synthesis to many folds (Li et al., 2019).

14.4 PROTEOMICS OF STRESS TOLERANCE IN MICROALGAE UNDER SALINITY

Salinization is a threat to current soil environment. Heterologuous genes are introduced to plants during cultivation as an innovative strategy to become accustomed to alkali-saline environment. Germplasm isolation or genetic resource isolation is the primary prerequisite adapted for this technique (Liu et al., 2019). Deserts, craters, salt lakes, polar regions are some of the extreme environments for algal diversity to be found. Therefore, the current research focus is for the investigation of these algal germplasm. Salt tolerances by the microalgae are notably cateogorized into three phases: alarm response, regulation, and acclimatization. Alarm response talks about water influx, resulting in increase in microalgal cell volume as well as size. The internal osmotic pressure is balanced due to the increase in Na^+ ions and osmolytes. The second phase includes activation/deactivation of enzymes, light harvesting complexes (LHC), and transporters and enzymatic modifications. In recent times in *Scenedesmus obliquus* XJ002 and *Chlorella* sp. 1 S30 under high saline conditions, downregulation of PS II system, antenna proteins, and chlorophyll content was reported (Ji et al., 2018; Liu et al., 2019). The third phase involves control of proteins and genes changes and $Ca2^+$ activation signalling (Borowitzka, 2018; Einali, 2018). Therefore, the mentioned process involves antioxidant mechanism commencement and activation of damaged cellular substances in presence of saline stress that gives microalga protection (Einali, 2018). Both the and osmo-protecting molecules aids in salinity stress of the microalgal cells, further which it instigates the lipids synthesis that acts as energy-profond storages (Xia et al., 2014; Wang et al., 2016). For instance, a salinity stress (1.0 g L^{-1} NaCl) contributed to increase the lipid content (23.4% by DCW) and the proportion of saturated fatty acids (9.2% by DCW) in a mixed microalgal culture. Inclusion of integrated wastewater treatment has given potential sustainability for salinity stress in biorefinery (Mohan and Devi, 2014). High quantity of sucrose and proline were seen due to the gathering of *Scenedesmus* sp. IITRIND2 and augmented lipid. Moreover, *Scenedesmus* sp. IITRIND2 shows signs of salt tolerance as proteomics data identified two unique proteins: (a) putative salt tolerant protein and (b) alfin-like protein. The possible homolog of salt-tolerant protein (STO), which is a putative salt tolerance protein, could be identified in *Arabidopsis thaliana* as a Na^+/H^+ antiporter, which has an over expression resulting in improved salt tolerance. The quenching of the ROS produced because of salt stress found to be key adaptive mechanism used by *Scenedesmus* sp. (Arora et al., 2019).

Scenedesmus IITRIND2 efficiently grown in ASW due to differential expression of three antioxidant proteins together with thioredoxin, glutathione peroxidise, and catalase. When exposed to salinity, *Scenedesmus* sp. restructured cellular reserves and adjusted its surface potential with secretion of extra polysaccharide to constrain the ion channels, retain homeostasis, and unravel the cellular damage solving (Fan et al., 2016). From the algal-omics research, an efficient salinity driven metabolic adjustment by the microalga beginning with the increase of negatively charged lipids, upregulation of sugars, and proline buildup (Chen et al., 2017) followed by direction of carbon and energy flux towards TAG synthesis. In addition, the omics experiment revealed that both lipid and de novo cycling pathways increases the triglycererols TAG accumulation in microalgal cells.

A resultant strain noted as *Chlorella* sp. S30 and stress tolerance method was tested by comparative transcriptomic analysis. Even though the evolved strain can sustain 30 g/L salt, high salinity resulted in photosynthesis deprivation, fatty acid biosynthesis oxidative, phosphorylation, and tyrosine metabolism (Li et al., 2018). The associated genes of amino acid biosynthesis, antioxidant enzymes, central carbon metabolism, CO_2 fixation, and ABC transporter proteins are upregulated. The CO_2 mechanism in microalgae having a negative response due to high salt affinity. The relative transcriptomic analysis for the dramatically changed genes in amino acid metabolism are studied, and the allied genes of amino acid metabolism of *Chlorella* sp. S30 in presence of high saline and BG11 mediums are found (Sun et al., 2018).

Addition of salts is a simple yet effective technique to maximize the lipid accumulation while minimizing the invasion of non-target algae or grazers, therefore improving the economic feasibility of algal biofuel. Although salinity stress is commonly used for marine microalgae, especially for *Dunaliella*, an innovative culturing method was coined as the salinity-gradient operation in combination with nitrogen depletion to increase the lipid accumulation in *Chlamydomonas* sp. JSC4 (Chen et al., 2017). A new glycosyl transferase gene (UGT85A5) has been identified in *Arabidopsis* subjected to salt stress. Moreover, the inductive expression of UGT85A5 into tobacco plants has shown to perk up the salt tolerance of the transgenic plants (Sun et al., 2013). The upregulation of glycosyl transferase was reported also in the proteome test of *A. platensis* under low temperature stress. Besides transportation of metals, nutrients, and regulation of cellular pH, ABC transporters (spot 7B) were developed in cells to protect against the oxidative stress caused from ROS (Tahara et al., 2018). Induction of ABC transporters and hypothetical proteins was reported more than the half of the total induced proteins in salt stressed *Synechocystis* sp. PRK was significantly accumulated by the Cu acclimated strain *A. doliolum* under high concentrations (Rai et al., 2013). Moreover, PRK was upregulated in the tolerant *C. reinhardtii* strain under the salinity stress (Sithtisarn et al., 2017). ACS was found in bacteria, archaea, and eukaryote (Gao et al., 2016). ACS has proved a vital role in growth and resistance to salt and other stresses (Quan et al., 2008). Additionally, Sithtisarn et al. (2017) suggested the possible role of ligase in salinity tolerance of *C. reinhardtii*.

14.5 GENOMICS ON STRESS TOLERANCE IN MICROALGAE UNDER HIGH SALINITY

Since microalgae implies important resources, some species can accumulate up to 30–70% of lipids under certain conditions (Sun et al., 2018). With the increase of world population and the demand for energy, finding a suitable renewable energy resource has become a critical issue. Microalgae have been renowned as a potential source of livestock feed, pharmaceuticals, and alternative fuels. Recently, research has focused on lipids and carotenoids (Sun et al., 2018). Microalgae can produce large quantities of neutral lipids, TAG, on exposure to unfavorable growth surroundings, namely, high salinity. Eukaryotic microalgae use some methods that can help to surmount salt stress (DeJaeger et al., 2018). Generally, stress-induction methods decrease the microalgae growth rate. Salinity will induce stress induction in microalgae, so under high-stress condition, some genetic mutation will occur through accession of useful phenotypes and subsequent positive selection (Sun et al., 2018). *Picochlorum* SENEW3 (SE3) has been collected from the brackish water and exposed to high seasonal salinity conditions. The transport proteins was put into the wide range of salt tolerance in Picochlorum SE3. The algae encodes six copies of the NHX8/salt overly sensitive 1 (SOS1) gene, which is very much helpful for the expulsion of sodium ions from the cell, which is important for salt tolerance (Foflonker et al., 2015). Salinity stress rapidly augmented the saturated fatty acids (SFAs) buildup in the microalgae. During this stress many genes were upregulated. In the present study, the genes involved in the biosynthesis of fatty acid and TAG were examined. The genes which are involved are PDH2, ACCase, MAT, KAS2, FAT1, GPD1, DGTT1, DGTT2,

DGTT3, and DGTT4. Genes such as PDH2, ACCase, MAT, KAS2, and FAT were involved in the fatty acid biosynthesis, while genes GPD1 and DGTT1–4 were involved in the Kennedy pathway for the TAG biosynthesis. These genes are activated in the microalgae during high saline condition. With the help of gene expression analysis, it was stated that high salinity activates the expression of genes responsible for the fatty acid biosynthesis. This speculation gave the explanation for the conversion of starch-to-lipid by high salinity. Salt-induced stress rapidly increased saturated fatty acid (Atikij et al., 2019). The correlation between the productions of omega-3 fatty acids with the high salinity condition of microalgae was studied. Omega-3 fatty acids are much needed for human nutrition and aquaculture feeds. Marine microalgae species *Tetraselmis* sp. M8 was used to study the effect of salinity on fatty acids and related gene expression. Eicosapentaenoic acid (EPA), eicosatetraenoic acid (ETA), and docosahexaenoic acid (DHA) are some of the omega-3 fatty acids. *Tetraselmis* sp. expression of 15 genes was seen in FA synthesis at high salinity. Out of this, four genes, encoding BKAS, Δ5D, Δ6E, and ACSace, were differentially expressed according to salinity. The gene BKAS, codes for an enzyme which is involved in the extension of long-chain fatty acids by adding two carbons to the fatty acid chain. Gene encoding Δ5D increased with the sequence of nutrient stress in all salinities. From this experiment it is inferred that salinity had no important outcome on the fatty acid synthesis. But under nutrition depletion, the genes which are implicated in the fatty acid synthesis are unregulated by activating the omega-3 pathway (Adarme-Vega et al., 2014). ROS-related genes are the important fundamentals of the microalgae *Dunaliella* salt stress response system. In *Dunaliella* various meta-genes such as APX, CLPB1, CLPD, LHL3, SHMT2, DVR1, and WD40 were found to be an important backbone of ROS and signalling network. APX was the only enzyme selected as meta genes in *Dunaliella* sp. APX even though different scavenging enzymes were up regulated in response to salt stress. *Dunaliella* species has founded complex ways to regulate and detoxify ROS in salt stress conditions.

This study recognized a gene whose expression was altered in response to salt stress. This must be very useful in the secondary metabolite production (Panahi et al., 2019). *Chlorella* sp. AE10 is an evolved strain that can withstand high environmental stress. This strain proposed can tolerate the nitrogen starvation and high concentration of CO_2 and salt. *Chlorella* sp. S30 has high salinity, and salt tolerance was also analyzed. About 73,135 unigenes were identified for *Chlorella* sp., out of which only 1,739 of them were annotated. High salinity can cause oxidative stress, which can root a great harm in the microalgae. Genes such as the ND4, ND5, COX1, and COX2 in the oxidative phosphorylation pathway were downregulated, and the genes such as NuoA, SDHA, and SDHB were upregulated. The down regulations of PsaA, PsaB, PsaC, PsaI, and PsaJ (PS I) and PsbA, PsbB, PsbC, PsbD, PsbE (PS II) shows that the photosynthesis activity was declined. From the current study, the salt stress caused negative effects in oxidative phosphorylation and photosynthesis (Li et al., 2018). *C. reinhardtii* is a type of microalgae which are specialized to form aggregated and hypertrophied cells called "palmelloids" as a defense response to stressful conditions. To find the cause for the lipid content decrement of salt-resistant strains, transcription of key genes in the starch-to-lipid biosynthesis switching was analyzed. PDC, ALDH, and ACS are enzymes in the PDH-bypass pathway, which are upregulated in JSC4 in response to salinity. Expression of all 4 ACC genes (BCCP, BC, aCT, and bCT) is also induced by salinity conditions in JSC4. Transcriptional analysis also showed that genes encoding ACC, PDC, ALDH, and ACS are all upregulated during high salinity. Key genes are expressed when the salinity is increased, which enhances the lipid accumulation in JSC4, and when these genes are inactivated, it may cause delayed starch degradation and decreased lipid content in the salt-resistant strains. This study suggests that stress caused by high salinity plays an important role in stimulation of lipid accumulation in the microalgae (Kato et al., 2017).

A marine microalga, *Schizochytrium* sp. is very much useful for the feedstock for the sustainable production of biofuels and food. Under high stress, variation can occur via acquisition of beneficial phenotypes by random genomic mutations and subsequent positive selection. Using high salinity along with adaptive laboratory evolution (ALE) was performed to improve the antioxidant system and lipid accumulation. ALE150 showed lower reactive oxygen species (ROS) levels compared to the starting strain. In *Dunaliella* and *Chlamydomonas* sp., high salinity increases the SFA content, and the PUFA decreases when the fermentation process ends. But in the case of ALE150 evolved strain, it does not show any PUFA reduction, which reveals that the strain ALE150 increase SFA percentage and decreased the PUFA percentage when compared to the starting strain. In *Dunaliella* and *Chlamydomonas* sp., due to high salinity the SFA percentage increases, and the PUFA percentage decreased at the end. Whereas ALE150 PUFA strain does not show a decrease in the percentage. This shows that SFA buildup was increased, and PUFA deprivation was reduced in ALE150 compared to the starting strain. At high salinity, ALE raised an increase of FAS gene expression, which helps in the fatty acid synthesis (FAS) in the *Schizochytrium* sp. ALE150 (Sun et al., 2018).

14.6 MERITS AND DEMERITS OF SALINE TOLERANCE IN MICROALGAE

These microalgae have their application in various fields viz. biofuels, health supplements, pharmaceuticals, wastewater treatment, etc. It was noted that increase of pH will increase the culture media salinity, which turns out to be harmful for certain algae cells (Khan et al., 2018). Salinity have both positive and negative effects over microalgae in various aspects, like biomass yield, over the metabolism of the organism, and in the physiological state.

Salinity is one of the most important factor for the survival of both marine and freshwater microalgae. Increased or decreased salinity than that of inherent habitats of microalgae can hand out as a stress because changes would happen in order to adjust themselves to the surrounding environment. Marine microalgae alter lipid production with respect to low salinity and vice versa for freshwater microalgae. *Schizochytrium limacinum* showed up an improved lipid content, specifically saturated fatty acid at lowered salinity conditions. In diatom *Nitzschialaevis* increases the salt concentration and also increases the degree of fatty acid unsaturation (Kim et al., 2014). Different microalgal species can tolerate different salt concentrations. In some organism excess salinity inhibits the photosynthesis process, which in turn is noted to have a negative impact over the biomass yield. In the production of various biofuels, salt concentration had a positive role. High salt concentration was known to have an impact in ethanol production (cyanobacterium and *A. maxima*). Salt stress has an influence in the composition of microalgae cells. Low molecular weight carbohydrates were produced in response to these salt stresses (Cheng and He, 2014).

Salinity influences the growth pattern, biochemical profiles, and nutrients requirements below 36 ppt. *Pseudanabaena limnetica* was dominant over *Picochlorum atomus* in lower salinity culture. *Picochlorum atomus* can be cultivated in various locations which vary in salinity level from 2–36 ppt, without any adverse effects on biochemical profiles. Increase in salinity appears to have a positive effect for delay in contamination which gives good quality of high biomass yield (Von-Alvensleben et al., 2013).

Salinity and nutrient resource availability vary among different places depending on their season and cultivation locations, which potentially have an impact on biomass productivity. Among the following four species, *Desmodesmus armatus, Scenedesmus quadricauda*, *Mesotaenium* sp., and *Tetraedron* sp., which had been cultured at salinities of 2–18 ppt. It was noted that *D. armatus* has highest halo tolerance, which had an active growth up to 18 ppt, while *Mesotaenium* sp. was the least salinity-tolerant species, which had decreased growth rates from 11 ppt. *Mesotaenium* sp. was known to have the highest biomass productivity at 2 and 8 ppt, while in *S. quadricauda* and *Tetraedron* sp., which increases in 1 the total lipid and FA contents which was due to nutrient depletion and also increases the salinity to boost up in the total lipid and FA contents (Von Alvensleben et al., 2016).

Effects due to salinity on microalgae were studied in marine microalgae, wherein the target was on the improvement in lipid content, which was further converted into biodiesel. Several studies revealed significant effect 1 on the growth and physiology of microalgae due to the salinity stress of the culture medium. *Amphora subtropica* and *Dunaliella* sp. were found to be halophilic strains. Pigment levels increased in *A. subtropica*, and *Dunaliella* sp. tolerates high salt concentration, which is known to increase salinity with increased lipid content. Chlorophyll a and b contents were high in *A. subtropica* at 1M NaCl and *Dunaliella* sp. at 3M NaCl. When the salt concentration increases continuously, which exceeds beyond the optimal growth concentration, chlorophyll a and b had a significant decrease, while in the case of total carotenoids content and (Car/Chl a) ratio would be vice versa with the increase of salt in *A. subtropica* (BenMoussa-Dahmen et al., 2016). Maintenance of saturated fatty acid methyl esters in elevated amounts was also facilitated by salinity stress, which in turn had improved fuel properties. Documentation of positive impact was showed upon lipid synthesis due to salinity stress (Mohan and Devi, 2014).

Recycling of microalgae culture media, which would be used for further production, was often altered with increase in salinity due to evaporation and fouling by particulate and dissolved matter. *Tetraselmis* sp. MUR 233 was grown in the culture medium, which was recycled to increase the salinity (Sing et al., 2014). Composition of lipid and fatty acid of the microalgae varies accordingly to various culture conditions. Elevation of salt concentration has its own drawbacks in ion toxicity, rate of respiration, distribution of minerals, photosynthetic rate, and permeability of the cell membrane (Asulabh et al., 2012).

14.7 FUTURE PERSPECTIVES ON SALINE TOLERANCE

Salinity and its effect is the main abiotic factor that occurs in our environment, which is affecting the productivity of certain crops present in the worldwide. It is important to comprehend the overall mechanism of saline tolerance in plants and the salinity, i.e., existing in the soil as well. By understanding the interconnectivity between them, it will be easier to suggest certain perspective ways that are in trend to control the salinity, and also we can improve the property of plant saline tolerance. The productivity of the crops can be increased worldwide.

The mechanism of saline tolerance in plants is one of the complex process, and certain genes are also involved over here. In general, a particular plant's physiological responses towards this saline tolerance or salinity are difficult to understand, and this poses as a major problem in designing and interpreting an actual experiment. In addition, there are several mechanisms that are existing in a plant towards this saline tolerance mechanism. There are three main saline tolerance mechanisms proposed elsewhere (Munns and Tester, 2008). These are the ion-exclusion, tissue tolerance, and shoot ion-dependent tolerance. The effect of salinity towards the plant growth should be quantified for the better suggestion of perspective ideas. This can be done through the destructive and nondestructive approach. Salinity often causes reduction in plant growth, thereby affecting the overall yield. In destructive approach, the plants will be separated into parts. Measurements such as length of the root, height of the stem, and area of the leaf must be measured and calculated.

This kind of approach doesn't require any expensive equipment but only requires the limited amount of space of the frequent number of samplings to take place. After samplings of the plant parts had done, the reduction in biomass is calculated. This could be achieved through the comparison between the stressed plants and the control plants. The duration of stress interval needs to be calculated. The growth reduction difference measurement can be accurate if the biomass increase had been analyzed during the stress intervals. The effect of the saline tolerance can be observed in the plants when parts of the plants are cultured at starting time of the salt stress and between the end point of the period of interest. The plants growth should be observed both before and after the salt stress conditions. Hence, the better effect of the saline tolerance in the plant can be studied. But some plants can accumulate more amount of salt in its shoot. In such cases, mass of the salt becomes a significant fraction of the total mass. Some extreme halophytes such as *Salicornia* do exist in this property. The nondestructive approach involves the dose response curves that can be applied only to specific genotypes in plants.

Apart from this, it is important to observe the salinity effects on plants based on their water relation, transpiration. Water relation can be referred as the potential energy of water relative to pure water, and this determines the movement of water from the region where there is higher water potential to the region where there is lower water potential. Other than that, transpiration and transpiration use efficiency (TUE) are two important components that also need to be considered to study the effect of salinity in plants. Since they define the movement of water in plants, thereby the capacity of saline tolerance can be studied. TUE depends on both the genotype and the environment on which the plant grows. Before getting into the future perspective trends on saline tolerance, the previously mentioned are the important parameters, and it is of vital importance to recognize the saline tolerance mechanism in plants to develop a better response in them towards this major abiotic stress issue. The idea of saline tolerance should be focused in different viewpoints such as biochemical, physiological parameters, molecular, etc. The different approaches towards improving the saline tolerance in plants are as follows.

Transcriptome and genomic approaches may provide a brief functional view of the saline tolerance in plants. Wang et al. (2018) have identified that during transcriptome analysis, a set of unigenes were identified, and they are differentially expressed under salt stress conditions. They stated that by using GO and KEGG, the potential mechanism for saline tolerance or salt stress tolerance were identified. Further, these analyses reported that lipid homeostasis and the phosphatidic acid acetate level regulations are having a key role in further improving the tolerance of salt stress. Further, Zhang et al. (2019) represented about the physiological and molecular responses of *Populus* sp. towards salinity. They had used the poplars as a model species to study the NaCl stress that is occurring on them.

The idea of using bio stimulants is yet another strategy to overcome the negativity of salinity. Zhang et al. (2019c) studied the effect of melatonin in plants towards their saline tolerance response. Since melatonin is responsible for regulating hormone metabolism through the upregulation of abscisic and gibberellic acid, which contains certain catabolism-responsible genes. Hence, the authors finally described that the finding of this plant melatonin receptor can open a new perspective way of studying its role towards saline tolerance in them. Further, they had explained that the melatonin metabolism increases the expression of certain stress-responsive genes, i.e., related to the effect of salinity.

Protein kinase, ion homeostasis, and ROS plays a major role towards this effect of salinity. The different families of protein kinase are being involved in the plant's adaptation to salt stress (Szymanska et al., 2019; Zhang et al., 2019b). Specifically, the SNF1-related protein kinase (SnRK2.4 and SnRK2.10) had a role in ROS homeostasis modulation that shows response towards salinity (Szymanska et al., 2019). This is being done by regulating the expression of several genes that are related to ROS generation within them.

Borowitzka (2018) studied and described the effect of salt stress in ion homeostasis. The authors further represented a brief overview about the role of high-affinity potassium-type transporter 1 (HKT1) and their importance in different plant species under the effect of salinity. HKT1-type transporters are playing a crucial role in Na+ homeostasis, thereby maintaining a pivotal importance for an optimal K^+/Na^+ balance in the cytoplasm that leads to response in salinity for a plant survival. They described the role of these HKT1-type transporters and their functional differences in glycophytes and halophytes as well.

Some of the authors suggested the Proteomics approach towards the effect of salinity in plant growth. This isobaric tags for relative and absolute quantization (iTRAQ) based proteomic technique which had been implied to make out the differentially expressed proteins in leaves of two rice genotypes that solely differs in their salinity tolerance. The iTRAQ protein profiling was identified in both rice genotypes. This revealed that the differentially expressed proteins was found to be involved in the regulation of salinity responses, further in oxidation-reduction responses, in carbohydrate metabolism, and in photosynthesis as well. Regarding their sub-cellular localization, it was found that most of them were predicted to get localized in the region of cytoplasm and chloroplasts (Gupta and Huang, 2014). The role of genetic engineering techniques would help us to study more about the intracellular and intercellular interaction of saline tolerance response in plants. Regarding future prospective, there is a lack of information regarding the proteomic, transcriptomic, genetic techniques; it would be better to focus on the genetic engineering aspects in order to develop a saline tolerant plant.

ACKNOWLEDGMENT

This work was supported by B.S. Abdur Rahman Crescent Institute of Science and Technology, Chennai through CSM scheme (CSD/CSM/2022/29).

REFERENCES

Abdulkhair, W. M., and Alghuthaymi, M. A. (2016). Plant pathogens. Plant Growth, 49. doi:10.5772/65325.

Adarme-Vega, T. C., Thomas-Hall, S. R., Lim, D. K., and Schenk, P. M. (2014). Effects of long chain fatty acid synthesis and associated gene expression in microalga Tetraselmis sp. Marine Drugs, 12(6), 3381–3398.

Alexander, M. M., and Cilia, M. (2016). A molecular tug-of-war: Global plant proteome changes during viral infection. Curr. Plant Biol., 5, 13–24. doi:10.1016/j.cpb.2015.10.003 15.

Alishah Aratboni, H., Rafiei, N., Garcia-Granados, R., Alemzadeh, A., and Morones-Ramírez, J. R. (2019). Biomass and lipid induction strategies in microalgae for biofuel production and other applications. Microbial Cell Fact., 18(1).

Almeyda, M. D., Bilbao, P. G. S., Popovich, C. A., Constenla, D., and Leonardi, P. I. (2020). Enhancement of polyunsaturated fatty acid production under low-temperature stress in Cylindrotheca closterium. J. Appl. Phycol. 32, 989–1001.

Aratboni, H. A., Rafiei, N., Garcia-Granados, R., Alemzadeh, A., and Morones-Ramírez, J. R. (2019). Biomass and lipid induction strategies in microalgae for biofuel production and other applications. Microbial Cell Fact., 18(1), 178.

Arora, N., Kumari, P., Kumar, A., Gangwar, R., Gulati, K., Pruthi, P. A., Prasad, R., Kumar, D., Pruthi, V., and Poluri, K. M. (2019). Delineating the molecular responses of a halotolerant microalga using integrated omics approach to identify genetic engineering targets for enhanced TAG production. Biotechnol. Biofuels, 12(1). doi:10.1186/s13068-018-1343-1.

Asulabh, K. S., Supriya, G., and Ramachandra, T. V. (2012). Effect of salinity concentrations on growth rate and lipid concentration in *Microcystis* sp., *Chlorococcum* sp. and *Chaetoceros* sp. In National Conference on Conservation and Management of Wetland Ecosystems, 61 School of Environmental Sciences, Mahatma Gandhi University, Kottayam, Kerala.

Atikij, T., Syaputri, Y., Iwahashi, H., Praneenararat, T., Sirisattha, S., Kageyama, H., and Waditee-Sirisattha, R. (2019). Enhanced lipid production and molecular dynamics under salinity stress in green microalga *Chlamydomonas reinhardtii* (137C). Marine Drugs, 17(8), 484.

Barton, I. S., Fuqua, C., and Platt, T. G. (2017). Ecological and evolutionary dynamics of a model facultative pathogen: Agrobacterium and crown gall disease of plants. Environ. Microbiol., 20(1), 16–29. doi:10.1111/1462-2920.13976.

BenMoussa-Dahmen, I., Chtourou, H., Rezgui, F., Sayadi, S., and Dhouib, A. (2016). Salinity stress increases lipid, secondary metabolites and enzyme activity in *Amphora subtropica* and *Dunaliella* sp. for biodiesel production. Biores. Technol., 218, 816–825.

Borowitzka, M. A. (2018). The "stress" concept in microalgal biology—homeostasis, acclimation and adaptation. J. Appl. Phycol. doi:10.1007/s10811-018-1399-0 17.

Chen, B., Wan, C., Mehmood, M. A., Chang, J. S., Bai, F., and Zhao, X. (2017). Manipulating environmental stresses and stress tolerance of microalgae for enhanced production of lipids and value-added products–A review. Bioresour. Technol., 244, 1198–1206.

Chen, C., Bai, L. H., Xu, H., Dong, G. L., Ruan, K., Huang, F., and Cao, Y. (2008). Cloning and expression study of a putative carotene biosynthesis related (cbr) gene from the halotolerant green alga *Dunaliella salina*. Mole. Biol. Rep., 35(3), 321–327.

Chen, H.-H., Xue, L.-L., Liang, M.-H., and Jiang, J.-G. (2019). Sodium azide intervention, salinity stress and two-step cultivation of *Dunaliella tertiolecta* for lipid accumulation. Enzyme Microbial Technol. 127: 1–5. doi:10.1016/j.enzmictec.2019.04.008.

Chen, J-h, Wei, D., Lim, P-E. (2020). Enhanced coproduction of astaxanthin and lipids by the green microalga Chromochloris zofingiensis: Selected phytohormones as positive stimulators. Bioresour Technol. 295, 122242.

Cheng, D., and He, Q. (2014). Assessment of environmental stresses for enhanced microalgal biofuel production–an overview. Front. Ener. Res., 2, 26.

Chokshi, K., Pancha, I., Ghosh, A., and Mishra, S. (2017). Salinity induced oxidative stress alters the physiological responses and improves the biofuel potential of green microalgae *Acutodesmus dimorphus*. Biores. Technol., 244, 1376–1383. doi:10.1016/j.biortech.2017.05.003.

Clish, C. B. (2015). 50 metabolomics: An emerging but powerful tool for precision medicine. Mol. Case Stud., 1(1), a000588. doi:10.1101/mcs.a000588 42.

de Jaeger, L., Carreres, B. M., Springer, J., Schaap, P. J., Eggink, G., Dos Santos, V. A. M., . . . Martens, D. E. (2018). *Neochloris oleoabundans* is worth its salt: Transcriptomic analysis under salt and nitrogen stress. PLoS One, 13(4), 10.

Ding, W., Cui, J., Zhao, Y. T. et al. 2019a. Enhancing Haematococcus pluvialis biomass and γ-aminobutyric acid accumulation by two-step cultivation and salt supplementation. Bioresource Technology, 285: 121334.

Einali, A. (2018). The induction of salt stress tolerance by propyl gallate treatment in green microalga *Dunaliella bardawil*, through enhancing ascorbate pool and antioxidant enzymes activity. Protoplasma, 255(2), 601–611.

Escobar, C., Barcala, M., Cabrera, J., and Fenoll, C. (2015). Overview of root-knot nematodes and giant cells. In: C. Escobar, C. Fenoll, (Eds.), Plant Nematode Interactions—A View on Compatible Interrelationships, pp. 1–32. CABI Digital Library. doi:10.1016/bs.abr.2015.01.001.

Fan, J., Xu, H., and Li, Y. (2016). Transcriptome-based global analysis of gene expression in response to carbon dioxide deprivation in the green algae *Chlorella pyrenoidosa*. Algal Res., 16, 12–19.

Foflonker, F., Price, D. C., Qiu, H., Palenik, B., Wang, S., and Bhattacharya, D. (2015). Genome of the halotolerant green alga Picochlorum sp. 58 reveals strategies for thriving under fluctuating environmental conditions. Environ. Microbial., 17(2), 412–426.

Gao, J., Tao, J., Liang, W., and Jiang, Z. (2016). Cyclic (di) nucleotides: The common language shared by microbe and host. Curr. Opin. Microbiol., 30, 79–87.

Ge, W., Zhang, Y., Sun, Z., Li, J., Liu, G., Ma, Y., and Gao, J. (2017). Physiological and anatomical responses of *Phyllostachys vivax* and *Arundinaria fortunei* (Gramineae) under salt stress. Brazilian J. Bot., 40(1), 79–91.

Gull, A., Ahmad Lone, A., and Ul Islam Wani, N. (2019). Biotic and abiotic stresses in plants. In: Alexandre Bosco de Oliveira (Ed.), Abiotic and Biotic Stress in Plants. Intechopen doi:10.5772/intechopen.85832.

Gupta, B., and Huang, B. (2014). Mechanism of salinity tolerance in plants: Physiological, biochemical, and molecular characterization. Int. J. Genomics, 2014. doi:10.1155/2014/701596.

Hamed, S. M., Hassan, S. H., Selim, S., Wadaan, MAM, Mohany, M., Hozzein, W.N., AbdElgawad, H. (2020). Differential responses of two cyanobacterial species to R-metalaxyl toxicity: Growth, photosynthesis and antioxidant analyses, Environmental Pollution, 258: 113681.

Han, S. I., Kim, S., Lee, C., and Choi, Y. E. (2019). Blue-red LED wavelength shifting strategy for enhancing beta-carotene production from halotolerant microalga, *Dunaliella salina*. J. Microbiol., 57(2), 101–106.

Hartmann, A., Glaser, K., Holzinger, A., Ganzera, M., Karsten, U. (2020). Klebsormidin A and B, two new UV-sunscreen compounds in green microalgal Interfilum and Klebsormidium Species (Streptophyta) from terrestrial habitats. Front Microbiol 11, 499.

Ho, S. H., Nakanishi, A., Kato, Y., Yamasaki, H., Chang, J. S., Misawa, N., and Kondo, A. (2017). Dynamic metabolic profiling together with transcription analysis reveals salinity induced starch-to-lipid biosynthesis in alga *Chlamydomonas* sp. JSC4. Sci. Rep., 7(1), 1–11.

Huang, J., Zhang, H., Wang, J., and Yang, J. (2003). Molecular cloning and characterization of rice 6-phosphogluconate dehydrogenase gene that is up-regulated by salt stress a. Molecular Biol. Rep., 30(4), 223–227.

Hussain, F., and Usman, F. (2019). Fungal biotic stresses in plants and its control strategy. In: Alexandre Bosco de Oliveira (Ed.), Abiotic and Biotic Stress in Plants. Intechopen. doi:10.5772/intechopen.83406.

Ji, X., Cheng, J., Gong, D., Zhao, X., Qi, Y., Su, Y., and Ma, W. (2018). The effect of NaCl stress on photosynthetic efficiency and lipid production in freshwater microalga—Scenedesmus obliquus XJ002. Sci. Total Environ., 633, 593–599.

Jiang, M., Ye, Z. H., Zhang, H. J., and Miao, L. X. (2019). Broccoli plants over expressing an ERF transcription factor gene BoERF1 facilitates both salt stress and sclerotinia stem rot resistance. J. Plant Growth Regul., 38(1), 1–13.

Kato, Y., Ho, S. H., Vavricka, C. J., Chang, J. S., Hasunuma, T., and Kondo, A. (2017). Evolutionary engineering of salt-resistant chlamydomonas sp. strains reveals salinity stress-activated starch-to-lipid biosynthesis switching. Bioresour. Technol., 245, 1484–1490.

Khan, M. I., Shin, J. H., and Kim, J. D. (2018). The promising future of microalgae: Current status, challenges, and optimization of a sustainable and renewable industry for biofuels, feed, and other products. Microb. Cell Fact., 17, 36.

Kim, G., Mujtaba, G., Rizwan, M., and Lee, K. (2014). Environmental stress strategies for stimulating lipid production from microalgae for biodiesel. Appl. Chem. Eng., 25(25), 553–558.

Kumar K. K and Das, A. K. (2019). Diversity and community analysis of plant parasitic nematodes associated with citrus at citrus research station, Tinsukia, Assam. *Journal of Entomology and Zoology Studies* 2019; 7(3): 187-189.

Kumar, K. K., and Das, A. K. (2019). Diversity and community analysis of plant parasitic nematodes associated with citrus at citrus research station, Tinsukia, Assam.

Kumar, R., Bohra, A., Pandey, A. K., Pandey, M. K., and Kumar, A. (2017). Metabolomics for plant improvement: Status and prospects. Front. Plant Sci., 8. doi:10.3389/fpls.2017.01302.

Li, X., Li, X., Han, B., Zhao, Y., Li, T., Zhao, P., and Yu, X. (2019). Improvement in lipid production in Monoraphidium sp. 60 QLY-1 by combining fulvic acid treatment and salinity stress. Biores. Technol., 122179. doi:10.1016/j.biortech.2019.122179.

Li, X., Yuan, Y., Cheng, D., Gao, J., Kong, L., Zhao, Q., and Sun, Y. (2018). Exploring stress tolerance mechanism of evolved freshwater strain Chlorella sp. S30 under 30 g/L salt. Biores. Technol., 250, 495–504.

Liu, C., Liu, J., Hu, S., Wang, X., Wang, X., and Guan, Q. (2019). Isolation and identification of a halophilic and alkaliphilicmicroalgal strain. PeerJ, 7, e7189. doi:10.7717/peerj.7189 30.

Lupatini, A. L., Colla, L. M., Canan, C., and Colla, E. (2017). Potential application of microalga Spirulina platensis as a protein source. J. Sci. Food Agri., 97(3), 724–732.

Mohan, S. V., and Devi, M. P. (2014). Salinity stress induced lipid synthesis to harness biodiesel during dual mode cultivation of mixotrophic microalgae. Bioresour. Technol., 165, 288–294.

Munns, R., and Tester, M. (2008). Mechanisms of salinity tolerance. Annu. Rev. Plant Biol., 59, 651–681.

Paliwal, C., Mitra, M., Bhayani, K., Bharadwaj, S. V. V., Ghosh, T., Dubey, S., and Mishra, S. (2017). Abiotic stresses as tools for metabolites in microalgae. Biores. Technol., 244, 1216–1226.

Panahi, B., Dumas, J., and Hejazi, M. (2019). Integration of cross species RNA-seq meta-analysis and machine learning models identifies the most important salt stress responsive pathways in microalga Dunaliella. Front. Genet., 10, 752.

Qiao, J., Huang, S., Te, R., Wang, J., Chen, L., and Zhang, W. (2013). Integrated proteomic and transcriptomic analysis reveals novel genes and regulatory mechanisms involved in salt stress responses in Synechocystis sp. PCC 6803. Appl. Microbiol. Biotechnol. 97, 8253–8264. doi: 10.1007/s00253-013-5139-8

Quan, L. J., Zhang, B., Shi, W. W., and Li, H. Y. (2008). Hydrogen peroxide in plants: A versatile molecule of the reactive oxygen species network. J. Integra. Plant Biol., 50(1), 2–18.

Rai, S., Pandey, S., Shrivastava, A. K., Singh, P. K., Agrawal, C., and Rai, L. (2013). Understanding 48 the mechanisms of abiotic stress management in cyanobacteria with special reference to proteomics. In: Srivastava, A. K., Rai, A. N., and Neilan, B. A. (Eds.), Stress Biology of Cyanobacteria: Molecular Mechanisms to Cellular Responses. CRC Press. Taylor and Francis, USA, pp. 93–112.

Sansone, C., Galasso, C., Orefice, I., Nuzzo, G., Luongo, E., Cutignano, A., et al. 2017. The green microalga Tetraselmis suecica reduces oxidative stress and induces repairing mechanisms in human cells. Scientific Reports, 7: 41215.

Sing, S. F., Isdepsky, A., Borowitzka, M. A., and Lewis, D. M. (2014). Pilot-scale continuous recycling of growth medium for the mass culture of a halotolerant Tetraselmis sp. in raceway ponds under increasing salinity: A novel protocol for commercial microalgal biomass production. Biores. Technol., 161, 47–54.

Sithtisarn, S., Yokthongwattana, K., Mahong, B., Roytrakul, S., Paemanee, A., Phaonakrop, N., and Yokthongwattana, C. (2017). Comparative proteomic analysis of *Chlamydomonas reinhardtii* control and a salinity-tolerant strain revealed a differential protein expression pattern. Planta, 246, 843–856.

Song, J., and Wang, B. (2015). Using euhalophytes to understand salt tolerance and to develop saline agriculture: *Suaeda salsa* as a promising model. Ann. Bot., 115(3), 541–553.

Srivastava, G., Nishchal, and Goud, V. V. (2017). Salinity induced lipid production in microalgae and cluster analysis (ICCB 16-BR_047). Biores. Technol., 242, 244–252.

Sun, S., Wen, D., Yang, W., Meng, Q., Shi, Q., and Gong, B. (2019). Overexpression of caffeic acid O-methyltransferase 1 (COMT1) increases melatonin level and salt stress tolerance in tomato plant. J. Plant Growth Regul., 1–15.

Sun, X., Wen, C., Zhu, J., Dai, H., and Zhang, Y. (2020). Apple columnar gene MdDMR6 increases the salt stress tolerance in transgenic tobacco seedling and Apple Calli. J. Plant Growth Regul., 1–10.

Sun, X. M., Ren, L. J., Bi, Z. Q., Ji, X. J., Zhao, Q. Y., and Huang, H. (2018). Adaptive evolution of microalgae Schizochytrium sp. 18 under high salinity stress to alleviate oxidative damage and improve lipid biosynthesis. Bioresour. Technol., 267, 438–444.

Sun, Y. G., Wang, B., Jin, S. H., Qu, X. X., Li, Y. J., and Hou, B. K. (2013). Ectopic expression of Arabidopsis glycosyltransferase UGT85A5 enhances salt stress tolerance in tobacco. PLoS One, 8, e59924.

Szymanska, K. P., Polkowska-Kowalczyk, L., Lichocka, M., Maszkowska, J., Dobrowolska, G. (2019). SNF1-related protein kinases SnRK2.4 and SnRK2.10 modulate ROS homeostasis in plant response to salt stress. Int. J. Mol. Sci., 20. doi:10.3390/ijms20010143 25.

Tahara, H., Fukai, S., Sambe, M., Kobayashi, M., Uchiyama, J., and Ohta, H. (2013). Characterization of the ABC transporter gene slr1045 involved in acid-stress tolerance of Synechocystis sp. PCC 6803. Environ Experiment. Bot., 147(2018), 63–74, 34.

Těšitel, J. (2016). Functional biology of parasitic plants: A review. Plant Ecol. Evol., 149(1), 5–20.

Varela, J. C., Pereira, H., Vila, M., and León, R. (2015). Production of carotenoids by microalgae: Achievements and challenges. Photosynthesis Res., 125(3), 423–436.

Vo, H. N. P., Ngo, H. H., Guo, W., Liu, Y., Chang, S. W., Nguyen, D. D., Nguyen, P. D., Bui, X. T., and Ren, J. (2018). Identification of the pollutants' removal and mechanism by microalgae in saline wastewater. Biores. Technol. 275, 44–52. doi:10.1016/j.biortech.2018.12.026.

von Alvensleben, N., Magnusson, M., and Heimann, K. (2016). Salinity tolerance of four freshwater microalgal species and the effects of salinity and nutrient limitation on biochemical profiles. J. Appl. Phycol., 28(2), 861–876.

von Alvensleben, N., Stookey, K., Magnusson, M., and Heimann, K. (2013). Salinity tolerance of *Picochlorum atomus* and the use of salinity for contamination control by the freshwater cyanobacterium *Pseudanabaena limnetica*. PloS One, 8(5).

Wang, J.-H., Zhang, T.-Y., Dao, G.-H., Xu, X.-Q., Wang, X.-X., and Hu, H.-Y. (2017). Microalgae-based advanced municipal wastewater treatment for reuse in water bodies. Appl. Microbiol. Biotechnol., 101(7), 2659–2675.

Wang, N., Qian, Z., Luo, M., Fan, S., Zhang, X., and Zhang, L. (2018). Identification of salt stress responding genes using transcriptome analysis in green alga *Chlamydomonas reinhardtii*. Int. J. Mol. Sci., 19(11), 26.

Wang, T., Tian, X., Liu, T., Wang, Z., Guan, W., Guo, M., . . . Zhuang, Y. (2016). Enhancement of lipid productivity with a novel two-stage heterotrophic fed-batch culture of Chlorella protothecoides and a trial of CO2 recycling by coupling with autotrophic process. Biomass Bioener., 95, 235–243.

Wase, N., Tu, B., Rasineni, G. K., Cerny, R., Grove, R., Adamec, J., Black, P. N., and DiRusso, C. C. (2019). Remodeling of Chlamydomonas metabolism using synthetic inducers results in lipid storage during growth. Plant Physiol. 181(3), 1029–1049. doi:10.1104/pp.19.00758.

Xia, L., Rong, J., Yang, H., He, Q., Zhang, D., and Hu, C. (2014). NaCl as an effective inducer for lipid accumulation in freshwater microalgae *Desmodesmus abundans*. Bioresour. Technol., 161, 402–409.

Yang, R., Deng, C., Ouyang, B., and Ye, Z. (2011). Molecular analysis of two salt responsive NAC-family genes and their expression analysis in tomato. Mole. Biol. Rep., 38(2), 857–863.

Yarra, R., He, S. J., Abbagani, S., Ma, B., Bulle, M., and Zhang, W. K. (2012). Overexpression of a wheat Na+/H+ antiporter gene (TaNHX2) enhances tolerance to salt stress in transgenic tomato plants (*Solanum lycopersicum* L.). 46 Plant Cell, Tissue Organ Cult. (PCTOC), 111(1), 49–57.

You, Z., Zhang, Q., Peng, Z., and Miao, X. (2019). Lipid droplets mediate salt stress tolerance in *Parachlorella kessleri*. Plant Physiol., 00666, 2019. doi:10.1104/pp.19.00666.

Zhang, L., Pei, H., Yang, Z., Wang, X., Chen, S., Li, Y., and Xie, Z. (2019a). Microalgae nourished by mariculture wastewater aids aquaculture self-reliance with desirable biochemical composition. Biores. Technol. 278, 205–213. doi:10.1016/j.biortech.2019.01.066.

Zhang, Y., Liu, L., Chen, B., Qin, Z., Xiao, Y., Zhang, Y., Yao, R., Liu, H., and Yang, H. (2019b). Progress in understanding the physiological and molecular responses of populus to salt stress. Int. J. Mol. Sci., 20, 1312. doi:10.3390/ijms20061312.

Zhang, Z. X., Hu, Q., Liu, Y. N., et al. (2019c). Strigolactone represses the synthesis of melatonin, thereby inducing floral transition in Arabidopsis thaliana in an FLC-dependent manner. Journal of Pineal Research, 67(2): e12582.

Zhao, H., Zhao, X., Li, M., Jiang, Y., Xu, J., Jin, J., and Li, K. (2018). Ectopic expression of *Limonium bicolor* (Bag.) Kuntze DREB (LbDREB) results in enhanced salt stress tolerance of transgenic *Populus ussuriensis* Kom. Plant Cell, Tissue Organ Cult (PCTOC), 132(1), 123–136.

Zhu, J. K. (2001). Plant salt tolerance. Trends Plant Sci., 6, 66–71.

Zhu, Z., Chen, J., and Zheng, H. L. (2012). Physiological and proteomic characterization of salt tolerance in a mangrove plant, *Bruguiera gymnorrhiza* (L.) Lam. Tree Physiol., 32(11), 1378–1388.

Zubia, M., Freile-Pelegrín, Y. and Robledo, D., 2014. Photosynthesis, pigment composition and antioxidant defences in the red alga Gracilariopsis tenuifrons (Gracilariales, Rhodophyta) under environmental stress. Journal of Applied Phycology, 26: 2001–2010.

15 Metabolic Engineering of Hydrocarbon Biosynthesis in *Chlamydomonas reinhardtii* to Enhance Fuel Precursor Accumulation

K. Jothibasu, J. Tharun Kumar, Dolly Wattal Dhar, and Suchitra Rakesh

ABBREVIATIONS

TAG	Triacylglycerols
CRISPR	Clustered regularly interspaced short palindromic repeats
C. reinhardtii	*Chlamydomonas reinhardtii*

15.1 INTRODUCTION

The best alternative to meet the future demand for fuel is to replace it with renewable biofuels. Biofuels not only serve as an alternative to fossil fuels but also prevent the emission of harmful gaseous pollutants (Chen et al. 2018). The carbon feedstocks for plant and lignocellulose-based biofuels are derived from carbon dioxide (CO_2) fixation through photosynthesis. The first and second-generation biofuels depend on the processing of biomass to fuel. The biofuel from photosynthetic microorganisms can directly convert CO_2 into fuel molecules or fuel precursors and eliminate the intermediate biomass stage (Ruffing 2013). Oleaginous microbes with lipid content more than 20% like yeast, algae, molds, and bacteria can be used for biofuel production (Li et al. 2007). Algae are classified as macroalgae and microalgae based on the size. Macroalgae (seaweed) are multicellular and visible to the naked eye due to their large size, while microalgae are microscopic, single-celled, and may be prokaryotic similar to cyanobacteria (Chloroxybacteria) or eukaryotic as green algae (Chlorophyta) (Barsanti et al. 2008). Microalgae are simple photosynthetic organisms that live in saline, freshwater, wastewater and can grow in places of enough sunlight (Suchitra and Karthikeyan 2019). These can tolerate a wide range of temperatures, salinities, pH, and different light intensities (Barsanti et al. 2008) and are a rich source of carbon compounds due to high photosynthetic efficiency, lipid content, and carbon dioxide reduction ability (Das et al. 2011; Olguín 2012; Suchitra et al. 2020). These can produce a wide range of fuel molecules and their precursors but need to improve cost-effective commercial production.

Metabolic engineering is a powerful tool to improve microalgal fuel production (Jothibasu et al. 2021). This chapter focuses on the metabolic engineering of fatty acid and terpenoid biosynthesis in microalgae for strain improvement and to assess the fuel precursor recovery for the economical production of biofuels. Apart from biofuel, microalgae produce valuable coproducts like health supplements, pharmaceuticals, enzymes, lipids favoring bio-refineries to reduce the biofuel production

DOI: 10.1201/9781003219156-19

cost (Jagadevan et al. 2018). The following approaches can be employed to enhance hydrocarbon-based fuel production.

1. **Metabolic engineering:** Genetic engineering to enhance lipid accumulation and ability to tolerate bioreactor growth parameters such as temperature and oxygen etc.
2. **Low-cost feedstock**: Optimizing low-cost wastewater-based media composition and growth parameters.
3. **Synergism:** Synergism of microalgae with oleaginous yeast to enhance robust growth and biofuel production.
4. **Advanced photobioreactor:** Designing advanced photobioreactor with high light penetration and continuous production of biofuel.
5. **Cost-efficient downstream processing:** Low-cost technologies for biomass harvesting, drying, and oil extraction.
6. **Biorefinery:** A plant capable of simultaneously producing biofuel and other coproducts, for example, dietary supplements, surfactants etc.

15.2 STAIN IMPROVEMENT BY REGULATION OF TAG SYNTHETIC PATHWAY

Microalgal lipids are classified into two groups, nonpolar lipids (acylglycerols, sterols, free fatty acids, wax, and steryl esters) and polar lipids (phosphoglycerides, glycosylglycerides, and sphingolipids) based on their structures. These lipids perform different roles in metabolism and growth of microalgae. The neutral lipids serve as the energy reserves, while polar lipids are constituents of cell organelles and membranes. The lipids like inositol, sphingolipids, and polyunsaturated fatty acids function as intermediates in cell signaling and sense changes in the environment (Chen et al. 2018). Microalgae accumulate and store neutral lipids in the form of triacylglycerols (TAGs). Lipids and polyglucans are the energy and carbon reserves in microalgal cells as these ensure the survival of microalgae under variable light intensities and help to carry out biological processes such as macromolecules synthesis and multiplication (Zhu et al. 2016).

Among lipids, triacylglycerides (TAGs) from oleaginous microalgae are considered an ideal feedstock for biofuel production (Kim et al. 2019). Glycerol-3-phosphate acyltransferase (GPAT) is the initial rate-limiting enzyme that catalyzes the TAG synthesis from glyceraldehyde-3-phosphate (G-3-P) (Li et al. 2018). The most important route to the biosynthesis of triacylglycerol is the *sn*-glycerol-3-phosphate or Kennedy pathway and dihydroxyacetone phosphate pathways. The main source of glycerol backbone in the Kennedy pathway is sn-glycerol-3-phosphate produced by catabolism of glucose (glycolysis). Figure 15.1 illustrates the triacylglycerol's biosynthesis in microalgae by common C3 precursors. Acetyl-CoA is the primary metabolite in triacylglycerol biosynthesis. Two main enzymes, which regulate the conversion of acetyl-CoA to fatty acid chains, are (1) acetyl-CoA carboxylase (ACC), which converts acetyl-CoA to malonyl-CoA, and (2) fatty acid synthase II (FAS II), which elongates malonyl-ACP to the saturated fatty acid chain by two units (Post-Beittenmiller et al. 1991, 1992). Acyl-acyl carrier protein (ACP) thioesterase (TEs) specifies the chain length of the fatty acid in plants and green algae (Voelker 1996). The polyunsaturated fatty acid (PUFA) is synthesized from saturated fatty acid by the enzyme polyketide synthase (PKS) (Mata et al. 2010).

Metabolic engineering of the specific target sites with different strategies like overexpression of genes, inhibition of alternative pathways, transcription factor engineering, and flux balance analysis are the key factors for increasing the biomass and lipid concentration in microalgae (Banerjee et al. 2016). For the effective outcome of the metabolic engineering techniques, it is important to understand the basic biosynthetic pathways of microalgae at a molecular level. The microalgae species have different types and quantity of basal level lipids (Zhu et al. 2016) that can be altered by using genetic and bioprocess engineering (Alishah et al. 2019). The following specific pathways and respective genes were targeted for tuning of a metabolic pathway to improve TAG accumulation by inhibition (Table 15.1).

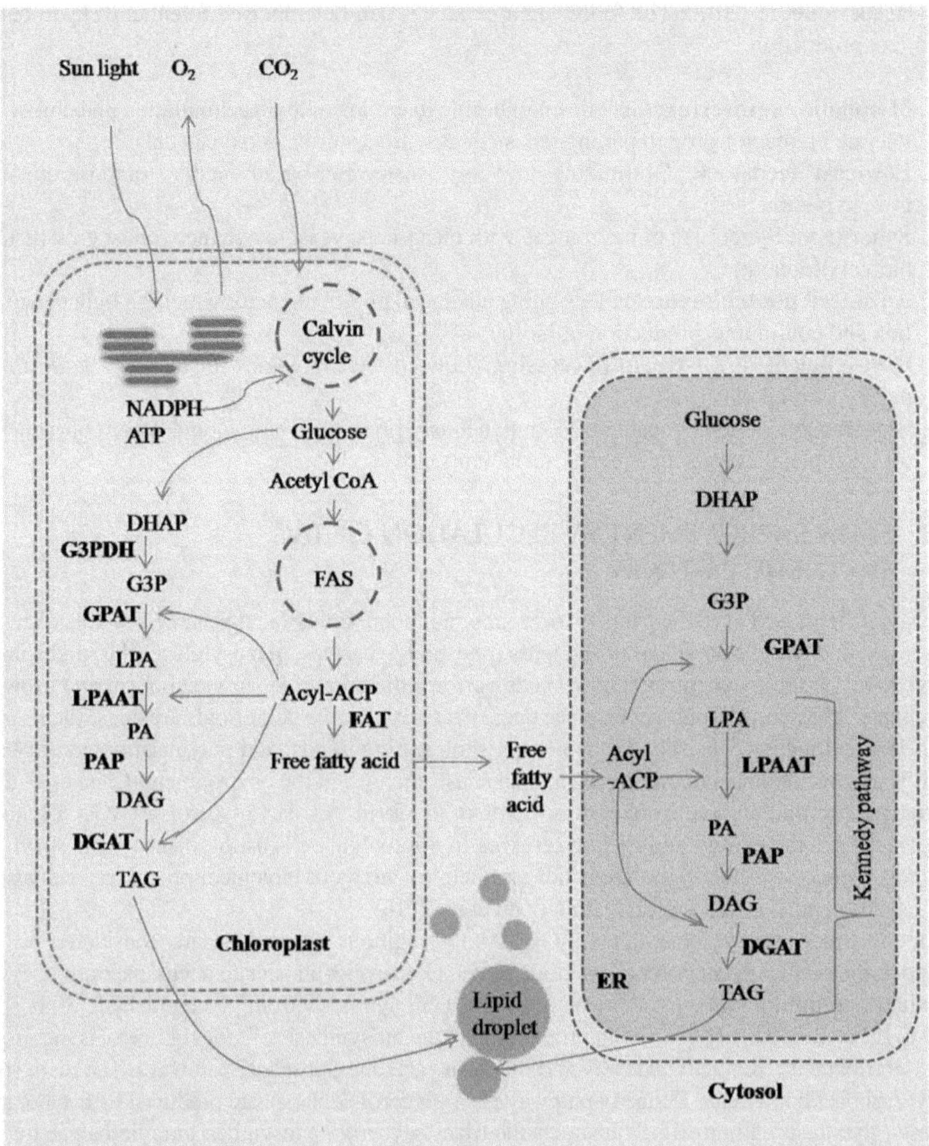

FIGURE 15.1 Triacylglycerol (TAG) synthesis in microalgae. NADPH, nicotinamide adenine dinucleotide phosphate; ATP, adenosine triphosphate; DHAP, dihydroxyacetone phosphate; G3P, glyceraldehyde 3-phosphate; G3pDH, G3P dehydrogenase; GPAT, glycerol 3-phosphate acyltransferase; PA/LPA/LPAAT/PAP, phosphatidic acid/Lyso-PA/LPA acyltransferase/PA phosphatase; DAG, diacylglycerol; DGAT, DAG acyltransferase; FAT, fatty acyl-ACP thioesterase; ACP, acyl-carrier protein; PC, phosphatidylcholine; PDAT, phospholipid diacylglycerol acyltransferase; ACCase, acetyl-CoA carboxylase; FAS, fatty acid synthase. The figure adapted and modified from Sharma et al. 2018; Goncalves et al. 2016; Banerjee et al. 2016; Bellou et al. 2014.

1. Competitive pathway: *CrPEPC1, CrPEPC2, CrCIS, AGPase, STA6-STA7*
2. TAG and lipid catabolism: *PLA2, CrDGAT2–4, CrACX2*
3. Enhancing photosynthetic efficiency: *LHC isoforms*
4. Transcription regulators: *NRR1*, *TAR1*, *CHT7*

TABLE 15.1
Summary of Genome Editing Studies in Microalgae to Enhance TAG Accumulation

Microalgae Species	Target Gene for Inhibition	Specific Gene Disruption Techniques	Outcome	Reference
CRISPRi				
C. reinhardtii	*PLA2*	Knock-out/ CRISPR/Cas9	64% increase in total lipid content	(Shin et al. 2019)
C. reinhardtii	*ΔCrPEPC1*	CRISPRi	+74.4% lipid content + 94.2% lipid productivity	(Kao and Ng 2017)
C. reinhardtii	Knock-down of PEPC enzyme with	CRISPRi	Enhanced lipid production up to 94%	(Serrano et al. 2009)
RNA interference				
C. reinhardtii	*ΔCrPEPC1*, *ΔCrPEPC2*	amiRNA	28.7~48.6% TFA content	(Wang et al. 2017)
C. reinhardtii	*ΔPEPC1*	RNAi	+ 20% TAG level	(Deng et al. 2014)
C. reinhardtii	Citrate synthase	Knock-down/RNAi	TAG level increased up to 169.5%	(Deng et al. 2013)
C. reinhardtii	Artificial silencing of *Diacylglycerol acyltransferase* *2–4* gene (*CrDGAT2–4*)	RNA interference	24%–34% increase in lipid content	(Deng et al. 2012)
C. reinhardtii	Repression of major lipid droplet protein (*MLDP*) gene expression	RNA Interference	40% increase in the average lipid droplet diameter	(Moellering and Benning 2010)
C. reinhardtii	*LHC isoforms*	Knock-down/RNAi	65% faster growth rate	(Mussgnug et al. 2007)
Mutation				
C. reinhardtii	*CrACX2*	Knock-out/random insertional mutagenesis	20% increase in oil content under N starvation	(Kong et al. 2017)
C. reinhardtii	*CHT7*	Knock-out/random insertional mutagenesis	90% reduction in TAG and lipid droplet accumulation	(Tsai et al. 2014)
C. reinhardtii	*TAR1*	Knock-out/random insertional mutagenesis	Unable to resume growth and recover from S and N deprivation	(Kajikawa et al. 2015)
C. reinhardtii	*NRR1*	Knock-out/random insertional mutagenesis	Reduction in TAG accumulation in defective mutant	(Boyle et al. 2012)
C. reinhardtii	*sta6 and sta7* mutations	Knock-out/random insertional mutagenesis	Tenfold enhancement in TAG accumulation in contrast to wild-type	(Work et al. 2010)

(Continued)

TABLE 15.1 (*Continued*)
Summary of Genome Editing Studies in Microalgae to Enhance TAG Accumulation

Microalgae Species	Target Gene for Inhibition	Specific Gene Disruption Techniques	Outcome	Reference
Chlamydomonas reinhardtii (starchless mutant)	*AGPase* (Adenosine diphosphoglucose pyrophosphorylase)	Knock-out/random insertional mutagenesis	Tenfold increase in TAG content	(Li et al. 2010)

Abbrevations: "+" increase in quantity percentile; *C. reinhardtii, Chlamydomonas reinhardtii;* TFA, Total fatty acids; TAG, triacylglycerol; CRISPRi, CRISPR interference; PEPC1, phosphoenolpyruvate carboxylase 1; PLA2, phospholipase A2; CrCIS, citrate synthase; CrDGAT2–4, acyl-CoA diacylglycerol acyltransferase; MLDP, major lipid droplet protein; LHC isoforms, light-harvesting antenna complexes; CrACX2, acyl-CoA oxidase; CHT7, compromised hydrolysis of triacylglycerols 7; TAR1, triacylglycerol accumulation regulator 1 kinase; AGPase, ADP-glucose pyrophosphorylase; STA6, ADP-glucose pyrophosphorylase; STA7–10, Isoamylase.

The following specific pathways were targeted for tuning of a metabolic pathway to improve TAG accumulation by overexpression (Table 15.2).

1. Enhancing fatty acid biosynthesis: *ACS*
2. Increasing intracellular reducing equivalents: *FDX*
3. Increasing TAG content: *GPAT, LPAAT, PAP, DGAT, PDAT*
4. Manipulation of transcription regulators: *DOF*
5. Modifying fatty acid profile: *SAD*

15.3 STAIN IMPROVEMENT BY REGULATION OF ISOPRENOID SYNTHETIC PATHWAY

Terpenoids are the most abundant and chemically diverse class of natural products synthesized in almost all living organisms. Isoprenoids also referred as terpenoids or terpenes comprise over 70,000 compounds. The diverse structures of terpenoids carry out different cellular functions such as electron transport, photosynthesis, pigments, membrane fluidity, signaling, and cell wall formation. In higher plants, the isoprenoids also serve as the precursor for different phytohormones, including cytokinins, gibberellins, abscisic acid, strigolactones, and brassinosteroids (Walter et al. 2010). Microalgal isoprenoids have gained great concern as means of sustainable energy to create a low-carbon society. It is used for biofuel purposes due to their lower hygroscopy, higher energy content, and good fluidity at low temperatures (Beller et al. 2015; Gupta and Phulara 2015; Vavitsas et al. 2018). It is classified according to the number of carbons as hemiterpenes (C5), monoterpenes (C10), sesquiterpenes (C15), diterpenes (C20), triterpenes (C30), and tetraterpenes (carotenoids, C40).

The central integrant of isoprenoid is the five-carbon skeleton precursors, namely, isopentenyl pyrophosphate (IPP) and its isomer dimethylallyl pyrophosphate (DMAPP). The synthesis of these precursors is carried out through two distinct pathways. Figure 15.2 depicts (a) the cytosolic mevalonate (MVA) pathway and (b) the plastidic non-mevalonatel-deoxy-D-xylulose-5-phosphate (DOXP) pathway (Lichtenthaler 1999). The mevalonate pathway typically occurs in the cytosol of eukaryotes and archaea, while DOXP pathway, also known as the methylerythritol (MEP) pathway occurs in the stroma of plastids in green algae and most bacteria (Baba and Shiraiwa 2013).

The primary metabolite of the mevalonate (MVA) pathway is the acetyl-CoA, which undergoes six enzymatic steps to form isopentenyl pyrophosphate (IPP) (Lohr et al. 2012). The IPP isomerase enzyme acts as a catalyst for isomerization of IPP to DMAPP resulting in two vital precursors for the production of a diverse group of isoprenoids. The rate-controlling enzyme of the MVA pathway

TABLE 15.2
Summary of Overexpression Studies in Microalgae to Enhance TAG Accumulation

Microalgae Apecies	Target Gene	Specific Pathway	Outcome	Reference
C. reinhardtii	*ACS*	Enhancing fatty acid biosynthesis	2.4 X TAG content	Rengel et al. 2018
C. reinhardtii	*FDX*	Increasing intracellular reducing equivalents	2.5 X lipid level	Huang et al. 2015
C. reinhardtii	*GPAT*	Increasing TAG content	1.5 X TAG content and above 50% improvement in TAG content	Boyle et al. 2012; Iskandarov et al. 2016
C. reinhardtii	*LPAAT*	Increasing TAG content	+20% TAG	Yamaoka et al. 2016
C. reinhardtii	*PAP*	Increasing TAG content	+7.5 to 21.8% lipid content	Deng et al. 2013
C. reinhardtii	*DGAT*	Increasing TAG content	2 X TAG content	Ahmad et al. 2015
C. reinhardtii	*DGAT2*	Increasing TAG content	29-fold improvement in TAG content	Iwai et al. 2014
C. reinhardtii	*DGAT2–1; DGAT2–5*	Increasing TAG content	20% and 44% improvement in lipid content, respectively	Deng et al. 2012
C. reinhardtii	*PDAT*	Increasing TAG content	32% increase in TAG content	Zhu et al. 2016
C. reinhardtii	*DOF*	Manipulation of transcription regulators	2 X Total lipid	Ibanez-Salazar et al. 2014
C. reinhardtii	*SAD*	Modifying fatty acid profile	2.7 X C 18:1 FA content	Hwangbo et al. 2013
C. reinhardtii	*CrCIS*	Competitive pathway	Decreased the TAG level by 45%	(Deng et al. 2013)

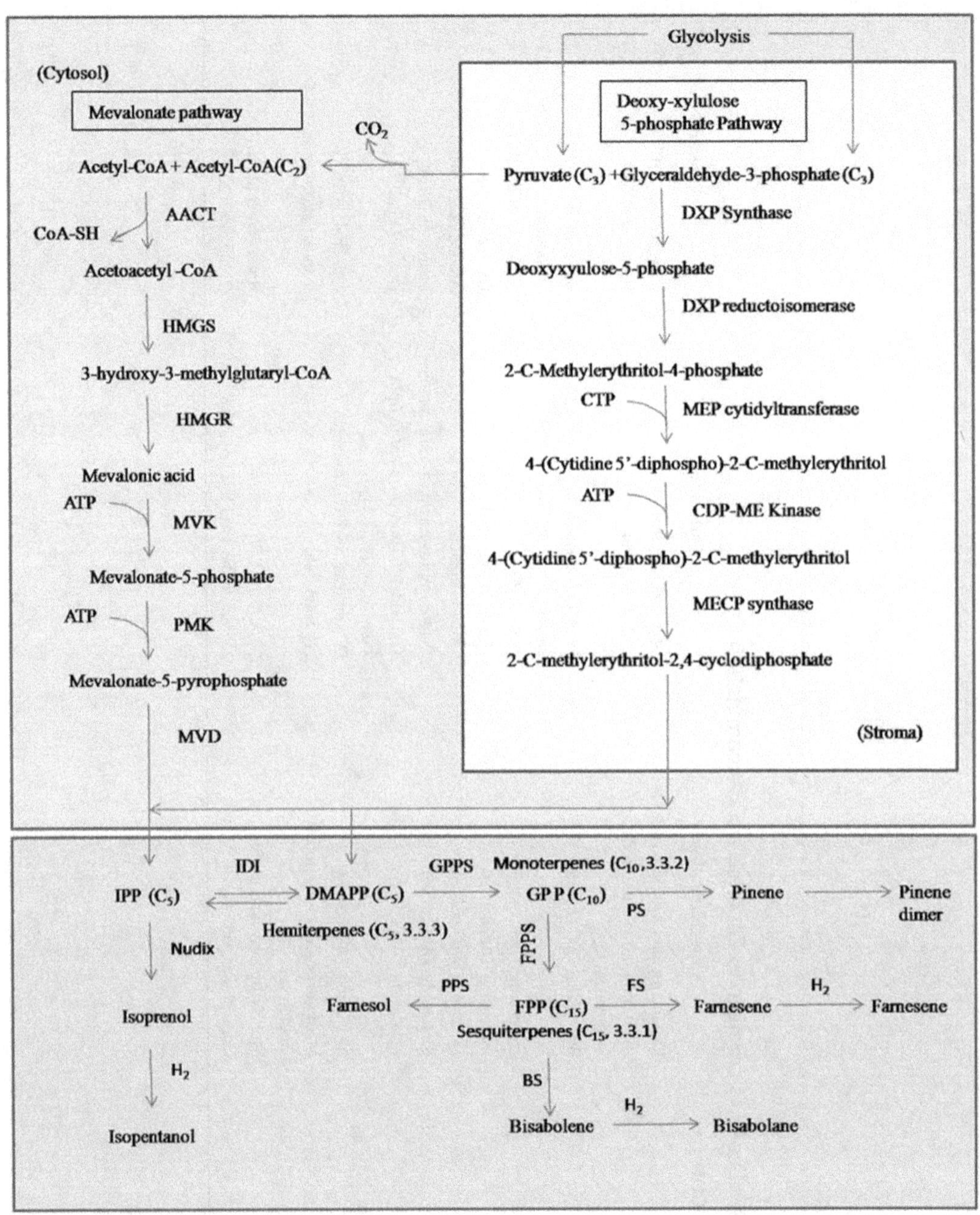

FIGURE 15.2 Isoprenoid biosynthesis pathways in microalgae. The mevalonate (MVA) pathway is from acetyl-CoA, and the deoxyxylulose 5-phosphate (DXP) pathway is from glyceraldehyde-3-phosphate and pyruvate. AACT, Acetoacetyl CoA transferase; HMGS, HMG CoA synthase; HMGR, HMG CoA reductase; MVK, mevalonate kinase; PMK, phosphomevalonate kinase; MVD, phosphomevalonate decarboxylase; IPP, isoprenyl diphosphate; DMAPP, dimethylallyl diphosphate; IDI, isoprenyl diphosphate isomerase; nudiX, nucleoside diphosphate hydrolases; GPP, geranyl pyrophosphate; FPP, farnesyl pyrophosphate; GPPS, geranyl pyrophosphate synthase; FPPS, farnesyl pyrophosphate synthase; PS, pinene synthase; FS, farnesene synthase; BS, bisabolene synthase; PPS, endogenous pyrophosphatase. The figure adapted and modified from Baba and Shiraiwa 2013; Phulara et al. 2016. Isoprenoid synthetic pathway, its occurrence and precursor are presented in Table 15.3.

TABLE 15.3
Isoprenoid Synthesis, Pathway, Occurrence, and Precursor

Pathway	Occurrence and Precursor	Reference
Mevalonate (MVA)	a) Starts from acetyl-CoA	(Rilling 1959)
	b) Presents in eukaryotes, archaea, and cytosol of higher plant	
Methylerythritol phosphate (MEP) pathway also known as non-mevalonate pathway	a) Starts from glyceraldehyde-3-phosphate (G3P) and pyruvate via deoxyxylulose 5-phosphate (DXP).	(Lange et al. 2000; Rodrıguez-Concepcion 2006).
	b) Present in prokaryotes, green algae. and chloroplasts of higher plants	
Glycolysis	a) Primary source of precursor	(Kanehisa and Goto 2000; Kanehisa et al. 2012)
Pentose pyrophosphate (PPP), entner–doudoroff pathway (EDP), fatty acid and amino acid metabolism	b) Also provide precursor	(Kanehisa and Goto 2000; Kanehisa et al. 2012)

TABLE 15.4
Sterol Contents of Different Microalgae

Microalgae Species	Phytosterol Content in (μg/g dw).	References
Chlorella vulgaris	13.5	Rzama et al (1994)
Scenedesmus sp.	23.5	
Nannochloropsis gaditana	13	Ryckebosch et al. (2014)
Haematococcus pulvialis	20	Bilbao et al. (2016)
Isochrysis galbana	14.9	
Phaeodactylum tricornutum	16.5	Randhir et al. (2020)
Tetraselmis suecica	10.9	
Thalassiosira pseudonana	34	
Eutreptiella eupharyngea	11.26	
Chaetoceros muelleri	22.86	

is the HMG-CoA reductase (Ruffing 2013). IPP and DMAPP undergo condensation and dephosphorylation to form polyprenyl pyrophosphates of various lengths which are catalyzed by different enzymes including farnesyl pyrophosphate synthase (FPPS), geranyl pyrophosphate synthase (GPPS), and geranylgeranyl pyrophosphate synthase (GGPPS) (Athanasakoglou et al. 2019). These polyprenyl phosphates then undergo several reactions including methylation, isomerization, and reduction to form sterols that include β-sitosterol, campesterol, stigmasterol, brassicasterol, and ergosterol (Randhir et al. 2020). The sterol content of different microalgae is shown in Table 15.4.

The mevalonate (MVA) pathway in plants is highly inhibited by mevinolin, which is a potent inhibitor of HMG-CoA reductase but does not affect the synthesis of carotenoid pigments and some isoprenes (Rohmer 1998). This led to the discovery of an alternative pathway, a mevalonate-independent pathway for isoprenoid synthesis. This pathway was first discovered in bacteria and was later found to be widespread amongst phototrophic eukaryotes (Rohmer 1999). The Methylerythritol pathway (MEP) is reported to take place in plastids for the precursors of plastidic isoprenoids such as carotenoids, phytol, phytohormones, and polyphenols for quinines (Schwender et al. 1996; Joyard et al. 2009). The primary metabolite of the methylerythritol pathway is the glyceraldehyde-3-phosphate and pyruvate, which undergo seven enzymatic steps to form Isopentenyl pyrophosphate (IPP) and Dimethylallyl pyrophosphate (DMAPP). The rate-controlling enzyme of the MEP pathway is 1-deoxy-D-xylulose-5-phosphate

TABLE 15.5
Carotenoid Contents of Different Microalgae

Microalgae Species	Carotenoids Content in μg/g dw	References
Eutreptiella eupharyngea	9.11	
Chaetoceros muelleri	58	(Hikihara et al. 2020)
Nannochloropsis sp.	14.4	
Isochrysis galbana	18.23	(Ryckebosch et al. 2014)
Porphyridium cruentum	26.9	(Guiheneuf and Stengel 2015)
Phaeodactylum tricornutum	62	(Wu et al. 2016)
Tetraselmis suecica	68.6	(Ahmed et al. 2014)
Golenkiniaaff sp.	63.8	(Rearte et al. 2020)
Scenedesmus bijugatus	46.8	(Minhas et al. 2020)
Desmodesmus subspicatus	118	(Eze et al. 2020)

(DXP) synthase (Ruffing 2013). Carotenoids in microalgae can be distinguished as primary and secondary carotenoids. The primary carotenoids are the accessory pigments with a central role as light-harvesting in photosynthetic machinery (Mulders et al. 2014). The secondary carotenoids, which include β-carotene, canthaxanthin, astaxanthin, lutein, zeaxanthin, fucoxanthin, among others, are some of the vital carotenoids harvested from microalgae that have a commercially high value (Henriquez et al. 2016). The carotenoid content of different microalgae is shown in Table 15.5.

The prenyl diphosphates formed from isopentenyl pyrophosphate (IPP) and dimethylallyl pyrophosphate (DMAPP) serve as the substrate for the synthesis of novel isoprenoid classes like geranyl pyrophosphate (GPP), which is a precursor for C10 monoterpenoids. Sequential addition of IPP forms farnesyl pyrophosphate (FPP) and geranylgeranyl pyrophosphate (GGPP), which are the precursors for C20 diterpenoids, C30 triterpenoids, C40 tetraterpenoids (cholesterol), and other sterols, carotenoids, etc. (Vranova et al. 2012). The isopernoid-based fuels have structural diversity to mimic petroleum-based fuels such as gasoline and diesel (Lee et al. 2008). Isoprenoids reported to be potential fuel candidates include the hemiterpene (C5) isoprene; monoterpenes (C10): terpinene, pinene, limonene, and sabinene; the sesquiterpene (C15) farnesene and their associated alcohols: isopentenyl, terpineol, geraniol, and farnesol (Liu et al. 2010; Rude and Schirmer 2009).

The increase in biomass and the selective enhancement of specific fuel precursors in microalgal cells are the obvious ways of making microalgal-based biofuel economically feasible. Conventional methods like nutrient deprivation, physical stress (temperature, salt, heavy metals, etc.) are reported to enhance the biomass in significant quantities in few classes of microalgae. However, these reduce the proliferation of cells and, therefore, limit the productivity of biomass and overall yield (Hu et al. 2008). Recent advances in omics studies (transcriptomics, proteomics, metabolomics, and lipidomics) have begun to unravel the molecular mechanisms and regulatory networks involved in microalgae fatty acid, triglycerides, and isoprenoid biosynthesis for biofuel production. This emphasizes the development of metabolic engineering strategies to increase fuel production (Lee et al. 2020). The summaries of terpenoid metabolic engineering in microalgae are presented in Table 15.6.

The endosymbiotic theory states that chloroplasts have originated from cyanobacteria over a billion years ago, in which an ancestral heterotrophic eukaryotic cell engulfed photosynthetic cyanobacteria (Bhattacharya et al. 2004; Gray 1993; Timmis et al. 2004). Boynton et al. (1988) demonstrated the first particle-bombardment-based stable chloroplast transformation. Although particle bombardment is the most preferred method of transformation in microalgae, it applies to only a limited number of microalgal species. Kindle et al. (1991) demonstrated glass-beads-based chloroplast transformation in *Chlamydomonas reinhardtii* by briefly agitating cells and DNA with glass beads. It is a cost-efficient approach for chloroplast genome engineering. Gan et al. (2018) successfully applied electroporation for chloroplast transformation in marine microalga *Nannochloropsis oceanica* by showing site-specific integration of fluorescence reporter gene and antibiotic resistance gene into the chloroplast

TABLE 15.6
Summary of Terpenoid Metabolic Engineering in *Chlamydomonas reinhardtii*

Sl. No.	Microalgae Species	Target Gene	Type of Manipulation	Outcome	Reference
1	*Chlamydomonas reinhardtii*	Terpene synthase	Overexpression	a) Engineered *Chlamydomonas reinhardtii* to produce sesquiterpene biodiesel precursor (E)-α-bisabolene b) Up to 10.3 ± 0.7mg bisabolene·g^{-1} cell dry weight produced in five days	(Wichmann et al. 2018)
2	*Chlamydomonas reinhardtii*	Patchoulol synthase	Overexpression	a) Patchoulol was produced up to 922 ± 242μgg^{-1} CDW in six days	(Lauersen et al. 2016)
4	*Chlamydomonas reinhardtii*	Squalene epoxidase	RNAi	a) 56–76% knock-down of mRNA Up to1.1 μg/mg DW of squalene	(Kajikawa et al. 2015)
4	*Chlamydomonas reinhardtii*	Squalene synthase	Nuclear overexpression	Squalene not detected	(Kajikawa et al. 2015)

TABLE 15.7
Comparison of Different Microalgal Transformation Methods

Transformation Methods	Predominant Type of Transformation	Requirement of Special Equipment	Required Cell Wall Removal	Essential Parameter/Advantages	Reference
Particle bombardment	Chloroplast	Complex	No	a) The bombardment device and costing DNA with microparticles (gold or tungsten) b) Chloroplast and mitochondria transformation c) Can transform genetic material to many algae including hard-walled diatoms d) Large DNA fragments can be delivered e) Delivery of multiple plasmids with high frequencies of cotransformation	Boynton et al. 1988
Glass beads	Nucleus	Simple	Yes	a) Preferred logarithmic growth of cell to increase exogenous DNA delivery in the nucleus b) Addition of membrane fusion agent polyethylene glycol (PEG) to increase transformation efficiency c) Simplicity and reproducibility	Kindle 1990
Electroporation	Nucleus and plastid	Simple	Yes	a) Preincubation, application of the electric pulse and postincubation at ~ 4°C to increase transformation efficiency	Gan et al. 2018

genome. The designed genome editing or overexpression construct can transform into microalgae for strain improvements. The selection of genetic transformation and predominant type of transformation either plastid or nuclear are essential to achieve high transformation efficiency. The comparison of different microalgal transformation methods is presented in Table 15.7.

15.4 CONCLUSION

The inhibition of genes, namely, PLA2, *CrPEPC1*, AGPase, major lipid droplet protein, citrate synthase, has enhanced TAG content above 60% as compared to wild type. Knock-out of these genes is directly related to enhance TAG content. Inhibition of citrate synthase expression led to the increased expression of TAG biosynthesis-related genes (Tables 15.1 and 15.2). In contrast, the overexpression of the Citrate synthase gene in *C. reinhardtii* showed a 45% reduction in TAG content (Deng et al. 2013). The inhibition of the target gene by mutation, CRISPRi, and RNA interference is presented in Table 15.1. Among the genome editing, the CRISPRi can minimize the off-target effect editing multiple genes at the same time and reducing the mutant screening process (Kumar et al. 2019). The CRISPR-Cas9 system has a plasmid and RNA-based approach, and both the components can transform microalgae by conventional transformation methods such as particle bombardment, electroporation, and glass bead. Different microalgae transformation methods are presented in Table 15.7. The isoprenoid metabolism has been engineered to enhance the production of fuel precursors like sesquiterpenes farnesene and bisabolene, monoterpenes pinene and limonene, and hemiterpenes isopentenyl and isopentanol (Table 15.6). The terpene synthase overexpression and amiRNA-based repression of geranylgeranyl pyrophosphate synthase (GGPPS) in *Chlamydomonas reinhardtii* led to a 15-fold increase in sesquiterpene biodiesel precursor (*E*)-α-bisabolene (Wichmann et al. 2018). Similarly, patchoulol synthase overexpression in *Chlamydomonas reinhardtii* reported up to $922 \pm 242 \mu g g^{-1}$ CDW Patchoulol production in six days (Lauersen et al. 2016). In contrast, the squalene synthase overexpression led to no significant improvement in squalene content (Kajikawa et al. 2015).

The current chapter concludes the inhibition of competitive pathway gene for triglycerides synthesis led to over 60% increase in TAG content in comparison to wild type. Arathi et al. (2020) discussed different qualitative and quantitative methods for algae lipid estimation. The CRISPR-Cas9 system is suitable for knock-in, knock-out, and knock-down of single or multiple genes in algae to enhance hydrocarbon content. The terpenoid biosynthesis pathway has been engineered for the production of specific fuel like bisabolene and patchoulol. Electroporation is a simple and reliable method of choice for both plastid and nuclear transformation to compare other conventional methods.

ACKNOWLEDGMENT

The authors thank the DST SERB Project (EEQ/2023/000530) for financial support.

REFERENCES

Ahmed F, Fanning K, Netzel M, Turner W, Li Y and Schenk PM. 2014. Profiling of carotenoids and antioxidant capacity of microalgae from subtropical coastal and brackish waters. Food Chemistry 165: 300–306. https://doi.org/10.1016/j.foodchem.2014.05.107.

Ahmad, I., Sharma, A. K., Daniell, H., & Kumar, S. (2015). Altered lipid composition and enhanced lipid production in green microalga by introduction of brassica diacylglycerol acyltransferase 2. Plant biotechnology journal, 13(4), 540–550. https://doi.org/10.1111/pbi.12278

Alishah AH, Rafiei N and Garcia GR. 2019. Biomass and lipid induction strategies in microalgae for biofuel production and other applications. Microbial Cell Factories 18: 178. https://doi.org/10.1186/s12934-019-1228-4.

Arathi S, Tharunkumar J, Jothibasu K, Karthikeyan S and Suchitra R. 2020. Qualitative and quantitative estimation of algal lipids for biofuel production. International Journal of Chemical Studies 8(4): 2451–2459. https://doi.org/10.22271/chemi.2020.v8.i4ab.10003.

Athanasakoglou A, Grypioti E, Michailidou S, Ignea C, Makris AM, Kalantidis K and Kampranis SC. 2019. Isoprenoid biosynthesis in the diatom *Hasleaostrearia*. New Phytologist 222(1): 230–243. https://doi.org/10.1111/nph.15586.

Baba M and Shiraiwa Y. 2013. Biosynthesis of lipids and hydrocarbons in algae. Photosynthesis 978–953. DOI:10.5772/56413.

Banerjee C, Dubey KK and Shukla P. 2016. Metabolic engineering of microalgal based biofuel production: Prospects and challenges. Frontiers in Microbiology 7: 432. https://doi.org/10.3389/fmicb.2016.00432.

Barsanti L. et al. (2008). Oddities and Curiosities in the Algal World. In: Evangelista, V., Barsanti, L., Frassanito, A.M., Passarelli, V., Gualtieri, P. (eds) Algal Toxins: Nature, Occurrence, Effect and Detection. NATO Science for Peace and Security Series A: Chemistry and Biology. Springer, Dordrecht. https://doi.org/10.1007/978-1-4020-8480-5_17.

Beller, H.R., Lee, T.S., Katz, L., 2015. Natural products as biofuels and bio-based chemicals: Fatty acids and isoprenoids. Nat. Prod. Rep. 32, 1508–1526. https://doi.org/10.1039/c5np00068h

Bellou S, Baeshen MN, Elazzazy AM, Aggeli D, Sayegh F and Aggelis G. 2014. Microalgal lipids biochemistry and biotechnological perspectives. Biotechnology Advances 32: 1476–1493. https://doi.org/10.1016/j.biotechadv.2014.10.003.

Bhattacharya, D., Yoon, H. S., & Hackett, J. D. (2004). Photosynthetic eukaryotes unite: endosymbiosis connects the dots. Bioessays, 26 (1), 50–60. https://doi.org/10.1002/bies.10376

Bilbao PGS, Damiani C, Salvador GA and Leonardi P. 2016. *Haematococcus pluvialis* as a source of fatty acids and phytosterols: Potential nutritional and biological implications. Journal of Applied Phycology 28(6): 3283–3294. DOI:10.1007/s10811–016–0899-z.

Boyle NR, Page MD, Liu B. 2012. Three acyltransferases and nitrogen-responsive regulator are implicated in nitrogen starvation-induced triacylglycerol accumulation in *Chlamydomonas*. The Journal of Biological Chemistry 287(19): 15811–15825. https://doi.org/10.1074/jbc.M111.334052.

Boynton JE, Gillham NW, Harris EH, Hosler JP, Johnson AM, Jones AR, Randolph-Anderson BL, Robertson D, Klein TM, Shark KB, Sanford JC (1988) Chloroplast transformation in Chlamydomonas with high velocity microprojectiles. Science 240, 1534–1538. https://doi.org/10.1126/science.2897716

Chen Z, Wang L, Qiu S, Ge S. 2018. Determination of microalgal lipid content and fatty acid for biofuel production. BioMed Research International, 1503126. https://doi.org/10.1155/2018/1503126.

Das P, Aziz SS and Obbard JP. 2011. Two phase microalgae growth in the open system for enhanced lipid productivity. Renewable Energy 36: 2524–2528. https://doi.org/10.1016/j.renene.2011.02.002.

Deng XD, Cai J and Fei X. 2013. Effect of the expression and knockdown of citrate synthase gene on carbon flux during triacylglycerol biosynthesis by green algae *Chlamydomonas reinhardtii*. BMC Biochemistry14: 38. https://doi.org/10.1186/1471-2091-14-38.

Deng XD, Cai J, Li Y and Fei X. 2014. Expression and knockdown of the PEPC1 gene affect carbon flux in the biosynthesis of triacylglycerols by the green alga *Chlamydomonas reinhardtii*. Biotechnology Letters 36: 2199–2208. https://doi.org/10.1007/s10529-014-1593-3.

Deng XD, Gu B, Li YJ, Hu XW, Guo JC and Fei XW. 2012. The roles of acyl-CoA: Diacylglycerol acyltransferase 2 genes in the biosynthesis of triacylglycerols by the green algae *Chlamydomonas reinhardtii*. Molecular Plant 5: 945–947. https://doi.org/10.1093/mp/sss040.

Eze CN, Aoyagi H and Ogbonna JC. 2020. Simultaneous accumulation of lipid and carotenoid in freshwater green microalgae *Desmodesmussubspicatus* LC172266 by nutrient replete strategy under mixotrophic condition. Korean Journal of Chemical Engineering 37(9): 1522–1529. https://doi.org/10.1007/s11814-020-0564-8.

Gan, Q., Jiang, J., Han, X., Wang, S., & Lu, Y. (2018). Engineering the chloroplast genome of oleaginous marine microalga Nannochloropsis oceanica. Frontiers in plant science, 9, 439. https://doi.org/10.3389/fpls.2018.00439

Goncalves EC, Wilkie AC, Kirst M, Rathinasabapathi, B. (2016). Metabolic regulation of triacylglycerol accumulation in the green algae: identification of potential targets for engineering to improve oil yield. Plant Biotechnology Journal, 14(8): 1649–1660. https://doi.org/10.1111/pbi.12523

Gray, M. W. (1993). Origin and evolution of organelle genomes. Current Opinion in Genetics and Development, 3(6), 884–890. https://doi.org/10.1016/0959-437X(93)90009-E

Guiheneuf F and Stengel DB. 2015. Towards the biorefinery concept: Interaction of light, temperature and nitrogen for optimizing the co-production of high-value compounds in *Porphyridium purpureum*. Algal Research 10: 152–163. https://doi.org/10.1016/j.algal.2015.04.025.

Gupta P, Phulara SC. 2015. Metabolic engineering for isoprenoid-based biofuel production. J. Appl. Microbiol. 119, 605–619. https://doi.org/10.1111/jam.12871

Henríquez V, Escobar C, Galarza J and Gimpel J. 2016. Carotenoids in microalgae. In: Carotenoids in Nature Vol 79: 219–237. Springer, Cham. https://doi.org/10.1007/978-3-319-39126-7_8.

Hikihara R, Yamasaki Y, Shikata T, Nakayama N, Sakamoto S, Kato S and Tanaka R. 2020. Analysis of phytosterol, fatty acid, and carotenoid composition of 19 microalgae and 6 bivalve species. Journal of Aquatic Food Product Technology 29(5): 461–479. https://doi.org/10.1080/10498850.2020.1749744.

Huang, L. F., Lin, J. Y., Pan, K. Y., Huang, C. K., & Chu, Y. K. (2015). Overexpressing Ferredoxins in Chlamydomonas reinhardtii Increase Starch and Oil Yields and Enhance Electric Power Production in a Photo Microbial Fuel Cell. International Journal of Molecular Sciences, 16(8), 19308–19325. https://doi.org/10.3390/ijms160819308.

Hu Q, Sommerfeld M, Jarvis E, Ghirardi M, Posewitz M, Seibert M and Darzins A. 2008. Microalgal triacylglycerols as feedstocks for biofuel production: Perspectives and advances. The Plant Journal 54(4): 621–639. https://doi.org/10.1111/j.1365-313X.2008.03492.x.

Hwangbo, K., Ahn, JW., Lim, JM. et al. Overexpression of stearoyl-ACP desaturase enhances accumulations of oleic acid in the green alga Chlamydomonas reinhardtii. Plant Biotechnol Rep 8, 135–142 (2013). https://doi.org/10.1007/s11816-013-0302-3.

Ibanez-Salazar A, Rosales-Mendoza S, Rocha-Uribe A, Ramirez-Alonso JI, Lara-Hernandez I, Hernandez-Torres A and Soria-Guerra RE. 2014. Over-expression of Dof-type transcription factor increases lipid production in *Chlamydomonas reinhardtii*. Journal of Biotechnology 184: 27–38. https://doi.org/10.1016/j.jbiotec.2014.05.003.

Iskandarov, U., Sitnik, S., Shtaida, N. et al. Cloning and characterization of a GPAT-like gene from the microalga Lobosphaera incisa (Trebouxiophyceae): overexpression in Chlamydomonas reinhardtii enhances TAG production. J Appl Phycol 28, 907–919 (2016). https://doi.org/10.1007/s10811-015-0634-1

Iwai, M., Ikeda, K., Shimojima, M., & Ohta, H. (2014). Enhancement of extraplastidic oil synthesis in C hlamydomonas reinhardtii using a type-2 diacylglycerol acyltransferase with a phosphorus starvation–inducible promoter. *Plant biotechnology journal, 12*(6), 808–819. DOI: 10.1111/pbi.12210

Jagadevan S, Banerjee A, Banerjee C, Guria C, Tiwari R, Baweja M, 2018. Biotechnology for Biofuels Recent developments in synthetic biology and metabolic engineering in microalgae towards biofuel production. Biotechnol. Biofuels 1–21. https://doi.org/10.1186/s13068-018-1181-1.

Jothibasu K, Dhar DW and Suchitra R. 2021. Recent developments in microalgal genome editing for enhancing lipid accumulation and biofuel recovery. Biomass and Bioenergy 150: 106093. https://doi.org/10.1016/j.biombioe.2021.106093.

Joyard J, Ferro M, Masselon C, Seigneurin-Berny D, Salvi D, Garin J and Rolland N. 2009. Chloroplast proteomics and the compartmentation of plastidial isoprenoid biosynthetic pathways. Molecular Plant 2(6): 1154–1180. https://doi.org/10.1093/mp/ssp088.

Kajikawa M, Kinohira S, Ando A, Shimoyama M, Kato M and Fukuzawa H. 2015. Accumulation of squalene in a microalga *Chlamydomonas reinhardtii* by genetic modification of squalene synthase and squalene epoxidase genes. PLoS One 10(3): e0120446. DOI:10.1371/journal.pone.0120446.

Kanehisa M and Goto S. 2000. KEGG: Kyoto encyclopedia of genes and genomes. Nucleic Acids Research 28: 27–30. DOI:10.1093/nar/28.1.27.

Kanehisa M, Goto S, Sato Y, Furumichi M and Tanabe M. 2012. KEGG for integration and interpretation of large-scale molecular data sets. Nucleic Acids Research 40: 109–114. https://doi.org/10.1093/nar/gkr988.

Kao PH and Ng IS. 2017. CRISPRi mediated phosphoenolpyruvate carboxylase regulation to enhance the production of lipid in *Chlamydomonas reinhardtii*. Bioresource Technology 245: 1527–1537. https://doi.org/10.1016/j.biortech.2017.04.111.

Kim M, Park BG, Kim EJ, Kim J and Kim BG. 2019. In silico identification of metabolic engineering strategies for improved lipid production in *Yarrowia lipolytica* by genome-scale metabolic modeling. Biotechnology for Biofuels 12: 187. https://doi.org/10.1186/s13068-019-1518-4.

Kindle K.L. 1990. High-frequency nuclear transformation of Chlamydomonas reinhardtii. Proc Natl Acad Sci USA 87, 1228–1232.

Kindle, K.L., Richards, K.L., Stern, D.B., 1991. Engineering the chloroplast genome: Techniques and capabilities for chloroplast transformation in Chlamydomonas reinhardtii. Proc. Natl. Acad. Sci. U. S. A. 88, 1721–1725. https://doi.org/10.1073/pnas.88.5.1721

Kong F, Liang Y, Légeret B, Beyly-Adriano A, Blangy S, Haslam RP, Napier JA, Beisson F, Peltier G and Li-Beisson Y. 2017. Chlamydomonas carries out fatty acid β-oxidation in ancestral peroxisomes using a bona fide acyl-CoA oxidase. Plant Journal 90: 358–371. https://doi.org/10.1111/tpj.13498.

Kumar V, Niraja P, Venkatesh S, Ajit P, Santanu S and Bhaskar D. 2019. CRISPR—Cas9 system for genome engineering of photosynthetic microalgae. Molecular Biotechnology 61: 541–561. https://doi.org/10.1007/s12033-019-00185-3.

Lange BM, Rujan T, Martin W and Croteau R. 2000. Isoprenoid biosynthesis: The evolution of two ancient and distinct pathways across genomes. Proceedings of the National Academy of Sciences USA 97: 13172–13177. https://doi.org/10.1073/pnas.240454797.

Lauersen KJ, Baier T, Wichmann J, Wördenweber R, Mussgnug JH, Hübner W, Huser T and Kruse O. 2016. Efficient phototrophic production of a high-value sesquiterpenoid from the eukaryotic microalga *Chlamydomonas reinhardtii*. Metabolic Engineering 38: 331–343. https://doi.org/10.1016/j.ymben.2016.07.013.

Lee SK, Chou H, Ham TS, Lee TS and Keasling JD. 2008. Metabolic engineering of microorganisms for biofuels production: From bugs to synthetic biology to fuels. Current Opinion in Biotechnology 19(6): 556–563. DOI:10.1016/j.copbio.2008.10.014.

Lee SY, Khoiroh I, Vo DVN, Kumar PS and Show PL. 2020. Techniques of lipid extraction from microalgae for biofuel production: A review. Environmental Chemistry Letters: 1–21. https://doi.org/10.1007/s10311-020-01088-5.

Li X, Liu P, Yang P, Fan C and Sun X. 2018. Characterization of the glycerol-3-phosphate acyltransferase gene and its real-time expression under cold stress in *Paeonia lactiflora* Pall. PLoS One 13(8): e0202168. https://doi.org/10.1371/journal.pone.0202168.

Li X, Xu H and Wu Q. 2007. Large-scale biodiesel production from microalga Chlorella protothecoides through heterotrophic cultivation in bioreactors. Biotechnology and Bioengineering 98(4): 764–771. https://doi.org/10.1002/bit.21489.

Li Y, Han D, Hu G, Dauvillee D, Sommerfeld M, Ball S and Hu Q. 2010. Chlamydomonas starchless mutant defective in ADP-glucose pyrophosphorylase hyper-accumulates triacylglycerol. Metabolic Engineering 12: 387–391. https://doi.org/10.1016/j.ymben.2010.02.002.

Lichtenthaler HK. 1999. The 1-deoxy-D-xylulose-5-phosphate pathway of isoprenoid biosynthesis in plants. Annual Review of Plant Biology 50(1): 47–65.

Liu T and Khosla C. 2010. Genetic engineering of Escherichia coli for biofuel production. Annual Review of Genetics 44: 53–69. https://doi.org/10.1146/annurev-genet-102209-163440.

Lohr M, Schwender J and Polle JE. 2012. Isoprenoid biosynthesis in eukaryotic phototrophs: A spotlight on algae. Plant Science 185: 9–22. https://doi.org/10.1016/j.plantsci.2011.07.018.

Mata TM, Martins AA and Caetano NS. 2010. Microalgae for biodiesel production and other applications: A review. Renewable and Sustainable Energy Reviews 14: 217–232.

Minhas AK, Barrow CJ, Hodgson P and Adholeya A. 2020. Microalga *Scenedesmusbijugus:* Biomass, lipid profile, and carotenoids production in vitro. Biomass and Bioenergy 142: 105749. https://doi.org/10.1016/j.biombioe.2020.105749.

Moellering ER and Benning C. 2010. RNA interference silencing of a major lipid droplet protein affects lipid droplet size in *Chlamydomonas reinhardtii*. Eukaryotic Cell 9: 97–106. https://doi.org/10.1128/EC.00203-09.

Mulders KJ, Lamers PP, Martens DE and Wijffels RH. 2014. Phototrophic pigment production with microalgae: Biological constraints and opportunities. Journal of Phycology 50(2): 229–242. https://doi.org/10.1111/jpy.12173.

Mussgnug JH, Thomas-Hall S, Rupprecht J, Foo A, Klassen V, McDowall A, Schenk PM, Kruse O and Hankamer B. 2007. Engineering photosynthetic light capture: Impacts on improved solar energy to biomass conversion. Plant Biotechnology Journal 5(6): 802–814. https://doi.org/10.1111/j.1467-7652.2007.00285.x.

Olguín EJ. 2012. Dual purpose microalgae-bacteria-based systems that treat wastewater and produce biodiesel and chemical products within a biorefinery. Biotechnology Advances 30(5): 1031–1046. https://doi.org/10.1016/j.biotechadv.2012.05.001.

Phulara SC, Chaturvedi P, Gupta P. 2016. Isoprenoid-based biofuels: homologous expression and heterologous expression in prokaryotes. Appl Environ Microbiol 82, 5730–5740. https://doi.org/10.1128/aem.01192-16

Post-Beittenmiller D, Jaworski JG and Ohlrogge JB. 1991. In vivo pools of free and acylated acyl carrier proteins in spinach. Evidence for sites of regulation of fatty acid biosynthesis. Journal of Biological Chemistry 266: 1858–1865.

Post-Beittenmiller D, Roughan G and Ohlrogge JB. 1992. Regulation of plant fatty acid biosynthesis: Analysis of acyl-coenzyme a and acyl-acyl carrier protein substrate pools in spinach and pea chloroplasts. Plant Physiology 100(2): 923–930. https://doi.org/10.1104/pp.100.2.923.

Randhir A, Laird DW, Maker G, Trengove R and Moheimani NR. 2020. Microalgae: A potential sustainable commercial source of sterols. Algal Research 46: 101772. https://doi.org/10.1016/j.algal.2019.101772.

Rearte TA, Figueroa FL, Gomez-Serrano C, Velez CG, Marsili S, Iorio ADF and Acién-Fernández FG. 2020. Optimization of the production of lipids and carotenoids in the microalga *Golenkiniaaff. brevispicula*. Algal Research 51: 102004. https://doi.org/10.1016/j.algal.2020.102004.

Rengel, R., Smith, R.T., Haslam, R.P., Sayanova, O., Vila, M., León, R., 2018. Overexpression of acetyl-CoA synthetase (ACS) enhances the biosynthesis of neutral lipids and starch in the green microalga Chlamydomonas reinhardtii. Algal Res. 31, 183–193. https://doi.org/https://doi.org/10.1016/j.algal.2018.02.009

Rilling HKB. 1959 On the mechanism of sqalene biogenesis from mevalonic acid. Journal of Biological Chemistry 234(6): 1424–1432.

Rodriguez-Concepcion M. 2006. Early steps in isoprenoid biosynthesis: Multilevel regulation of the supply of common precursors in plant cells. Phytochemistry Reviews 5: 1–15. https://doi.org/10.1007/s11101-005-3130-4.

Rohmer M. 1998. Isoprenoid biosynthesis via the mevalonate-independent route, a novel target for antibacterial drugs? In Progress in Drug Research: 135–154. Birkhäuser, Basel.

Rohmer M. 1999. The discovery of a mevalonate-independent pathway for isoprenoid biosynthesis in bacteria, algae and higher plants. Natural Product Reports 16(5): 565–574.

Rude MA and Schirmer A. 2009. New microbial fuels: A biotech perspective. Current Opinion in Microbiology 12(3): 274–281. https://doi.org/10.1016/j.mib.2009.04.004.

Ruffing AM. 2013. Metabolic engineering of hydrocarbon biosynthesis for biofuel production. Liquid, Gaseous and Solid Biofuels–Conversion Techniques: 263–299. DOI:10.5772/52050.

Ryckebosch E, Bruneel C, Termote-Verhalle R, Muylaert K and Foubert I. 2014. Influence of extraction solvent system on extractability of lipid components from different microalgae species. Algal Research 3: 36–43. https://doi.org/10.1016/j.algal.2013.11.001.

Rzama A, Dufourc EJ and Arreguy B. 1994. Sterols from green and blue-green algae grown on reused waste water. Phytochemistry 37(6): 1625–1628. https://doi.org/10.1016/S0031-9422(00)89579-5.

Schwender J, Seemann M, Lichtenthaler HK and Rohmer M. 1996. Biosynthesis of isoprenoids (carotenoids, sterols, prenyl side-chains of chlorophylls and plastoquinone) via a novel pyruvate/glyceraldehyde 3-phosphate non-mevalonate pathway in the green alga *Scenedesmus obliquus*. Biochemical Journal 316(1): 73–80. https://doi.org/10.1042/bj3160073.

Serrano G, Herrera-Palau R, Romero JM, Serrano A, Coupland G and Valverde F. 2009. *Chlamydomonasconstans* and the evolution of plant photoperiodic signaling. Current Biology 19: 359–368. https://doi.org/10.1016/j.cub.2009.01.044.

Sharma PK, Saharia M, Srivstava R, Kumar S and Sahoo L (2018) Tailoring Microalgae for Efficient Biofuel Production. Front. Mar. Sci. 5:382. doi: 10.3389/fmars.2018.00382.

Shin YS, Jeong J, Nguyen THT, Kim JYH, Jin ES and Sim SJ. 2019. Targeted knockout of phospholipase A2 to increase lipid productivity in *Chlamydomonas reinhardtii* for biodiesel production. Bioresource Technology 271: 368–374. https://doi.org/10.1016/j.biortech.2018.09.121.

Suchitra R, Karthikeyan S. 2019. Co-cultivation of microalgae with oleaginous yeast for economical biofuel production. Journal of Farm Sciences 32: 125–130.

Suchitra R, Tharunkumar J, Bhavya S, Jothibasu K and Karthikeyan S. 2020. Sustainable microalgae harvesting strategies for the production of biofuel and oleochemicals. Highlights BioSciences 3: 1–8. https://doi.org/10.36462/H.BioSci.20209.

Tsai CH, Warakanont J, Takeuchi T, Sears BB, Moellering ER and Benning C. 2014. The protein compromised hydrolysis of triacylglycerols 7 (CHT7) acts as a repressor of cellular quiescence in *Chlamydomonas*. Proceedings in the National Academy of Sciences 111: 15833 LP—15838. https://doi.org/10.1073/pnas.1414567111.

Timmis, J., Ayliffe, M., Huang, C. et al. 2004. Endosymbiotic gene transfer: organelle genomes forge eukaryotic chromosomes. Nat Rev Genet 5, 123–135 (2004). https://doi.org/10.1038/nrg1271

Vavitsas, K., Fabris, M., & Vickers, C. E. (2018). Terpenoid metabolic engineering in photosynthetic microorganisms. Genes, 9(11), 520. https://doi.org/10.3390/genes9110520.

Voelker T. 1996. Plant acyl-ACP thioesterases: Chain-length determining enzymes in plant fatty acid biosynthesis. In Genetic Engineering: 111–133. Springer, Boston, MA.

Vranová E, Coman D and Gruissem W. 2012. Structure and dynamics of the isoprenoid pathway network. Molecular Plant 5(2): 318–333. https://doi.org/10.1093/mp/sss015.

Walter MH, Floss DS and Strack D. 2010. Apocarotenoids: Hormones, mycorrhizal metabolites and aroma volatiles. Planta 232(1): 1–17. DOI:10.1007/s00425-010-1156-3.

Wang C, Chen X, Li H, Wang J and Hu Z. 2017. Biotechnology for biofuels artificial miRNA inhibition of phosphoenolpyruvate carboxylase increases fatty acid production in a green microalga *Chlamydomonas reinhardtii*. Biotechnology for Biofuels: 1–11. https://doi.org/10.1186/s13068-017-0779-z.

Wichmann J, Baier T, Wentnagel E, Lauersen KJ and Kruse O. 2018. Tailored carbon partitioning for phototrophic production of (E)-α-bisabolene from the green microalga *Chlamydomonas reinhardtii*. Metabolic Engineering 45: 211–222. https://doi.org/10.1016/j.ymben.2017.12.010.

Work VH, Radakovits R, Jinkerson RE, Meuser JE, Elliott LG, Vinyard DJ, Laurens LML, Dismukes GC and Posewitz MC. 2010. Increased lipid accumulation in the *Chlamydomonas reinhardtiista* 7–10 starchless isoamylase mutant and increased carbohydrate synthesis in complemented strains. Eukaryotic Cell 9: 1251–1261. https://doi.org/10.1128/EC.00075-10.

Wu H, Li T, Wang G, Dai S, He H and Xiang W. 2016. A comparative analysis of fatty acid composition and fucoxanthin content in six *Phaeodactylum tricornutum* strains from different origins. Chinese Journal of Oceanology and Limnology 34(2): 391–398. https://doi.org/10.1007/s00343-015-4325-1.

Yamaoka, Y., Achard, D., Jang, S., Legéret, B., Kamisuki, S., Ko, D., ... & Lee, Y. (2016). Identification of a Chlamydomonas plastidial 2-lysophosphatidic acid acyltransferase and its use to engineer microalgae with increased oil content. *Plant biotechnology journal, 14*(11), 2158–2167. https://doi.org/10.1111/pbi.12572.

Zhu LD, Li ZH and Hiltunen E. 2016. Strategies for lipid production improvement in microalgae as a biodiesel feedstock. BioMed Research International, 8792548. https://doi.org/10.1155/2016/8792548.

16 Review of Algal Nanotechnology
Guidelines to Research Scholars

G. Rani and Ramachandran Uma

16.1 INTRODUCTION

Reviews on Algal nanotechnology has been already provided by many authors (Aishwarya *et al.*, 2015; Oscar *et al.*, 2016; Azhar *et al.*, 2018); however the aim of the present review is to critically examine the state-of-art and express the informed views, discuss the problems faced by research scholars on the technologies, provide guidance on the government-funded facilities for nanotechnology work. Due to the development of science and technology, researchers are merging green chemistry/green synthesis and green engineering with nanotechnology resulting in a new field "Green Nano" (Sharda, 2019). Currently researchers are into synthesizing biologically active nanoparticles due to spread of infectious diseases and increased drug resistance among microbes. Improvement of the eco-friendly nanotechnology procedure for the synthesis of nanoparticles is on the rise (Anju *et al.*, 2018). Globally biological synthesis has gained much importance because the chemical methods are a laborious process, and chemicals used are not eco-friendly. Hence, green synthesis of nanoparticles using algae has attracted the focus of scientists from all fields. Since many research scholars of the other fields of science are also focusing on algal nanotechnology, we have provided the image of all algal source of nanoparticles in this review. Algae is a group of organism of thallophyta of the plant kingdom which are autotrophic (synthesize starch) due to the presence of the green pigment chlorophyll. It occurs in fresh water, marine, soil, on other plants as epiphytes, endophytes, etc. Morphological organization of algae varies from micro unicellular, colonial, and filamentous to macroalgae.

16.2 HISTORY OF NANOTECHNOLOGY

The history of nanoparticle usage dates back to the 9th century when the artisans of Mesopotamia used nanoparticles to generate glittering effects to pots (Giljohann *et al.*, 2010). Chaudhary (2011) reported that Ayurvedic bhasmas are in nanometer dimensions and are considered as nanomedicine free from toxicity in therapeutic doses. Bhasmas produced by biological methods of nanoparticles is prescribed with several medicines in Ayurveda. This is one of the most ancient applications of traditional nanomedicine.

Nanotechnology, of the 21st-century frontier, is defined as the understanding and control of matter at dimensions between 1 and 100 nm where unique phenomena enable novel applications. Modern nanotechnology was the brainchild of Richard Feynman, the 1965 Nobel Prize Laureate in physics (Drexler, 1992). Almost 15 years after Feynman's lecture, a Japanese scientist, Norio Taniguchi, was the first to use nanotechnology "to describe semiconductor processes that occurred on the order of a nanometer" (Taniguchi, 1974). According to the consumer product inventory of April 2013, there are 1,266 commercialized nanoproducts with 714 products in health and fitness, 104 in food and beverages, and in 28 products for children. The analysis revealed silver nanoparticles as the most commercialized material worldwide.

DOI: 10.1201/9781003219156-20

A vast majority of these nanomaterials find their potential application as antimicrobial agent. However, their extensive application may lie in biotechnology, biomedicine, cancer therapy, pharmaceutical, catalysis, medical imaging, and environmental sector (Wiechers and Musee, 2010). Nanoparticles have entered our day to day life as they are used in many cosmetics, therapeutics (Czupryna and Tsourkas, 2006), drug delivery to specific sites in the body and gene delivery systems (Jin and Ye, 2007), biocompatible replacements for body parts and fluids, material for bone and tissue regeneration, and biosensors (Prow *et al.*, 2006).

16.3 ALGAL NANOTECHNOLOGY

Various prokaryotic and eukaryotic systems like plants, plant products, algae, fungi, yeast, bacteria, and even viruses (Thakkar *et al.*, 2010) are regularly used as biogenerators of nanoparticles. However, microalgae and macroalgae (seaweeds) in its fresh form and dry form have proved to be "bionanofactories" synthesising the nanoparticles (Davis *et al.*, 1998). Hence algae-mediated biosynthesis of nanoparticles is treated as a new branch, "phyconanotechnology" (Thakkar *et al.*, 2010; Sharma *et al.*, 2015). This review article is focused on the biosynthesis and characterization of nanoparticles from Cyanophyceae, Chlorophyceae, Bacillariophyceae, Phaeophyceae, and Rhodophyceae (Table 16.1), which are found to synthesize different types of metallic NPs, such as silver, gold, ferrihydrite, and palladium, and their usage in different fields. The article also throws light on the government of India initiatives to develop the research on nanotechnology and establishing centers for nanotechnology with instrumentation facilities.

16.4 MECHANISM OF BIOSYNTHESIS OF ALGAL NANOPARTICLES

Basically the mechanism of biosynthesis of metallic nanoparticles by algae is similar with slight variations in different methods, depending upon the reducing agents in the algae. In the first step algal extract either with the fresh algae or dried powder of algae is extracted with water or with organic solvent by heating or boiling. In the next step, molar solution of ionic metal compound is prepared and mixed with the algal extract. According to Dahoumane *et al.*, (2012), Dahoumane *et al.*, 2014a), depending upon the algae, biosynthesis of nanoparticles can be extracellular or intracellular. It is observed that if the reducing agents in algal aqueous solution is polysaccharides, reducing sugars, peptides, proteins, pigments, then extracellular formation of metallic nanoparticles are formed due to the reduction of metal ions (Greene *et al.*, 1986; Barwal *et al.*, 2011; Abdel-Raouf *et al.*, 2013; Dahoumane *et al* ., 2012; Dahoumane *et al.*, 2014b). In intracellular formation of metallic nanoparticles, algal metabolism such as photosynthesis, respiration may be responsible for the reduction of metallic ions (Sicard *et al.*, 2010; Barwal *et al.*, 2011; Dahoumane *et al* ., 2012, 2014a; Jeffryes *et al.*, 2015). Reduction of metallic ions in blue-green algae into metallic nanoparticles is either by the reducing agents NADPH or NADPH dependent reductase of electron transport system (ETS) of photosynthesis or respiration (Sicard *et al.*, 2010; Focsan *et al.*, 2011; Dahoumane *et al.*, 2012; Dahoumane et al., 2014a; Dahoumane *et al.*, 2014b; Eroglu *et al.*, 2013; Jeffryes *et al.*, 2015). The various mechanisms of the reduction of metallic ions into metallic nanoparticles are represented in Figure 16.1(a). Nitrogenase enzyme is reported to be responsible for the reduction of intracellular gold (Au) ions into Au nanoparticles (Oza *et al* ., 2012). Stabilizing and capping the metal nanoparticles in aqueous solutions is done by proteins through amino groups or cysteine residues and sulfated polysaccharides (Singaravelu *et al.*, 2007). Synthesis of nanoparticles using algae takes comparatively shorter time period than the other biosynthesizing methods (Thakkar *et al.*, 2010; Rauwel *et al.*, 2015).

In general, the biosynthesis of algae-mediated nanoparticles is subjected to the method of treatment, as shown in Figure 16.1(b), for identifying the type of nanoparticle (silver/gold NP) synthesized. Biosynthesis of silver NPs in the algal extract treated with 0.1 mM AgNO3 solution turns

TABLE 16.1
Biosynthesis and characterization of nanoparticles from Cyanophyceae, Chlorophyceae, Bacillariophyceae, Phaeophyceae, and Rhodophyceae

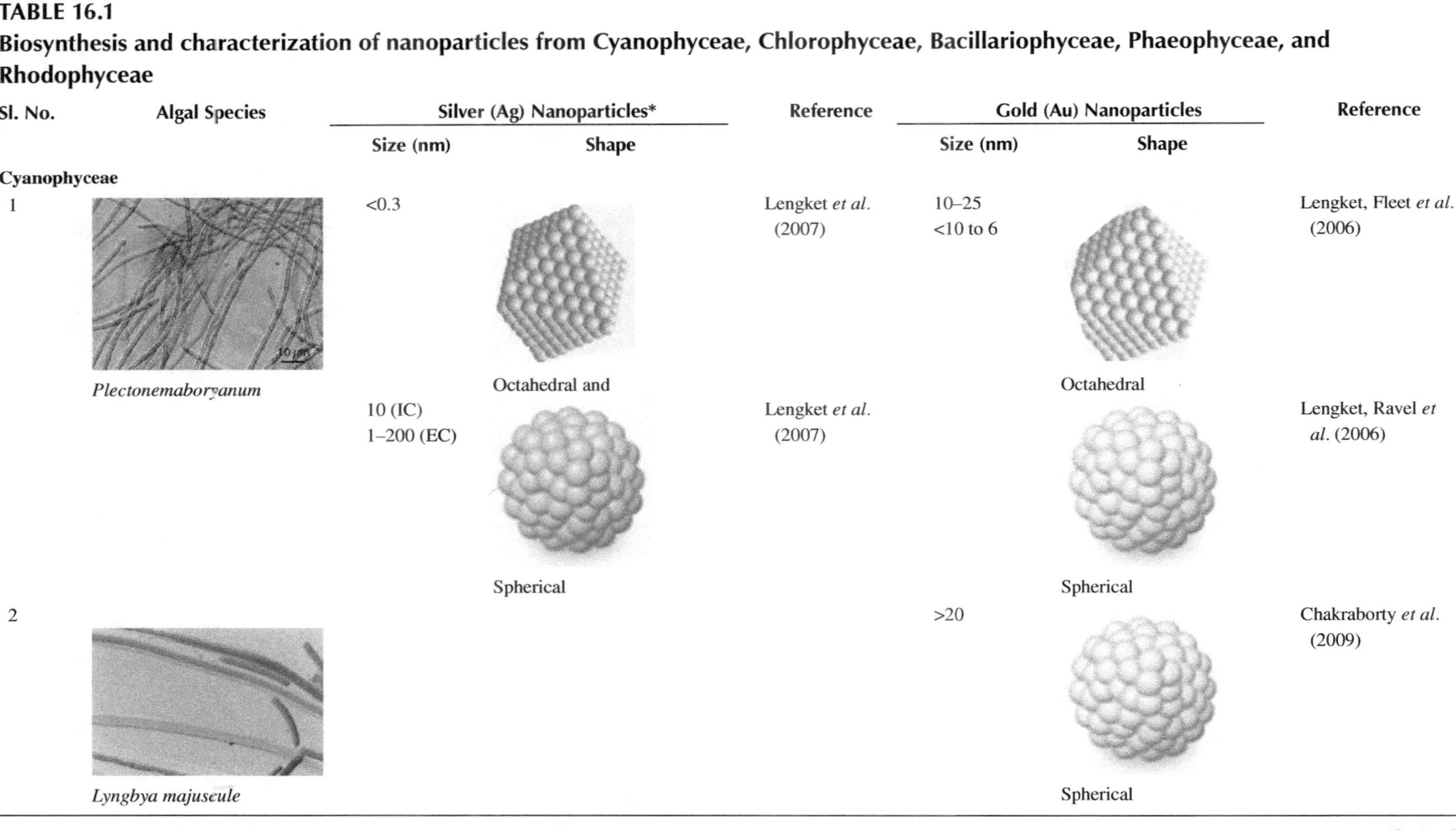

Sl. No.	Algal Species	Silver (Ag) Nanoparticles*		Reference	Gold (Au) Nanoparticles		Reference
		Size (nm)	Shape		Size (nm)	Shape	
Cyanophyceae							
1	*Plectonemaboryanum*	<0.3	Octahedral and	Lengket *et al.* (2007)	10–25 <10 to 6	Octahedral	Lengket, Fleet *et al.* (2006)
		10 (IC) 1–200 (EC)	Spherical	Lengket *et al.* (2007)		Spherical	Lengket, Ravel *et al.* (2006)
2	*Lyngbya majuscule*				>20	Spherical	Chakraborty *et al.* (2009)

(Continued)

TABLE 16.1 *(Continued)*

Sl. No.	Algal Species	Silver (Ag) Nanoparticles*		Reference	Gold (Au) Nanoparticles		Reference
		Size (nm)	Shape		Size (nm)	Shape	
3	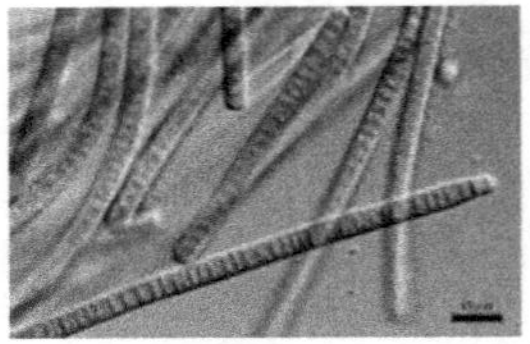 *Microcoleus* sp	44–79	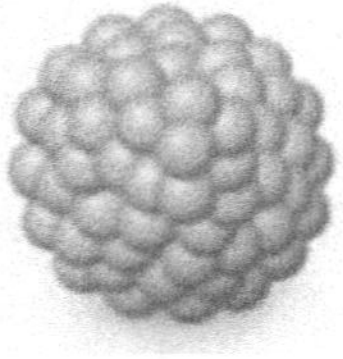 Spherical	Sudha *et al.* (2013)			
4	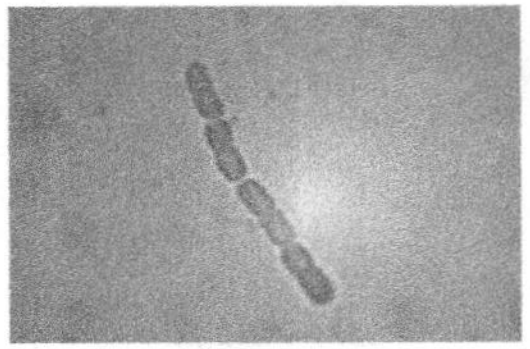 *Nostoc ellipsosporum*				137–209 (length) 33–69 (diameter)	 Nanorod	Parial *et al.* (2012a)
5	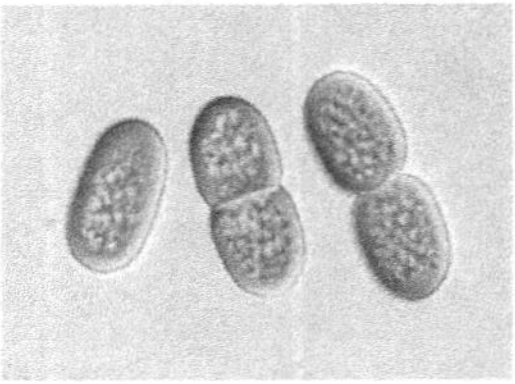 *Synechocystis* sp.				13±2	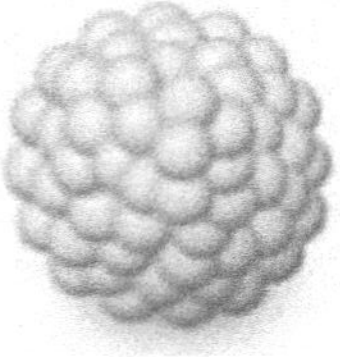 Spherical	Focsan *et al.* (2011)

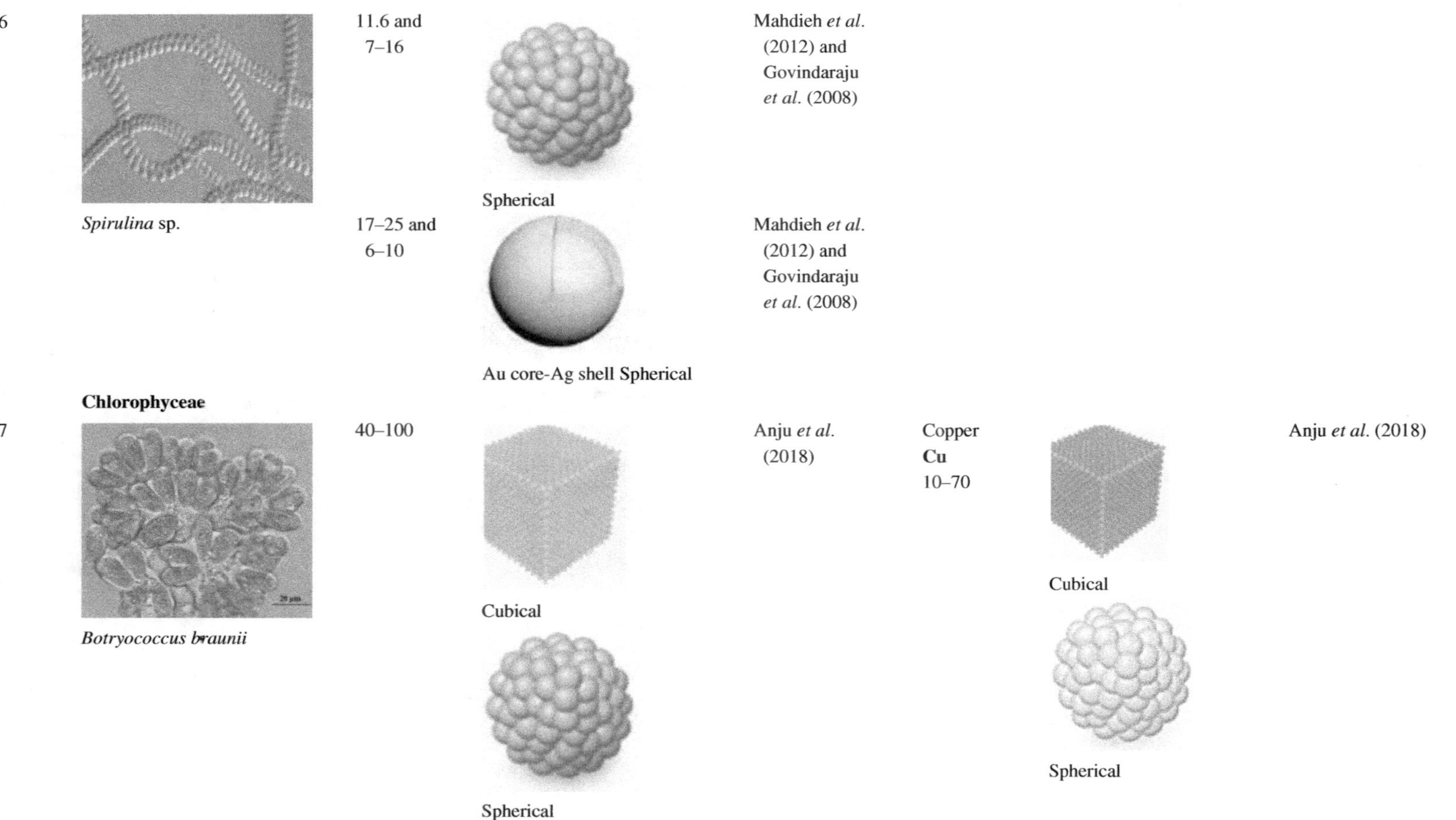

6	*Spirulina* sp.	11.6 and 7–16	Spherical	Mahdieh *et al.* (2012) and Govindaraju *et al.* (2008)			
		17–25 and 6–10	Au core-Ag shell Spherical	Mahdieh *et al.* (2012) and Govindaraju *et al.* (2008)			
	Chlorophyceae						
7	*Botryococcus braunii*	40–100	Cubical Spherical	Anju *et al.* (2018)	Copper **Cu** 10–70	Cubical Spherical	Anju *et al.* (2018)

(Continued)

TABLE 16.1 *(Continued)*

Sl. No.	Algal Species	Silver (Ag) Nanoparticles*		Reference	Gold (Au) Nanoparticles		Reference
		Size (nm)	Shape		Size (nm)	Shape	
			 Triangular			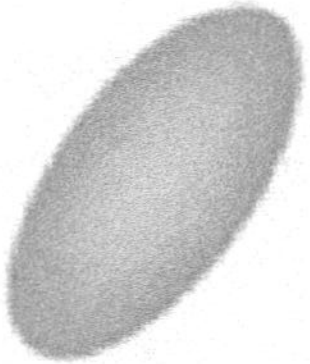 Elongated	
8	 *Chlamydomonas reinhardtii*	5–15 *in vitro* 5–35 *in vivo*	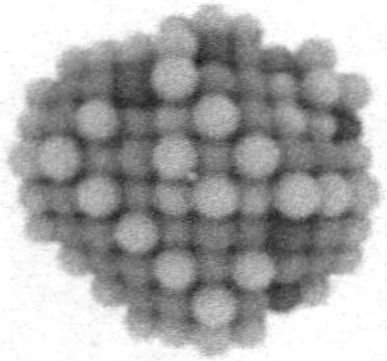 Rounded and  Rectangular	Barwal *et al.* (2011)			
			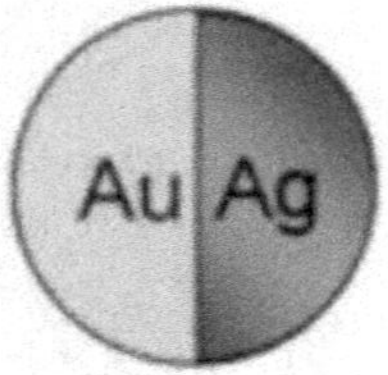 Bimetallic Ag-Au Nanoalloy Spherical/round	Dahoumane, Wijesekera *et al.* (2014)			

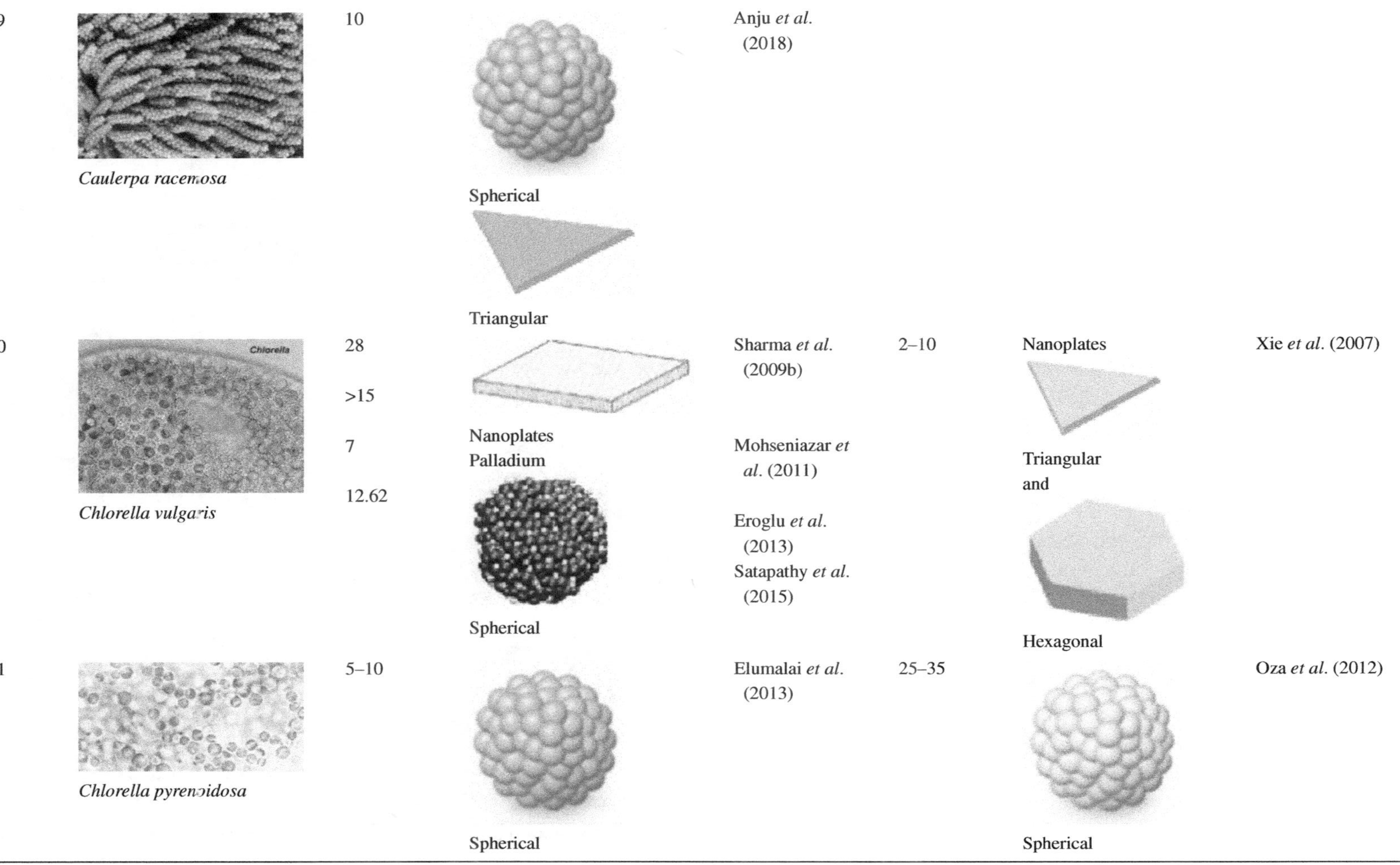

9	*Caulerpa racemosa*	10	Spherical Triangular	Anju *et al.* (2018)			
10	*Chlorella vulgaris*	28 >15 7 12.62	Nanoplates Palladium Spherical	Sharma *et al.* (2009b) Mohseniazar *et al.* (2011) Eroglu *et al.* (2013) Satapathy *et al.* (2015)	2–10	Nanoplates Triangular and Hexagonal	Xie *et al.* (2007)
11	*Chlorella pyrenoidosa*	5–10	Spherical	Elumalai *et al.* (2013)	25–35	Spherical	Oza *et al.* (2012)

(Continued)

TABLE 16.1 *(Continued)*

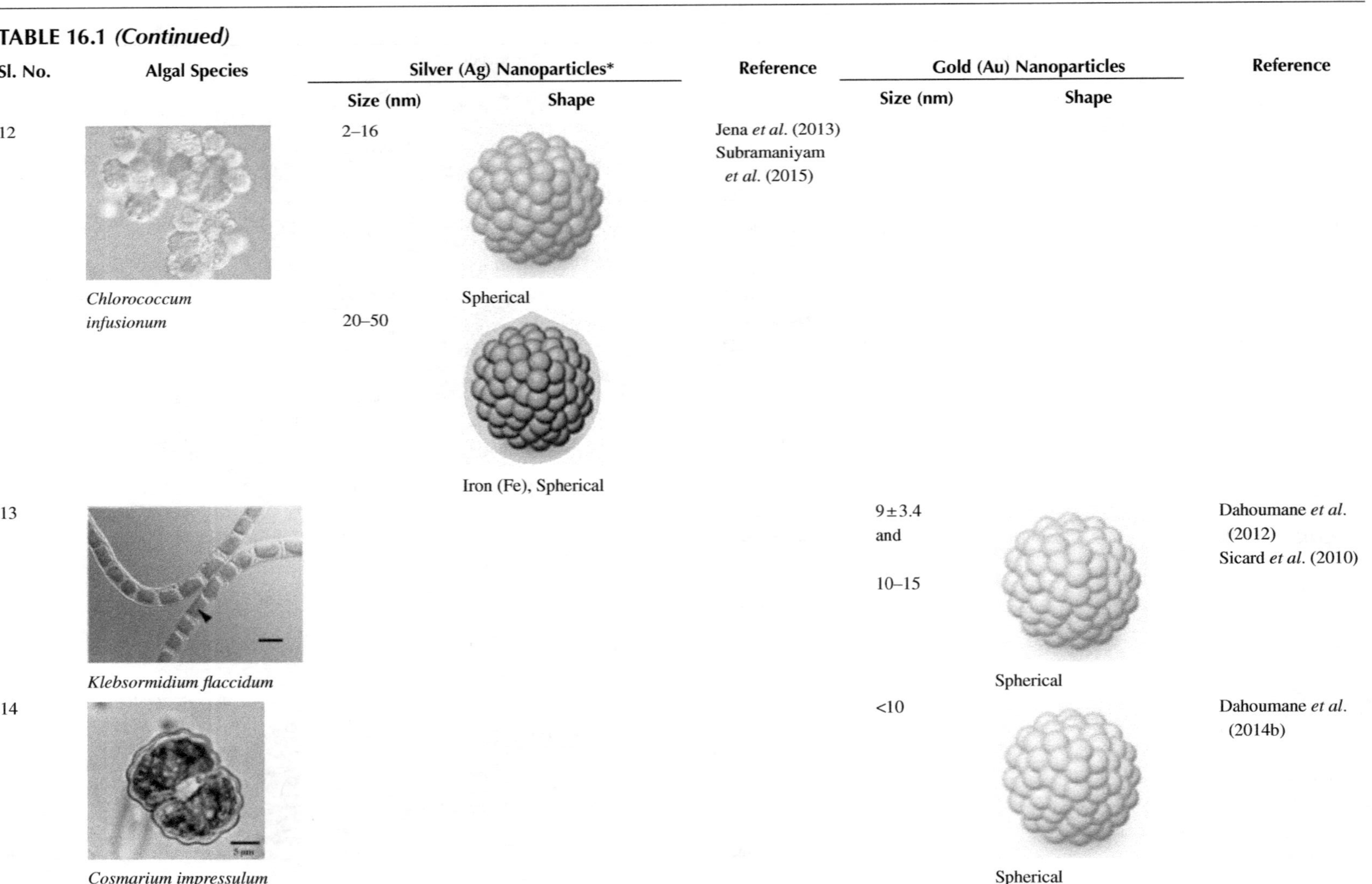

Sl. No.	Algal Species	Silver (Ag) Nanoparticles*		Reference	Gold (Au) Nanoparticles		Reference
		Size (nm)	Shape		Size (nm)	Shape	
12	*Chlorococcum infusionum*	2–16	Spherical	Jena *et al.* (2013)			
		20–50	Iron (Fe), Spherical	Subramaniyam *et al.* (2015)			
13	*Klebsormidium flaccidum*				9±3.4 and 10–15	Spherical	Dahoumane *et al.* (2012) Sicard *et al.* (2010)
14	*Cosmarium impressulum*				<10	Spherical	Dahoumane *et al.* (2014b)

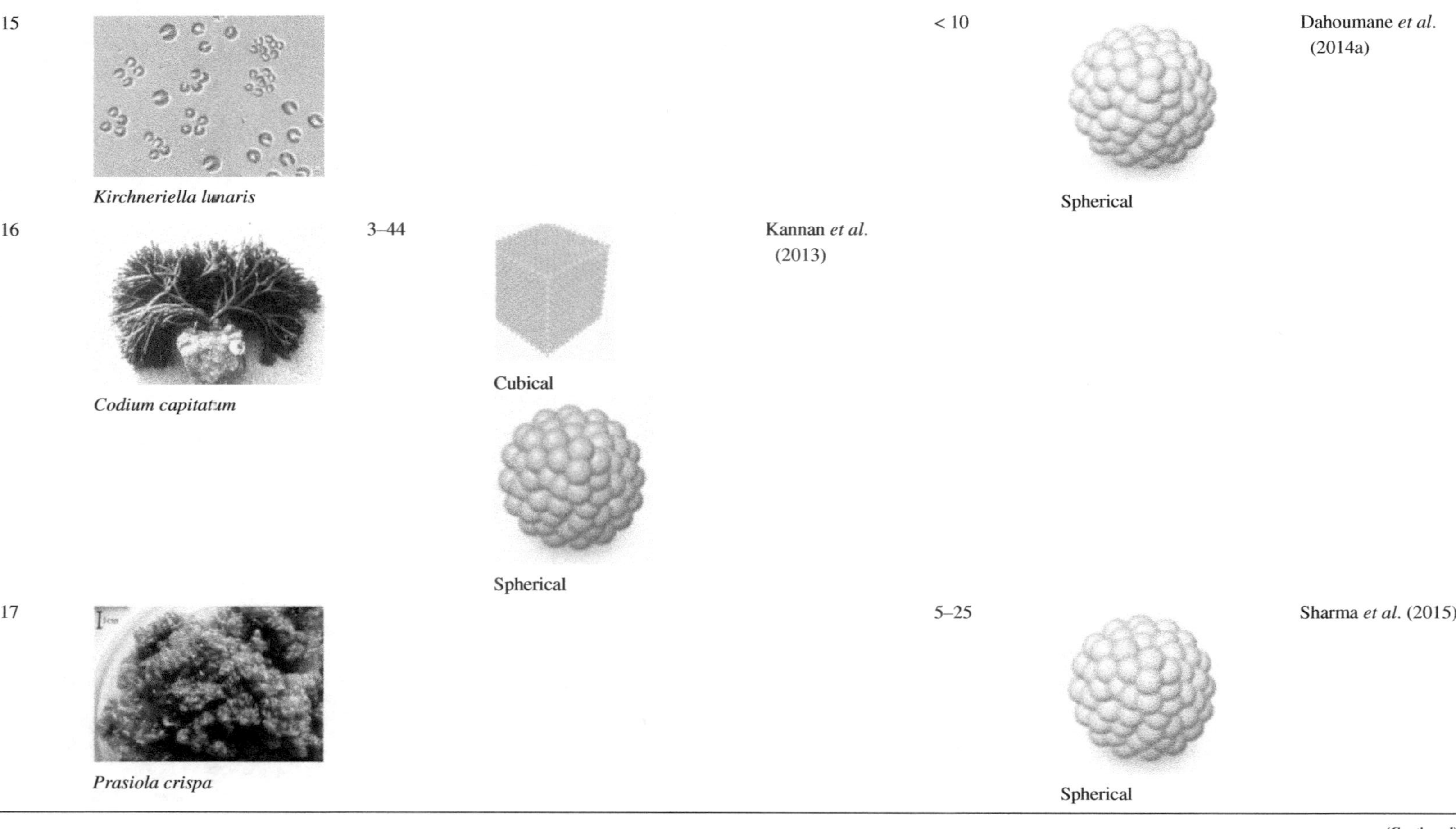

15	*Kirchneriella lunaris*				< 10	Spherical	Dahoumane *et al.* (2014a)
16	*Codium capitatum*	3–44	Cubical Spherical	Kannan *et al.* (2013)			
17	*Prasiola crispa*				5–25	Spherical	Sharma *et al.* (2015)

(Continued)

TABLE 16.1 *(Continued)*

Sl. No.	Algal Species	Silver (Ag) Nanoparticles*		Reference	Gold (Au) Nanoparticles		Reference
		Size (nm)	Shape		Size (nm)	Shape	
18	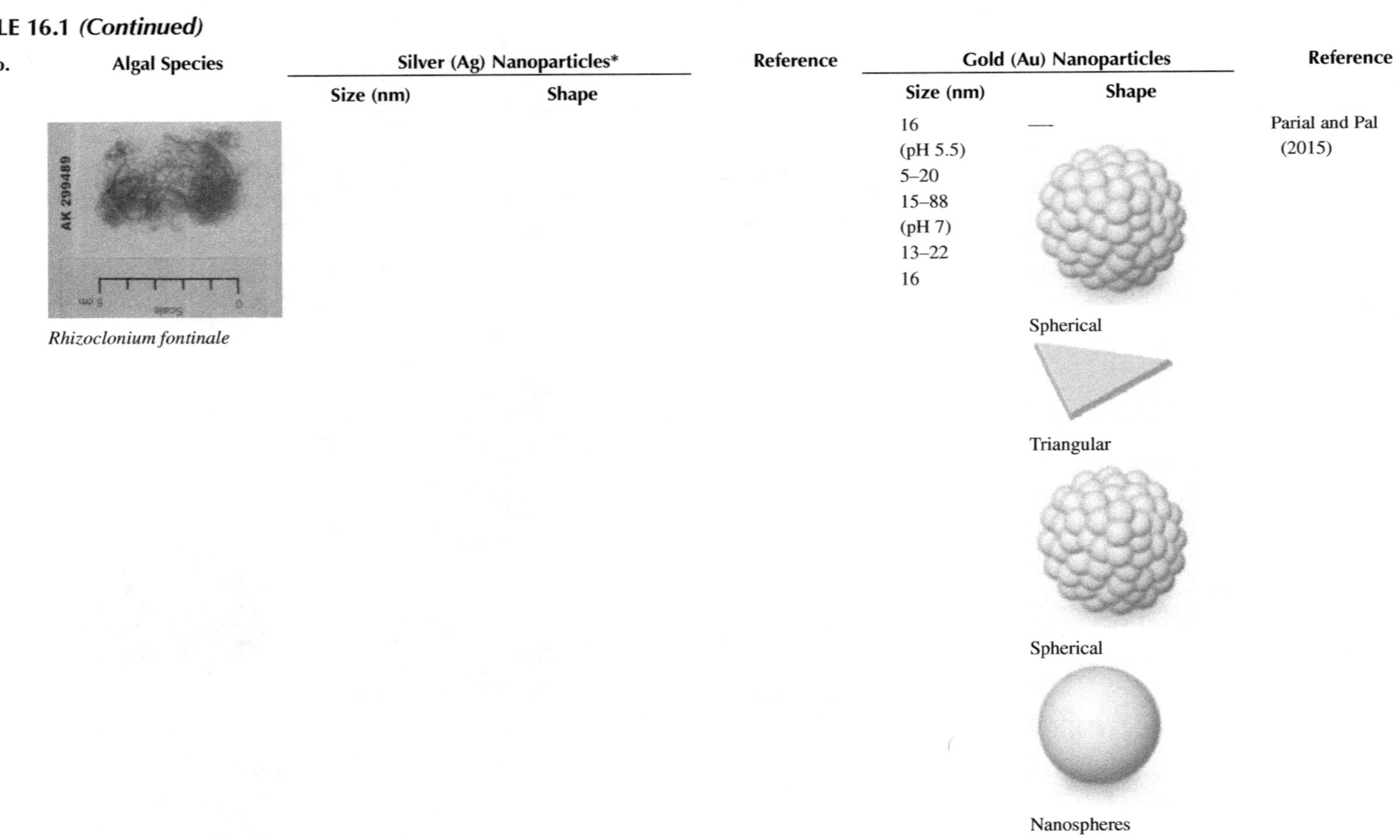 *Rhizoclonium fontinale*				16 (pH 5.5) 5–20 15–88 (pH 7) 13–22 16	— Spherical Triangular Spherical Nanospheres	Parial and Pal (2015)

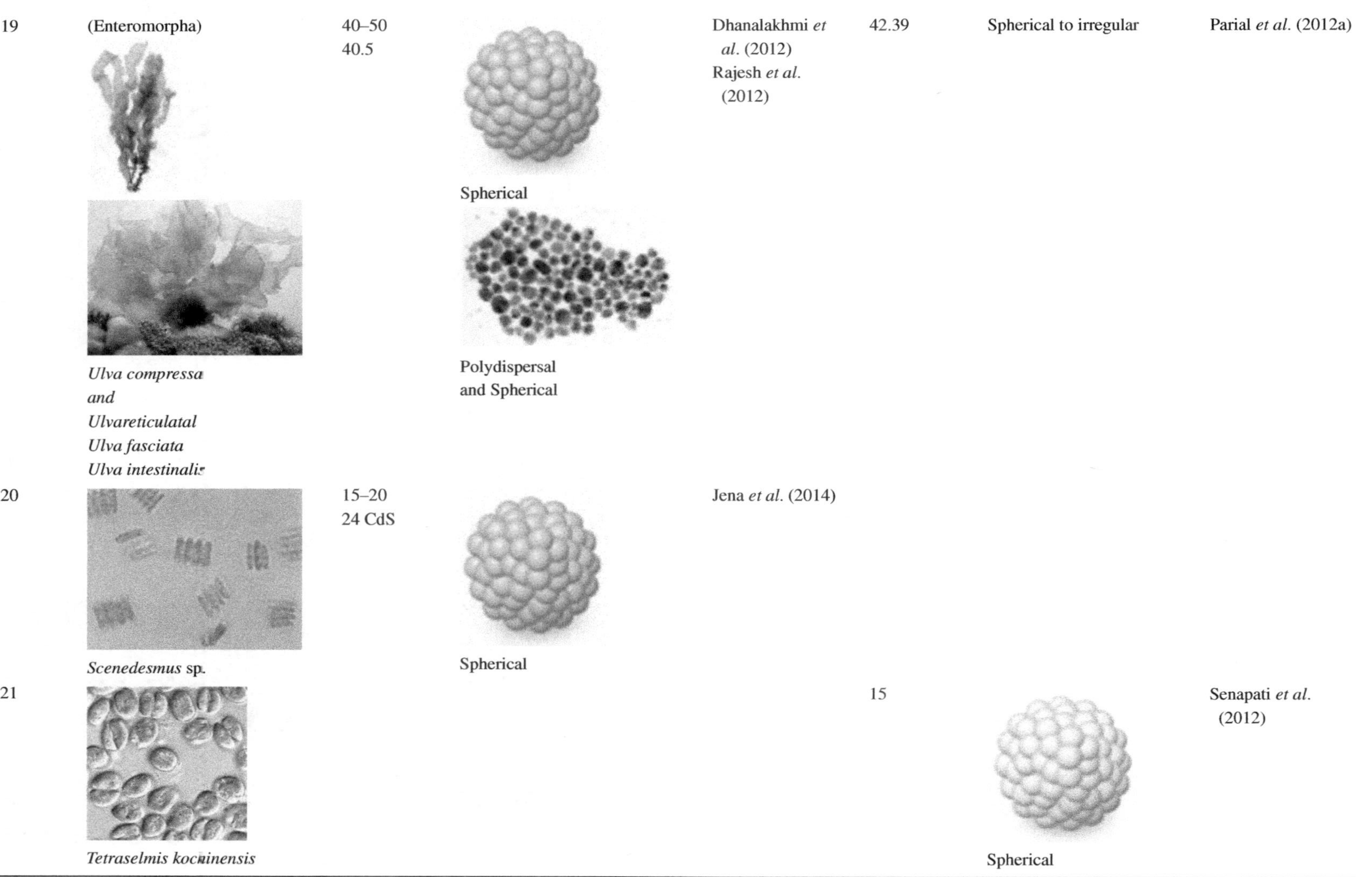

19	(Enteromorpha) *Ulva compressa and Ulvareticulatal Ulva fasciata Ulva intestinalis*	40–50 40.5	Spherical Polydispersal and Spherical	Dhanalakhmi *et al*. (2012) Rajesh *et al*. (2012)	42.39	Spherical to irregular	Parial *et al*. (2012a)
20	*Scenedesmus* sp.	15–20 24 CdS	Spherical	Jena *et al*. (2014)			
21	*Tetraselmis kochinensis*				15	Spherical	Senapati *et al*. (2012)

(*Continued*)

TABLE 16.1 *(Continued)*

Sl. No.	Algal Species	Silver (Ag) Nanoparticles*		Reference	Gold (Au) Nanoparticles		Reference
		Size (nm)	Shape		Size (nm)	Shape	
22	Ochrophyta 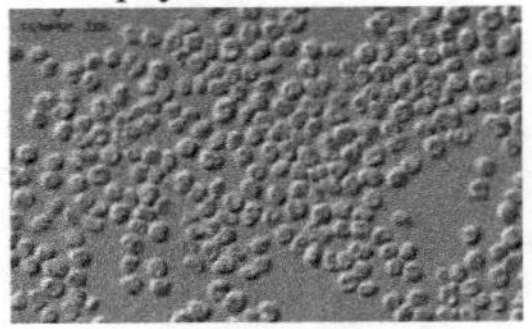 *Nannochloropsis oculata*	>15 57.25	Cubical 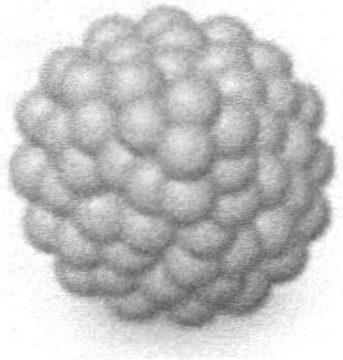 Spherical  Hexagonal	Gnanakani *et al.* (2019)			
23	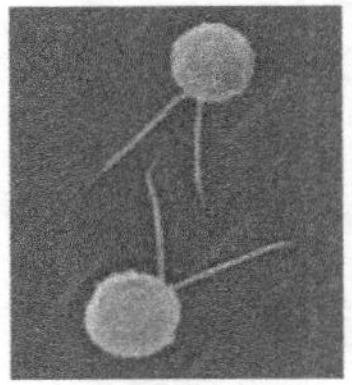 *Isochrysis galbana*	73.9	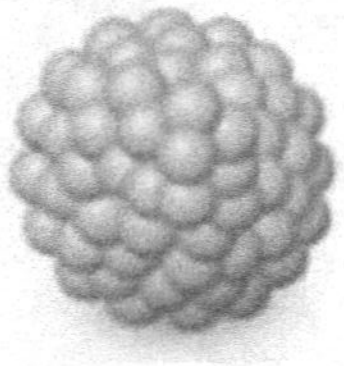 Spherical	Merin *et al.* (2010)			

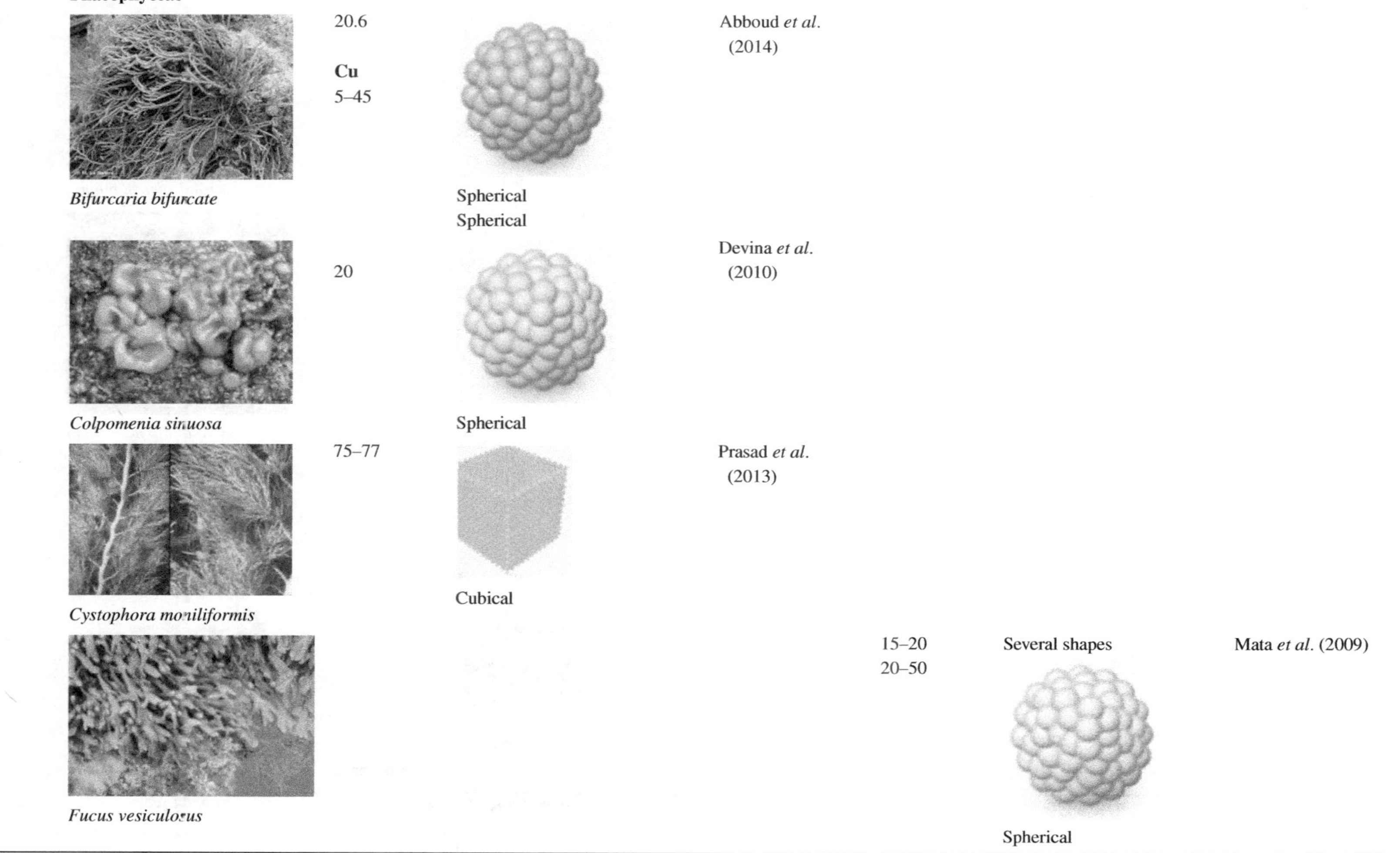

No.	Alga	Size (nm)	Shape	Reference	Size (nm)	Shape	Reference
	Phaeophyceae						
24	*Bifurcaria bifurcate*	20.6 **Cu** 5–45	Spherical Spherical	Abboud *et al.* (2014)			
25	*Colpomenia sinuosa*	20	Spherical	Devina *et al.* (2010)			
26	*Cystophora moniliformis*	75–77	Cubical	Prasad *et al.* (2013)			
27	*Fucus vesiculosus*				15–20 20–50	Several shapes Spherical	Mata *et al.* (2009)

(Continued)

TABLE 16.1 *(Continued)*

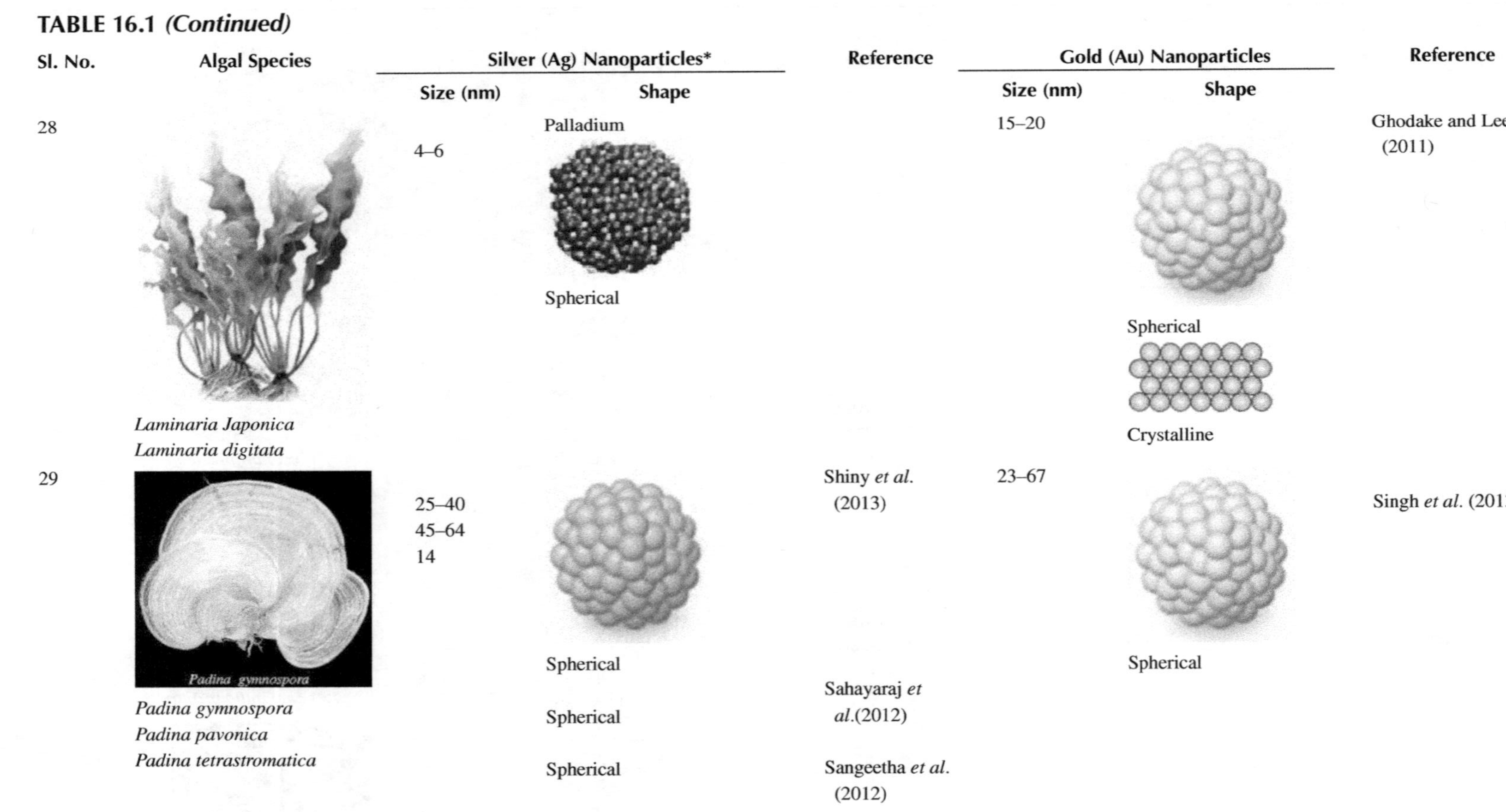

Sl. No.	Algal Species	Silver (Ag) Nanoparticles*		Reference	Gold (Au) Nanoparticles		Reference
		Size (nm)	Shape		Size (nm)	Shape	
28	*Laminaria Japonica* *Laminaria digitata*	4–6	Palladium Spherical		15–20	Spherical Crystalline	Ghodake and Lee (2011)
29	*Padina gymnospora* *Padina pavonica* *Padina tetrastromatica*	25–40 45–64 14	Spherical Spherical Spherical	Shiny *et al.* (2013) Sahayaraj *et al.*(2012) Sangeetha *et al.* (2012)	23–67	Spherical	Singh *et al.* (2013)

30	 *Sargassum cinereum* *Sargassum muticum* *Sargassum plagiophyllum* *Sargassum vulgare* *Sargassum wightii*	45–76 5–15 and 30–57 18±4 20–50 7 18–42 ~10	ZnO Fe_3O_4  Spherical AgCl	Mohandass *et al.* (2013) Azizi *et al.* (2013) Azizi *et al.* (2014) Mahdavi *et al.* (2013) Dhanalakhmi *et al.* (2012) Dhas *et al.* (2014) Govindaraju *et al.* (2015)	8–12	 Nanoplates	Singaravelu *et al.* (2007)
31	 *Stoechospermum marginatum*				62.5	 Spherical	Rajathi *et al.*(2012)
32	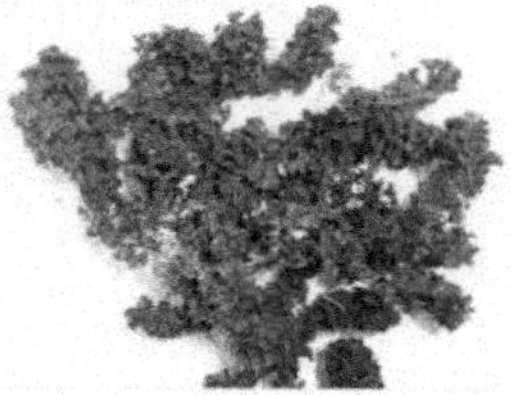 *Turbinaria conoides*	2–19	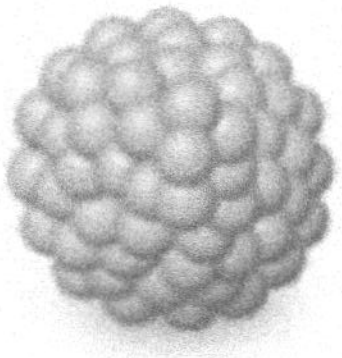 Spherical	Rajeshkumar *et al.* (2013)	60	 Triangular  Rectangular	Rajeshkumar *et al.* (2013)
					2–17	Square	Vijayan *et al.* (2014)

(Continued)

TABLE 16.1 *(Continued)*

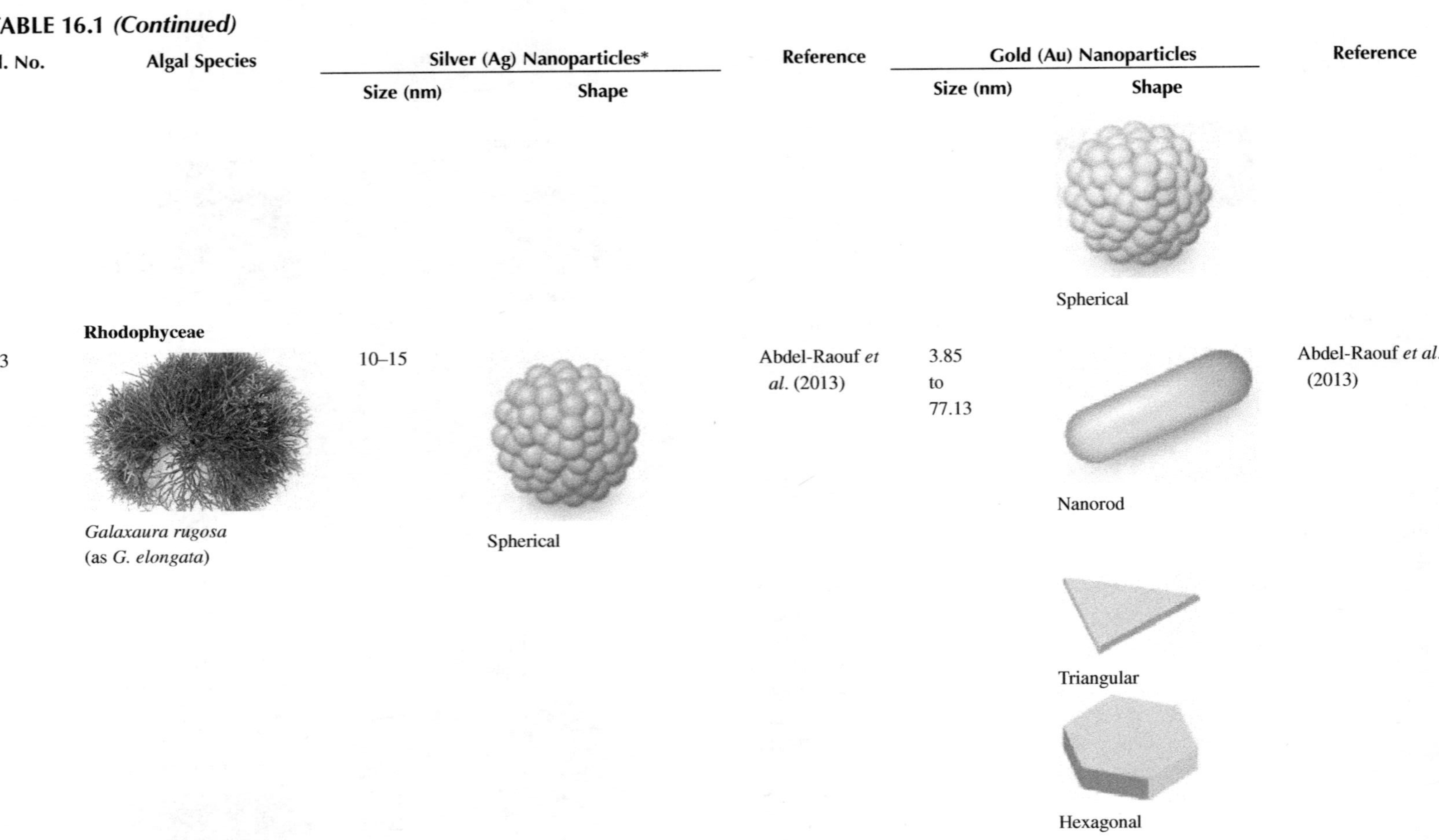

Sl. No.	Algal Species	Silver (Ag) Nanoparticles*		Reference	Gold (Au) Nanoparticles		Reference
		Size (nm)	Shape		Size (nm)	Shape	
						Spherical	
	Rhodophyceae						
33	*Galaxaura rugosa* (as *G. elongata*)	10–15	Spherical	Abdel-Raouf *et al.* (2013)	3.85 to 77.13	Nanorod Triangular Hexagonal	Abdel-Raouf *et al.* (2013)

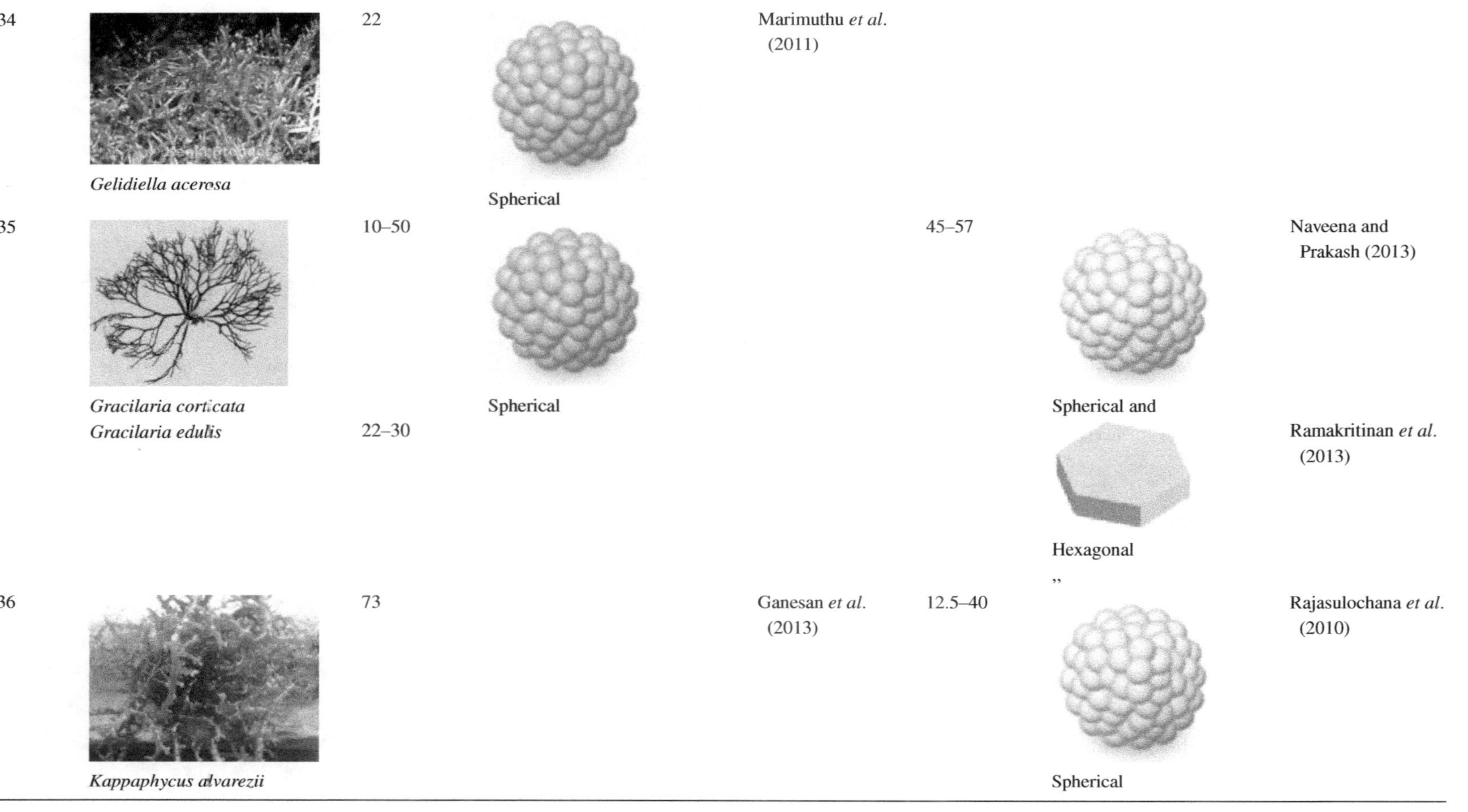

34	*Gelidiella acerosa*	22	Spherical	Marimuthu *et al.* (2011)			
35	*Gracilaria corticata*	10–50	Spherical		45–57	Spherical and	Naveena and Prakash (2013)
	Gracilaria edulis	22–30				Hexagonal	Ramakritinan *et al.* (2013)
						”	
36	*Kappaphycus alvarezii*	73		Ganesan *et al.* (2013)	12.5–40	Spherical	Rajasulochana *et al.* (2010)

(Continued)

TABLE 16.1 *(Continued)*

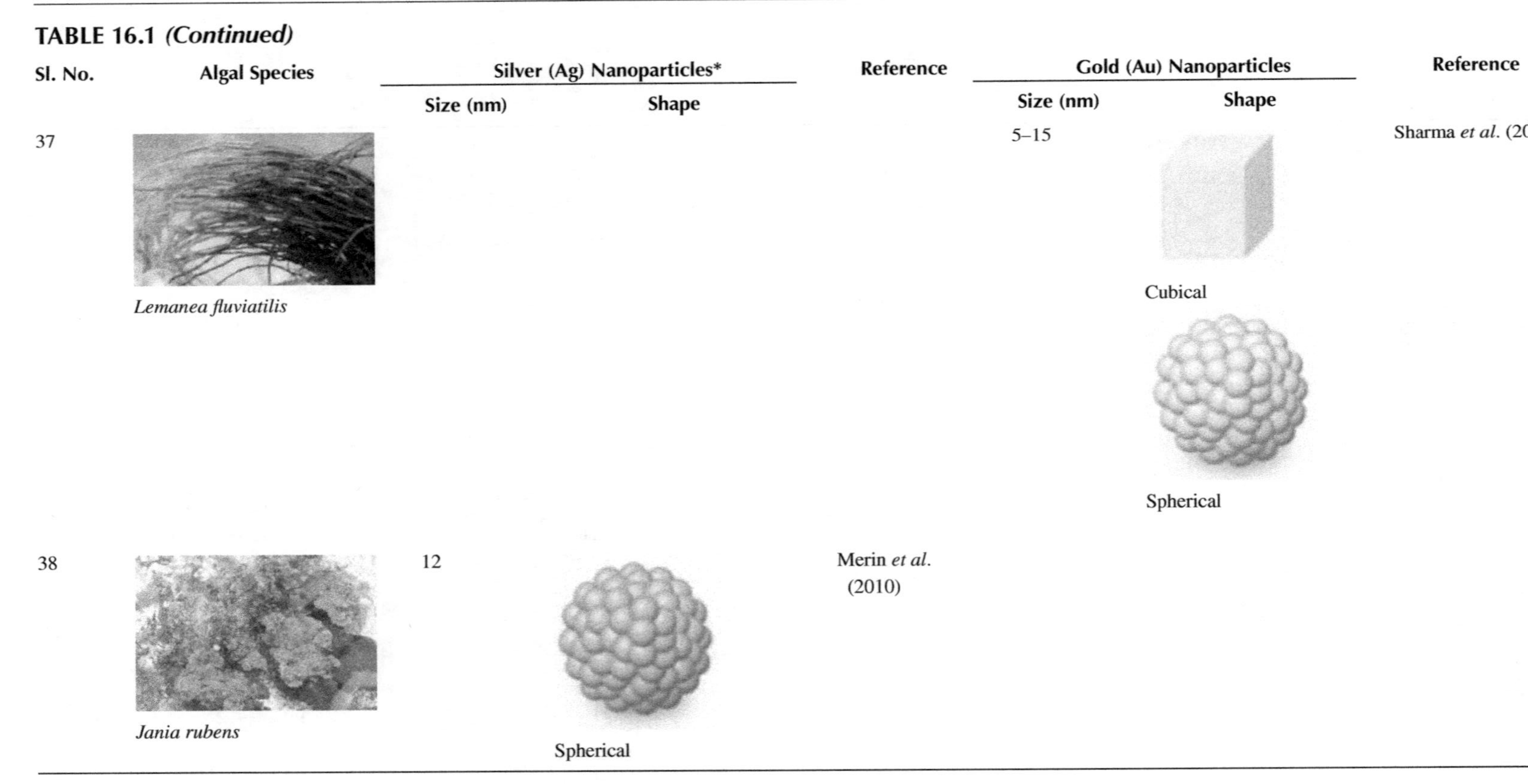

Sl. No.	Algal Species	Silver (Ag) Nanoparticles*		Reference	Gold (Au) Nanoparticles		Reference
		Size (nm)	Shape		Size (nm)	Shape	
37	*Lemanea fluviatilis*				5–15	Cubical Spherical	Sharma *et al.* (2014)
38	*Jania rubens*	12	Spherical	Merin *et al.* (2010)			

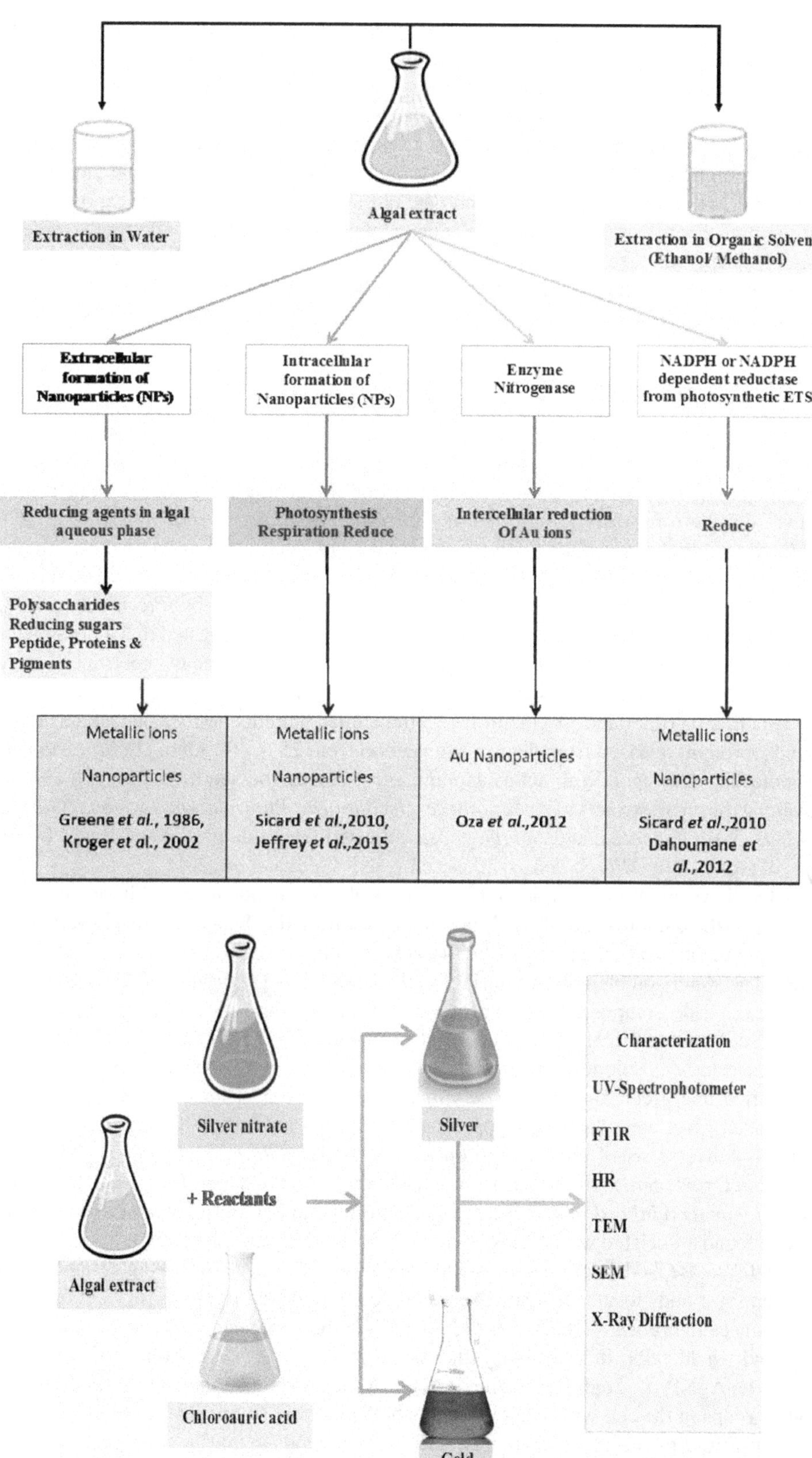

FIGURE 16.1 (a) Biosynthesis (b) characterization of algal nanoparticles.

from yellow solution to brown. Algal extract that synthesize gold NPs changes from yellow to ruby pink when treated with 1 mM aqueous $HAuCl_4$ solution.

Initially the reduction of metallic ions to silver nanoparticles or gold nanoparticles is monitored visually. Further characterization of the nanoparticles to study their size, shape, and other characterization are analyzed by UV-Spectrometer, high resolution transmission electron microscope (HRTEM), scanning electron microscope (SEM), and X-ray diffraction. Physical parameters such as temperature, pH, initial concentration, and type of the metals, duration of the exposure, type and concentration of the reducing agents in the aqueous phase can be altered selectively to determine the desirable shape and size of the metal nanoparticles as well as to prevent the aggregation and agglomeration of the nanometals (Oza *et al* ., 2012; Dahoumane, Yépremian *et al.*, 2014; Parial and Pal, 2015).

16.5 BIOSYNTHESIS OF SILVER (AG) NANOPARTICLES

Biosynthesis of AgNPs can be accomplished by physical, chemical, and green synthesis; however, synthesis via biological precursors has shown remarkable outcomes (Ahamed *et al.*, 2019). A variety of algae that include cyanobacteria are used for synthesis of AgNPs viz., *Plectonema boryanum, Oscillatoria willei, Arthrospira (Spirulina) platensis, Aphanothece, Oscillatoria, Phormidium, Lyngbya, Gloeocapsa, Microcoleus, Synechococcus*, and *Aphanocapsa* (Singh *et al.*, 2014; Chakraborty *et al* ., 2012; Mahdieh *et al* ., 2012; Govindarajulu *et al.*, 2008). Intracellular and extracellular biosynthesis of spherical and octahedral AgNPs by *Plectonema boryanum* was >10 nm and up to 200 nm respectively. Lengket *et al.* (2006a) suggested that a possible mechanism for the intracellular bioreduction of $AgNO_3$ at 25°C could be the cyanobacterial metabolic processes, utilizing nitrate (NO_3) by reducing nitrate (NO_3-) to nitrite (NO_2-) and ammonium (NH_4+), which is incorporated in glutamine. Extracellular bioreduction was attributed to organic compounds (protein) released from dead cyanobacteria from 25 to 100°C. In *Oscillatoria willei* the protein molecules are reported to act as capping agent during the synthesis of AgNPs (Ali *et al.*, 2011). Among the algal extracts of *Aphanothece, Oscillatoria, Phormidium, Lyngbya, Gloeocapsa, Microcoleus, Synechococcus*, and *Aphanocapsa* collected from mangroves and tested for the biosynthesis of AgNp, only *Microcoleus* synthesised AgNP (Sudha *et al.*, 2013).

Microalgae have an important role in the synthesis of silver nanoparticles (Sharma *et al.*, 2009b; Merin *et al.*, 2010; Barwal *et al.*, 2011; Mohseniazar *et al.*, 2011; Jena *et al.*, 2013; Satpathy *et al.*, 2015). Among the micro green algae *Chlorella vulgaris* plays a significant role in the biosynthesis of AgNPs. The water soluble polysaccharide of algal extract of *Chlorella vulgaris* showed a complex monosaccharide composition (fructose, maltose, lactose, and glucose), sulfate, uronic acids, total protein content, and total carbohydrate. Silver nanoparticles (AgNPs) were synthesized using soluble polysaccharides solution of *Chlorella vulgaris* (Noura *et al.*, 2020). Sharma, Yngard *et al.* (2009) reported that green alga, *Chlorella vulgaris*, was a catalyst for the reduction of Ag+ ions and synthesized silver nanoplates with appreciable yield. Merin *et al.* (2010) reported the synthesis of AgNPs in a diverse set of marine chlorophyceae members, *Tetraselmis gracilis* and *Chlorella salina*, diatom *Chaetoceros calcitrans*, and the haptophyte *Isochrysis galbana*. AgNPs synthesized by *Chlamydomonas reinhardtii* were located in the peripheral cytoplasm and basal body of the flagella. It was found associated with ATP synthase, RUBP carboxylase, ferredoxin NADP+ reductase, superoxide dismutase, sedoheptulose-1, 7-bisphosphatase, and oxygen-evolving enhancer proteins. The change in size and biosynthesis rate of NPs in protein-depleted fractions further confirmed that protein enhances the process (Barwal *et al.*, 2011). The fresh extracts of *Chlorococcum humicola* (in vitro) and whole cells (in vivo) biosynthesized AgNPs (Jena *et al.*, 2013), and the binding of proteins to the AgNPs through free amine groups, cysteine residue, and electrostatic attraction of carboxylic groups in the cell wall was reported, which probably stabilizes the AgNPs.

Synthesis of AgNPs was carried out in two different methods in marine microalgae *Chaetoceros calcitrans, Chlorella salina, Isochryis galbana*, and *Tetraselmis gracilis*. In the first method, silver nitrate solution was added to the algal culture in the shaker, and in the second method, algal culture in its exponential phase was subjected to microwave irradiation for 5 seconds and 15 seconds every five minutes and then treated with $AgNo_3$ solution for the synthesis of AgNPs (Merin *et al.*, 2010). Silver nanoparticles syntheses were observed in normal and microwave irradiated microalgae and screened against human pathogens for the presence of antimicrobials. Thus, special focus is placed on AgNP synthesis and microwave-assisted synthesis methods. The chapter on biosynthesis of nanomaterials using algae by Rahman *et al.* (2020) describes and reviews green chemistry-based approaches on the biosynthesis of nanomaterials using algae. Detailed literature review was utilized to evaluate the present state-of-the-art methods in meeting all the requirements for such synthesis. Studies demonstrated synthesis of silver nanoparticles using green algae (*Botryococcus braunii*), which in turn is used for synthesis of biologically important benzimidazoles. In *Botryococcus braunii* biogenic synthesis of copper nanoparticles was also observed along with AgNPs (Anju *et al.*, 2018). Green synthesis of silver nanoparticles using freshwater green alga *Pithophora oedogonia* was considerably rapid, and silver nanoparticles were generated within few minutes of silver ions coming in contact with the algal extract. It is believed that phytochemicals present in the extract of *P. oedogonia* reduced the silver ions into metallic nanoparticles (Sinha *et al.*, 2015).

Production of silver nanoparticles by the diatom *Phaeodactylum tricornutum* showed different sizes and chemical composition associated with the diatom frustules and extracellular polymeric substances. While the silver ions inhibit the *P. tricornutum* growth, the cells were able to generate AgNPs with a diameter below 200 nm. The "green synthesis" method conducted at ambient temperature and pressure conditions can be considered as an environmentally friendly approach and a low-cost technique for the production of AgNPs (Wishkerman *et al.*, 2017). In a recent report, the aqueous extract of a diatom *Amphora*-46 was used for the light-induced biosynthesis of polycrystalline AgNPs, in which fucoxanthin, a photosynthetic pigment, was responsible for the reduction of Ag ion (Jena *et al.*, 2015).

Synthesis of silver nanoparticles (AgNPs) using water-soluble polysaccharides extracted from four marine macroalgae, namely, *Pterocladia capillacae*, *Jania rubins*, *Ulva faciata*, and *Colpmenia sinusa*, are considered as reducing agents for silver ions as well as stabilizing agents for the synthesized AgNPs (El-Rafiea *et al.*, 2013). Green macroalgae that synthesised AgNPs are *Caulerpa racemosa* (Karthikeyan *et al.*, 2015), *Codium capitatum, Ulva (Enteromorpha) compressa, Ulva fasciata, Ulva reticulate, Padina gymnospora, Padina tetrastromatica, Padina pavonica, Sargassum cinereum, Sargassum muticum, Sargassum plagiophyllum, Sargassum longifolium* (Shanmugam *et al.*, 2014), *Turbinaria conoides* (Shanmugam *et al* ., 2012) are some of the Phaeophyceae members from which AgNPs are synthesized. Red macroalgae *Gelidiella acerosa, Gracilaria dura, Gracilaria corticata, Kappaphycus alvarezii* also synthesized AgNPs.

Synthesis of (AgNPs) by the reduction of aqueous solutions of silver nitrate (AgNO3) with dry powder, fresh extract, and chloroform extracts of *Padina pavonia* showed a ratio of converted AgNPs as 88.5, 86.2, and 90.5% respectively (Abdel-Raouf *et al.*, 2019). Kathiravan *et al.* (2015) studied green synthesis of silver nanoparticles using marine algae *Caulerpa racemosa* from Gulf of Mannar and suggested that peptides may play an important role in the reduction of AgNO3 into Ag nanoparticles. They have further concluded that the extract of marine seaweed *C. racemosa* is capable of producing Ag nanoparticles extracellularly, and these nanoparticles are quite stable in solution due to capping likely by the proteins present in the extract.

Brown seaweed mediates the synthesis of silver nanomaterials using extract of *Sargassum longifolium*. Some kinetic studies such as time incubation and pH were conducted for improved production of silver nanomaterials. The pH and reaction time range were changed, and the absorbance was taken for the characterization of the nanoparticles at various time intervals. High pH level showed increased absorbance due to increased nanoparticles synthesis (Rajeshkumar *et al.*, 2014). Biosynthesis of silver nanoparticles (AgNPs) using the aqueous extract of brown seaweed *Padina tetrastromatica* leaf extract showed complete reduction of silver ions after 72 h of reaction at 300°C under shaking conditions (Jegadeeswaran *et al* ., 2012). *Sargassum wightii* and *Fucus vesiculosus* have been used for synthesizing AgNPs of different sizes and shapes. Extracellular AgNPs synthesis is reported in *Sargassum wightii* (Govindaraju *et al.*, 2009) and also from *Sargassum plagiophyllum, Ulva reticulata*, and *Enteromorpha compressa* (Dhanalakshmi *et al* ., 2012). Aqueous extract of the red algae *Gelidiella acerosa* biosynthesized an average of 22 nm, spherical-shaped AgNPs capped with aromatic compounds or alkanes or amines (Marimuthu *et al.*, 2011). Agar extracted from *Gracilaria dura* reduce $AgNO_3$ solution thus synthesizing spherical AgNPs (Shukla *et al* ., 2012).

16.6 BIOSYNTHESIS OF GOLD NANOPARTICES

Biosynthesis of gold nanoparticles is obtained by the chemical reduction of tetrachloroauric acid by sodium citrate or NaBH4. The biogenic synthesis of AuNPs using algae is a simple, low-cost, environmental-friendly, nontoxic, reliable, and safe approach that can be used for a range of applications (Khan and Cho, 2018). Cyanophyceae members *Plectonema boryanum, Spirulina platensis, Calothrix* sp., *Phormidium valderianum, Microcoleus chthonoplastes* synthesize AuNPs. The cyanobacteria *Phormidium valderianum, P. tenue*, and *Microcoleus chthonoplastes* and the green algae *Rhizoclonium fontinale, Ulva intestinalis, Chara zeylanica*, and *Pithophora oedogoniana* were exposed to hydrogen tetrachloroaurate solution and were screened for their suitability for producing nano-gold and found that *Chara zeylanica* and *Pithophora oedogoniana* could not synthesize gold nanoparticles. Hence, choice of algae is important in the study of nanoparticle synthesis (Parial *et al.*, 2012b). However, *Pithophora oedogoniana* synthesize AgNPS very efficiently. Gold nanorods were synthesized (together with gold nanospheres) by *Phormidium valderianum* at acidic pH (pH 5), and at pH 7 and pH 9 they produced spherical, triangular, and hexagonal AuNPs. Govindaraju *et al.* (2008) reported biosynthesis of AuNPs and AgNPs as well as bimetallic nanoalloy (Au core-Ag shell NPs) utilizing the cyanobacterium *Spirulina platensis*. When the cyanobacterial biomass was added to 10^{-3} M solution of $HAuCl^4$ and AgNO3, alone or in combination for 120 h, the color of the reaction mixture changed to brown in the case of AgNO3 and ruby red in the case of $HAuCl^4$, whereas in case of bimetallic NPs, the color of the reaction mixture changed from purple to brown. Parial *et al.* (2012a) reported biosynthesis of Au nanorods ranging from 137 to 209 nm in length and 33 to 69 nm in diameter using *Nostoc ellipsosporum*, after incubating the growing filaments with Au^{3+} (15 mg L−1) solution at pH 4.5 and 20°C for 48 h. Cyanobacterial species *Lyngbya majuscule* and *Spirulina subsalsa* and a green alga *Rhizoclonium hieroglyphicum* were exposed to Au solution, and different percentages of gold uptake were observed with maximum uptake in *R. hieroglyphicum* (40–80%) followed by *L. majuscule* (35–60%) and *S. subsalsa* (20–35%) (Chakraborty *et al.*, 2009). Focsan *et al.* (2011) observed synthesis of intracellular AuNPs in *Synechocystis* sp., which formed brown spots in the cytoplasm, thylakoids, cell wall, and plasma membrane.

Biosynthesis of gold nanoparticles by microalgae, mainly green algae, have been employed (Xie *et al.*, 2007; Shakibaie *et al.*, 2010; Sicard *et al.*, 2010; Dahoumane *et al* ., 2012; Oza *et al* ., 2012; Senapati *et al* ., 2012). The green microalgae that synthesized AuNPs are *Chlorella pyrenoidusa, C. vulgaris, Chlamydomonas reinhardtii, Klebsormidium flaccidum, Cosmarium impressulum, Kirchneriella lunaris.* The most dominating parameters for the synthesis of AuNPs were pH 8, 100°C, and 100 ppm aurochlorate salt (Oza *et al* ., 2012). *C. vulgaris* has strong binding ability towards tetrachloroaurate ions to form algal-bound gold reducing into Au(O). Approximately 88% of algal-bound gold attained metallic state, and the crystals of gold were accumulated in the inner and outer parts of cell surfaces with tetrahedral, decahedral, and icosahedral structures (Jianping *et al.*, 2007). Dahoumane, Yéprémian *et al.* (2014) reported the intracellular biosynthesis of spherical AuNPs in *Cosmarium impressulum* and *Kirchneriella lunaris*, where the intracellular polysaccharides controlled the shape, size, and macromolecule stabilization of AuNPs in the thylakoid membrane. Diatoms (*Navicula atomus* and *Diadesmus gallica*) have the ability to synthesize gold nanoparticles, gold, and silica-gold ionanocomposites (Mubarak Ali *et al.*, 2013). *Navicula atomus* and *Diadesmus gallica* synthesized different size of AuNPs, and *N. atomus* produced homogenous

AuNPs (Schrofel *et al*., 2011). Senapati *et al* . (2012) have reported the intracellular synthesis of gold nanoparticles (5–35 nm) using *Tetraselmis kochinensis*. Nanoparticles produced are more concentrated on the cell wall than on the cytoplasmic membrane, possibly due to the reduction of metal ions by the enzymes present in the cell wall and the cytoplasmic membrane.

Macroalgae have been investigated widely for the synthesis of AuNPs, including green algae (Parial *et al*., 2012a; Parial and Pal, 2015; Sharma *et al*., 2015), brown algae (Singaravelu *et al*., 2007; Mata *et al*., 2009; Rajasulochana *et al*., 2010; Ghodake and Lee, 2011; Rajathi *et al* ., 2012; Ganesan *et al*., 2013; Singh *et al*., 2013), and red algae (Naveena and Prakash, 2013; Ramakritinan *et al*., 2013). The green algae *Rhizoclonium fontinale* and *Ulva intestinalis* have been reported to produce AuNPs intracellularly, and green algae *Chara zeylanica* and *Pithophora oedogonium* were unable to produce AuNPs (Parial *et al*., 2012a). Parial and Pal (2015) reported the synthesis of monodispersed AuNPs using *Rhizoclonium fontinale* with maximum yield by incubating the algae in 15 mg L−1HAuCl.XH2O (pH 9) for 72 h. Another freshwater green alga, *Prasiola crispa*, has also been reported to produce nearly spherical AuNPs of size ranging from 5 to 25 nm within 12 h at room temperature (Sharma *et al*., 2015).

In *Padina gymnospora* extracellular biosynthesis of gold nanoparticles has been attempted, and rapid formation of gold nanoparticles was achieved in a short time frame, and FTIR spectra confirmed that hydroxyl groups present in the algal polysaccharides were involved in the gold bioreduction (Singh *et al*., 2013). Extracellular biosynthesis of gold nanoparticles in a marine alga, *Sargassum wightii*, was able to form high-density and extremely stable gold nanoparticles (8–12 nm) in a short timespan (Singaravelu *et al*., 2007). Dead biomass of *Fucus vesiculosus* was found to reduce Au(III)(gold hydroxide) to Au(0) (gold oxide) with varying sizes and shapes under the influence of different pH ranges from 2.0 to 11 and suggested that algal pigment fucoxanthin, which is rich in hydroxyl groups, to be responsible for this bioreduction process (Mata *et al*., 2009). The brown alga *Laminaria japonica* has also been reported to biosynthesize AuNPs (Ghodake and Lee, 2011). The reaction mixture was quickly converted to red when different concentrations of algal extracts were incubated with HAuCl4 (2 mM) solution at 37°C, which is suggested to be due to polysaccharides in the cell wall. The Au+ (90–95%) was converted to AuNPs with size of 15–20 nm within 10–20 min intracellularly. Gold nanoparticles are synthesized by the brown algae *Stoechospermum marginatum*, *Turbinaria conoides*, *Ecklonia cava* (Rajathi *et al* ., 2012; Rajeshkumar *et al*., 2013; Venkatesan *et al*., 2014).

Extracellular biosynthesis of AuNPs using the red marine alga *Kappaphycus alvarezii* (Rajasulochana *et al*., 2010) is influenced by the sulfated polysaccharides of the algae. Rapid biosynthesis of AuNPs by the red alga *Galaxaura elongata* using either aqueous or ethanolic algal extracts is reported to be due to Alloaromadendrene oxide, hexadecanoic acid, gallic acid, 11-eicosenoic acid, andrographolide, oleic acid, catechin, epigallocatechin, and epicatechin acting as reducing, capping, and stabilizing agent (Abdel-Raouf *et al*., 2013). Ramakritinan *et al*. (2013) employed *Gracilaria* sp. to form nanoparticles of Ag, Au, and even bimetallic Ag-Au nanoalloys. Naveena and Prakash (2013), Suganthi and Rani (2014) reported biosynthesis of AuNPs in *Gracilaria corticata*. In *Lemanea fluviatilis*, algal proteins assisted in the synthesis of AuNPs by acting as reducing and stabilizing agents (Sharma *et al*., 2014).

16.7 APPLICATION OF SILVER NANOPARTICLES

AgNPs have been reported to be associated with many potential applications such as antifungal (Marimuthu *et al*., 2011), antimicrobial (Sharma, Yngard *et al*., 2009), wound healing (Silver *et al*., 2006; Tian *et al*., 2007), anticancer (Boca *et al*., 2011; Govindaraju *et al*., 2015) therapies, as well as catalytic activity (Mohanpuria *et al*., 2008) and having good conductivity (Kang and Guo, 2007; Liu and Yu, 2011). These properties have led to an increase in their usefulness in the field of drug delivery, enzyme-released systems, biosensor designing, and ultrasound medical imaging (Ngeontae *et al*., 2009; Vaidyanathan *et al*., 2009). AgNPs are also used in textile industry for creating cotton fabric which has antibacterial properties (Zhang *et al.*, 2021). They are also used as biosensors for detecting glucose (Ngeontae *et al.*, 2009). Due to their sensitivity towards surface absorption of the metal, these NPs possess applications in sensing and imaging (Lee and El-Sayed, 2006). AgNPs biosynthesized by blue-green algae showed significant activity against several bacterial strains, including several human pathogens such as *Escherichia coli, Bacillus subtilis, Corynebacterium, Proteus vulgaris, Salmonella typhi, Staphylococcus aureus*, and *Vibrio cholera* (Oscar *et al.*, 2016). Green marine microalgae *Tetraselmis gracilis* and *Chlorella salina*, the diatom *Chaetoceros calcitrans*, and the haptophyte *Isochrysis galbana* synthesized AgNPs, which were found to be potent antimicrobial agents with antibacterial activity against human pathogens such as *E. coli, Klebsiella* sp., *P. vulgaris*, and *Pseudomonas aeruginosa* (Merin *et al.*, 2010). Another possible biomedical application is suggested for silver NPs synthesized from the aqueous extract of the diatom Amphora-46.

Monica Terracciano *et al.* (2018) reviewed recent progress made on diatom biosilica-based system applications for drug delivery applications. Among all the available nanomaterials for drug delivery applications, porous silica NPs have been investigated in numerous studies due to unique properties of diatoms. The authors emphasize that nature has provided us with the three-dimensional (3-D) porous structures, the single-celled photosynthetic diatom. Feng *et al.* (2019) reviewed the effects of Diatom nanoparticles on algae in terms of adsorption, distribution, and ecotoxicity. According to him, ecotoxicity of nanoparticles (NPs) and their potential hazards to the environment has recently been a subject of great concern due to rapid development of nanotechnology and widespread use of nanoproducts. When NPs pass the cell membrane of algae by endocytosis or passive diffusion, the carrier of NPs in the cell membrane may be involved. Reported studies in the review tend to explain the damage of NPs to organelles. The concentration of NPs in the water environment is low, and NPs may accumulate in organisms of higher trophic level and produce significant toxic effects by the stepwise delivery or enrichment of the food chain.

Green synthesis of silver nanoparticles using marine algae *Caulerpa racemosa* showed best antibacterial activity against human pathogens such as *Staphylococcus aureus* and *Proteus mirabilis* (Kathiravan *et al.*, 2015). Biosynthesized silver nanoparticles from *Turbinaria conoides* were found to be highly toxic against gram-positive bacteria *Bacillus subtilis* (MTCC3053) and gram-negative bacteria *Klebsiella planticola* (MTCC2277) (Rajeshkumar *et al.*, 2013). Silver nanoparticles synthesized from *Sargassum longifolium* exhibited antifungal activity against the clinical pathogenic fungi *Aspergillus fumigatus, Candida albicans*, and *Fusarium* sp. This green process gives the greater potential biomedical applications of silver nanoparticles. The synthesized silver nanoparticles from *Pithophora oedogonia* exhibited strong potential inhibitory antibacterial activity against both gram-negative and gram-positive pathogenic bacteria. Extract of green alga *Botryococcus braunii* synthesized copper and silver nanoparticles, which was highly toxic against two gram-negative bacterial strains, *Pseudomonas aeruginosa* (MTCC 441) and *Escherichia coli* (MTCC 442); two gram-positive bacterial strains, *Klebsiella pneumoniae* (MTCC 109) and *Staphylococcus aureus* (MTCC 96); and a fungal strain, *Fusarium oxysporum* (MTCC 2087). Thus, Anju *et al.* (2018) highlight that currently researchers are into synthesizing biologically active nanoparticles due to spread of infectious diseases and increased drug resistance among microbes. Several studies propose that AgNPs may attach to the surface of the cell membrane, disturbing permeability and respiration functions of the cell (Morones *et al.*, 2005; Kvitek *et al.*, 2008). It is also possible that AgNPs not only interact with the surface of membrane but can also penetrate inside the bacteria (Sondi and Sondi, 2007). Prashant Agarwal *et al.* (2019) reviewed literature on recent advances and have provided an overview of wastewater treatment processes by applying an amalgamation of nanoparticles and microalgae synthesis. The authors recognize that microalgae culture is gaining tremendous attention, providing combined benefit of treating wastewater as a growth medium and algal biomass production, which can be used for several livestock purposes. The ubiquitous microalgae being extremely diverse are capable of accumulating toxic contaminants and heavy metals from wastewater, making them superior contender to become a powerful nanofactory.

16.8 APPLICATION OF GOLD NANOPARTICLES

The AuNPs are used for targeting the nucleus, which in turn helps in therapeutic delivery and diagnostics at intracellular levels (Tkachenko *et al.*, 2003). In medical application, they have been used for targeting in computed tomography and in cancer therapy (Kim *et al.*, 2010). Gold NP synthesis by *Nitzschia* sp. (diatoms) were successfully coupled with the antibiotics penicillin and streptomycin and showed an increased antibacterial activity (Oscar *et al.*, 2016). The highest activity was observed against *E. coli* and *Enterobacter aerogenes*, moderate activity against *S. aureus*, with the lowest activity against *Enterococcus faecalis*, indicating effectiveness of the coupled complex against gram-negative bacteria (Naveena and Prakash, 2013). AuNPs was synthesized using Porphyran as a reducing agent. Porphyran is a sulfated carbohydrate derived from the red algae *Porphyra* (red seaweed). The synthesized AuNPs are used as a carrier for the delivery of an anticancer drug. Porphyran-capped AuNPs enhanced the cytotoxicity on the human glioma cell line (LN-229) compared to native porphyran. The AuNPs were also used as a carrier for the delivery of the anticancer drug doxorubicin hydrochloride (DOX). The DOX-loaded AuNPs demonstrated higher cytotoxicity to the LN-229 cell line compared to that of an equal dose of a native DOX solution. The biosynthesized AuNPs have potential use as a carrier for anticancer drug delivery (Venkatpurvar, 2011). The chemical constituents of the algal extract of *Galaxaura elongate* (red algae) are andrographolide, alloaromadendrene oxide, glutamic acid, hexadecanoic acid, oleic acid, 11-eicosenoic acid, stearic acid, gallic acid, epigallocatechin catechin, and epicatechin gallate, which act as a reducing, stabilizing, and capping agent in the synthesis of AuNPs. These AUNPs showed antibacterial activity against *Escherichia coli, Klebsiella pneumonia, Staphylococcus aureus*, and *Pseudomonas aeruginosa* (Shamaila *et al.*, 2016). Green synthesis of AuNPs from aqueous extract of the seaweed *Turbinaria conoides* was examined for its antibiofilm activity against marine biofilm-forming bacteria (Vijayan *et al.*, 2014). Biological screening of AuNPs synthesized by *Chlorella vulgaris* against the human pathogen *Candida albicans* and *Staphylococcus aureus* revealed them to be susceptible to synthesized aqueous AuNPs. Therefore, AuNPs by *Chlorella vulgaris* extract can be used as an effective drug (Annamalai and Nallamuthu, 2015). AuNPs of *Tetraselmis kochinensis* can be utilized in drug delivery, biomedical application, and catalysis (Senapathi *et al* ., 2012). *Padina gymnospora* synthesized AuNPs, which is used in cosmetics, food, and consumer goods (Singh *et al.*, 2013). AuNPs synthesized by *Cystosiera baccata*, which is referred as Au@CB, was examined against apoptotic activity and found that Au@CB has a significant potential for the treatment of rectal cancer (Gonzalez *et al.*, 2017). *Ecklonia cava* synthesized AuNPs, which showed good antimicrobial properties and biocompatibility with the human keratinocyte cell line and hence can be applied in biomedical, drug delivery, tissue engineering, and biosensors (Venkatesan *et al.*, 2014). AuNPs synthesized by *Padina terastromatica* were assessed for their in vitro cytotoxic activity on human liver cancer (Hep G2) and lung cancer (A549) cell lines at different concentrations by a comparison with that of the standard drug cyclophosphamide (Rajeshkumar *et al.*, 2017). The analysis of application of algal nanoparticles are as shown in Figure 16.2.

16.9 NANOTECHNOLOGY INITIATIVES IN INDIA

Nanotechnology is already addressing key economic sectors and can provide solutions to some of the world's most critical development problems. In 2010, Department of Science and Technology appointed a task force which has been asked to advice Nano Mission Council to develop a regulatory body for nanotechnology in India. National nanotechnology centers have been created that allows access to industry for R&D and also provide academia-industry linkage. The Nano-Manufacturing, Industry Liaison, and Innovation (NILI) Working Group are a key institution for promoting and facilitating nanotechnology innovation and to improve technology transfer to industry. It also promotes interagency cooperation in the areas of standards, nomenclature, nano-manufacturing research, and use of programs that encourage innovation in small business. Apart from this, federal

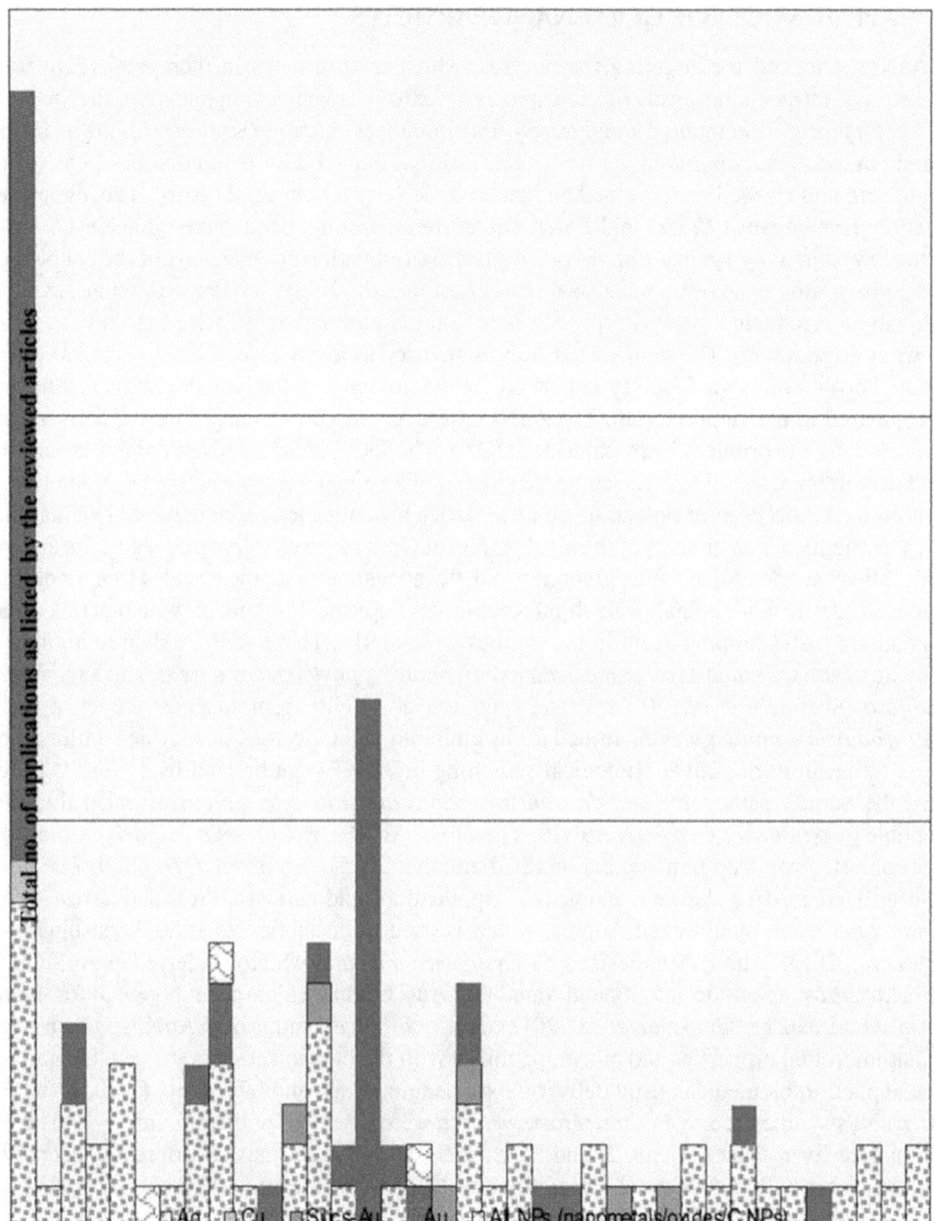

FIGURE 16.2 Analysis of application of algal nanoparticles.

governments created specialized centers for promoting technology transfer, for example: The Robert C. Byrd National Technology Transfer Center (NTTC), The Federal Laboratory Consortium for Technology Transfer, The Agricultural Research Service are actively involved in nanotechnology transfer. Industry has taken its own initiative for commercialization of nanotechnology by creating Nano Business Commercialization Association.

The Indian nanotechnology initiative is a multiagency effort and has strong similarity with US multiagency model. The key agencies that have undertaken major initiatives for capacity creation are the Department of Science and Technology (DST) and Department of Information Technology (DIT). Other agencies showing major involvement are the Department of Biotechnology (DBT), Council of Scientific and Industrial Research (CSIR), Ministry of

New and Renewable Energy (MNRE), Ministry of Health and Family Welfare (MoHWF), Indian Council of Agricultural Research (ICAR), Indian Space Research Organisation (ISRO), Department of Atomic Energy (DAE), and Defence Research and Development Organisation (DRDO). Nanotechnology as a distinct area of government research started with NSTI (Nano Science and Technology Initiative) in the Xth plan period (2002–2007) with an allocation of rupees 60 crores (approx. USD 12 million). NSTI was initiated and implemented by DST. NSTI helped in establishing units for developing research excellence in nanoscience, centers for nanotechnology each aimed at application development and two national instrumentation/characterization facilities. In all, 14 national institutions, including seven IITs and ten universities have been supported under the NSTI. The other major program that complemented the nanotechnology initiatives is the National Program for Smart Materials (NPSM) launched in 2002. Public-funded research organizations have been the major stakeholder in developing the knowledge capacity in the country. CSIR, ICAR, DRDO, ISRO, ARCI, National Institute of Pharmaceuticals Education and Research (NIPER), Bhabha Atomic Research Centre (BARC) are playing a major role in nanotechnology research (Bhattacharya *et al* ., 2012).

16.10 ECONOMICS AND CHOICE OF ALGAL NANOTECHNOLOGY RESEARCH

It is to be noted that, while most of nanoparticles have very specific uses, quite a few of them, like Intracellular NPs, Extracellular NPs, EC Polysaccharides, etc., are meant for several other applications. However, there is a clear indication that most of the researchers tend to choose studies focusing on antibacterial, antifungal, or antimicrobial properties. This is attributed to low costs, as this does not involve a huge burden of expenses in terms of chemicals. Equipment used for characterization of NPs, such as SEM, HRTEM, FTIR, X-ray diffraction, etc., pose a barrier to researchers' studies due to heavy expense per sample. Nevertheless, such studies are possible only through dedicated funding set aside for researchers by reliable organizations or institutions or government organizations that can lead to innovative fascinating discoveries.

16.11 CONCLUSION AND WAY FORWARD

- The biogenic synthesis of metal nanoparticles can be a promising process for production of other metal and metal oxide nanoparticles which can have environmental, pharmaceutical, medical, and biotechnological applications.
- Investigation on the antibacterial effect of nanosized silver colloidal solution against human pathogens reveals high efficacy of silver nanoparticles as a strong antimicrobial agent.
- The characteristics of diatoms are well suited for nanocomposites preparation that has great importance for future applications. It is also expected that silica-gold and EPS-gold bionanocomposites have potentially a great value for various applications and should be further studied.
- *Spirulina* biomass with gold nanoparticles synthesized may be used for medical, pharmaceutical, and technological purposes.
- Nanoparticles in combination with commercially available antibiotics could be used as an antimicrobial agent after further trials on experimental animals. The cited eco-friendly synthesis procedure of NPs could be easily scaled up in future for the industrial and therapeutic needs.
- Several potential disease conditions such as chronic immune responses of inflammation, allergy, respiratory disorders, gastrointestinal-related disorders, neurological disorders, several types of cancers resulting from oxidative damage to DNA and tissue damages are emerging with the increased use of nanomaterials.
- It is recommended to have a regular health monitoring program and periodical medical surveillance of pulmonary, renal, liver, and hematopoietic functions while working with human pathogenic microbes and carcinogenic materials.

- Although studies have elucidated the significant biomedical potential of biogenic metallic nanoparticles (MNPs), it is very important to explore the hazards associated with the use of biogenic MNPs. Evidence indicates that genetic toxicity causes mutation, carcinogenesis, and cell death (Barabadi *et al.*, 2019).
- India has a stretch of about 7,500 km coastline having a rich flora of diversified seaweeds which can be further exploited by the researchers for the biosynthesis of AgNps and AuNPs and explore their applications in all fields.
- Competency in this area of research is an immense challenge as it is a knowledge intensive area requiring development of advanced R&D infrastructure, significant R&D investment, skilled manpower having inter-disciplinary competence, access/development of sophisticated instruments, entrepreneurship, and synergy among a divergent set of stakeholders.

REFERENCES

Abboud Y, Saffaj T, Chagraoui A, Bouari AE, Brouzi K, Tanane O, Ihssane B (2014) Biosynthesis, characterization and antimicrobial activity of copper oxide nanoparticles (CONPs) produced using brown alga extract (*Bifurcaria bifurcata*). Appl Nanosci 4:571–576.

Abdel-Raouf N, Al-Enazi NM, Ibraheem IBM (2013) Green biosynthesis of gold nanoparticles using *Galaxaura elongata* and characterization of their antibacterial activity. Arab J Chem. doi:10.1016/j.arabjc.2013.11.044.

Abdel-Raouf N, Nouf MAE, Ibraheem BMI, Reem MA, Manal MA (2019) Biosynthesis of silver nanoparticles by using of the marine brown alga *Padina pavonia* and their characterization. Saudi J Biol Sci 26:1207–1215.

Ahmad S, Munir S, Zeb N, Ullah A, Khan B, Ali J, Bilal M, Omer M, Alamzeb M, Salman SM, Ali S (2019) Green nanotechnology: A review on green synthesis of silver nanoparticles—an ecofriendly approach. Int J Nanomedicine 14:5087–5107. https://doi.org/10.2147/IJN.S200254.

Aishwarye S, Sruti S, Kuldeep S, Siva KC, Amit V, Pushpa S, Ravindra K, Brijesh R, Veena A (2015) Algae as crucial organisms in advancing nanotechnology: A systematic review. J Appl Phycol 28(3):1759–1774.

Ali DM, Sasikala M, Gunasekaran M, Thajuddin N (2011) Biosynthesis and characterization of silver nanoparticles using marine cyanobacterium *Oscillatoria willei* NTDM01. Dig J Nanomater Bios 6:385–390.

Anju A, Khushbu G, Tejapl SC, Dipti V (2018) Biogenic synthesis of copper and silver nanoparticles using green algae *Botryococcus braunii* and its antimicrobial activity. Bioinorg Chem Appl 2018. Article ID 7879403. https://doi.org/10.1155/2018/7879403.

Annamalai J, Nallamuthu T (2015) Characterization of biosynthesis gold nanoparticles from aqueous extract of *Chlorella vulgaris* and their antipathogenic properties. Appl Nanosci 5:603–607.

Ashiqur R, Shishir K, Tabish N (2020) Biosynthesis of nanomaterials using algae. In *Microalgae Cultivation for Biofuels Production*, pp. 265–279.

Azhar UK, Masudulla K, Nazia M, Moo HC, Mohammad MK (2018) Critical review: Recent progress of algae and blue–green algae-assisted synthesisof gold nanoparticles for various applications. Bioprocess Biosyst Eng. https://doi.org/10.1007/s00449-018-2012-2.

Azizi S, Ahmad MB, Namvar F, Mohamad R (2014) Green biosynthesis and characterization of zinc oxide nanoparticles using brown marine macroalga *Sargassum muticum* aqueous extract. Mater Lett 116:275–277.

Azizi S, Namvar F, Mahdavi M, Ahmad M, Mohamad R (2013) Biosynthesis of silver nanoparticles using brown marine macroalga, *Sargassum muticum* aqueous extract. Materials 6:5942–5950.

Barabadi H, Najafi M, Samadian H, Azarnezhad A, Vahidi H, Mahjoub MA, Koohiyan M, Ahmadi A. (2019) A Systematic Review of the Genotoxicity and Antigenotoxicity of Biologically Synthesized Metallic Nanomaterials: Are Green Nanoparticles Safe Enough for Clinical Marketing? Medicina (Kaunas). 55(8):439. doi: 10.3390/medicina55080439. PMID: 31387257; PMCID: PMC6722661.

Barwal I, Ranjan P, Kateriya S, Yadav SC (2011) Cellular proteins of *Chlamydomonas reinhardtii* control the biosynthesis of silver nanoparticles oxido-reductive. J Nanobiotechnol 9:1–12.

Bhattacharya S, Jayanthi AP, Shilpa S (2012) Nanotechnology development in India: Investigating ten years of India's efforts in capacity building. CSIR-NISTADS Strategy Paper on Nanotechnology, No. I, July, NISTADS, India.

Boca SC, Potara M, Gabudean A-M, Juhem A, Baldeck PL, Astilean S (2011) Chitosan-coated triangular silver nanoparticles as a novel class of biocompatible, highly effective photothermal transducers for in vitro cancer cell therapy. Cancer Lett 311:131–140.

Chakraborty N, Banerjee A, Lahiri S, Panda A, Ghosh AN, Pal R (2009) Biorecovery of gold using cyanobacteria and an eukaryotic alga with special reference to nanogold formation—a novel phenomenon. J Appl Phycol 21:145–152.

Chaudhary A (2011) Ayurvedic bhasma: Nanomedicine of ancient India—its global contemporary perspective. J Biomed Nanotechnol 7(1):68–69. doi:10.1166/jbn.2011.1205. PMID: 21485807.

Czupryna J, Tsourkas A (2006) Suicide gene delivery by calcium phosphate nanoparticles: A novel delivery by calcium phosphate nanoparticles. Cancer Biol Ther 5:1691–1692.

Dahoumane SA, Djediat C, Yepremian C, Coute A, Fievet F, Coradin T, Brayner R (2012) Recycling and adaptation of *Klebsormidium flaccidum* microalgae for the sustained production of gold nanoparticles. Biotechnol Bioeng 109:284–288.

Dahoumane SA, Wijesekera K, Filipe CDM, Brennan JD (2014a) Stoichiometrically controlled production of bimetallic gold-silver alloy colloids using micro-alga cultures. J Colloid Interface Sci 416:67–72.

Dahoumane SA, Yéprémian C, Djédiat C, Couté A, Fiévet F, Coradin T, Brayner R (2014b) A global approach of the mechanism involved in the biosynthesis of gold colloids using micro-algae. J Nanopart Res 16:2607.

Davis SA, Patel HM, Mayes EL, Mendelson NH, Franco G, Mann S (1998). Brittle bacteria: A biomimetic approach to the formation of fibrous composite materials. Chem Mater 10:2516–2524.

Dhanalakshmi PK, Riyazulla A, Rekha R, Poonkodi S, Thangaraju N (2012) Synthesis of silver nanoparticles using green and brown seaweeds. Phykos 42(2):39–45.

Dhas TS, Kumar VG, Karthick V, Angel KJ, Govindaraju K (2014) Facile synthesis of silver chloride nanoparticles using marine alga and its antibacterial efficacy. Spectrochim Acta A 120:416–420.

Drexler KE (1992) *Nanosystems: Molecular Machinery, Manufacturing, and Computation.* Wiley Interscience, London, pp. 311–312 and 398–403.

El-Rafiea HM, El-Rafieb MH, Zahranc MK (2013) Green synthesis of silver nanoparticles using polysaccharides extracted from marine macro algae. Carbohydr Polym 96:403–410.

Elumalai S, Santhose BI, Devika R, Revathy S (2013) Collection, isolation, identification, and biosynthesis of silver nanoparticles using microalga *Chlorella pyrenoidosa.* Nanomechanics Sci Technol Int J 4:59–66.

Eroglu E, Chen X, Bradshaw M, Agarwal V, Zou J, Stewart SG, Duan X, Lamb RN, Smith SM, Raston CL, Iyer KS (2013) Biogenic production of palladium nanocrystals using microalgae and their immobilization on chitosan nanofibers for catalytic applications. RSC Adv 3:1009–1012.

Felix LO, Sasikumar V, Manivel A, Nooruddin T, Dharumadurai D, Chari N (2016) Algal nanoparticles: Synthesis and biotechnological potentials intechopen. doi:10.5772/62909. www.intechopen.com/books/algae-organisms-for-imminent-biotechnology/algal-nanoparticles-synthesis-and-biotechnological-potentials.

Feng W, Wen G, Ling X, Zhongyang D, Haile M, Anzhou M, Norman T (2019) Effects of nanoparticles on algae: Adsorption, distribution, ecotoxicity and fate. Appl Sci 9:1534. doi:10.3390/app9081534. www.mdpi.com/journal/applsci.

Focsan M, Ardelean II, Craciun C, Astilean S (2011) Interplay nanoparticle biosynthesis and metabolic activity of cyanobacterium *Synechocystis* sp. PCC 6803 between gold. Nanotechnology 22:1–8.

Ganesan V, Aruna Devi J, Astalakshmi A, Nima P, Thangaraja A (2013) Eco-friendly synthesis of silver nanoparticles using a sea weed, *Kappaphycus alvarezii* (Doty) Doty ex P.C. Silva. Int J Eng Adv Tech 2:559–563.

Ghodake G, Lee DS (2011) Biological synthesis of gold nanoparticles using the aqueous extract of the brown algae *Laminaria japonica.* J Nanoelectron Optoelectron 6:1–4.

Giljohann DA, Seferos DS, Daniel WL, Massich MD, Patel PC, Mirkin CA (2010) Gold nanoparticles for biology and medicine. Angewandte Chemie Int Ed (English) 49:3280–3294.

Gonzalez-Ballesteros N, Prado-Lopez S, Rodriguez-Gonzalez JB, Lastra M, Rodriguez-Arguelles MC (2017) Green synthesis of gold nanoparticles using brown algae *Cystoseira baccata*: Its activity in colon cancer cells. Colloids Surf B Biointerfaces 153:190–198.

Govindaraju KV, Basha SK, Kumar VG, Singaravelu G (2008) Silver, gold and bimetallic nanoparticles production using single-cell protein (*Spirulina platensis*) Geitler. J Mater Sci 43:5115–5122.

Govindaraju KV, Kiruthiga V, Ganesh Kumar V, Singaravelu G (2009) Extracellular synthesis of silver nanoparticles by a marine alga, *Sargassum wightii* Grevilli and their antibacterial effects. J Nanosci Nanotechnol 9:5497–5501.

Govindaraju KV, Krishnamoorthy K, Alsagaby SA, Singaravelu G, Premanathan M (2015) Green synthesis of silver nanoparticles for selective toxicity towards cancer cells. IET Nanobiotechnol. doi:10.1049/iet-nbt.2015.0001.

Greene B, Hosea M, McPherson R, Henzl M, Alexander MD, Darnall DW (1986) Interaction of gold(I) and gold(III) complexes with algal biomass. Environ Sci Technol 20:627–632.

Jeffryes C, Agathos SN, Rorrer G (2015) Biogenic nanomaterials from photosynthetic microorganisms. Cur Opin Biotechnol 33:23–31.

Jegadeeswaran P, Rajaeswari S, Venckatesh R (2012) Green synthesis of silver nanoparticles from extract of *Padina tetrastromatica* leaf. Digest J Nanomat Biostruct 7(3):991–998.

Jena J, Pradhan N, Dash BP, Sukla LB, Panda PK (2013) Biosynthesis and characterization of silver nanoparticles using microalga *Chlorococcum humicola* and its antibacterial activity. Int J Nanomater Bios 3:1–8.

Jena J, Pradhan N, Nayak RR, Dash BP, Sukla LB, Panda PK, Mishra BK (2014) Microalga *Scenedesmus* sp.: A potential low-cost green machine for silver nanoparticle synthesis. J Microbiol Biotechnol 24:522.

Jena J, Pradhan N, Nayak RR, Dash BP, Sukla LB, Panda PK, Mishra BK (2015) Pigment mediated biogenic synthesis of silver nanoparticles using diatom Amphora sp. and its antimicrobial activity. J Saudi Chem Soc 19:661.

Jianping X, Jim YL, Daniel ICW, Yen PT (2007). Identification of active biomolecules in the high-yield synthesis of single-crystalline gold nanoplates in algal solutions. Small 3(4):668–672.

Jin S, Ye K (2007) Nanoparticle-mediated drug delivery and gene therapy. Biotechnol Prog. 23(1):32–41. doi:10.1021/bp060348j. PMID: 17269667.

Kang MG, Guo LJ (2007) Nanoimprinted semitransparent metal electrodes and their application in organic light-emitting diodes. Adv Mater 19:1391–1396.

Kannan RRR, Stirk WA, Staden JV (2013) Synthesis of silver nanoparticles using the seaweed *Codium capitatum* P.C. Silva (Chlorophyceae). S Afr J Sci Bot 86:1–4.

Karthikeyan T, Sundaramanickam A, Shanmugam N, Balasubramanain T (2015) Green synthesis of silver nanoparticles using marine algae *Caulera racemosa* and their antibacterial activity against some human pathogens. Appl Nanosci 5:499–504.

Kathiraven T, Sundaramanickam A, Shanmugam N, Balasubramanian T (2015) Green synthesis of silver nanoparticles using marine algae *Caulerpa racemosa* and their antibacterial activity against some human pathogens. Appl Nanosci 5:499–504.

Khan MM, Cho MH (2018) Positively charged gold nanoparticles for hydrogen peroxide detection. BioNanoScience 8(2):537–543.

Kim D, Jeong YY, Jon S (2010) A drug-loaded aptamer–gold nanoparticle bioconjugate for combined CT imaging and therapy of prostate cancer. ACS Nano 4:3689–3696.

Kroger N, Lorenz S, Brunner E, Sumper M (2002) Self-assembly of highly phosphorylated silaffins and their function in biosilica morphogenesis. Science 298:584–586.

Kvitek L, Panacek A, Soukupova J, Kolar M, Vecerova R, Prucek R (2008) Effect of surfactant and polymers on stability and antibacterial activity of silver nanoparticles (NPAS). J Phys Chem C 112:5825–5834.

Lee KS, El-Sayed MA (2006) Gold and silver nanoparticles in sensing and imaging: sensitivity of plasmon response to size, shape, and metal composition. J Phys Chem B.5;110(39):19220-5. doi: 10.1021/jp062536y.

Lengket MF, Fleet ME, Southam G (2006a) Morphology of gold nanoparticles synthesized by filamentous cyanobacteria from gold(I)-thiosulfate and gold(III)-chloride complexes. Langmuir 22:2780–2787.

Lengket MF, Fleet ME, Southam G (2007) Biosynthesis of silver nanoparticles by filamentous cyanobacteria from a silver(I) nitrate complex. Langmuir 23:2694–2699.

Lengket MF, Ravel B, Fleet ME, Wanger G, Gordon RA, Southam G (2006b) Mechanisms of gold bioaccumulation by filamentous cyanobacteria from gold(III)-chloride complex. Envi Sci Technol 40:6304–6309.

Liu C, Yu X (2011) Silver nanowire-based transparent, flexible, and conductive thin film. Nanoscale Res Lett 6. doi:10.1186/1556-276X-6-75.

Mahdavi M, Namvar F, Ahmad MB, Mohamad R (2013) Green biosynthesis and characterization of magnetic iron oxide (Fe_3O_4) nanoparticles using seaweed (*Sargassum muticum*) aqueous extract. Molecules 18:5954–5964.

Mahdieh M, Zolanvari A, Azimee AS, Mahdieh M (2012) Green biosynthesis of silver nanoparticles by *Spirulina platensis*. Scientia Iranica 19:926–929.

Marimuthu V, Palanisamy SK, Sesurajan S, Sellappa S (2011) Biogenic silver nanoparticles by *Gelidiella acerosa* extract and their antifungal effects. Avicenna J Med Biotechnol 3:143–148.

Mata YN, Torres E, Blazquez ML, Ballester A, González F, Munoz JA (2009) Gold(III) biosorption and bioreduction with the brown alga *Fucus vesiculosus*. J Hazard Mater 166:612–618.

Merin DD, Prakash S, Bhimba BV (2010) Antibacterial screening of silver nanoparticles synthesized by marine micro algae. Asian Pac J Trop Med 3:797–799.

Mohandass C, Vijayaraj AS, Rajasabapathy R, Satheeshbabu S, Rao SV, Shiva C, De-Mello I (2013) Biosynthesis of silver nanoparticles from marine seaweed *Sargassum cinereum* and their antibacterial activity. J Invest Dermatol 78:206–209.

Mohanpuria P, Rana NK, Yadav SK (2008) Biosynthesis of nanoparticles: Technological concepts and future applications. J Nanoparticle Res 10:507–517.

Mohseniazar M, Barin M, Zarredar H, Alizadeh S, Shanehbandi D (2011) Potential of microalgae and Lactobacilli in biosynthesis of silver nanoparticles. Bio Impacts 1:149–152.

Monica T, Luca DS, Ilaria R (2018) Diatoms green nanotechnology for biosilica-based drug delivery systems. Pharmaceutics 10:242. doi:10.3390/pharmaceutics10040242 www.mdpi.com/journal/pharmaceutics.

Morones JR, Elechiguerra JL, Camacho A, Holt K, Kouri JB, Ramírez JT, et al. (2005) The bactericidal effect of silver nanoparticles. Nanotechnol 16:2346–2353.

Mubarak Ali D, Sasikala M, Gunasekaran M, Thajuddin N (2013). Biosynthesis and characterization of silver nanoparticles using marine cyanobacterium, *Oscillatoria willei* NTDM01. Dig J Nanomater Biostruct 6:385–390.

Naveen BE, Prakash S (2013) Biological synthesis of gold nanoparticles using algae *Gracilaria corticata* and its application as a potent antimicrobial and antioxidant agent. Asian J Pharm Clin Res 6(2):179–182.

Ngeontae W, Janrungroatsakul W, Maneewattanapinyo P, Ekgasit S, Aeungmaitrepirom W, Tuntulani T (2009) Novel potentiometric approaching glucose biosensor using silver nanoparticles as redox marker. Sens Actuators B 137:320–326.

Noura EAEN, Mervat HH, Sami ASD, Shimaa RD (2020) Production, extraction and characterization of *Chlorella vulgaris* soluble polysaccharides and their applications in AgNPs biosynthesis and biostimulation of plant growth. Sci Rep 10:3011. https://doi.org/10.1038/s41598-020-59945-w 2.

Oza G, Pandey S, Mewada A, Kalita G, Sharon M (2012) Facile biosynthesis of gold nanoparticles exploiting optimum pH and temperature of fresh water algae *Chlorella pyrenoidusa*. Adv Appl Sci Res 3:1405–1412.

Parial D, Pal R (2015) Biosynthesis of monodisperse gold nanoparticles by green alga *Rhizoclonium* and associated biochemical changes. J Appl Phycol 27:975–984.

Parial D, Patra HK, Dasgupta AK, Pal R (2012a) Screening of different algae for green synthesis of gold nanoparticles. Eur J Phycol 47:22–29.

Parial D, Patra HK, Roychoudhury P, Dasgupta AK, Pal R (2012b) Gold nanorod production by cyanobacteria—a green chemistry approach. J Appl Phycol 24:55–60.

Prasad TNVKV, Kambala VSR, Naidu R (2013) Phyconanotechnology synthesis of silver nanoparticles using brown marine algae *Cystophora moniliformis* and their characterisation. J Appl Phycol 25:177–182.

Prashant A, Ritika G, Neeraj A (2019) Advances in synthesis and applications of microalgal nanoparticles for wastewater treatment. J Nanotechnol 2019:1–9. Article ID 7392713.

Prow T, Grebe R, Merges C, Smith J, Mcleod D, Leary J, Lutty G (2006). Nanoparticle tethered biosensors for autoregulated gene therapy in hyperoxic endothelium. Nanomedicine-nanotechnology Biology and Medicine. Nanomed. Nanotechnol Biol Med 2:276–276. doi:10.1016/j.nano.2006.10.046.

Rahman, Ashiqur & Kumar, Shishir & Nawaz, Tabish. (2020). Biosynthesis of Nanomaterials Using Algae. In book: Microalgae Cultivation for Biofuels Production (pp.265–279) DOI: 10.1016/B978-0-12-817536-1.00017-5

Rajasulochana P, Dhamotharan R, Murugakoothan P, Murugesan S, Krishnamoorthy P (2010) Biosynthesis and characterization of gold nanoparticles using the alga *Kappaphycus alvarezii*. Int J Nanosci 9:511–516.

Rajathi FAA, Parthiban C, Kumar VG, Anantharaman P (2012) Biosynthesis of antibacterial gold nanoparticles using brown alga, *Stoechospermum marginatum* (kutzing). Spectrochim Acta Part A Mol Biomol Spectrosc 99:166–173.

Rajesh S, Raja DP, Rathi JM, Sahayaraj K (2012). Biosynthesis of silver nanoparticles using *Ulva fasciata* (Delile) ethyl acetate extract and its activity against *Xanthomonas campestris* pv. *Malvacearum*. J Biopest 5:119–128.

Rajeshkumar S, Kumar SV, Malaakodi C, Vanaja M, Paulkumar K, Annadurai G (2017) Optimized synthesis of gold nanoparticles using green chemical process and its in vitro anticancer activity against HepG2 and A549 cell lines. Mech Mater Sci Eng. ISSN 2412–5954.

Rajeshkumar S, Malarkodi C, Paulkumar K, Vanaja M, Gnanajobitha G, Annadurai G (2014) Algae mediated green fabrication of silver nanoparticles and examination of its antifungal activity against clinical pathogens. Int J Met. doi:10.1155/2014/692643.

Rajeshkumar S, Malarkodi C, Vanaja M, Gnanajobitha G, Paulkumar K, Kannan C, Annadurai G (2013) Antibacterial activity of algae mediated synthesis of gold nanoparticles from *Turbinaria conoides*. Der Pharma Chemica 5:224–229.

Ramakritinan CM, Kaarunya E, Shankar S, Kumaraguru AK (2013) Antibacterial effects of Ag, Au and bimetallic (Ag-Au) nanoparticles synthesized from red algae. Solid State Phenom 201:211–230.

Rauwel P, Küünal S, Ferdov S, Rauwel E (2015). A review on the green synthesis of silver nanoparticles and their morphologies studied via TEM. Adv Mater Sci Eng 682749:9.

Sahayaraj K, Rajesh S, Rathi JM (2012) Silver nanoparticles biosynthesis using marine alga *Padina pavonica* (linn.) and its microbicidal activity. Dig J Nanomater Biostruct 7:1557–1567.

Sangeetha N, Manikandan S, Singh M, Kumaraguru AK (2012) Biosynthesis and characterization of silver nanoparticles using freshly extracted sodium alginate from the seaweed *Padina tetrastromatica* of Gulf of Mannar, India. Curr Nanosci 8:697–702.

Satapathy S, Shukla SP, Sandeep KP, Singh AR, Sharma N (2015) Evaluation of the performance of an algal bioreactor for silver nanoparticle production. J Appl Phycol 27:285–291.

Schrofel A, Kratosova G, Bohunicka M, Dobrocka E, Vavra I (2011) Biosynthesis of gold nanoparticles using diatoms—silica-gold and EPS-gold bionanocomposite formation. J Nanoparticle Res 13:3207–3216.

Senapati S, Syed A, Moeez S, Kumar A, Ahmad A (2012) Intracellular synthesis of gold nanoparticles using alga *Tetraselmis kochinensis*. Mater Lett 79:116–118.

Shakibaie M, Forootanfar H, Mollazadeh-Moghaddam K, Bagherzadeh Z, Nafissi-Varcheh N, Shahverdi AR, Faramarzi MA (2010) Green synthesis of gold nanoparticles by the marine microalga *Tetraselmis suecica*. Biotechnol Appl Biochem 57(2):71–75.

Shamaila S, Zafar N, Riaz S, Sharif R, Nazir J, Naseem S. Gold Nanoparticles: An Efficient Antimicrobial Agent against Enteric Bacterial Human Pathogen. Nanomaterials (Basel). 2016 Apr 14;6(4):71. doi: 10.3390/nano6040071. PMID: 28335198; PMCID: PMC5302575.

Shanmugam R, Chelladurai M, Kanniah P, Mahendran V, Gnanadas G, Gurusamy A (2014) Algae mediated green fabrication of silver nanoparticles and examination of its antifungal activity against clinical pathogens. Int J Metal 2014:1–8. Hindawi Publishing Corporation.

Shanmugam R, Chellapandian K, Gurusamy A (2012) Green synthesis of silver nanoparticles using marine brown algae *Turbinaria conoides* and its antibacterial activity. Int J Pharma Bio Sci Oct; 3(4):(P) 502–510.

Sharda S (2019) Safe nano is green nano, Chapter-2 in Green synthesis, characterization and applications of nanoparticles micro and nano technologies, 27–36. https://doi.org/10.1016/B978-0-08-102579-6.00002-2.

Sharma B, Purkayastha DD, Hazra S, Gogoi L, Bhattacharjee CR, Ghosh NN, Rout J (2015) Biosynthesis of gold nanoparticles using a fresh water green alga, *Prasiola crispa*. Mater Lett 116:94–97.

Sharma B, Purkayastha DD, Hazra S, Thajamanbi M, Bhattacharjee CR, Narendra Nath Ghosh NN, Rout J (2014) Biosynthesis of fluorescent gold nanoparticles using an edible freshwater red alga, *Lemanea fluviatilis* (L.) C.Ag. and antioxidant activity of biomatrix loaded nanoparticles. Bioprocess Biosyst Eng 37:2559–2565.

Sharma VK, Park K, SrinivasaRao M (2009a). Colloidal dispersion of gold nanorods: Historical background, optical properties, seed-mediated synthesis, shape separation and self-assembly. Mater Sci Eng R 65:1–3.

Sharma VK, Yngard RA, Lin Y (2009b) Silver nanoparticles: Green synthesis and their antimicrobial activities. Adv Colloid Interf Sci 145:83–96.

Shiny PJ, Mukherjee A, Chandrasekaran, N (2013) Marine algae mediated synthesis of the silver nanoparticles and its antibacterial efficiency. Int J Pharm Sci 5:239–241.

Shukla MK, Singh RP, Reddy CRK, Jha B (2012) Synthesis and characterization of agar-based silver nanoparticles and nanocomposite film with antibacterial applications. Bioresour Technol 107:295–300.

Sicard C, Brayner R, Margueritat J, Hemadi M, Coute A, Yepremian C, Djediat C, Aubard J, Fievet F, Livage J, Coradin T (2010) Nanogold biosynthesis by silica-encapsulatedmicro-algae: A Bliving biohybrid material. J Mater Chem 20:9342–9347.

Silver S, Phung LT, Silver G (2006) Silver as biocides in burn and wound dressings and bacterial resistance to silver compounds. J Ind Microbiol Biotechnol 33:627–634.

Singaravelu GJS, Arockiamary C, Ganesh Kumarb V, Govindaraju K (2007) A novel extracellular synthesis of monodisperse gold nanoparticles using marine alga, *Sargassum wightii* Greville. Colloids Surf B: Biointerfaces 57:97–101.

Singh MR, Kalaivani S, Manikandan N, Sangeetha A, K. Kumaraguru (2013) Facile green synthesis of variable metallic gold nanoparticle using *Padina gymnospora*, a brown marine macroalga. Appl Nanosci 3:145–151.

Singh MR, Kumar M, Manikandan S, Chandrasekaran N, Mukherjee A, Kumaraguru AK (2014) Drug delivery system for controlled cancer therapy using physico-chemically stabilized bioconjugated gold nanoparticles synthesized from marine macroalgae *Padina gymnospora*. J Nanomed Nanotechol S5:009.

Sinha SN, Paul D, Halder N (2015) Green synthesis of silver nanoparticles using fresh water green alga *Pithophora oedogonia* (Mont.) Wittrock and evaluation of their antibacterial activity. *Appl Nanosci* 5:703–709.

Sondi I, Salopek-Sondi B (2007) Silver nanoparticles as antimicrobial agent: A case study on *E. coli* as a model for gram-negative bacteria. J Colloid Interface Sci 275:77–82.

Subramaniyam V, Subashchandrabose SR, Thavamani P, Megharaj M, Chen Z, Naidu R (2015) *Chlorococcum* sp. MM11—a novel phyco-nanofactory for the synthesis of iron nanoparticles. J Appl Phycol 27:1861–1869.

Sudha SS, Rajamanickam K, Rengaramanujam J (2013) Microalgae mediated synthesis of silver nanoparticles and their antibacterial activity against pathogenic bacteria. Indian J Exp Biol 51:393–399.

Suganthi S, Rani G (2014) Biosynthesis and characterization of gold nannoparticles from *Gracilaria corticata*—Nannoscience and nanotechnology. An Indian J 8(12):475–481.

Taniguchi N (1974) On the basic concept of nano technology. Proceedings of the International Conference on Production Engineering. Tokyo, Part, Japan Society of Precision Engineering 18–23 technological concepts and future applications. J Nanoparticle Res 10:507–517.

Thakkar KN, Mhatre SS, Parikh RY (2010) Biological synthesis of metallic nanoparticles. Nanomed 6:257–262.

Tian J, Wong KKY, Ho CM, Lok CN, Yu WY, Che CM, Chiu JF, Tam PKH (2007) Topical delivery of silver nanoparticles promotes wound healing. Chem Med Chem 2:129–136.

Tkachenko AG, Xie H, Coleman D, Glomm W, Ryan J, Anderson MF, Franzen S, Feldheim DL (2003) Multifunctional gold nanoparticle –peptide complexes for nuclear targeting. J Am Chem Soc 125:4700–4701.

Venkatesan J, Manivasagan P, Kim S, Vishnu Kirthi A, Marimuthu S, Rahuman AA (2014) Marine algae-mediated synthesis of gold nanoparticles using a novel Ecklonia cava. Bioprocess Biosyst Eng 37:1591–1597.

Venkatpurwar V (2011) Porphyran capped gold nanoparticles as a novel for delivery of anticancer drugs: In vitro cytotoxicity study. Int J Pharm 409:314–320.

Vijayan SR, Santhiyagu P, Singamuthu M, Ahila NK, Jayaraman R, Ethiraj K (2014) Synthesis and characterization of silver and gold nanoparticles using aqueous extract of seaweed, *Turbinaria conoides*, and their antimicrofouling activity. Sci World J. doi:10.1155/2014/938272.

Wiechers JW, Musee N. (2010)Engineered inorganic nanoparticles and cosmetics: facts, issues, knowledge gaps and challenges. J Biomed Nanotechnol. (5):408-31. doi: 10.1166/jbn.2010.1143. PMID: 21329039.

Wishkerman A, Arad (Malis) S (2017) Production of silver nanoparticles by the diatom *Phaeodactylum tricornutum* proceedings of SPIE. Vol.10248: 102480W-1 to 102480W-7pp.

Xie J, Lee JY, Wang DIC, Ting YP (2007) Identification of active biomolecules in the high-yield synthesis of single-crystalline gold nanoplates in algal solutions. Small 3:672–682.

Zhang, S, Zhang,T, He Jinxin, Xia, Dong Cellulose (2021) Effect of AgNP distribution on the cotton fiber on the durability of antibacterial cotton fabrics. Scholarly Journal 14: 9489–9504. DOI:10.1007/s10570-021-04113-0

17 Microalgae-Based Nanoparticles and Biocomposites for Biomedical Applications

Rajamohamed Beema Shafreen

17.1 INTRODUCTION

Nanotechnology deals with the design, synthesis, and manipulation of matter at nanoscale dimension (1 to 100 nm). Various metals such as silver, gold, copper, palladium, and platinum have been widely used for synthesis of nanoparticles. Among these, silver nanoparticles offer an innumerable amount of application due to their biocompatibility and low-toxicity nature. Nanomaterials generated either through bottom-up or top-down approaches possesses unique physicochemical properties such as high surface area and surface energy. Due to this exclusive property, they are used in diverse areas such as medicine, drug delivery, waste remediation, water treatment, agriculture, health care, cosmeceuticals, electronics, and nanomaterials-based sensors. Nanomaterials can be produced through physical, chemical, and biological methods. However, nanomaterials synthesized through greener routes are cost-effective, nontoxic, and eco-friendlier than nanomaterials synthesized through physicochemical methods, as the latter influences an environmental and toxicity issues. Besides, physicochemical methods require a lot of energy, organic solvents, and toxic-reducing agents that results in the generation of high levels of hazardous wastes. Due to the high operational cost and low yield, these methods are unsuitable for scale-up productions. On the other hand, it has strong tendency to form agglomerates. In order to overcome all these barriers, scientific community targeted biomaterials for nanoparticle synthesis. Biomaterials such as bacteria, algae, fungi, virus, and plants are used for nanoparticle production. Not all organisms are capable of producing metallic nanoparticles. The organisms which have the ability to reduce metal ions are used for nanomaterial synthesis. Intriguingly, biomaterials from marine source plays an important role in nano-based drugs for improving the human life. This is due to the fact that marine organisms have the potentiality to reduce inorganic elements present in the sea. Moreover, agglomerate formation is highly averted owing to the production of stabilizing molecules such as phytochelatins, heavy metal binding peptides, which stabilize the synthesized nanoparticles from forming agglomerates. The carbonyl group from the amino acid residues of proteins and peptides bind metal and thereby forms a shield covering the metal nanoparticles, a process known as capping. By forming the layer, it stabilizes the generated nanoparticles and prevent agglomeration. Further, biogenic production of nanoparticles extends the applications of nanotechnology in various domains, including photo-imaging, catalysis, biosensors, drug delivery, diagnostic, and therapeutic purposes. Most important applications of microalgae in biotechnology field are represented in Figure 17.1. Biosynthesis of nanoparticles using algae has gained widespread attraction due to the presence of rich diverse metabolites including polysaccharides, carotenoids (astaxanthin, fucoxanthin, neoxanthin, violaxanthin, carotene, zeaxanthin, siphonaxanthin, canthaxanthin, cryptoxanthin, and leutin), vitamins (A, B, C, D, E and K), polyphenols (flavonoids, flavones, isoflavone, flavonols, dihydroflavonols,

DOI: 10.1201/9781003219156-21

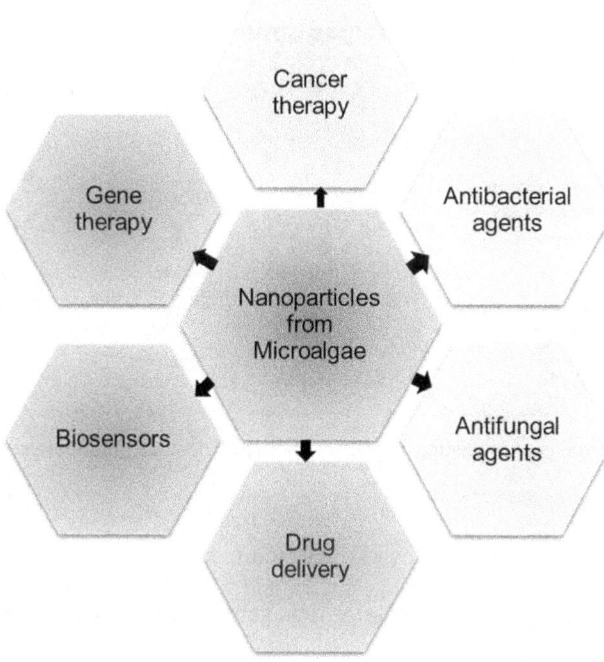

FIGURE 17.1 Applications of algal synthesized nanoparticles in biotechnology sector.

flavanones, and proanthocyanidins), omega-3 fatty acids (docosahexaenoic acid (DHA), eicosapentaenoic acid (EPA), and alpha-linolenic acid (ALA)), sterols (beta-sitosterol, ergosterol, stigmasterol, campesterol, and brassicasterol), peptides, amino acids, mycosporine-like amino acids (MAAs), and minerals.

In particular, marine algae are of significant importance due to the production of different polysaccharides, proteins, fats, polyphenols, pigments (carotene, xanthophyll, chlorophylls, phycocyanin, and phycoerythrin), antioxidants, tocopherol, antibiotics, chemopreventive, antidiabetic, and other pharmaceutically valuable compounds. However, studies on unexplored marine algae could be useful in the discovery of novel metabolites with high biomedical value. The present chapter describes the algae-mediated biosynthesis of nanoparticles, biocomposites, and their potential applications in different spheres.

17.2 ALGAL SOURCE FOR NANOPARTICLE PRODUCTION

Algae become a great resource for nanoparticle synthesis due to high growth rate, hyper accumulation of heavy metal ions, and biomass yield. Algae are classified as micro and macroalgae. Several group of algae such as Chlorophyceae, Phaeophyceae, Rhodophyceae, and Cyanophyceae have been reported for biosynthesis of thermodynamically stable nanoparticles with unique size and shape. Although intracellular and extracellular synthesis of nanoparticles have been described, biosynthesis of nanoparticles from microalgae can be classified into four methods (Figure 17.2). First method involves the use of biomolecules to form disrupted cells of microalgae, second method make use of cell free supernatant in which the cells are removed through centrifugation and filtration process, third method uses harvested whole cells, and the fourth method involves use of live algal cells. Each alga has the ability to synthesize unique nanoparticles due to the presence of different reducing agents. In general, the synthesized nanoparticles were characterized using UV-Vis spectroscopy, Fourier transform infrared (FTIR) spectroscopy, scanning electron microscopy

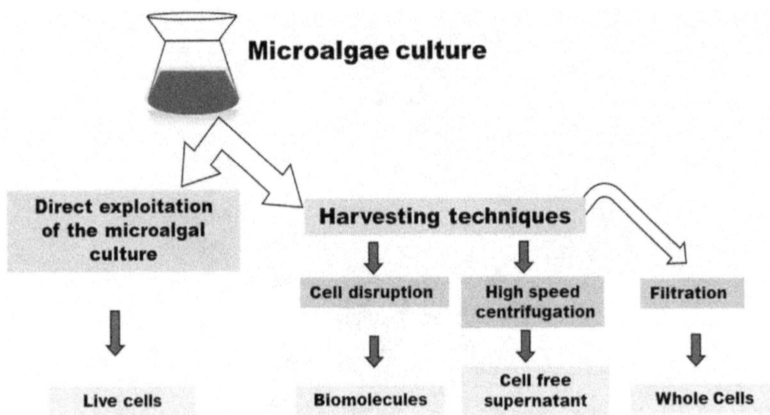

FIGURE 17.2 Four methods of biogenic synthesis of nanoparticles from microalgae.

(SEM), transmission electron microscopy (TEM), atomic force microscopy (AFM), energy-dispersive X-ray (EDX), dynamic light scattering (DLS), and X-ray diffraction (XRD). The synthesized nanoparticles are commonly cubical, spherical, rectangular, decahedral, or polygonal in shape.

17.3 IMPORTANCE OF USING MICROALGAE IN NANOTECHNOLOGY

Synthesis of nanoparticles through physical and chemical processes are highly expensive. Apart, the chemicals used for reducing and capping process are toxic and aggressive. Thus, green chemistry approaches play an important role in nanotechnologies exclusively when these nanoparticles are integrated with medical applications such as drug delivery, tissue repair, biomaterials, surgical devices, and sanitizers. The green biosynthesis of nanoparticles using plants serves with dual role, as capping and reducing agent; however, the slower growth rate limits its potential application in various sectors. Interestingly, microalgae can be grown in laboratory conditions with good yield and low cost. Hence, nanoparticle synthesis with microalgae is suggested to be the valuable alternative for physical and chemical methods.

17.4 DIFFERENT TYPES OF NANOPARTICLES

17.4.1 Silver Nanoparticles

Silver nanoparticles is the most extensively studied nanoparticles due to their unusual beneficial properties, including antimicrobial, surface chemistry, and particle stability. Since surface plasmon resonance peaks of silver nanoparticles fall between 450 and 530 nm, the generated silver nanoparticles will differ in size, shape, volume, and surface property. Therefore, it is widely used as antibacterial and antifungal agents in healthcare industries. Many products incorporated with silver nanoparticles have been approved by Food and Drug Administration (FDA, USA) and are commercially available. Studies on alcoholic extract of *Chlorella* sp. to reduce silver metal ions for nanoparticle biosynthesis (Ag/AgCl) resulted in the production 10–20 nm sized particles with antibacterial activity against *E. coli* and *Bacillus* spp. (Kashyap et al., 2019). Recently, microwave-assisted green synthesis of nanoparticles have been demonstrated as microwave-assisted biosynthesis has been reported to control the shape and morphology of the generated nanostructures. Exposure of dried-biomass of *C. vulgaris* to microwave radiation (180 W) resulted in an average size of 24 nm sized nanoparticles. Similarly, the aqueous extract of *Trichodesmium erythraeum* for biosynthesis of silver nanoparticles resulted in an average size of 26.5 nm. The synthesized nanoparticles exhibited an antibacterial activity against drug-resistant bacteria strains including *E. coli*, *Staphylococcus aureus*, and *Streptococcus pneumoniae*. The study extended in clinical strains of *Proteus mirabilis and S. aureus* substantiated the antibacterial activity of nanoparticles. In addition, the generated silver nanoparticles exhibited an antioxidant and anticancer (against HeLa and MCF-7 cell lines) activities (Sathishkumar et al., 2019).

17.4.2 Gold Nanoparticles

Gold nanoparticles can absorb and scatter light efficiently. By controlling the particle size, shape, and surface chemistry, optical and electronic properties of gold nanoparticles can be tuned, which allows for promising applications in different domains. Due to the attractive physicochemical properties including localized surface plasmon resonance, radioactivity, large surface-volume ratio, optoelectronic, fluorescence quenching, and biocompatibility, gold nanoparticles have been employed in imaging, sensory probes, electronic devices, photothermal therapy of cancer, catalysis, diagnostic, and therapeutic purposes. Besides, gold nanoparticles are attractive tools for gene therapy and drug delivery due to large surface area-to-volume ratio as well as controlled and sustained releasing properties. Importantly, gold nanoparticles absorb light and can turn up heat, resulting in increased localized temperature and an acoustic wave generation. This dual functional agent widens gold nanoparticles applications in photothermal and photoacoustic imaging of cancer. It supports highly useful function for cancer theragnosis (combined diagnosis and therapeutics) as well as monitoring the responses to therapy. As it is projected to reduce the cost and risks associated with cancer treatment, the gold nano particle has become highly attractive for cancer therapy.

Synthesis of gold nanoparticles have been described in algal species. Remarkably, biomolecules from microalgae have been reported to serve as both reducing as well as protecting agents. Such dual role-playing properties have been described in *Egregia* sp., in which the algal extract has been shown to have the potential to act as a reducing agent for the gold ions. Most importantly, the formation of ligand shell around the nanoparticles for stability aids in biocompatibility of gold nanoparticles (Colin et al., 2018). In another study, the ethanolic extract and dry matter of *Galaxaura elongata* treated with chloroauric acid has been reported to form rapid and highly stable gold nanoparticles with antibacterial activities against *Klebsiella pneumonia*, *Pseudomonas aeruginosa*, *E. coli*, methicillin-resistant *S. aureus* (MRSA), and *S. aureus* (Abdel-Raouf et al., 2017). Gold nanoparticles synthesized from marine algae, *Gelidiella acerosa* was found to exhibit an enzyme inhibitory activity against α-amylase and α-glucosidase enzyme, designating the antidiabetic potential of the synthesized gold nanoparticles. In addition, it exhibited a strong antioxidant and antibacterial activity against *E. coli*, *S. marcescens*, *K. pneumonia*, and *B. subtilis* (Senthilkumar et al., 2019).

17.4.3 Copper Oxide Nanoparticles

Copper (Cu) is a key trace element required for the growth of higher organisms. The daily requirement of copper is about 2–4 mg, which can be obtained through dietary supplements. The essential role of copper is to act as a cofactor for the enzymes and involved in the synthesis of neuropeptides. Also, it has an essential role in cell signaling and immune cell regulation, to kill the pathogens. Copper is essential to mediate an antioxidant defense and involved in the functions of the immune system (Abdul, W et al., 2021). Due to the potentiality of copper, it has been assisted in the development of copper and copper oxide (CuO)-based nanoparticle synthesis. Among several metal oxides, CuO nanoparticles have received much attention because of their multifarious applications. CuO has a monoclinic structure and a p-type semiconductor with a narrow bandgap of 1.7 eV. The extract from *Anabaena cylindrica* has been reported for the synthesis of CuO nanoparticles. In which the extract was added in batch into copper sulfate solution with constant stirring (500–1000 rpm), and solution was maintained at 60°C with the pH of 6.2–10.2. The synthesis of CuO oxide nanoparticle was confirmed through color change and the generated CuO nanoparticles were analyzed using different characterization techniques (Zetasizer, XPS, EDS, FESEM, TEM, and FTIR). The analysis reveals nanocrystals with rod-like morphology with a dimension of 40–60 nm (Waris et al., 2021). In general, CuO nanoparticles exhibit antibacterial activity by generating reactive oxygen species (ROS) or by disrupting the cell membrane and damage DNA, proteins, and proton efflux pumps (Sutradhar et al., 2014; Akintelu et al., 2020).

17.4.4 Other Metal Oxide Nanoparticles

Apart from metals, metal oxides such as zinc oxide (ZnO), iron oxide (Fe_2O_3), magnesium oxide (MgO), titanium dioxide (TiO_2), nitric oxide (NO), and aluminum oxide are used for nanoparticle synthesis. In application view, the transition metal oxides are greatly used in sensors and manufacture of personal care products due to their semiconductor and catalytic properties.

17.5 APPLICATIONS OF NANOPARTICLES FROM MICROALGAE

17.5.1 Antimicrobial Activity

In general, silver and gold nanoparticles show antibacterial activity against both gram-positive and gram-negative bacteria. An evaluation of antibacterial activity using silver nanoparticles synthesized from *C. vulgaris* against foodborne pathogens such as *Salmonella enterica* and *S. aureus* showed remarkable antibacterial activity in *S. enterica* than *S. aureus* (Torabfam andYüce, 2020). It is hypothesized that gram-negative bacteria harboring thick lipopolysaccharide layer (negatively charged) are more likely to interact with the nanoparticles (positively charged), and moreover it facilitates the entry of nanoparticles into the cell that eventually leads to intracellular damage. However, in the case of gram-positive bacteria, peptidoglycan layer protects the cells from nanoparticle destruction and thus shows slight resistant activity in comparison to gram-negative bacteria (Figure 17.3). Soluble polysaccharides extracted from *C. vulgaris* have been used to synthesize silver nanoparticles, and evaluation of biosynthesized nanomaterials revealed the antimicrobial activity against *Bacillus* sp., *Erwinia* sp., and *Candida* sp. (El-Naggar et al., 2020). The biofabricated

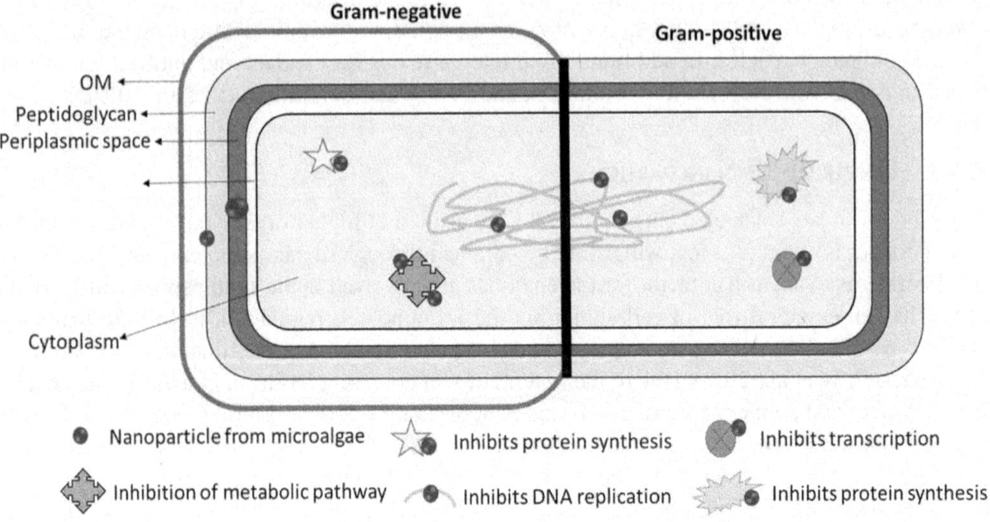

FIGURE 17.3 Antibacterial mechanisms of action of nanoparticles.

CuO nanoparticles also possess vigorous antibacterial activities against gram-positive and gram-negative bacterial strains and therefore gaining attention as antibacterial agent due to their unique morphologies and size.

Nanoparticles synthesized from microalgae have been reported for antifungal activity against *Candida* species. These nanoparticles are considered to exert an antifungal activity by disrupting the cell membrane integrity and thereby inhibiting the budding process (Kim et al., 2009). Antifungal activity has been reported from the silver nanoparticles biosynthesized from an ethanolic extract of three novel freshwater microalgae (*Dictyosphaerium* sp. strain HM1, *Dictyosphaerium* sp. strain HM2, and *Pectinodesmus* sp. strain HM3) against *Candida albicans*. These nanoparticles also exhibited significant antibacterial (against 14 bacterial strains), antitumor (liver and breast cancer) and antiviral (Newcastle disease virus, NDV) activities (Khalid et al., 2017).

17.5.2 Antibiofilm Activity

The widespread prevalence of antimicrobial resistance and the ability of the microorganisms to form biofilm on inert and biological surfaces poses a serious threat to the healthcare sector. The significance of metal nanoparticles, derived from microalgae, in quenching biofilm as well as antimicrobial properties has become an increasingly attractive and alternative treatment against microbial infections. Anti-biofilm and anti-quorum sensing potential of green synthesized nanoparticles have been reported from microalgae. Figure 17.4 represents the schematic illustration of antibiofilm mechanism of action of microalgae nanoparticles. A study from Vishwakarma and Sirisha (2020) reported that synthesis of silver nanoparticles from sulfated polysaccharides of *Chlamydomonas reinhardtii* and its application on planktonic and biofilm growth of *S. enterica* and *Vibrio harveyi* revealed the inhibition of bacterial cell attachment to the surface by reducing the cell surface hydrophobicity. In addition, the synthesized silver nanoparticles significantly disrupted the preformed biofilms of *S. enterica* and *V. harveyi* by considerably reducing the exopolysaccharide (EPS) and extracellular DNA (eDNA) contents of biofilm.

17.5.3 Biostimulation of Plants

Nanoparticles and nanomaterials at smaller concentration are reported to promote plant growth in agricultural and horticultural sector (Juárez-Maldonado et al., 2019). Synthesis of biogenic silver

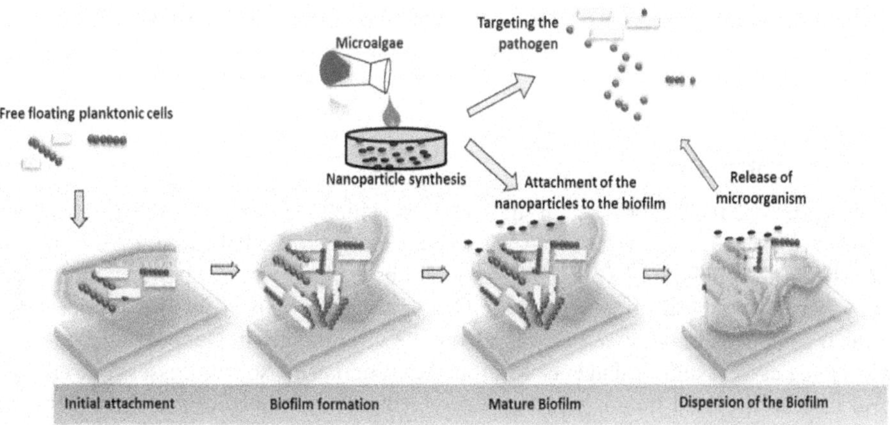

FIGURE 17.4 Schematic illustration of biofilm formation and antibiofilm mechanisms of action of nanoparticles.

nanoparticles from soluble polysaccharide extracts of *C. vulgaris* enhanced the growth of *Triticum vulgare* and *Phaseolus vulgaris* plants by augmenting the seedling growth, root and shoot length, leaf area, chlorophyll pigments, total phenolic contents, total carbohydrate and protein contents, fresh and dry biomass, and antioxidant activities (El-Naggar et al., 2020). Besides acting as a plant biostimulant, it provides greener and successful alternative solution for disease management in agronomical crops. A study reported that biosynthesis of silicon nanoparticles from the lysed cells of cyanobacteria *Oscillatoria agardhii* and its application in wheat improved the plant resistance to diseases, environment stress, irrigation quality, plant growth, yield productivity, and enhanced the quality and quantity of the wheat crops. Apart, it increased the levels of antioxidant enzymes such as catalase, peroxidase, superoxide dismutase, and improved the wheat quality parameters, including flour protein and glutamine (Haggag et al., 2018).

17.5.4 Anticancer Activity

Cancer is one of the most dreaded diseases of humans and metastasis is considered to be the major reason for mortality rates. According to the survey from International Agency for Research on Cancer (IARC) on global cancer burden, 19.3 million new cases of cancer is estimated with 10 million deaths from cancer worldwide in 2020 (Ferlay et al., 2021). Although several bioactive components are effective against treating cancer cells, it becomes unrealistic during therapeutic administration. To make it most efficient and successful, nano-based therapeutic approach are being employed. Numerous novel bioactives from microalgae source are reported for anticancerous property due to their excellent diversity in chemical constituent (Table 17.1). Importantly, the bioactives from microalgae induces apoptotic cell death through caspase-dependent or independent pathway. Among the microalgae, cyanobacteria are potentially exploitable for bioactive substances due to their efficacy in killing cancer cells through induction of apoptosis (Figure 17.5). Although they are effective in killing cancer cells, not even 10% of microalgal bioactive constituents have been employed for commercial use due to solubility issues. However, nano-based drugs provide a novel method to overcome poor solubility of hydrophobic bioactive compounds. Nanocarrier formulations are successfully employed for therapeutic and diagnostic purposes in cancer. Nanotechnology provides unique advantages including protection of the bioactive compound from enzymatic attack or metabolizing agents, prolongs the circulation time in bloodstream, increases the specific-target efficacy, successful accumulation at target site, improves the permeability as well as exhibits controlled and sustained drug release. Importantly, nanotechnology-based approaches such as nanoparticles, dendrimers, liposomes, carbon nanotubes, and micelles can be used to overcome the biological barriers, including cellular, immune, and blood-brain barriers for successful drug delivery. Although numerous bioactives from microalgae have been reported for anticancer property, there are only few reports on microalgal nanoparticles in cancer applications (Hussein et al., 2020; Bajpai et al., 2018).

17.5.5 Antiviral Activity

Viruses have become the major threat to the universe, which lead to mass epidemic and pandemic outbreaks. Few viruses, such as SARS-CoV, influenza, hepatitis, herpes simplex virus etc., can cause deadly diseases in human. Due to lack of facilities and preventive measures, these viruses are potentially harmful and alarming worldwide. Though vaccines are available, proper sanitization and personal hygiene is imperative to escape from the viral infection. Nanoparticles are promising candidates to be used as antiviral agents. Microalgae from marine source possess extremely diverse and novel metabolites, and thus it signifies a promising source for antiviral agents. Interestingly, there are several microalgal extracts showing antiviral activity. From *Haematococcus pluvialis* and *Dunaliella salina*, antiviral compounds were extracted using pressurized liquid extraction

TABLE 17.1
Anticancer and Anti-Tumorigenic Activity of Bioactive Constituents from Microalgae

Bioactives	Function	Cancer Cell Line	Algal Source	Reference(s)
Cryptophycin	Antiproliferative	Solid tumor and human tumor cell lines	*Nostoc* sp.	Shih and Teicher, 2001
Borophycin	Cytotoxicity	KB and LoVo cell lines	*Nostoclinckia* and *N. spongiaeforme* var. tenue	Banker and Carmeli, 1998
Calothrixins A and B	Cytotoxicity	Human cervical cancer cell line, HeLa cells	*Calothrix* sp.	Xu et al., 2016
Scytonemin	Antiproliferative	Fibroblast and endothelial cells	*Stigonema* sp.	Pathak et al., 2019
Coibamide-A	Cytotoxicity, antiproliferative G1 cell cycle arrest	Lung cancer cell lines, glioblastoma cells, breast cancer cells	*Leptolyngbya* sp.	Serrill et al., 2016
Phycocyanin	DNA fragmentation, upregulates the expression of Fas and ICAM, downregulates the expression of Bcl-2, activates caspases	Breast cancer, colorectal cancer, hepatoma, leukemia, human cervical cancer	*Spirulina* sp., *Aphanizomenon* sp., *Phormidium* sp., *Lyngbya* sp., *Synechocystis* sp., and *Synechococcus* sp.	Jiang et al., 2017
Astaxanthin	Antiproliferative, arrest cell cycle at G0/G1 phase, upregulates the expression of p27 and regulates extracellular signal-regulated kinase	KATO-III and SNU-1 and gastric cancer cell lines	*Chlorella* sp., *Chlorococcum* sp., *Scenedesmus* sp., and *Haematococcus pluvialis*	Joo et al., 2009
Fucoxanthin	Cell cycle in the G0/G1 stage, Fragmentation of DNA and induction of apoptosis, cytotoxicity, activates caspase-3 and downregulates the expression of Bax and Bcl-2	Leukemic HL-60 cells, prostate cancer cells	*Phaeodactylum tricornutum*	Kotake-Nara et al., 2005
Neoxanthin	Induction of apoptosis, DNA fragmentation, cleavages of caspase-3 and PARP	Prostate cancer cells	*Phaeodactylum tricornutum*	Kotake-Nara et al., 2005
Docosahexaenoic acid (DHA) and EPA	Induce apoptosis, generate ROS, and activates caspase pathway	HepG2 cancer cells, gastric cancer cells, and pancreatic cancer cells	*Schizochytrium* sp.	Park and Kim, 2017

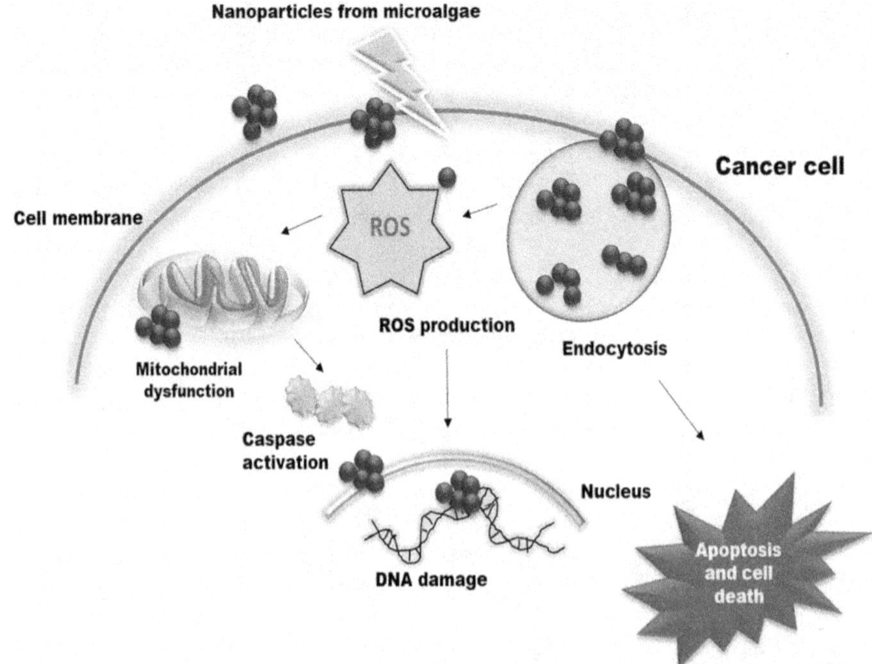

FIGURE 17.5 Possible cellular mechanism of anticancer activity of microalgae nanoparticles.

(PLE). The hexane, ethanol, and water extract were evaluated for antiviral activity against the herpes simplex virus type 1 (HSV-1). The extract with antiviral activity from *H. pluvualis* was subjected to GC-MS analysis was identified with short-chain fatty acids, a key molecule to inhibit HSV-1. Few compounds such as β-ionone, neophytadiene, phytol, palmitic acid, and α-linolenic acid from *D. salina* extracts were found to possess antiviral activity. The marine microalgae *Gyrodiniumim pudicum* contains unique sulfated polysaccharide p-KG03 (comprising of sugar units conjugated with uronic acid and sulfate groups), and it was reported as the novel antiviral compound. The compound has shown activity against the tumor cell growth and infection by encephalomyocarditis virus. Similarly, *Navicula directa* (W. Smith) Ralfs, a diatom collected from the deep sea, has been shown to possess antiviral activities against HSV-1 and 2 and influenza A virus. The compound responsible for antiviral activity was identified as the sulfated polysaccharide and named as naviculan. Though, several extracts of microalgae are reported with potential antiviral activity, there are no reports related to nanoparticles from microalgae to exhibit antiviral activity. Thus, this review will be an eye-opener to use marine microalgal species for nanoparticle synthesis and further applications as antiviral activity.

17.6 BIONANOCOMPOSITES FROM MICROALGAE

Microalgae have been reported for several other applications such as synthesis of semiconductor nanoparticles, silicon nanoparticles, bimetallic nanoparticles (alloy nanoparticles), other metallic nanoparticles, and metal oxide nanoparticles. Silicon nanoparticles have been successfully employed as bioindicators for detection of toxic compounds from industrial waste, as well as used as biostimulators, nanopesticides, nanofertilizers, and nanoherbicides in agricultural sector. In recent years, algae-based nanocomposites have been the focus of the scientific community due to its positive impact on environment, including carbon sequestration and production of biodegradable products at lower cost.

In general, composite are materials which consist of a matrix of natural fibers. The natural fibers are obtained from crops, like cotton, hemp, and flax. Because of the unique properties such as biodegradability, recyclable, low cost, and light weight, biocomposites are considered to be the promising candidates in several industries, and their importance is growing day by day. These materials are incorporated with other inorganic nanoparticles, such as carbon fibers to maintain the strength and enhance the extensive nature. Also, the inorganic part provides thermal properties and physical stability, which is very much required for improving the functional properties of the materials. These materials allow various applications in various disciplines, for example, biosensors, biomedical engineering, tissue engineering, and drug delivery. These compounds are versatile and most reliable due to the ease of processing and the variety of starting materials. Microalgae are of extensive demand, and thus biocomposites and bionanocomposites can assist in various applications. For biomedical applications, these materials are designed in such a manner that they are incorporated into the human system and remain until they are needed and dissolved later.

Preparation of bionanocomposites from tetrachloroaurate reduction by diatoms (*Navicula atomus, Diadesmis gallica*) has been reported valuable, efficient, and cost-effective. Besides, preparation is very reliable and eco-friendly, compared to other chemical methods that are toxic to the environment. Due to their remarkable properties, silica-gold and EPS-gold bionanocomposites have been valued for several applications. Similarly, silicon-germanium (Si-Ge) oxide nanocomposite in the diatom *Stauroneis* sp. was synthesized by using a two-stage cultivation process, and the characterization process revealed the incorporation of germanium into diatom cells.

17.7 CONCLUSION

Overall, this chapter details the synthesis of nanoparticles and bionanocomposites from microalgae and their importance in biomedical field. Despite that several applications of microalgae are described, microalgae associated with medicinal applications are largely booming due to their medicinal value as well as ecofriendly, low cost, and reproducible nature. It has attracted the researchers in the field to identify several novel compounds especially from marine source to treat incurable diseases such as diabetes, cancer, neurodegenerative disorders, etc.

REFERENCES

Abdel-Raouf, N., Al-Enazi, N. M., and Ibraheem, I. B. (2017). Green biosynthesis of gold nanoparticles using *Galaxaura elongata* and characterization of their antibacterial activity. Arabian Journal of Chemistry, 10, S3029–S3039.

Abdul, W., Misbahud, D., Asmat, A., Muhammad, A., Shakeeb, A., Abdul, B., and Atta, U. K. (2021). A comprehensive review of green synthesis of copper oxide nanoparticles and their diverse biomedical applications. Inorganic Chemistry Communications, 123, 108369.

Akintelu, S. A., Folorunso, A. S., Folorunso, F. A., and Oyebamiji, A. K. (2020). Green synthesis of copper oxide nanoparticles for biomedical application and environmental remediation. Heliyon, 6(7), e04508.

Bajpai, V. K., Shukla, S., Kang, S. M., Hwang, S. K., Song, X., Huh, Y. S., and Han, Y. K. (2018). Developments of cyanobacteria for nano-marine drugs: Relevance of nanoformulations in cancer therapies. Marine Drugs, 16(6), 179. https://doi.org/10.3390/md16060179.

Banker, R., and Carmeli, S. (1998). Tenuecyclamides A– D, cyclic hexapeptides from the cyanobacterium *Nostoc spongiaeforme* var. tenue. Journal of Natural Products, 61(10), 1248–1251.

Colin, J. A., Pech-Pech, I. E., Oviedo, M., Águila, S. A., Romo-Herrera, J. M., and Contreras, O. E. (2018). Gold nanoparticles synthesis assisted by marine algae extract: Biomolecules shells from a green chemistry approach. Chemical Physics Letters, 708, 210–215.

El-Deeb, N. M., Abo-Eleneen, M. A., Al-Madboly, L. A., Sharaf, M. M., Othman, S. S., Ibrahim, O. M., and Mubarak, M. S. (2020). Biogenically synthesized polysaccharides-capped silver nanoparticles: Immunomodulatory and antibacterial potentialities against resistant *Pseudomonas aeruginosa*. Frontiers in Bioengineering and Biotechnology, 8, 643.

El-Naggar, N. E. A., Hussein, M. H., Shaaban-Dessuuki, S. A., and Dalal, S. R. (2020). Production, extraction and characterization of *Chlorella vulgaris* soluble polysaccharides and their applications in AgNPs biosynthesis and biostimulation of plant growth. Scientific Reports, 10(1), 1–19.

Ferlay, J., Colombet, M., Soerjomataram, I., Parkin, D. M., Piñeros, M., Znaor, A., and Bray, F. (2021). Cancer statistics for the year 2020: An overview. International Journal of Cancer. 10.1002/ijc.33588. Advance online publication.

Haggag, W. M., Hoballah, M. M. E., and Ali, R. R. (2018). Applications of nano biotechnological microalgae product for improve wheat productivity in semai-aird areas. International Journal of Agricultural Technology, 14(5), 675–692.

Hussein, H. A., Mohamad, H., Ghazaly, M. M., Laith, A. A., and Abdullah, M. A. (2020). Anticancer and antioxidant activities of *Nannochloropsis oculata* and *Chlorella* sp. extracts in co-application with silver nanoparticle. Journal of King Saud University-Science, 32(8), 3486–3494.

Jiang, L., Wang, Y., Yin, Q., Liu, G., Liu, H., Huang, Y., and Li, B. (2017). Phycocyanin: A potential drug for cancer treatment. Journal of Cancer, 8(17), 3416–3429.

Joo, M., Park, J. J., Lee, B., Hong, S., Hwang, J., Chung, Y., Kim, J., Yeon, J., Kim, J., Byun, K., and Bak, Y. T. (2009). Anticancer effects of astaxanthin in stomach cancer cell lines. Journal of Gastroenterology and Hepatology, 24.

Juárez-Maldonado, A., Ortega-Ortíz, H., Morales-Díaz, A. B., González-Morales, S., Morelos-Moreno, Á., Sandoval-Rangel, A., . . . Benavides-Mendoza, A. (2019). Nanoparticles and nanomaterials as plant biostimulants. International Journal of Molecular Sciences, 20(1), 162.

Kashyap, M., Samadhiya, K., Ghosh, A., Anand, V., Shirage, P. M., and Bala, K. (2019). Screening of microalgae for biosynthesis and optimization of Ag/AgCl nano hybrids having antibacterial effect. RSC Advances, 9(44), 25583–25591.

Khalid, M., Khalid, N., Ahmed, I., Hanif, R., Ismail, M., and Janjua, H. A. (2017). Comparative studies of three novel freshwater microalgae strains for synthesis of silver nanoparticles: Insights of characterization, antibacterial, cytotoxicity and antiviral activities. Journal of Applied Phycology, 29(4), 1851–1863.

Kim, K. J., Sung, W. S., Suh, B. K., Moon, S. K., Choi, J. S., Kim, J. G., and Lee, D. G. (2009). Antifungal activity and mode of action of silver nano-particles on *Candida albicans*. Biometals, 22(2), 235–242.

Kotake-Nara, E., Asai, A., and Nagao, A. (2005). Neoxanthin and fucoxanthin induce apoptosis in PC-3 human prostate cancer cells. Cancer Letters, 220(1), 75–84.

Park, M., and Kim, H. (2017). Anti-cancer mechanism of docosahexaenoic acid in pancreatic carcinogenesis: A mini-review. Journal of Cancer Prevention, 22(1), 1–5.

Pathak, J., Pandey, A., Maurya, P. K., Rajneesh, R., Sinha, R. P., and Singh, S. P. (2019). Cyanobacterial secondary metabolite scytonemin: A potential photoprotective and pharmaceutical compound. Proceedings of the National Academy of Sciences, India Section B: Biological Sciences, 1–15.

Sathishkumar, R. S., Sundaramanickam, A., Srinath, R., Ramesh, T., Saranya, K., Meena, M., and Surya, P. (2019). Green synthesis of silver nanoparticles by bloom forming marine microalgae *Trichodesmium erythraeum* and its applications in antioxidant, drug-resistant bacteria, and cytotoxicity activity. Journal of Saudi Chemical Society, 23(8), 1180–1191.

Senthilkumar, P., Surendran, L., Sudhagar, B., and Kumar, D. R. S. (2019). Facile green synthesis of gold nanoparticles from marine algae *Gelidiella acerosa* and evaluation of its biological potential. SN Applied Sciences, 1(4), 1–12.

Serrill, J. D., Wan, X., Hau, A. M., Jang, H. S., Coleman, D. J., Indra, A. K., Alani, A. W., McPhail, K. L., and Ishmael, J. E. (2016). Coibamide A, a natural lariat depsipeptide, inhibits VEGFA/VEGFR2 expression and suppresses tumor growth in glioblastoma xenografts. Investigational New Drugs, 34(1), 24–40.

Shih, C., and Teicher, B. A. (2001). Cryptophycins: A novel class of potent antimitotic antitumor depsipeptides. Current Pharmaceutical Design, 7(13), 1259–1276.

Sutradhar, P., Saha, M., and Maiti, D. (2014). Microwave synthesis of copper oxide nanoparticles using tea leaf and coffee powder extracts and its antibacterial activity. Journal of Nanostructure in Chemistry, 4(1), 86.

Torabfam, M., and Yüce, M. (2020). Microwave-assisted green synthesis of silver nanoparticles using dried extracts of *Chlorella vulgaris* and antibacterial activity studies. Green Processing and Synthesis, 9(1), 283–293.

Vishwakarma, J., and Sirisha, V. L. (2020). Unraveling the anti-biofilm potential of green algal sulfated polysaccharides against *Salmonella enterica* and *Vibrio harveyi*. Applied Microbiology and Biotechnology, 104, 6299–6314.

Waris, A., Din, M., Ali, A., Ali, M., Afridi, S., Baset, A., and Khan, A. U. (2020). A comprehensive review of green synthesis of copper oxide nanoparticles and their diverse biomedical applications. Inorganic Chemistry Communications, 123(2021), 108369.

Xu, S., Nijampatnam, B., Dutta, S., and Velu, S. E. (2016). Cyanobacterial metabolite calothrixins: Recent advances in synthesis and biological evaluation. Marine Drugs, 14(1), 17.

18 Recent Biotechnological Advances in Understanding the Omega-3 Fatty Acid Accumulation in Algal Lipids

Ayushi Dalmia and Ajay W Tumaney

18.1 INTRODUCTION

Omega-3 (ω-3) polyunsaturated fatty acids (PUFAs) have recently gained augmented academic and industrial interest due to its myriad health benefits. The health-promoting role of ω-3 PUFAs, particularly in the maintenance of cardiovascular, neurological, and visual health is well established. There is mounting evidence highlighting the importance of dietary ω-3 fatty acid consumption throughout the life cycle, especially during infancy, pregnancy, and old age. The ω-3 fatty acid supplementation is proved to be beneficial against various diseases such as cardiovascular diseases, dyslipidemia, depression, diabetes, and cancer. Anti-inflammatory properties of ω-3 fatty acids help in conferring positive effect against conditions such as rheumatoid arthritis, Crohn's disease, asthma, eczema, and psoriasis (Ghasemi Fard et al., 2019; Gogus and Smith, 2010; MacLean et al., 2006; Shahidi and Ambigaipalan, 2018; Simopoulos, 1991; Swanson et al., 2012; Yashodhara et al., 2009). The major fatty acids in ω-3 series include α-linolenic acid (ALA; 18:3 $\Delta^{9, 12, 15}$), stearidonic acid (SDA; 18:4 $\Delta^{6, 9, 12, 15}$) and a very long chain of polyunsaturated fatty acids (VLCPUFAs), namely, eicosapentaenoic acid (EPA; 20:5 $\Delta^{5, 8, 11, 14, 17}$), docosapentaenoic acid (DPA; 22:5 $\Delta^{7, 10, 13, 16, 19}$), and docosahexaenoic acid (DHA; $\Delta^{4, 7, 10, 13, 16, 19}$) (Prasad et al., 2020; Sijtsma and De Swaaf, 2004). Nutritionally, VLCPUFAs are the most biologically active form of ω-3 fatty acids. Currently, marine fish is the major commercial source of EPA and DHA for human consumption. However, there has been rising concern on fish utilization due to declining fish stock, heavy metal toxicity, and inconsistency in fatty acid content. The global demand for EPA and DHA is projected to be valued at $2,614 million by the year 2021 (www.frost.com). Rising global demand of ω-3 fatty acid, in addition to decreased sustainability of fish sources, calls for exploring alternative sources (Adarme-Vega et al., 2014). Microalgae are the primary producers of ω-3 fatty acids. Various microalgal species are known to accumulate large amounts of EPA and DHA. There has been increased interest in utilizing microalgal lipids rich in ω-3 fatty acids for human consumption. Various advantages associated with microalgae include use of nonarable land, year-round production, rapid growth rate, and very high product density. Additionally, microalgae also are a vegetarian source of ω-3 fatty acids (Chisti, 2007).

Various industries that extract ω-3 rich oil from microalgae have come up. However, microalgae can also be used as a source of genes for genetic engineering to further augment ω-3 content. There can be two plausible approaches to enhance ω-3 production by employing genetic engineering, one aiming towards ameliorating the quality or quantity of ω-3 in lipids in the ω-3 producing strain and the other being generation of novel transgenic sources of ω-3 PUFAs. To accomplish this, it is important to have a thorough understanding of the ω-3 fatty acid metabolic pathway. The understanding of ω-3 fatty acid synthesis pathway of ω-3 producing microalgae is well established (Xiao, 2003). However, understanding of the microalgal genes that partition ω-3 fatty acids into storage

DOI: 10.1201/9781003219156-22

lipid molecules is still in infancy. Hence we present the current understanding of the genes/mechanism of ω-3 PUFA accumulation in algal biosystems.

18.2 MICROALGAE AS PRODUCERS OF ω-3 FATTY ACIDS

Marine microalgal species belonging to various classes have been identified as a prodigious source of ω-3 PUFAs. Table 18.1 shows the PUFA content in various microalgal species. *Pavlova lutheri, Pavlova viridis, Phaeodactylum tricornutum, Thraustochytrium* sp., *Nannochloropsis* sp., *Cryptocodinium conhii, Schizochytrium* sp., *Thalasiossira pseudonana, Nannochloropsis oceanica* are few examples of microalgal species rich in ω-3 lipids (Guihéneuf et al., 2009; C. Hu et al., 2008; H. Hu and Gao, 2003; Jiang and Chen, 2000; Patil et al., 2007; Qu et al., 2013; Scott et al., 2011; Tonon et al., 2005). Microalgal species belonging to *Isochrysis, Chaetoceros, Nannochloropsis, Pavlova, Gymnodinum*, and *Phaeodactylum* genus are commonly cultured for aquaculture feed to enrich zooplankton and fish larve with ω-3 fatty acids (Reitan et al., 1997). Lately, several industries based on microalgal ω-3 lipids have emerged aiming for VLCPUFA production for supplements and infant food formulations. *Cryptocodinium conhii* is industrially used for the production of DHA rich oil (Kyle, 2001). *Nannochloropsis* sp. is cultured commercially for EPA production by a few companies; Israel-based Sembiotic, Nekton-Algafuel in Portugal and LGem in Netherlands (Khozin-Goldberg et al., 2011). USA-based company Omega Tech produces DHA rich oil from *Schizochytrium* sp. marketed as "DHA Gold" (Katiyar and Arora, 2020).

18.3 DISTRIBUTION OF ω-3 FATTY ACIDS IN MICROALGAL LIPIDS

An understanding of PUFA distribution in cells gives insights on the underlying PUFA accumulating pathway. Therefore, it becomes crucial to know the PUFA distribution in the cells, whether it is present in the membrane lipids such as those of phospholipid or glycolipid or the cytosolic lipids in the form of TAG. A major amount of EPA is found in MGDG in *Pavlova lutheri* and *Nannochloropsis oculata. Monodus subterraneus* and *Phaeodactylum tricornutum* accumulate most of its EPA in monogalactosyl diacylglycerol (MGDG), phosphatidylglycerol (PG), and digalactosyldiacylglycerol (DGDG). Growing *Pavlova lutheri* cells in semicontinuous mode had 45% of EPA in MGDG whereas 33% in TAG (Mühlroth et al., 2013). In algae PUFA accumulation in TAG generally occur when stress is induced like in case of cell division arrest. In the stationary phase of *Nannochloropsis oculata*, TAG contributed to 90% of total fatty acid and 68% of the total EPA share. It is interesting to note that the fatty acid partitioning into the end storage molecule varies widely with algal species as well as growth phase or culture conditions. *Nannochloropsis oculata* and *Phaeodactylum tricornutum* both are EPA producers, but the amount of EPA partitioned into TAG is much higher in *Nannochloropsis oculata* than in *Phaeodactylum tricornutum. Phaeodactylum tricornutum* does not accumulate any DHA in TAG. The bacillariophyceae *Thalassiosira pseudonana* and the haptophyte *Pavlova lutheri* were found to accumulate higher amounts of EPA and DHA in TAG during stationary phase compared to the log phase (Tonon et al., 2002). The findings of this study highlight the fact that the EPA and DHA accumulation trend is varied in different species and cannot be extrapolated. In case of *Phaeodactylum tricornutum*, not much variation in fatty acid composition was seen with change in growth phase (Alonso et al., 2000). However, with transitioning from log to stationary phase, EPA content in TAG increased from 3% to 40% (Tonon et al., 2002). In *Thraustochytrium aureum*, TAG is the major lipid which contains around 40% of DHA (Hur et al., 2002). The previously mentioned studies highlight that ω-3 fatty acid accumulation pattern is different for different algae. It is desirable to have PUFA in TAG for nutritional and extraction applications. The partitioning of PUFA into TAG in various algae and their growth phase is performed by different enzymes. Therefore, elucidation of enzymes that channel PUFA in neutral lipids can help in selecting targets for genetic engineering techniques.

18.4 BIOCHEMICAL ENGINEERING TO INCREASE LIPIDS/PUFA

Biochemical engineering refers to controlling of nutritional factors to yield enhanced lipid levels. Stress in the form of nutrient starvation has been observed to direct the flux of carbon towards lipid production and result in increased lipid content. There are various methods of inducing stress such as phosphorus starvation, iron starvation, high salinity, which are known to induce lipid production. The most common means of increasing lipid accumulation is through nitrogen starvation. Increased lipid levels in nitrogen limitation conditions has been reported in many microalgal species such as *Chlamydomonas reinhardtii*, *Aurantiochytrium* sp., and *Chlorella* sp. (Jakobsen et al., 2008; Illman et al., 2000; Boyle et al., 2012). Low dissolved oxygen levels are favorable for DHA accumulation in *Aurantiochytrium* sp. (Jakobsen et al., 2008). The increase in lipid due to nitrogen starvation may be related to upregulation of various genes of lipid synthesis pathway. Some regulatory genes are also upregulated in nitrogen starvation. During stress conditions the carbon flux is channelled to lipid synthesis, and cell growth is negatively affected. Therefore, in case of biochemical engineering, though lipid content can be increased, the biomass accumulation is compromised. For industrial use however an improved ability to accumulate lipids without hindering biomass accumulation is favorable. Genetic engineering can help increase ω-3 in TAG without having a negative impact on biomass, if appropriate genes are exploited. In case of nutrient stress, lipid biosynthesis genes are upregulated; the genes however are unspecific to fatty acid. Metabolic engineering can facilitate selective increase of ω-3 fatty acids in lipids by using genes specific to ω-3 fatty acid accumulation. In order to achieve this, a thorough understanding of microalgal lipid synthesis genes is required.

18.5 TRIACYLGLYCEROL (TAG) SYNTHESIS ENZYMES

TAG synthesis pathway, or Kennedy pathway, was discovered in the 1950s by Eugene Kennedy and colleagues. The Kennedy pathway takes place in the endoplasmic reticulum (ER) and involves sequential addition of fatty acyl-CoA building blocks onto a glycerol-3-phosphate (G3P) backbone to form TAG. The pathway begins with acylation of G3P by the enzyme glycerol-3-phosphate acyltransferase (GPAT) to form lysophosphatidic acid (LPA). The enzyme GPAT has the least specificity and is also considered rate limiting. Further acylation of LPA is catalyzed by lysophosphatidic acid acyltransferase (LPAT) to produce phosphatidic acid (PA). Phosphatidic acid phosphatase (PAP) then dephosphorylate PA to form diacylglycerol (DAG). Finally, the enzyme acyl CoA: diacylglycerol acyltransferase (DGAT) performs acylation of DAG to form TAG. DGAT is considered crucial for TAG synthesis pathway and is the most committed step (Y. Xu et al., 2018). DGAT activity and specificity greatly define the quantity of TAG and its fatty acid composition. The pathway is illustrated in Figure 18.1. TAG formation from DAG can alternatively happen by the acyl-CoA independent pathway where phosphatidylcholine acts as an acyl donor, and the reaction is catalyzed by the enzyme phospholipid: diacylglycerol acyltransferase (PDAT). The function of PDAT is well established in higher plant. However, compared to DGAT, the enzyme PDAT is less studied. DGAT and PDAT are considered key players in determining the lipid quality and quantity. Although our understanding of algal TAG synthesis pathway is based on those pathways found in other systems, specially those in higher plants, recent studies have highlighted some differences in the plant and algal TAG biosynthesis. The green algae *Chlamydomonas reinhardtii* uses chloroplast specific acyltansferases for synthesis of TAG from DAG which is in contrast to plants where TAG synthesis occurs at the ER (Fan et al., 2011). The following section describes the current understanding of the key microalgal enzymes of TAG synthesis pathway.

18.5.1 GPAT

The enzyme GPAT is crucial for initiation of TAG synthesis, and therefore, its genetic manipulation is of great importance. The enzyme is regulated at multiple levels, including transcriptional and posttranscriptional level (Courchesne et al., 2009). GPAT and other four genes from yeast when overexpressed in *Chlorella minutissima* resulted in a considerable increase in lipid content (Hsieh et al., 2012). In plants GPATs are found both in ER and in plastid, whereas in case of various algae including *Cyanidioschyzon merolae*, *Phaeodactylum tricornutum*, *Thalassiosira pseudonana*, *Chlamydomonas reinhardtii*, and *Ostreococcus tauri*, GPAT was found only in plastid (Lykidis and Ivanova, 2014). Sequence similarity between plant and algal DGAT was only 50–60%. However, secondary structure was conserved among *Chlamydomonas reinhardtii*, *Arabidopsis thaliana*, and *Glycine max* as revealed by 3D structure analysis. Active site of various algal GPATs that interact with the fatty acyl motifs and G3P were also predicted. In case of *Chlamydomonas reinhardtii*, potential amino acid targets for improving the stability of GPAT were identified. These studies thus not only strengthen our understanding of algal GPATs and provides leads for further genetic manipulation but also highlights the need for further studies to have a better know-how of GPAT's role in algal ω-3 metabolism (Misra and Panda, 2013). The gene phospholipid acyltransferase 2 (PLAT2) when disrupted in the protist *Aurantiochytrium limacinum* F26-b showed decrease in GPAT activity. Overexpression of PLAT 2 resulted in an increased GPAT activity and a concomitant rise in DHA containing LPA. The study concluded that PLAT2 exhibited GPAT activity in *Aurantiochytrium limacinum* and showed preference for the ω-3 fatty acid DHA (Nutahara et al., 2019).

18.5.2 DGAT

The enzyme DGAT is considered very important for TAG synthesis pathway and is the rate limiting enzyme. There are three classes of DGAT: DGAT1 and DGAT2, which are membrane bound and are the major contributor for TAG synthesis, and third class is DGAT3, which is cytosolic. All the three DGAT classes have very little sequence similarity and belong to different gene families (Y. Xu et al., 2018). In conditions that promote TAG accumulation in algae such as nitrogen starvation, expression of DGAT has been found to increase, thereby highlighting the importance of DGATs in lipid accumulation. Unlike plants and animals, microalgae are known to have multiple isoforms of DGAT. *Chlamydomonas reinhardtii*, *Chlamydomonas zofingiensis, Nannochloropsis oceanica, Tetraselmis chui* are few algal species that have been reported to have multiple copies of DGATs (Xu et al., 2013; Liu et al., 2016; Mao et al., 2019; Zienkiewicz et al., 2017; Úbeda-Mínguez et al., 2017; La Russa et al., 2012). There are various algal genes related to TAG synthesis which are annotated but lacks functional characterization. However, predicted enzymes need molecular and biochemical examination. The most preferred method of functional characterization is by expression in quadruple mutant yeast strain (*Saccharomyces cerevisae*, H1246) devoid of TAG to see for complementation of DGAT activity. Other popular ways are to knock out or overexpress gene in the host. Additionally, the efficiency and the substrate specificities can only be understood by experimentation. DGATs from various algal species have been functionally characterized, including those from *Chlamydomonas reinhardtii, Haematococcus pluvialis, Lobosphaera incisa, Tetraselmis chui, Chlorella vulgaris* (Nguyen et al., 2019; Sitnik et al., 2018; Úbeda-Mínguez et al., 2017; Kirchner et al., 2016; Hung et al., 2013). Table 18.2 shows the studies done on algal DGATs. DGAT2, otherwise known as DGAT type 2 (DGTT), is the most abundant DGAT found in microalgae. Multiple copies of DGTT has been reported in various algal species, such as *Phaeodactylum tricornutum*, *Chlamydomonas reinhardtii, Nannochloropsis oceanica, Chlamydomonas zofingiensis* (Hung et al., 2013; Zienkiewicz et al., 2017; Mao et al., 2019; Dinamarca et al., 2017). Not all putative DGATs appear to be functionally active, for instance DGTT5 from *Chlamydomonas reinhardtii* did not show biochemical activity. Overexpression of some DGATs did not affect the lipid profile; for instance, when three Type 2 DGAT enzymes from *Chlamydomonas reinhardtii* were overexpressed, no alteration in the fatty acid or the TAG profile was seen (La Russa et al., 2012). Indicating role of some regulatory mechanism in controlling lipid accumulation. Functional characterization aided in identifying isoforms with a stronger activity than others, an example being DGTT2 from *Chlamydomonas reinhardtii*, which had a very strong activity and led to a ninefold increase in TAG levels in yeast cells (Liu et al., 2016).

Similarly, *Nannochloropsis oceanica* DGTT5 was reported to be a potential tool for genetic engineering owing to its activity (Zienkiewicz et al., 2017).

Various studies have explored the expression of algal DGATs in plants and studied the effect in terms of oil yield and fatty acid composition. Expression of *Chlorella ellipsoidea* DGAT 1 in *Brassica napus* and *Arabidopsis thaliana* increased the total oil content and also the lipid content per 1,000 seeds and did not have an effect on the fatty acid profile (Guo et al., 2017). DGAT2 from the PUFA producing protist *Thraustochytrium aureum* when expressed in *Arabidopsis thaliana* was found to selectively incorporate the MUFA oleic acid in TAG stores (Zhang et al., 2013). Some plant DGATs have also been expressed in algal species to see its effect on TAG production. DGAT2 from *Brassica napus* was expressed in *Chlamydomonas reinhardtii* and resulted in increase in MUFA by 7% and the increase in ALA by 12% (Ahmad et al., 2015).

Phylogenetic analysis shows that *Nannochloropsis oceanica* has DGAT2s from different origin, namely, from green algae, red algae, and from an ancestral eukaryotic heterotrophic host. Figure 18.2 shows the various DGAT2s from *Nannochloropsis oceanica*. Endosymbiosis often leads to gene transfer in algae and might have resulted in DGATs of different origin. DGAT2A is localized in the ER and is responsible for the flux of saturated fatty acid (SFA), DGAT2C prefers PUFA substrate and is localized in the chloroplast, whereas DGAT2D is cytosolic and catalyzes addition of monounsaturated fatty acids (MUFA) into TAG. Therefore, all the three DGATs are localized at different organelles and have varied fatty acyl-CoA preference (Xin et al., 2017). DGTT from *Chlamydomonas reinhardtii* also showed variability in fatty acid selectivity, DGTT1 preferred PUFA, DGTT2 exhibited preference for monounsaturated fatty acids (MUFA), whereas DGTT3 preferred hexadecanoic acid (Liu et al., 2016). The gene isoforms have complementary roles and work in synergy to yield the final fatty acid composition of TAG. The heterogeneity of DGAT activity observed further highlights the importance of functional characterization and shows how each DGAT2 isoform can be used to yield specific fatty acid composition in TAG (Xin et al., 2017).

The knowledge of algal DGAT3 is in infancy as DGAT3 from very few algae have been studied. DGAT3 in algae was first reported by Cui et al. (2013) in *Phaeodactylum tricornutum*. The enzyme had high sequence similarity to bacterial DGAT3 and further functional characterization of the same confirmed DGAT activity. The enzyme was found to prefer 18 carbon PUFAs. DGAT3 from *Chlamydomonas reinhardtii* was also functionally characterized. *In silico* prediction of subcellular localization suggest that most of the DGAT3 in green algae are localized in the chloroplast indicative of a soluble *de novo* TAG synthesis pathway (Bagnato et al., 2017). Some DGATs are also known to have dual function exhibiting wax synthase (WS) and DGAT activity. *Phaeodactylum tricornutum* WS/DGAT had positive impact on lipid and TAG accumulation (Cui et al., 2018). The fact that so many of the putative DGATs when subjected to experimental analysis did not show activity highlights the importance of biochemical characterization. It was interesting to note that in most cases, different isoforms had different fatty acyl-CoA preference and also enzymes from PUFA producing organisms did not necessarily exhibit preference for accumulating PUFAs.

18.5.3 PDAT

PDAT is an important enzyme in TAG biosynthesis. Understanding of microalgal PDAT is limited and needs deeper understanding. *Chlamydomonas reinhardtii PDAT* was found to have other functions apart from glycerolipid/phospholipid: diacylglycerol acyltransferase activity and includes lipase and DAG: DAG transacylase. The enzyme PDAT is important for phospholipid turnover and its importance is even more crucial in case of stress conditions where it helps in degradation of membrane lipids and synthesis of TAG. Under stress condition PDAT from *Chlamydomonas reinhardtii* was found to incorporate fatty acids from membrane glycerolipids into TAG (Yoon et al., 2012). In *Nannochloropsis gaditana*, partitioning of EPA in nitrogen replete and nitrogen deplete condition among lipid class showed that EPA was higher in the membrane lipids in the nitrogen sufficient condition and on nitrogen starvation was found to increase in TAG fraction, hinting a plausible role of PDAT in channelling the flux of fatty acids into neutral lipids from membrane lipids (Banerjee et al., 2017).

18.5.4 Multigene Approach

Some efforts in exploring the effect of simultaneous overexpression of multiple key genes of TAG biosynthesis have been made. Multigene and single gene overexpression of the Kennedy pathway genes (GPAT, LPAT, and DGAT) in the algae *Neochloris oleobundans* resulted in increased lipid content. Overexpression of single gene resulted in up to 1.3 fold increase in the total fatty acids, whereas 1.4 fold increase in TAG content. In case of multiple gene overexpression, the growth rate and the photosynthetic activity was negatively affected, and the total fatty acid increased by only 1.2 fold (Muñoz et al., 2019).

Recent academic efforts to understand microalgal lipid metabolism has helped in strengthening our knowledge of lipid metabolism in algal cells and facilitated the identification of potential targets for genetic manipulation to augment ω-3 fatty acids in lipids.

18.6 CONCLUSION AND FUTURE PERSPECTIVE

ω-3 PUFAs are known to have multiple health benefits and are of nutritional importance. EPA and DHA are the most biologically active ω-3 fatty acids. There has been a constant increase in the demand of ω-3 fatty acid and therefore, the need of alternate sustainable sources to meet supply. Microalgae, especially those of the marine environment, are primary producers of ω-3 PUFAs, and there has been an increased interest in utilizing microalgae for industrial applications. Lately, there have been enormous efforts in augmenting the ω-3 PUFA content in microalgae by various means. Biochemical engineering approaches have been extensively worked upon to improve lipid yield in microalgae, and various nutritional stress such as those of phosphorus and nitrogen are well studied. It is now an established fact that nutrient deprivation increases lipid content but negatively affects biomass accumulation. It is noteworthy to mention that during nutrient deprivation, TAG synthesis genes and some regulatory genes are upregulated. Using genetic engineering approach by manipulating ω-3 specific target genes can facilitate an increase in ω-3 accumulation in algal lipids. The first step is to identify the potential targets for enhancing ω-3 fatty acid accumulation in algal lipids. Key genes of TAG synthesis pathway have been identified in algae. Advancements in microalgal biotechnology facilitate isolation and utilization of genes for genetic transformation. Recent studies have revealed a lot about ω-3 lipid metabolism in algae and have brought in light many potential target genes for genetic engineering. The genes of the glycerolipid pathway, including GPAT, DGAT, and PDAT appear to be promising candidates for bioengineering. These studies thus not only strengthen our understanding of algal enzymes and provide leads for further genetic manipulation but also set the stage for further research to have better understanding of the role of algal enzymes in ω-3 metabolism.

REFERENCES

Adarme-Vega, T. C., Thomas-Hall, S. R., and Schenk, P. M. (2014). Towards sustainable sources for omega-3 fatty acids production. *Current Opinion in Biotechnology*, *26*, 14–18. https://doi.org/10.1016/j.copbio.2013.08.003.

Ahmad, I., Sharma, A. K., Daniell, H., and Kumar, S. (2015). Altered lipid composition and enhanced lipid production in green microalga by introduction of brassica diacylglycerol acyltransferase 2. *Plant Biotechnology Journal*, *13*(4), 540–550. https://doi.org/10.1111/pbi.12278.

Alonso, D. L., Belarbi, E. H., Fernández-Sevilla, J. M., Rodríguez-Ruiz, J., and Grima, E. M. (2000). Acyl lipid composition variation related to culture age and nitrogen concentration in continuous culture of the microalga *Phaeodactylum tricornutum*. *Phytochemistry*, *54*(5), 461–471. https://doi.org/10.1016/S0031-9422(00)00084-4.

Bagnato, C., Prados, M. B., Franchini, G. R., Scaglia, N., Miranda, S. E., and Beligni, M. V. (2017). Analysis of triglyceride synthesis unveils a green algal soluble diacylglycerol acyltransferase and provides clues to potential enzymatic components of the chloroplast pathway. *BMC Genomics*, *18*(1), 1–23. https://doi.org/10.1186/s12864-017-3602-0.

Banerjee, A., Maiti, S. K., Guria, C., and Banerjee, C. (2017). Metabolic pathways for lipid synthesis under nitrogen stress in *Chlamydomonas* and *Nannochloropsis*. *Biotechnology Letters*, *39*(1), 1–11. https://doi.org/10.1007/s10529-016-2216-y.

Boyle, N. R., Page, M. D., Liu, B., Blaby, I. K., Casero, D., Kropat, J., Cokus, S. J., Hong-Hermesdorf, A., Shaw, J., Karpowicz, S. J., Gallaher, S. D., Johnson, S., Benning, C., Pellegrini, M., Grossman, A., and Merchant, S. S. (2012). Three acyltransferases and nitrogen-responsive regulator are implicated in nitrogen starvation-induced triacylglycerol accumulation in *Chlamydomonas*. *Journal of Biological Chemistry*, *287*(19), 15811–15825. https://doi.org/10.1074/jbc.M111.334052.

Chisti, Y. (2007). Biodiesel from microalgae. *Biotechnology Advances*, *25*(3), 294–306. https://doi.org/10.1016/j.biotechadv.2007.02.001.

Courchesne, N. M. D., Parisien, A., Wang, B., and Lan, C. Q. (2009). Enhancement of lipid production using biochemical, genetic and transcription factor engineering approaches. *Journal of Biotechnology*, *141*(1–2), 31–41. https://doi.org/10.1016/j.jbiotec.2009.02.018.

Cui, Y., Zhao, J., Wang, Y., Qin, S., and Lu, Y. (2018). Characterization and engineering of a dual-function diacylglycerol acyltransferase in the oleaginous marine diatom *Phaeodactylum tricornutum*. *Biotechnology for Biofuels*, *11*(1), 1–13. https://doi.org/10.1186/s13068-018-1029-8.

Cui, Y., Zheng, G., Li, X., Lin, H., Jiang, P., and Qin, S. (2013). Cloning and characterization of a novel diacylglycerol acyltransferase from the diatom *Phaeodactylum tricornutum*. *Journal of Applied Phycology*, *25*(5), 1509–1512. https://doi.org/10.1007/s10811-013-9991-9.

Dinamarca, J., Levitan, O., Kumaraswamy, G. K., Lun, D. S., and Falkowski, P. G. (2017). Overexpression of diacylglycerolnacyltrasferase gene in *Phaeodactylum tricornutum* directs carbon towards lipid biosynthesis. *Journal of Phycology*, *53*(2), 405–414. https://doi.org/10.1111/jpy.12513.

Fan, J., Andre, C., and Xu, C. (2011). A chloroplast pathway for the de novo biosynthesis of triacylglycerol in *Chlamydomonas reinhardtii*. *FEBS Letters*, *585*(12), 1985–1991. https://doi.org/10.1016/j.febslet.2011.05.018.

Ghasemi Fard, S., Wang, F., Sinclair, A. J., Elliott, G., and Turchini, G. M. (2019). How does high DHA fish oil affect health? A systematic review of evidence. *Critical Reviews in Food Science and Nutrition*, *59*(11), 1684–1727. https://doi.org/10.1080/10408398.2018.1425978.

Gogus, U., and Smith, C. (2010). n-3 omega fatty acids: A review of current knowledge. *International Journal of Food Science and Technology*, *45*, 417–436. https://doi.org/10.1111/j.1365-2621.2009.02151.x.

Guihéneuf, F., Mimouni, V., Ulmann, L., and Tremblin, G. (2009). Combined effects of irradiance level and carbon source on fatty acid and lipid class composition in the microalga *Pavlova lutheri* commonly used in mariculture. *Journal of Experimental Marine Biology and Ecology*, *369*(2), 136–143. https://doi.org/10.1016/j.jembe.2008.11.009.

Guo, X., Fan, C., Chen, Y., Wang, J., Yin, W., Wang, R. R. C., and Hu, Z. (2017). Identification and characterization of an efficient acyl-CoA: Diacylglycerol acyltransferase 1 (DGAT1) gene from the microalga *Chlorella ellipsoidea*. *BMC Plant Biology*, *17*(48), 1–16. https://doi.org/10.1186/s12870-017-0995-5.

Hsieh, H.-J., Su, C.-H., and Chien, L.-J. (2012). Accumulation of lipid production in Chlorella minutissima by triacylglycerol biosynthesis-related genes cloned from *Saccharomyces cerevisiae* and *Yarrowia lipolytica*. *Journal of Microbiology*, *50*(3), 526–534. https://doi.org/10.1007/s12275-012-2041-5.

Hu, C., Li, M., Li, J., Zhu, Q., and Liu, Z. (2008). Variation of lipid and fatty acid compositions of the marine microalga *Pavlova viridis* (Prymnesiophyceae) under laboratory and outdoor culture conditions. *World Journal of Microbiology and Biotechnology*, *24*(7), 1209–1214. https://doi.org/10.1007/s11274-007-9595-0.

Hu, H., and Gao, K. (2003). Optimization of growth and fatty acid composition of a unicellular marine picoplankton, *Nannochloropsis* sp., with enriched carbon sources. *Biotechnology Letters*, *25*(5), 421–425. https://doi.org/10.1023/A:1022489108980.

Hung, C. H., Ho, M. Y., Kanehara, K., and Nakamura, Y. (2013). Functional study of diacylglycerol acyltransferase type 2 family in *Chlamydomonas reinhardtii*. *FEBS Letters*, *587*(15), 2364–2370. https://doi.org/10.1016/j.febslet.2013.06.002.

Hur, B.-K., Cho, D.-W., Kim, H.-J., Park, C.-I., and Suh, H.-J. (2002). Effect of culture conditions on growth and production of docosahexaenoic acid (DHA) using *Thraustochytrium aureum* ATCC 34304. *Biotechnology and Bioprocess Engineering*, *7*(1), 10–15. https://doi.org/10.1007/bf02935873.

Illman, A. M., Scragg, A. H., and Shales, S. W. (2000). Increase in *Chlorella* strains calorific values when grown in low nitrogen medium. *Enzyme and Microbial Technology*, *27*(8), 631–635. https://doi.org/10.1016/S0141-0229(00)00266-0.

Jakobsen, A. N., Aasen, I. M., Josefsen, K. D., and Strøm, A. R. (2008). Accumulation of docosahexaenoic acid-rich lipid in thraustochytrid *Aurantiochytrium* sp. Strain T66 : Effects of N and P starvation and O 2 limitation. *Applied Microbial and Cell Physiology*, *80*(I), 297–306. https://doi.org/10.1007/s00253-008-1537-8.

Jiang, Y., and Chen, F. (2000). Effects of temperature and temperature shift on docosahexaenoic acid production by the marine microalga *Crypthecodinium cohnii. JAOCS, Journal of the American Oil Chemists' Society*, *77*(6), 613–617. https://doi.org/10.1007/s11746-000-0099-0.

Katiyar, R., and Arora, A. (2020). Health promoting functional lipids from microalgae pool: A review. *Algal Research*, *46*, 101800. https://doi.org/10.1016/j.algal.2020.101800.

Khozin-Goldberg, I., Iskandarov, U., and Cohen, Z. (2011). LC-PUFA from photosynthetic microalgae: Occurrence, biosynthesis, and prospects in biotechnology. *Applied Microbiology and Biotechnology*, *91*(4), 905–915. https://doi.org/10.1007/s00253-011-3441-x.

Kirchner, L., Wirshing, A., Kurt, L., Reinard, T., Glick, J., Cram, E. J., Jacobsen, H. J., and Lee-Parsons, C. W. T. (2016). Identification, characterization, and expression of diacylgylcerol acyltransferase type-1 from *Chlorella vulgaris. Algal Research*, *13*, 167–181. https://doi.org/10.1016/j.algal.2015.10.017.

Kyle, D. J. (2001). The large-scale production and use of a single-cell oil highly enriched in docosahexaenoic acid. *ACS Symposium Series*, *788*(Figure 1), 92–107. https://doi.org/10.1021/bk-2001-0788.ch008.

La Russa, M., Bogen, C., Uhmeyer, A., Doebbe, A., Filippone, E., Kruse, O., and Mussgnug, J. H. (2012). Functional analysis of three type-2 DGAT homologue genes for triacylglycerol production in the green microalga *Chlamydomonas reinhardtii. Journal of Biotechnology*, *162*(1), 13–20. https://doi.org/10.1016/j.jbiotec.2012.04.006.

Liu, J., Han, D., Yoon, K., Hu, Q., and Li, Y. (2016). Characterization of type 2 diacylglycerol acyltransferases in *Chlamydomonas reinhardtii* reveals their distinct substrate specificities and functions in triacylglycerol biosynthesis. *Plant Journal*, *86*(1), 3–19. https://doi.org/10.1111/tpj.13143.

Lykidis, A., and Ivanova, N. (2014). Genomic prospecting for microbial biodiesel production. *Bioenergy*, 407–418. https://doi.org/10.1128/9781555815547.ch31.

MacLean, C. H., Newberry, S. J., Mojica, W. A., Khanna, P., Issa, A. M., Suttorp, M. J., Lim, Y.-W., Traina, S. B., Hilton, L., Garland, R., and Morton, S. C. (2006). Effects of omega-3 fatty acids on cancer risk: A systematic review. *Journal of Americal Medical Association*, *295*(4), 404–414. https://doi.org/10.1016/j.atherosclerosis.2006.02.012.

Mao, X., Wu, T., Kou, Y., Shi, Y., Zhang, Y., and Liu, J. (2019). Characterization of type I and type II diacylglycerol acyltransferases from the emerging model alga *Chlorella zofingiensis* reveals their functional complementarity and engineering potential. *Biotechnology for Biofuels*, *12*(1), 1–17. https://doi.org/10.1186/s13068-019-1366-2.

Misra, N., and Panda, P. K. (2013). In search of actionable targets for agrigenomics and microalgal biofuel production: Sequence-structural diversity studies on algal and higher plants with a focus on GPAT protein. *OMICS A Journal of Integrative Biology*, *17*(4), 173–186. https://doi.org/10.1089/omi.2012.0094.

Mühlroth, A., Li, K., Røkke, G., Winge, P., Olsen, Y., Hohmann-Marriott, M. F., Vadstein, O., and Bones, A. M. (2013). Pathways of lipid metabolism in marine algae, co-expression network, bottlenecks and candidate genes for enhanced production of EPA and DHA in species of chromista. *Marine Drugs*, *11*, 4662–4697. https://doi.org/10.3390/md11114662.

Muñoz, C. F., Weusthuis, R. A., Adamo, S. D., and Wijffels, R. H. (2019). Effect of single and combined expression of lysophosphatidic acid acyltransferase, glycerol-3- phosphate acyltransferase, and diacylglycerol acyltransferase on lipid accumulation and composition in *Neochloris oleoabundans. Frontiers in Plant Science*, *10*(November), 1–11. https://doi.org/10.3389/fpls.2019.01573.

Nguyen, T., Xu, Y., Abdel-Hameed, M., Sorensen, J. L., Singer, S. D., and Chen, G. (2019). Characterization of a type-2 diacylglycerol acyltransferase from *Haematococcus pluvialis* reveals possible allostery of the recombinant enzyme. *Lipids*. https://doi.org/10.1002/lipd.12210.

Nutahara, E., Abe, E., Uno, S., Ishibashi, Y., Watanabe, T., Hayashi, M., Okino, N., and Ito, M. (2019). The glycerol-3-phosphate acyltransferase PLAT2 functions in the generation of DHA-rich glycerolipids in *Aurantiochytrium limacinum* F26-b. *PLoS One*, *14*(1), 1–21. https://doi.org/10.1371/journal.pone.0211164.

Patil, V., Källqvist, T., Olsen, E., Vogt, G., and Gislerød, H. R. (2007). Fatty acid composition of 12 microalgae for possible use in aquaculture feed. *Aquaculture International*, *15*(1), 1–9. https://doi.org/10.1007/s10499-006-9060-3.

Prasad, P., Anjali, P., and Sreedhar, R. V. (2020). Plant-based stearidonic acid as sustainable source of omega-3 fatty acid with functional outcomes on human health. *Critical Reviews in Food Science and Nutrition*, 1–13. https://doi.org/10.1080/10408398.2020.1765137.

Qu, L., Ren, L. J., and Huang, H. (2013). Scale-up of docosahexaenoic acid production in fed-batch fermentation by Schizochytrium sp. Based on volumetric oxygen-transfer coefficient. *Biochemical Engineering Journal*, *77*, 82–87. https://doi.org/10.1016/j.bej.2013.05.011.

Reitan, K. I., Rainuzzo, J. R., Øie, G., and Olsen, Y. (1997). A review of the nutritional effects of algae in marine fish larvae. *Aquaculture*, *155*(1–4), 207–221. https://doi.org/10.1016/S0044-8486(97)00118-X.

Scott, S. D., Armenta, R. E., Berryman, K. T., and Norman, A. W. (2011). Use of raw glycerol to produce oil rich in polyunsaturated fatty acids by a thraustochytrid. *Enzyme and Microbial Technology*, *48*(3), 267–272. https://doi.org/10.1016/j.enzmictec.2010.11.008.

Shahidi, F., and Ambigaipalan, P. (2018). Omega-3 polyunsaturated fatty acids and their health benefits. *Annual Review of Food Science and Technology*, *9*, 345–381. https://doi.org/10.1146/annurev-food-111317-095850.

Sijtsma, L., and De Swaaf, M. E. (2004). Biotechnological production and applications of the ω-3 polyunsaturated fatty acid docosahexaenoic acid. *Applied Microbiology and Biotechnology*, *64*(2), 146–153. https://doi.org/10.1007/s00253-003-1525-y.

Simopoulos, A. P. (1991). Omega-3 fatty acids in health and disease and in growth and development. *American Journal of Clinical Nutrition*, *54*(3), 438–463. https://doi.org/10.1093/ajcn/54.3.438.

Sitnik, S., Shtaida, N., Guihéneuf, F., Leu, S., Popko, J., Feussner, I., Boussiba, S., and Khozin-Goldberg, I. (2018). DGAT1 from the arachidonic-acid-producing microalga *Lobosphaera incisa* shows late gene expression under nitrogen starvation and substrate promiscuity in a heterologous system. *Journal of Applied Phycology*, *30*(5), 2773–2791. https://doi.org/10.1007/s10811-017-1364-3.

Swanson, D., Block, R., and Mousa, S. A. (2012). Omega-3 fatty acids EPA and DHA : Health. *Journal of Advanced Nutrition*, *3*, 1–7. https://doi.org/10.3945/an.111.000893.Omega-3.

Tonon, T., Harvey, D., Larson, T. R., and Graham, I. A. (2002). Long chain polyunsaturated fatty acid production and partitioning to triacylglycerols in four microalgae. *Phytochemistry*, *61*(1), 15–24. https://doi.org/10.1016/S0031-9422(02)00201-7.

Tonon, T., Sayanova, O., Michaelson, L. V., Qing, R., Harvey, D., Larson, T. R., Li, Y., Napier, J. A., and Graham, I. A. (2005). Fatty acid desaturases from the microalga *Thalassiosira pseudonana*. *FEBS Journal*, *272*(13), 3401–3412. https://doi.org/10.1111/j.1742-4658.2005.04755.x.

Úbeda-Mínguez, P., García-Maroto, F., and Alonso, D. L. (2017). Heterologous expression of DGAT genes in the marine microalga *Tetraselmis chui* leads to an increase in TAG content. *Journal of Applied Phycology*, *29*(4), 1913–1926. https://doi.org/10.1007/s10811-017-1103-9.

Xiao, Q. (2003). Biosynthesis of docosahexaenoic acid (DHA, 22:6–4, 7,10,13,16,19): Two distinct pathways. *Prostaglandins Leukotrienes and Essential Fatty Acids*, *68*, 181–186. https://doi.org/10.1016/S0952-3278(02)00268-5.

Xin, Y., Lu, Y., Lee, Y. Y., Wei, L., Jia, J., Wang, Q., Wang, D., Bai, F., Hu, H., Hu, Q., Liu, J., Li, Y., and Xu, J. (2017). Producing designer oils in industrial microalgae by rational modulation of co-evolving type-2 diacylglycerol acyltransferases. *Molecular Plant*, *10*(12), 1523–1539. https://doi.org/10.1016/j.molp.2017.10.011.

Xu, J., Kazachkov, M., Jia, Y., Zheng, Z., and Zou, J. (2013). Expression of a type 2 diacylglycerol acyltransferase from *Thalassiosira pseudonana* in yeast leads to incorporation of docosahexaenoic acid β-oxidation intermediates into triacylglycerol. *FEBS Journal*, *280*(23), 6162–6172. https://doi.org/10.1111/febs.12537.

Xu, Y., Caldo, K. M. P., Pal-Nath, D., Ozga, J., Lemieux, M. J., Weselake, R. J., and Chen, G. (2018). Properties and biotechnological applications of Acyl-CoA: Diacylglycerol acyltransferase and phospholipid: Diacylglycerol acyltransferase from terrestrial plants and microalgae. *Lipids*, *53*(7), 663–688. https://doi.org/10.1002/lipd.12081.

Yashodhara, B. M., Umakanth, S., Pappachan, J. M., Bhat, S. K., Kamath, R., and Choo, B. H. (2009). Omega-3 fatty acids: A comprehensive review of their role in health and disease. *Postgraduate Medical Journal*, *85*(1000), 84–90. https://doi.org/10.1136/pgmj.2008.073338.

Yoon, K., Han, D., Li, Y., Sommerfeld, M., and Hu, Q. (2012). Phospholipid: Diacylglycerol acyltransferase is a multifunctional enzyme involved in membrane lipid turnover and degradation while synthesizing triacylglycerol in the unicellular green microalga *Chlamydomonas reinhardtii*. *The Plant Cell*, *24*, 3708–3724. https://doi.org/10.1105/tpc.112.100701.

Zhang, C., Iskandarov, U., Klotz, E. T., Stevens, R. L., Cahoon, R. E., Nazarenus, T. J., Pereira, S. L., and Cahoon, E. B. (2013). A thraustochytrid diacylglycerol acyltransferase 2 with broad substrate specificity strongly increases oleic acid content in engineered *Arabidopsis thaliana* seeds. *Journal of Experimental Botany*, *64*(11), 3189–3200. https://doi.org/10.1093/jxb/ert156.

Zienkiewicz, K., Zienkiewicz, A., Poliner, E., Du, Z. Y., Vollheyde, K., Herrfurth, C., Marmon, S., Farré, E. M., Feussner, I., and Benning, C. (2017). *Nannochloropsis*, a rich source of diacylglycerol acyltransferases for engineering of triacylglycerol content in different hosts. *Biotechnology for Biofuels*, *10*(1), 1–20. https://doi.org/10.1186/s13068-016-0686-8. www.frost.com/news/press-releases/omega-3-epadha-ingredients-manufacturers-promote-advanced-applications-expand-reach/.

Zhang, C., [illegible], Klar, [illegible] [illegible]

Zhang, K., [illegible] [illegible]

Theme IV

Large-Scale Bioprocesses for Algal Cultivation

Constraints and Challenges

19 Neural-Network Approach in Seaweed Research

An Emerging Field for Prediction and Modelling of Critical Parameters

Vaibhav A. Mantri, Dineshkumar Ramalingam, Mudassar Anisoddin Kazi, and M. Vignesh

19.1 INTRODUCTION

Seaweed "although misnomer" are integral part of human well-being from time immemorial. They are marine renewable resources, ubiquitously distributed in all the world oceans. Several of them are considered as key stone species due to their ability of structuring marine habitats, providing nursery and breeding grounds to the marine vertebrate and invertebrate taxa and ecosystem services (Bustamante et al. 2017). It is interesting to note that, technically there are diverse groups of organisms artificially classified based on their pigment, food reserves, cell wall polysaccharides, and flagella orientation and construction. About 15,000 taxa have been described and segregated into three major groups, namely, red (Rhodophyta), brown (Ochrophyta), and green (Chlorophyta) seaweeds (Vuong et al. 2017). It has been shown that they have different evolutionary lineages and polyphyletic origin. Their utilization has been reported from prehistoric civilization through archaeological evidence from places like China, Japan, Costa Rica, and Egypt (Dillehay et al. 2008). Further, medicinal properties of these marine forms was known in ancient medicinal literature, such as use of *Sargassum* for the treatment of goiter, *Gelidium* as a remedy for afflictions in intestine, and *Laminaria* in complicated childbirths for the dilation of the cervix wall (Dawson 1966). The burgeoning population growth and need for viable and sustainable supply of resources has brought the attention back to this seemingly plentiful and comparatively cheap resource.

19.2 APPLICATIONS OF SEAWEED IN INDUSTRY AND CLIMATE RESEARCH

The harvesting of seaweeds for commercial purpose has attend new milestone registering 31.2 million tones year^{-1} production (of which 95% accounts to farming) with a trade value of US$ 11.7 billion (FAO 2018). It may also be noted that, the larger proportion of which (~83%) is going inedible sector, either as direct food stuff such as "nori" from *Porphyra/Pyropia* spp., "kombu" from *Saccharina japonica* and "wakame" from *Undaria pinnatifida*. Further, it may be noted that, among the economically important seaweed species, 145 species are directly used for food while 101 are for phycocolloid production. These include 125 Rhodophyta, 64 Ochrophyta, and 32 Chlorophyta (Hurtado 2022). Nevertheless, seaweed resources are traditionally harvested for chemical constituents such as hydrocolloids, namely, agar, alginate, and carrageenan, besides, food, feed, and agriculture stimulants. The global seaweed based hydrocolloid industry has already crossed 90,000 metric tons of production with a market value of US$ 1.2 billion in 2015 (Porse and Rudolph 2017). Seaweed ocean farms are considered more sustainable when compared to their counterpart in land-based agriculture. This is essentially because their cultivation does not require freshwater irrigation, chemical fertilizer, or pesticide inputs. Emerging applications of seaweed in the energy sector have been well

DOI: 10.1201/9781003219156-24

received in recent years. It has been estimated that the annual harvest of 500 million dry tons of seaweed with 50% carbohydrate content could produce about 1.25 billion megawatt-hours' worth of methane or liquid fuel. As per estimate about 85 billion megawatt-hours of energy from fossil fuels has been used by the world in 2012. Considering this statistical trend, the energy production from the seaweeds would equate roughly to 1.5% of the world energy requirement from fossil fuels (IEA 2014). Similarly, seaweed based aquaculture offers critical opportunity to mitigate and adapt to climate change. Seaweed farms can act as a CO_2 sink. It may be noted that seaweed aquaculture contributes to climate change adaptation by dampening wave energy and protecting shorelines, and by elevating pH and supplying oxygen to the waters, thereby reducing the effects of ocean acidification and deoxygenation. As per the estimate, increasing the growth of seaweed biomass through farming up to 14% year^{-1} would generate 500 MT dry weight by 2050, which would add directly or indirectly to about 10% to the world's present supply of food. This amount of seaweed will be able to absorb about 135 MT of carbon, which in terms of carbon credit will help in growth and profitability of seaweed businesses (World Bank Group 2016).

19.3 SEAWEED CULTIVATION

Among the domestication and farming sector, Asian countries lead with highest proportion of both red seaweeds (*Eucheuma*, *Kappaphycus*, *Gracilaria*, *Porphyra*/*Pyropia*, and *Gelidium*) and brown (*Saccharina* and *Undaria*) seaweeds. Among the several taxa cultivated commercially, species of *Saccharina* and *Undaria* farmed in subtemperate to temperate zones such as China PR, Japan, and the Republic of Korea dominated. However, since the beginning of 2010 the scenario has been changed. It may be noted that currently species of *Kappaphycus* and *Eucheuma* cultivated in subtropical to tropical waters of Indonesia, the Philippines, Tanzania, Malaysia, and India are the top producers. Among Western countries, commercial large-scale farming not implemented yet with few exceptions such as *Gracilaria* and *Macrocystis* cultivation in Chile, *Undaria pinnatifida*, *Palmaria palmata*, and *Pyropia umbilicalis* in France and *Saccharina latissima* in Canada along the integrated multi-trophic aquaculture (IMTA). Nevertheless, countries such as Denmark, Ireland, Norway, Portugal, and Spain have made appreciable progress in implementing regional seaweed farming initiatives (Hurtado 2022).

It is also interesting to note that besides open sea farming, land-based pond or tank cultivation has also received lot of attention due to its advantage of obtaining uniform product, protection of natural calamities, and ability to maintain controlled ambient conditions. The taxa such as *Chondrus*, *Asparagopsis*, *Palmaria*, *Hydropuntia*, *Gracilaria*, *Gelidium*, and *Porphyra*/*Pyropia* of Rhodophyta and *Codium*, *Monosroma*, *Caulerpa*, and *Ulva* of Chlorophyta are farming by these techniques. It may also be noted that, currently, land-based techniques are more advanced in terms of mechanization and sophistication. These systems are largely engaged in primary seeding of life history stages of specific seaweed necessary step prior to out planting of ropes or nets in open-sea cultivation practice, rather than biomass production. The economic viability although seen in terms of considerable reduction in labor costs than open sea farming compensating technological investments, the commercial feasibility of land-based production systems will come when high-value products of niche applications can be developed through these systems such as specialty chemicals, cosmeceuticals, secondary metabolites, and drug precursors. Hurtado (2022) has enlisted 11 genera, 25 species, and two varieties of Rhodophyta; 7 genera and 12 species of Ochrophyta; and 5 genera and 10 species and one variety of Chlorophyta, which are commercially farmed.

19.4 FUTURE SEAWEED INDUSTRY NEEDS

In their review paper titled "The Evolution Road of Seaweed Aquaculture: Cultivation Technologies and the Industry 4.0" authors provided several opportunities and challenges faced by seaweed growers (García-Poza et al. 2020). It has been clearly shown that the multidisciplinary approach for taking forward the seaweed aquaculture domain is highly essential and promising. There is an urgent need for the optimization of seaweed production and processing-based technologies for enhanced output coupled with lower cost of production. Establishing seaweed based industry either cultivation or processing needs proper planning from the beginning. Therefore, collecting data on aspect of these systems is necessary. García-Poza (2020) lamented the need of multidisciplinary approach for success of seaweed-related industry and listed couple of domains: (i) computational models, (ii) computational fluid dynamics (CFD), (iii) mechanical and chemical engineering, (iv) informatics and electrotechnical engineering, and (v) biological sciences and engineering. Industry 4.0 concept enables utilization of smart digital technology, big data, and machine learning for achieving holistic and well-connected functional ecosystem more focused on manufacturing as well as supply chain management. This concept in seaweed industry can make use of more computing power with logarithmic and artificial intelligence for better productivity and efficiency, thereby reducing the overall costs. The recent years have seen research efforts in this direction, thus preparing strong foundation stone for the industry to grow.

19.5 BIOLOGICAL COMPUTATION

It may be interesting to note that, upon the evolution of biological knowledge and massive growth in availability of data, there is an opportunity available for analyzing and evaluating such massive datasets in biological perspective. The machine learning and deep learning techniques are thus evolved and subsequently applied for enabling the data management. The various probabilistic and statistic tools, machine learning algorithms have been developed to model multiple biological processes, optimize the data, and predict the results (Manisekhar et al. 2020; Shastry and Sanjay 2020). These algorithms consider raw features within comparatively large and annotated dataset and analyze them based on hidden patterns to deliver the predictive tool. These techniques are used for working on the complicated biological problems that are complex normal computing methods to comprehend. Further, these computational tools can perform a wide variety of roles, such as integrating data and knowledge generated from a particular experiment, furthering understanding of process, testing of certain hypothesis, interpreting experimental data, helping to generate novel insights, tracing chains of causation, performing sensitivity analyses, and developing new approaches (Brodland 2015). These computational based models essentially make use of observation and manipulation exactly in similar fashion as physical experiments, since in both these cases their ultimate goal is the same. However, it may be noted that these models cannot replace routine laboratory experiments; similarly, they cannot prove the mechanism (but can disprove it). But models can definitely help in offering different perspectives than those gained through experiments. The use of model-based research in biology helps in advancing the progress of the work much faster than without them. The computing models are increasingly becoming a standard methodology in several scientific investigations and becoming nearly obligatory (Brodland 2015). As statistics did a few decades ago, computational modelling is adopted into biological science these days in a similar way. Indeed, biological computation and modelling has an important role to play in future investigations of critical importance.

19.5.1 Artificial Neural Network

Artificial neural network (ANN) model is a machine learning model, and it is used to solve complex and nonlinear processes in many applications, including prediction, pattern recognitions, modeling and simulation, and optimization purposes (Dhanarajan et al. 2014; Mohamed 2019). A typical ANN architecture essentially consists of three layers: input layer that gathers data, output layer that generates required data after computation and hidden layer(s) that connects the input and output layers. The fundamental processing unit of neural network is termed as neurons, which has two main roles, such as collection of input factors and generation of output. In most of the process, the set of input and output data given for ANN model will be split into 70% for training, 15% for validating, and another 15% for testing the network, i.e., being developed (Dineshkumar, Dhanarajan et al. 2015; Huang et al. 2007).

The neuronal links in each layer are termed by weights and bias, and generally, it will be trained using back propagation algorithm. This algorithm attempts to reduce the error between given output and trained output by adjusting the weights and bias until the error value is negligible by repeatedly feeding the error in the network (Dhanarajan et al. 2017; Nayak et al. 2018). The performance indices such as correlation coefficient (R) and mean squared error (MSE) (Eqn. 1) are often employed in training the neural network. While executing the model, ANN understands the cause-effect correlation between the input and output factors by updating its weights and biases. ANN will implement an activation function (Eqn. 2) while training is in process, in which each input is multiplied by weights, and this product (weight and input of each neuron) will be added with bias and then subsequently transferred to generate an output through the activation function, as shown in Figure 19.1.

$$MSE = \sum \frac{(Experimental - Predictedvalue)^2}{n} \quad (1)$$

$$Y_j = \sum (x_i w_{ij}) + b_j \quad (2)$$

where, Y_j = activation function, w_{ij} = weight connection between neurons i and j, x_i is the input at neuron i, and b_j is the bias of neuron j.

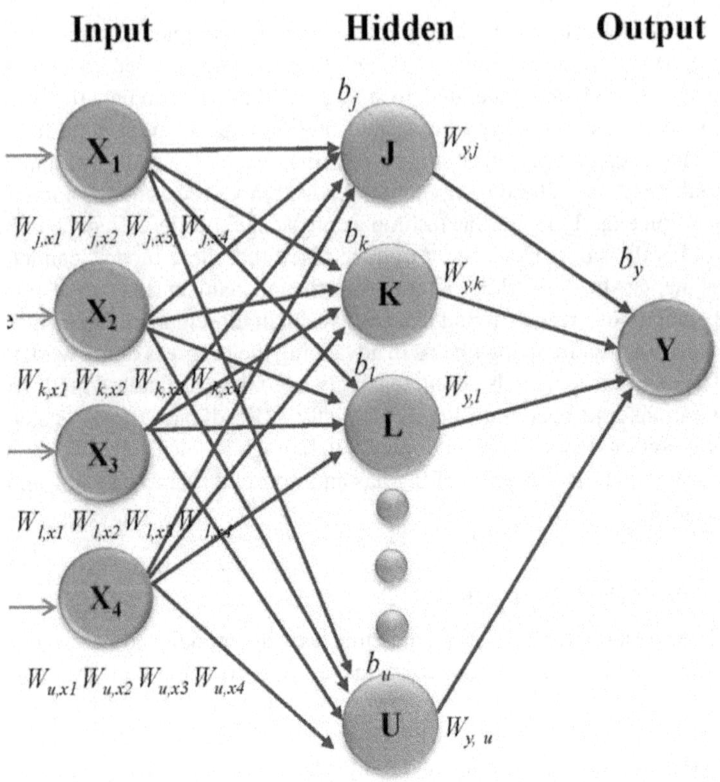

FIGURE 19.1 Schematic representation of a neural network architecture.

In order to obtain an effective application of ANN, it is essential to choose the best neural network topology. This can be achieved by optimizing the required number of neurons in the hidden layer through trial and error basis, wherein a maximum R-value and minimum MSE-value is expected (Dhanarajan et al. 2017; Dineshkumar, Dash et al. 2015). This optimal ANN topology will eliminate data overfitting, and the selected topology needs to be validated experimentally to corroborate its efficacy in prediction or modelling. The computation of ANN shall be performed in software tools such as Mathworks Inc., Natick, USA.

The performance comparison of ANN with response to surface methodology (RSM) have been analyzed by many studies (Mohamed 2019; Sargent 2001; Shi et al. 2012; Vignesh et al. 2020). Overall, ANN showed superiority in terms of prediction, pattern recognition, and optimization purposes. It can be reiterated that the statistical optimization methods like RSM can fail to describe complex nonlinearity processes as it uses second order polynomial equations, whereas ANN can be practically employed in almost all research fields for prediction of the nonlinear processes due to its simulation performance of biological neural network. Once the ANN model is successfully developed, it can be then combined with optimization algorithms such as genetic algorithm (GA) or particle swarm optimization (PSO) through a fitness function for identifying optimal input factors which can provide the best output (Mookherjee et al. 2018; Vignesh et al. 2020).

19.5.2 Genetic Algorithm

Genetic algorithm (GA) is a modern optimization tool that imitates the theory of natural selection where the fittest one can survive (Figure 19.2). It is a heuristic search that works by repeating the set of processes—the selection of data, attrition, and reproducing the potential solutions with the operators like crossing over and mutation to obtain the optimal one. It can be implemented easily when compared to other machine learning algorithms. It is widely used in various medical and medical-related fields to classify and optimize massive datasets containing a wide range of unsorted data. It is used in analyzing data of microarrays, mutations in genetic structures. Genetic algorithm-based software is used in the detection of QRS complexes as a part of the automatic interpretation of ECG (Ahmed et al. 2020).

19.5.3 Particle Swarm Optimization

The traditional form of particle swarm optimization (PSO) was first defined by J. Kennedy and R. C. Eberhart. It is the analogy of bird flocking, where, a bird is compared to the particle-containing solution. Likewise, each particle keeps on moving its position by adjusting the velocity (according to equations 3 and 4) and interacting with other particles to obtain a better seat in the dimension of space (Kennedy and Eberhart 1995). Each particle stores the memory of the previous location and

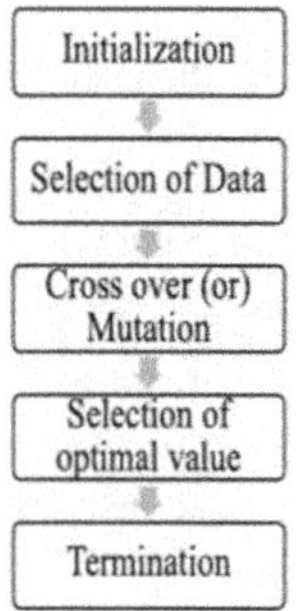

FIGURE 19.2 Flowchart depicting the steps involved in genetic algorithm for identifying the optimal value.

adjusts to the position of the best particle. Ultimately, the whole swarm is moved upon finding the best values in each iteration. This iterative process repeats until reaching the termination criteria.

$$V_i^k = W^{(k-1)}V_i^{(k-1)} + C_1R_1(L_i^{(k-1)} - P_i^{(k-1)}) + C_2R_2(G_i^{(k-1)} - P_i^{(k-1)}) \quad (3)$$

$$P_i^k = P_i^{(k-1)} + V_i^k \quad (4)$$

Here, k is the iteration of the algorithm;

V_i^{k-1}—Velocity of the particle I at the iteration k-1;
P_i^{k-1}—Position of the particle I at the iteration k-1;
L_i^{k-1}—Local best solution of the swarm;
W^{k-1}—Inertia weight;

C_n and R_n—Learning factors and random variables respectively.

It is also used as a learning algorithm in ANN for optimization purposes, that is, in identifying global optimum values. Recently, DNA barcode sets were constructed based on PSO (Wang et al. 2018).

19.5.4 Studies on Artificial Neural Network Modelling in Seaweeds

In recent times, ANN has been employed successfully in many biological systems for analyzing data. However, its application in seaweeds remains limited. A comparative analysis of ANN and empirical models was performed to predict the thermophysical properties of *Saccharina latissima* (Kelp) important in the process of drying and temperature-controlled storage (Sappati et al. 2019). In this study, the performance of ANN prediction was superior in comparison to the conventional regression model and the Choi and Okos' model. It was observed that the thermophysical properties mainly correlated to moisture contents than to other proximate and effect of seasonal variation. In last few years' massive shoals of pelagic *Sargassum* has become a major ecological and economical concern for the Mexican, Caribbean, and African coastal regions (Arellano-Verdejo et al. 2019). In order to have early detection of *Sargassum* bloom, Arellano-Verdejo et al. (2019) have developed a detection system ERISNet using deep neural network modeling based on

TABLE 19.1
Application of ANN Models in Seaweeds

Seaweed Species	Machine Learning Algorithms	Application	Reference
Saccharina latissima	Artificial Neural Network (ANN)	To predict the thermophysical properties	Sappati et al. (2019)
Sargassum sp.	Convolutional Neural Network (CNN) and Recurrent Neural Network (RNN)	For early detection of *Sargassum* bloom	Arellano-Verdejo et al. (2019)
Sargassum sp.	ANN and Support Vector Machine (SVM)	To predict the pyrolytic conversion	Saleem and Ali (2017)
Sargassum filipendula	ANN	To study the biosorption potential	Fagundes-Klen et al. (2007)
Gracilaria dura	ANN and Particle Swarm Optimization (PSO)	To optimize different physiochemical parameters to improve seedling production	Vignesh et al. (2020)

convolutional neural network (CNN) and recurrent neural network (RNN). To train and test the ERISNet, a dataset of pixel values with and without *Sargassum* was built by using Aqua-MODIS imagery. ERISNet was able to classify pixel values with and without *Sargassum* with maximum probability of 90.08% showing its capability in detecting arrival of algal blooms. Two machine learning algorithms, ANN and Support Vector Machine (SVM), were used to predict the pyrolytic conversion of *Sargassum* species (Saleem and Ali 2017). The accuracy of ANN model of prediction was higher than SVM with lower error rate and higher coefficient of determination (0.998). ANN was also tested to fit the equilibrium data of the binary mixture of cadmium-zinc ions biosorption by the *Sargassum filipendula* by Fagundes-Klen et al. (2007). The performance of ANN was found better in comparison to conventional biosorption isotherm models, and the study reported that the biomass of *S. filipendula* has higher affinity for zinc ions. Recently, the ANN model was applied for optimization of different physiochemical parameters for seedling production in an important agarophyte, *Gracilaria dura* (Vignesh et al. 2020). ANN was combined with particle swarm optimization (PSO) to optimize different physiochemical parameters to achieve accurate regeneration strategy in seedling production of *G. dura*. The results showed that the prediction accuracy of ANN-PSO was better than response surface methodology (RSM) with prediction error in optimum seedling regeneration 1.25% and 13.75% respectively. These studies certainly demonstrated the efficiency and possible wide applications of ANN approach to overcome the technological problems and optimization in complex biological system such as seaweed for their economical and ecological role.

ACKNOWLEDGMENTS

Authors would like to thank the Council of Scientific and Industrial Research (CSIR), New Delhi, for its financial support. Thanks are also due to Director, CSIR-CSMCRI, for the facilities.

REFERENCES

Ahmed, Z., Mohamed, K., Zeeshan, S., Dong, X., 2020. Artificial intelligence with multi-functional machine learning platform development for better healthcare and precision medicine. *Database*. doi:10.1093/database/baaa010.

Arellano-Verdejo, J., Lazcano-Hernandez, H.E., Cabanillas-Terán, N., 2019. ERISNet: Deep neural network for *Sargassum* detection along the coastline of the Mexican Caribbean. *Peer J*, 7, e6842.

Brodland, G.W., 2015. How computational models can help unlock biological systems. *Seminars in Cell and Developmental Biology*, 47–48, 62–73.

Bustamante, M., Tajadura, J., Díez, I., Saiz-Salinas, J.I., 2017. The potential role of habitat-forming seaweeds in modeling benthic ecosystem properties. *Journal of Sea Research*, 130, 123–133.

Dawson, E.Y., 1966. *Marine Botany, an Introduction*. New York: Holt, Rinehart and Winston, p. 371.

Dhanarajan, G., Mandal, M., Sen, R., 2014. A combined artificial neural network modeling– particle swarm optimization strategy for improved production of marine bacterial lipopeptide from food waste. *Biochemical Engineering Journal*, 84, 59–65.

Dhanarajan, G., Rangarajan, V., Bandi, C., Dixit, A., Das, S., Ale, K., Sen, R., 2017. Biosurfactant-biopolymer driven microbial enhanced oil recovery (MEOR) and its optimization by an ANN-GA hybrid technique. *Journal of Biotechnology*, 256, 46–56.

Dillehay, T.D., Ramírez, C., Pino, M., Collins, M.B., Rossen, J., Pino-Navarro, J.D., 2008. Monte Verde: Seaweed, food, medicine, and the peopling of South America. *Science*, 320, 784–786.

Dineshkumar, R., Dash, S.K., Sen, R., 2015b. Process integration for microalgal lutein and biodiesel production with concomitant flue gas CO_2 sequestration: A biorefinery model for healthcare, energy and environment. *RSC Advances*, 5:90, 73381–73394.

Dineshkumar, R., Dhanarajan, G., Dash, S.K., Sen, R., 2015a. An advanced hybrid medium optimization strategy for the enhanced productivity of lutein in *Chlorella minutissima*. *Algal Research*, 7, 24–32.

Fagundes-Klen, M.R., Ferri, P., Martins, T.D., Tavares, C.R.G., Silva, E.A., 2007. Equilibrium study of the binary mixture of cadmium–zinc ions biosorption by the Sargassum filipendula species using adsorption isotherms models and neural network. *Biochemical Engineering Journal*, 34:2, 136–146.

FAO, 2018. The global status of seaweed production, trade and utilisation. *Goldfish Research Programme*, 124. Rome. 120 pp.

García-Poza, S., Leandro, A., Cotas, C., Cotas, J., Marques, J.C., Pereira, L., Gonçalves, A.M.M., 2020. The evolution road of seaweed aquaculture: Cultivation technologies and the industry 4.0. *International Journal of Environmental Research and Public Health*, 17, 6528.

Huang, J., Mei, L.H., Xia, J., 2007. Application of artificial neural network coupling particle swarm optimization algorithm to biocatalytic production of GABA. *Biotechnology & Bioengineering*, 96:5, 924–931.

Hurtado, A.Q., 2022. *Genetic Resources for Farmed Seaweeds—Thematic Background Study*. Rome: FAO. https://doi.org/10.4060/cb7903en.

IEA, 2014. *Key World Energy Statistics 2014*. Paris: International Energy Agency.

Kennedy, J., Eberhart, R., 1995. Particle swarm optimization. In: Proceeding of the IEEE international conference on neural networks, Perth, Australia, pp. 1942–1948.

Manisekhar, S.R., Siddesh, G.M., Manvi, S.S., 2020. Introduction to bioinformatics. In: *Statistical Modelling and Machine Learning Principles for Bioinformatics Techniques, Tools, and Applications. Algorithms for Intelligent Systems*, eds. Srinivasa, K., Siddesh, G., Manisekhar, S. Singapore: Springer, pp. 3–9.

Mohamed, Z.E., 2019. Using the artificial neural networks for prediction and validating solar radiation. *Journal of the Egyptian Mathematical Society*, 27, 47.

Mookherjee, A., Dineshkumar, R., Kutty, N.N., Agarwal, T., Sen, R., Mitra, A., Maiti, T.K., Maiti, M.K., 2018. Quorum sensing inhibitory activity of the metabolome from endophytic Kwoniella sp. PY016: Characterization and hybrid model-based optimization. *Applied Microbiology and Biotechnology*, 102:17, 7389–7406.

Nayak, M., Dhanarajan, G., Dineshkumar, R., Sen, R., 2018. Artificial intelligence driven process optimization for cleaner production of biomass with co-valorization of wastewater and flue gas in an algal biorefinery. *Journal of Cleaner Production*, 201, 1092–1100.

Porse, H., Rudolph, B., 2017. The seaweed hydrocolloid industry: 2016 updates, requirements, and outlook. *Journal of Applied Phycology*, 29, 2187–2200.

Saleem, M., Ali, I., 2017. Machine learning based prediction of pyrolytic conversion for red seaweed. 7th International Conference on Biological, Chemical & Environmental Sciences (BCES-2017), September 6–7, Budapest (Hungary).

Sappati, P.K., Nayak, B., Van Walsum, G.P., 2019. Thermophysical properties prediction of brown seaweed (*Saccharina latissima*) using artificial neural networks (ANNs) and empirical models. *International Journal of Food Properties*, 22:1, 1966–1984.

Sargent, D.J., 2001. Comparison of artificial neural networks with other statistical approaches. *Cancer*, 91, 1636–1642.

Shastry, K.A., Sanjay, H.A., 2020. Machine learning for bioinformatics. In: *Statistical Modelling and Machine Learning Principles for Bioinformatics Techniques, Tools, and Applications. Algorithms for Intelligent Systems*, eds. Srinivasa, K., Siddesh, G., Manisekhar, S. Singapore: Springer, pp. 25–39.

Shi, H.Y., Lee, K.T., Lee, H.H., Ho, W.H., Sun, D.P., Wang, J.J., Chiu, C.C., 2012. Comparison of artificial neural network and logistic regression models for predicting in-hospital mortality after primary liver cancer surgery. *PLoS One*, 7:4, e35781.

Vignesh, M., Kazi, M.A., Rathore, M.S., Kavale, M.G., Dineshkumar, R., Mantri, V.A., 2020. Artificial neural network modelling for seedling regeneration in *Gracilaria dura* (Rhodophyta) under different physiochemical conditions. *Plant Cell, Tissue and Organ Culture*. https://doi.org/10.1007/s11240-020-01943-x.

Vuong, D., Kaplan, M., Lacey, H.J., Crombie, A., Lacey, E., Piggott, A.M., 2017. A study of the chemical diversity of macroalgae from South Eastern Australia. *Fitoterapia*, 126, 53–64.

Wang, B., Zheng, X., Zhou, S., Zhou, C., Wei, X., Zhang, Q., Wei, Z., 2018. Constructing DNA barcode sets based on particle swarm optimization. *IEEE/ACM Transactions on Computational Biology and Bioinformatics*, 15:3, 999–1002.

World Bank Group, 2016. Seaweed aquaculture for food security, income generation and environmental health in tropical developing countries. https://elibrary.worldbank.org/doi/full/10.1596/24919.

20 Constraints and Challenges on Large-Scale Cultivation of Economically Important Algae

V. Veeragurunathan, Dineshkumar Ramalingam, Apoorva Bhayani, and P. Gwen Grace

20.1 INTRODUCTION

Microalgal biomass and its products are considered as potential candidates for its application in the field of pharmaceuticals, nutraceuticals, renewable energy, natural colourant, poultry feed, and aqua feed applications. After develpoing a proof of concept in lab conditions, it is very essential to scale up the entire process in order to commercialise it. For the commercial production of microalgal metabolites, generally microalgae have been cultivated either in open pond systems (e.g., raceway ponds) or in closed systems (e.g., photobioreactors (PBRs)).

As compared to the PBRs, open pond cultivation confers several advantages like less capital investment, lower cultivation cost, less periodical maintaince, and cleaning and reduction inoperational energy consumption (Mutanda et al. 2011). Whereas closed cultivation systems requires high capital and operating costs including maintenace. According to Chisti (2007), it is estimated that the cost for production of a kilogram of microalgal biomss would be USD 2.95 for photobioreactor and USD 3.80 for raceway ponds, assuming no cost availability of carbon dioxide. The aspiration behind process scale-up is to increase the production capacity with similar or increased product quality economically. While scaling-up the microalgal cultivation process, there are many physicochemical and biological factors which hurdle the higher rate of biomass production in open pond systems. Hence, this chapter emphasizes the constraints and challenges in large-scale cultivation of microalgae.

Seaweeds have been utilized for centuries worldwide mainly for food purpose to the coastal communities. Seaweeds are also utilized for production of pigments, polysaccharides, minerals, and vitamins (Trivedi et al. 2016). Among sulfated polysaccharides, agar, carrageenan, and alginates are major industrially exploited products due to their unique gelling, stabilizing, and emulsifying characteristics. Carrageenan is mainly used in pet food, dairy and meat industries, agar is mainly used as foodstuff in Asian countries, and alginic acid is used in textile printing industries (FAO 2018). In a global scenario, biomass production through seaweed farming was about 31.2 million tons, which showed a sale value of USD 11.7 billion in 2016 and recorded 27.3% of the whole marine aquaculture production (FAO 2018). In a world scenario, only 221 species have commercial value, and among them, onlyten species, such as *Caulerpa* spp., *Enteromorpha clathrata*, *Eucheuma* spp., *Gracilaria* spp., *Kappaphycus alvarezii*, *Monostroma nitidum*, *Porphyra* spp., *Saccharina japonica*, *Sargassum fusiforme*, and *Undaria pinnatifida*, are being commercially cultivated.

Though 844 seaweed species are reported in India, only few seaweed species, such as *Gelidiella acerosa*, *Gracilaria* spp., *Sargassum* spp., and *Turbinaria* spp., are commercially utilized for agar and alginate production. Indiscriminate and repetitive harvesting in natural seaweed beds leads fast depletion of natural resources. Central Salt and Marine Chemicals Research Institute (CSIR-CSMCRI) developed viable and successful cultivation technology available for three economically important agarophytes, *Gelidiella acerosa* (Ganesan et al. 2011), *Gracilaria debilis* (Veeragurunathan et al. 2019), *Gracilaria dura* (Veeragurunathan et al. 2015), and *Gracilaria*

DOI: 10.1201/9781003219156-25

edulis (Ganesan et al. 2011b) and *Kappaphycus alvarezii* (Eswaran et al. 2002). Commercialization of *Kappaphycus* cultivation technology was initiated in 2000 in India, and there are more than 1,000 households in Tamil Nadu alone engaged in seaweed farming as alternative livelihood and earning an income of ₹ 15,000/per month (Mantri et al. 2017). The seaweed cultivation has now emerged as a viable option for livelihood improvement of the low-income coastal community in the country. There are many constraints and challenges while taking algal cultivation at commercial scale. In this book chapter, we elaborated the constraints and challenges on large-scale commercial cultivation of economically important algae in India.

20.2 FACTORS AFFECTING THE LARGE-SCALE CULTIVATION OF MICROALGAE

Microalgal cultivation at large scale can be improved by optimizing the various physical and chemical parameters like temperature, light, culture mixing, pH, etc. Such parameters not only influence the biomass concentration but it also affects the biochemical composition and metabolic activity of the microalgal cells. So it becomes very essential to regulate them during large-scale cultivation. The factors given in the next section have a direct impact on microalgal growth and subsequent product synthesis.

20.2.1 Physical Factors

20.2.1.1 Temperature

Out of all the physical factors, temperature plays a very important role in large-scale microalgal cultivation. Each microalgal specices has particular optimal temperature range, in which they show maximum biomass production and metabolite production along with other cultivation conditions. But it would be very difficult to maintain such optimum temperature conditions during large-scale cultivation, especially in outdoor cultivation. During outdoor cultivation, there are a lot of seasonal shifts, which cause fluctuation in temperature and ultimately leads to the decrease in total concentration of biomass. According to a study carried out by Bechet et al. (2010), it was observed that most of the commercially used microalgal species would tolerate the temperature range between 35°C and 40°C. In outdoor cultivations during summertime, temperature exceeding greater than optimum temperature occurs very rapidly. Such upsurge in temperature causes a decrease in growth, photosynthesis, and respiration of the microalgae because of denaturation and inactivation of photosynthetic proteins and disproportion between ATP production and energy demand (Raven and Geider 1988). It shall be noted that at low temperatures, there would be reduced activity of carboxylase enzyme, which in turn affects the growth rate of microalgae.

In outdoor cultivation using PBRs, reactor surface gets overheated. To reduce the temperature, cool water should be sprayed onto the reactor surface to exchange the heat. While opting for open pond cultivation during wintertime, it is advisable to mix lukewarm water periodically to resist the decrease of culture temperature. In such cases, ponds situated near industries having hot water boiler outlets may be used for economic benefits.

20.2.1.2 Light

Light is another important growth-limiting physical factor. Even though light is required for photosynthesis, elevated or decreased amount of light can exhibit significant effect on microalgal growth. Light provided at proper wavelength, intensity, and duration can improve the microalgal growth in PBRs. Exorbitant intensity of light may cause photo oxidation and inhibition, while low light can cause growth limitation (Carvalho et al. 2011).

While dealing with large volume of microalgal cultivation, culture density is a drawback in terms of light penetration. Soeder (1980) has explained the concept of self-shading and areal density. High areal density causes low biomass productivity because of light limitation. In addition, due to photo oxidation or photo inhibition, there will be detrimental effects on microalgal growth and product formation. Orientation of cultivation system is also very important for light availability. Inappropriate orientation may cause shadow on algal culture during light cycle.

20.2.1.3 Culture Mixing

In large-scale cultivation, microalgal culture needs to be gently agitated to avoid sedimentation or accumulation of algal cells at the bottom surface of the pond. Proper mixing of the culture increases the availability of light and gas flow. In general, an increase in gas flow rate and light availability increases the biomass productivity (Gao et al. 2018). Insufficient mixing would cause changes in local concentration of pH and nutrients, which can hamper the biomass production. While mixing, hydrodynamic forces should be minimal in order to resist cell death due to the turbulence force or shear stress. Open or close cultivation system needs to be designed in a way that the dead zone effect can be reduced. In such dead zone, there is no turbulence or mixing, so it leads to the accumulation of dead or sedimentation of cells and eventually interrupts the entire flow channel.

20.2.2 Chemical Factors

20.2.2.1 pH

Microalgae are quite responsive towards the pH shifts. Every microalgal species shows maximum growth at optimum pH along with other optimum condition. In large-scale cultivation, it is essential to maintain optimal pH range. Shift in pH can cause effect on microalgal biomass production. In cultivation medium, increased pH may provoke precipitation of phosphate, and this may be prevented by lowering the pH in a way that it does not affect the carbon assimilation (Yaakob et al. 2014). pH has impact on the accessibility of inorganic carbon (Azov 1982). Efficiency to utilize external CO_2 also decreases with decreased pH (Wang et al. 2018). Study done by Qiu et al. (2017) on *Chlorella sorokiniana* reported that an increase in pH enhances the protein content, whereas C/N ratio decreases. While scale-up, it is economically not advisable to use buffer systems. In such case, pH fluctuations can be reduced by providing nitrogen sources like ammonium and nitrate in certain combinations (Scherholz and Curtis 2013). Choi et al. (2017) proposed the cost-effectivness of the bicarbonate buffer system prepared by dissolving CO_2 in KOH solution. Such buffer system shows buffering capasity at 10% CO_2 concentration as well. It also provides carbon and phosphates to the cell.

20.2.2.2 Salinity

Salinity is one of the critical factors while cultivating microalgae at large-scale. Due to the temperature effect, salinity tends to be gradually increased due to continuous evaporation (Ishika et al. 2018). So periodically, evaporation lost needs to be compensated with the water. In microalgae, increase in salinity above the optimum range can cause reduction in growth rate (Ben-Amotz et al. 1985). Changes in salinity cause ion stress, osmotic stress in microalgae. It also affects the fatty acid profile as well (Xu and Beardall 1997). Freshwater microalgae cultivation always requires a huge amount of freshwater supplementation periodically. In the past few years, freshwater availability has become an issue. Generally, freshwater strains have less tolerance towards increased salinity concentration. Hence, the use of fresh water for microalgae cultivation is cumbersome in the regions where fresh water is a limited factor (Ishika et al. 2017). In spite of the use of freshwater species, marine strains are highly recommended. Such marine strains can grow into a range of salinity concentrations, which confers the selective advantages like reduction in contamination, low water requirement for evaporation compensation, higher lipid production, etc.

20.2.3 Biological Factors

Microalgal culture on large-scale cultivation may be contaminated by bacteria, fungi, other algae, zooplanktons, etc. Biological contamination is one of the major factors responsible for low algal biomass productivity (Wang et al. 2013).

Bacterial contamination can either induce or reduce microalgae growth (Haines and Guillard 1974). Closed cultivation systems like photobioreactors can be used to reduce the contamination problem to a certain extent (Fulbright et al. 2018). It is impractical to cultivate bacteria-free microalgae at large scale cultivation, specifically in open pond cultivation. Sometimes bacteria growing with microalgae can induce growth-promoting substances like vitamins, which helps to improve microalgae growth. However, Baker and Herson's (1978) study depicted that several bacteria release toxins that are heat labile, have relatively high molecular weight, and probably protein structures, which imparts the algicidal effects and inhibits the microalgae growth. In addition, bacteria are able to hamper microalgal growth in different methods like (1) they can kill directly or indirectly by releasing lytic composites (2) by modifying the microenvironment of microalgae (3) by competing for available nutrients (Wang et al. 2016).

Zooplanktons were also reported to affect the microalgal growth in open pond cultivation systems. Moreno-Garrido and Canavate (2001) indicated that zooplanktons have great growth capacity and high grazing abilities. Hence, they can graze dense culture of *Dunaliella* in 48–72h. They found that some taxonomically unidentified ciliates are gifted to eat dense algal cultures in 48h. In the remedy, 10 mg/l quinine can wipe out ciliates very promptly, whereas it can allow *Dunaliella* cells to regain the density (Moreno-Garrido and Canavate 2001).

20.3 CONSTRAINTS ON LARGE-SCALE CULTIVATION OF SEAWEEDS

20.3.1 Requirement of Initial Seed Material

Two thousand floating rafts can be used for hectare level cultivation of economically important seaweeds. For initiating large-scale cultivation of *Gracilaria edulis*, 2 tons fr.wt initial seed material is required, whereas for *Gracilaria debilis* 8 tons fr.wt initial seed material is required (Table 20.1). For large-scale cultivation of *Kappaphycus alvarezii*, 80 tons fr.wt initial seed material is required. So maintaining initial seed material is the main constraint for initiating one hectare level cultivation.

20.3.2 Harvest Cycle Variations

Indian Sea conditions are not unique and common in all over India. The favorable period for seaweed cultivation varied in each maritime states of India. In Tamil Nadu, maximum harvest cycle/year is five growth cycle or harvest cycles, whereas in Gujarat and Andhra Pradesh, it is three to four harvest cycles. These variations are attributed to differences in habit and habitat variations in seashore. In Tamil Nadu costal region, the shallow and lagoon region is available, whereas other parts of India, it is with open sea.

TABLE 20.1
Initial Seed Material Requirement for Hectare Level Cultivation

S. No.	Name of Algae	Initial Seed Material/Raft (kg.fr.wt)	No. of Rafts/ Hectare	Total Initial Seed Requirement for Hectare Level Cultivation (tons fr.wt)	References
1	*Gracilaria debilis*	4.5	2,000	9	Veeragurunathan et al. (2019)
2	*Gracilaria edulis*	1	2,000	2	Veeragurunathan et al. (2016)
3	*Gracilaria dura*	1.5	2,000	3	Veeragurunathan et al. (2015)
4	*Kappaphycus alvarezii*	40	2,000	8	Eswaran et al. 2002

20.3.3 Grazing and Drifting of Seaweeds

Mostly fully grown seaweed material in the raft was densely grazed by grazers even though the lower portion was covered by a fish net. *Pteroscirtes mitratus*, *Siganus canaliculatus*, *Siganus javus*, *Pelates quadrilineatus*, *Monodactylus kottelati*, *Terapon puta*, *Scarus ghobban*, *Acanthurus gahhm*, and small crabs are the common grazers in seaweed farming area (Veeragurunathan et al. 2015). Young juvenile of above grazing fishes entered in the raft and ate the grown material of seaweeds and enlarged in size. About 200–300 g of young juveniles were observed in fully grown rafts of *Gracilaria edulis, Gracilaria debilis*, and *Gracilaria dura* (Figure 20.1).

FIGURE 20.1 Grazers found in fully grown *G. dura* raft.

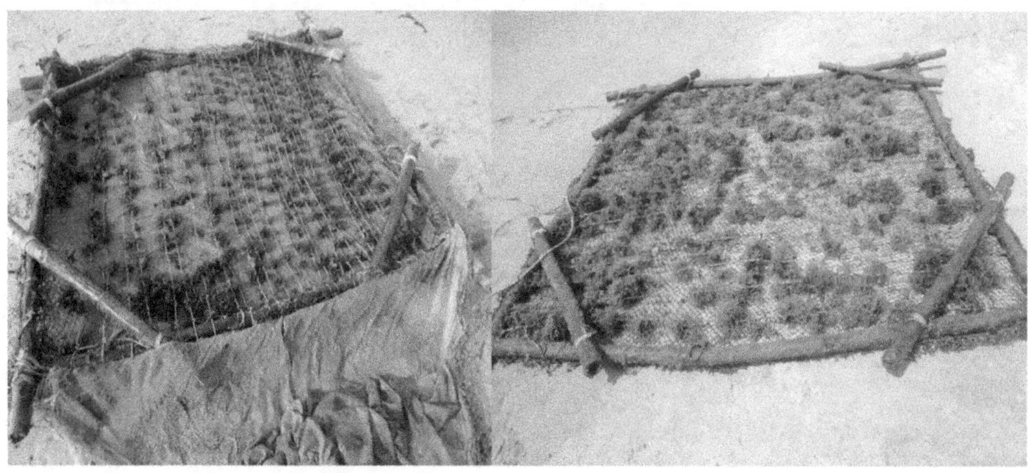

FIGURE 20.2 Drifting of fully grown material due to monsoon effect: 2a. *G. dura* 2b. *G. debilis*.

During monsoon season, the sea became very rough with high wave action. So fully grown algae were detached from the main rope and drifted towards seashore even though the lower portion was covered by fish net. Nearly 35–60% of fully grown material of *Gracilaria edulis*, *G. debilis*, and *G. dura* was lost due to heavy winds coupled with strong water currents and high waves (Figure 20.2a, 2b).

20.3.4 Epiphytism and Drifting Material Load

Buschmann et al. (1999) stated that persistent use of the same cultivated area triggers the development of pests that affect *Gracilaria* production. During summer months, severe epiphytism was observed in seaweed farmings. *Lyngbhya*, a blue-green alga attached in cultivation ropes and later it reached algal surface. It smoothened algal surface, and algal tissue became soft and bleached conditions which ultimately detached from main cultivation ropes. About 25% to 50%

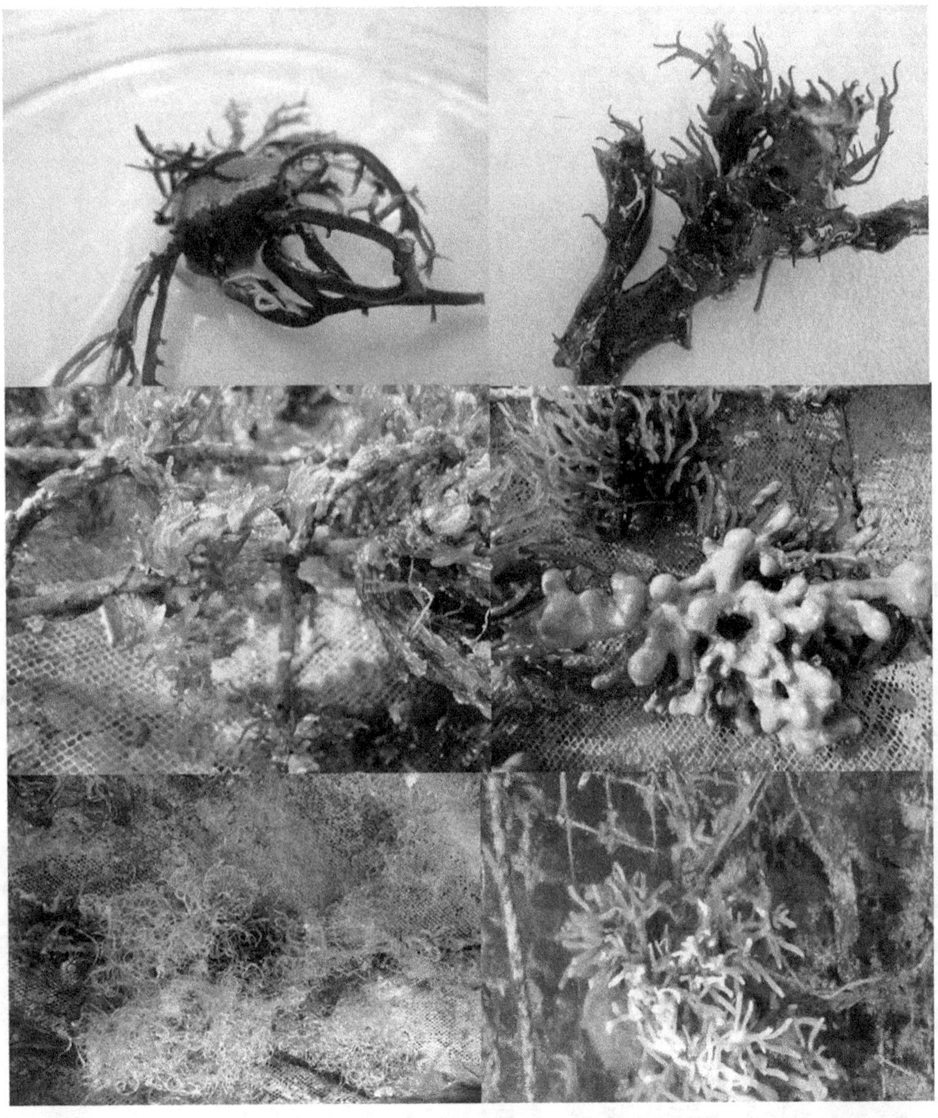

FIGURE 20.3A–3F Epifaunal attachment and 3e–3f epiphytic attachment on fully grown *G. dura* in raft. 3a–3d epifaunal attachment: (a) *Echinochalina* sp., (b) *Botryllus* sp., (c) *Modiolus* sp., (d) *Didemnum* sp., (e) *Chaetomorpha crassa*, (f) *Lyngbya* sp.

FIGURE 20.4 Deposition of *Sargassum* spp., *Turbinaria* spp., and seagrasses in seeded raft of *Gracilaria edulis*.

of loss of material was observed in *Gracilaria edulis*, *G. debilis*, and *G. dura*. Irrespective of any seasons, epifaunal attachments, particularly polychaetes, isopods, decapods, gastropod, ophiuroidea, ectoprocta, maxillopod, acidiacea, and anthozoa, were severe on cultivated algal material. It reduced the available nutrient contents and light intensity and resulted in retarded growth (Figure 20.3a–3d). Epiphytes also added the weight on the host alga, thereby promoting their detachment (Figure 20.3e, 3f).

In Mandapam coastal regions, during monsoon season, several algae, such as *Sargassum* spp., *Turbinaria* spp., and seagrasses drifted from islands of Gulf of Mannar and deposited in the rafts (Figure 20.4). It reduced light intensity and nutrient availability to growing algae and also reduced volume-surface ratio, which ultimately retarded the algal growth in seaweed farming. The accumulated *Sargassum* and *Turbinaria* spp. entangled in main cultivation ropes and resulted in drifting the seeded material from the main rope, and 30% loss of material was observed in *Gracilaria edulis*, *G. debilis*, and *G. dura*. Accumulation of other unwanted seaweeds in raft triggered sediment deposition, which also resulted in retarded growth.

20.3.5 Unawareness of Seaweed Farming by Local Fishermen

In all the maritime villages in India, onshore fishing is the most important activity, and only 10–20% of the coastal population in particular coastal villages preferred deep-sea fishing since it required sophisticated boat facilities. Many fishermen think that fishing and their boat movement will be disturbed if commercial seaweed farming is done in a large-scale level in onshore or open sea conditions in their villages, and they will be forced to involve seaweed farming activities. Because of this misconception in the fishermen community, most of them are against seaweed farming activities.

20.3.6 Loss of Genetic Vigor in the Seed Material

Significant reduction in biomass yield of seaweeds over time due to loss of vigor has been observed after one to two years of higher harvested biomass, and this was reported in Chile for *G. chilensis*

(Buschmann et al. 2001), in Lakshadweep for *G. edulis* (Kaladharan et al. 1996), and for *Gelidiella acerosa* (Rama Rao and Subbaramaiah 1977) and for *Hypnea musciformis* (Ganesan et al. 2006) in India. Since most of the seaweed farming activity depends on vegetative seed material, after several years, biomass yield was reduced due to loss of vigor.

20.3.7 Unfavorable Weather Conditions

Though the east coast of India has a lengthy coastline, particularly coastal areas in Andhra Pradesh and Odisha state's coast, most of them are cyclone-passing areas. Periodical cyclones/typhoons are reported in these areas. During monsoon period, both the states received higher rainfall, which resulted in lower salinity (10–15 ppt) in seawater in onshore area up to 3 km from seashore. During cyclone period, the sea became very rough, and turbulent waves formed, which resulted in severe loss to commercial farming of seaweeds. In 2013, mass mortality of *Kappaphycus alvarezii* was observed due to unfavorable weather conditions, especially higher surface seawater temperature (33–35°C), which was coupled with bacterial infection. As a result, biomass of *K. alvarezii* became white in color and colloidal in nature and oozed out into the sea (Figure 20.5a–5c).

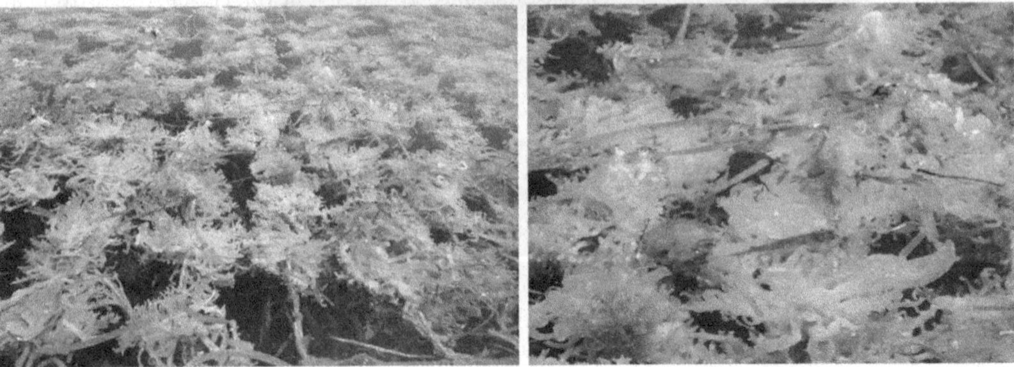

FIGURE 20.5 Thallus softening and bleaching in *Kappaphycus alvarezii* due to bacterial infection coupled with extreme seawater temperature. (a) Raft showing bleached and softened thallus of *Kappaphycus alvarezii*, (b–c) Close view of infected thallus.

20.4 CONCLUSION

There are many physicochemical and biological factors which are impediments to the higher rate of biomass production in microalgal culture. These constraints may be solved by detailed analyzing of impact of individual parameters. However, for constraints in seaweeds cultivtion, implementation of new farming methods in onshore and offshore cultivation, selection of ideal strain for particular habitat, developing superior strain for commercially important algae (heat stress tolerant, sailinity tolerant, robust growth nature) by treating with biostimulant or fertilizers, tissue culture, protoplasmic fusion, or any conventional method will solve the problems. Both marine micro- and macroalgal culture certainly paved the employment opportunity to coastal rural population in India. Production of many value-added products from a single resource also triggered the need for implementation of large-sacle commercial farming of algae in India.

ACKNOWLEDGMENT

Financial support from SERB-DST, New Delhi **(CRG/2019/005304)** and CSIR, New Delhi, is greatly acknowledged. Authors express their heartfelt thanks to Dr. S. Kannan, director, CSMCRI, for his encouragement to pursue this book chapter. We also thank Dr. Vaibhav Mantri, divisional chair, Division of Applied Phycology and Biotechnology, for his suggestions to prepare this book chapter.

REFERENCES

Azov Y (1982) Effect of pH on inorganic carbon uptake in algal cultures. Applied and Environmental Microbiology 43(6): 1300–1306.

Baker KH, Herson DS (1978) Interactions between the diatom Thallasiosira pseudonanna and an associated pseudomonad in a mariculture system. Applied and Environmental Microbiology 35(4): 791–796.

Bechet Q, Shilton A, Fringer OB, Munoz R, Guieysse B (2010) Mechanistic modeling of broth temperature in outdoor photobioreactors. Environmental Science & Technology 44(6): 2197–2203.

Ben-Amotz A, Tornabene TG, Thomas WH (1985) Chemical profile of selected species of microalgae with emphasis on lipids 1. Journal of Phycology 21(1): 72–81.

Buschmann AH, Correa JA, Westermeier R, Hernández-González MC, Norambuena R (1999) Mariculture of red algae in Chile. World Aquaculture 30: 41–45.

Buschmann AH, Troell M, Kautsky N (2001) Integrated algal farming: a review. Cah Biol Mar 42: 83–90.

Carvalho AP, Silva SO, Baptista JM, Malcata FX (2011) Light requirements in microalgal photobioreactors: An overview of biophotonic aspects. Applied Microbiology and Biotechnology 89(5): 1275–1288.

Chisti Y (2007) Biodiesel from microalgae. Biotechnology Advances 25(3): 294–306.

Choi YY, Joun JM, Lee J, Hong ME, Pham HM, Chang WS, Sim SJ (2017) Development of large-scale and economic pH control system for outdoor cultivation of microalgae *Haematococcus pluvialis* using industrial flue gas. Bioresource Technology 244: 1235–1244.

FAO (2018) The global status of seaweed production, trade and utilization. Globefish Research Programme. Vol 124. Rome. pp. 1–210.

Fulbright SP, Robbins-Pianka A, Berg-Lyons D, Knight R, Reardon KF, Chisholm ST (2018) Bacterial community changes in an industrial algae production system. Algal Research 31: 147–156.

Ganesan M, Thiruppathi S, Jha B (2006) Mariculture of *Hypnea musciformis* (Wulfen) Lamouroux in south east coast of India. Aquaculture 256: 201–211.

Ganesan M, Thiruppathi S, Eswaran K , Reddy CRK, Jha B (2011) Development of an improved method of cultivation to obtain high biomass of the red alga Gelidiella acerosa (Gelidiales, Rhodophyta) in the open sea. Biomass and Bioenergy 35(7): 27–29.

Ganesan M, Nivedita Sahu, Eswaran K (2011b) Raft culture of Gracilaria edulis in open sea along the south-eastern coast of India. Aquaculture 321: 141–151.

Gao X, Kong B, Vigil RD (2018) Multiphysics simulation of algal growth in an airlift photobioreactor: Effects of fluid mixing and shear stress. Bioresource Technology 251: 75–83.

Haines KC, Guillard RR (1974) Growth of vitamin b12-requiring marine diatoms in mixed laboratory cultures with vitamin b12-producing marine bacteria 1 2. Journal of Phycology 10(3): 245–252.

Ishika T, Bahri PA, Laird DW, Moheimani NR (2018) The effect of gradual increase in salinity on the biomass productivity and biochemical composition of several marine, halotolerant, and halophilic microalgae. Journal of Applied Phycology 30(3): 1453–1464.

Ishika T, Moheimani NR, Bahri PA (2017) Sustainable saline microalgae co-cultivation for biofuel production: A critical review. Renewable and Sustainable Energy Reviews 78: 356–368.

Kaladharan P, Vijayakumaran K, Chennubhotla VSK (1996) Optimization of certain physical parameters for the mariculture of *Gracilaria edulis* (Gmelin) Silva in Minicoy lagoon (*Luccadive archipelago*). Aquaculture 139: 265–270.

Mantri VA, Eswaran K, Shanmugam M, Ganesan M, Veeragurunathan V, Thiruppathi S, Reddy CRK, Abhiram S (2017) An appraisal on commercial farming of *Kappaphycus alvarezii* in India: Success in diversification of livelihood and prospects. Journal of Applied Phycology 29: 335–357.

Moreno-Garrido I, Canavate JP (2001) Assessing chemical compounds for controlling predator ciliates in outdoor mass cultures of the green algae *Dunaliella salina*. Aquacultural Engineering 24(2): 107–114.

Mutanda T, Ramesh D, Karthikeyan S, Kumari S, Anandraj A, Bux F (2011) Bioprospecting for hyper-lipid producing microalgal strains for sustainable biofuel production. Bioresource Technology 102(1): 57–70.

Qiu R, Gao S, Lopez PA, Ogden KL (2017) Effects of pH on cell growth, lipid production and CO2 addition of microalgae *Chlorella sorokiniana*. Algal Research 28: 192–199.

Rama Rao K, Subbaramaiah K (1977) Regeneration and re-growth of Gelidiella acerosa (Forssk.) Feldmann etHamel at Kilakkarai southeastern shores of India. Indian Journal of Marine Sciences 6: 175–177.

Raven JA, Geider RJ (1988) Temperature and algal growth. New Phytologist 110(4): 441–461.

Scherholz ML, Curtis WR (2013) Achieving pH control in microalgal cultures through fed-batch addition of stoichiometrically-balanced growth media. BMC Biotechnology 13(1): 39.

Soeder CJ (1980) Massive cultivation of microalgae: Results and prospects. Hydrobiologia 72(1–2): 197–209.

Trivedi N, Baghel R, Bothwell J, Vishal Gupta, Reddy CRK, Lali AM, Jha B (2016) An integrated process for the extraction of fuel and chemicals from marine macroalgal biomass. Sci Rep 6, 30728.

Veeragurunathan V, Eswaran K, Malarvizhi J, Gobalakrishnan M (2015) Cultivation of *Gracilaria dura* in the open sea along the southeast coast of India. Journal of Applied Phycology 27: 2353–2365.

Veeragurunathan V, Kamalesh P, Malar Vizhi J, Nripat S, Meena R, Mantri VA (2019) *Gracilaria debilis* cultivation, agar characterization and economics: Bringing new species in the ambit of commercial farming in India. Journal of Applied Phycology 31: 2609–2621.

Veeragurunathan V, Kamlesh P, Nripat S, Malarvizhi J, Mandal SK, Mantri VA (2016) Growth and biochemical characterization of green and red strains of the tropical agarophytes *Gracilaria debilis* and *Gracilaria edulis* (Gracilariaceae, Rhodophyta). Journal of Applied Phycology 28(6): 3479–3489.

Wang H, Hill RT, Zheng T, Hu X, Wang B (2016) Effects of bacterial communities on biofuel-producing microalgae: Stimulation, inhibition and harvesting. Critical Reviews in Biotechnology 36(2): 341–352.

Wang H, Zhang W, Chen L, Wang J, Liu T (2013) The contamination and control of biological pollutants in mass cultivation of microalgae. Bioresource Technology 128: 745–750.

Wang Z, Wen X, Xu Y, Ding Y, Geng Y, Li Y (2018) Maximizing CO2 biofixation and lipid productivity of oleaginous microalga Graesiella sp. WBG1 via CO2-regulated pH in indoor and outdoor open reactors. Science of the Total Environment 619: 827–833.

Xu XQ, Beardall J (1997) Effect of salinity on fatty acid composition of a green microalga from an antarctic hypersaline lake. Phytochemistry 45(4): 655–658.

Yaakob Z, Kamarudin KF, Rajkumar R, Takriff MS, Badar SN (2014) The current methods for the biomass production of the microalgae from wastewaters: An overview. World Applied Sciences Journal 31(10): 1744–1758.

21 Harnessing the Blue-Green Gold

Scale-Up and Bio-Integrations

Sai Kishore Butti, M. Madhumala, Nivedita Sahu, and S. Sridhar

21.1 INTRODUCTION: MICROALGAL PHOTOSYNTHESIS

Photosynthesis appears to be an expedient solution for a sustainable future at a time when the world is on the verge of enervation. Photosynthesis is a naturally occurring bioprocess that has the ability to utilize the always available solar energy for the production of essential compounds. Photosynthesis has been one of the major life-giving sources on Earth since the beginning of evolution, and it is still one up to today [1]. The rapid utilization of the scarcely available fossil reserves to derive energy and commodities has enforced us to look back at the earliest process of photosynthesis for reliable solutions. The biological process by which solar energy is transformed into biomass, bio-products, and biofuel is known as photosynthesis. Originally found in gas-fermenting archaeal bacteria, this process has evolved into the complex sugar production found in higher organisms as a major natural process that makes efficient use of solar power. Now, photosynthesis has developed into a sustainable source of feedstock reserves available for primitive life-forms, primary consumers, and also mankind (Figure 21.1).

To effectively placate the ever-increasing demand for sustainable energy, detrimental causes of environmental pollution, and negate climate change, it is an essential to elucidate and utilize the working potential of the photosynthetic process [2–4]. Photosynthesis is a sequential energy transfer mechanism beginning with light harvesting at photosystems progressing towards carbon assimilation at rubisco. Drawback of photosynthesis is its energy conversion efficiency where most of the harvested light energy is diffused as non-photochemical quenching, CO_2 sequestration, and NAD^+ regeneration to produce carbohydrates and charge separation between the cascading proteins (Figure 21.2a, b).

Global photosynthetic researchers are working with some of the most advanced bio-techniques like metabolomics/fluxomics, transcriptomics, mi-sequencing, microbial chemistry, biophysics, etc. to elucidate the intricate metabolic pathways for regulation and optimization. In this chapter the CO_2 sequestration mechanism, nutritional mode, spectrum of biobased products synthesized, and their ability to photo-symbiosis with other bioprocesses like acidogenesis, bioelectrogenesis, bioanoxygenesis, etc. of different photo-biocatalysts is elaborated. Understanding their metabolic versatility and applications in establishing a bioeconomy platform are essentially addressed, providing future research scope.

Microalgae are a promising part in many ecosystems. They are the primary producers responsible for organic carbon synthesis and oxygen supply to next trophic levels in the food chain. Microalgae have been fostering humanity and sustaining life through multiple dimensions like bioenergy and biofuels generation, mitigating climate change through CO_2 sequestration, producing edible substances and industrially important chemicals like nutraceutical. Microalgae employ photosynthetic machinery to transform solar energy and CO_2 into high-value bio-based products. The advantages of growing microalgae include efficient photosynthesis, simple nutritional demands, and biotechnological attributes like rapid growth rates, accumulation of primary and secondary

DOI: 10.1201/9781003219156-26

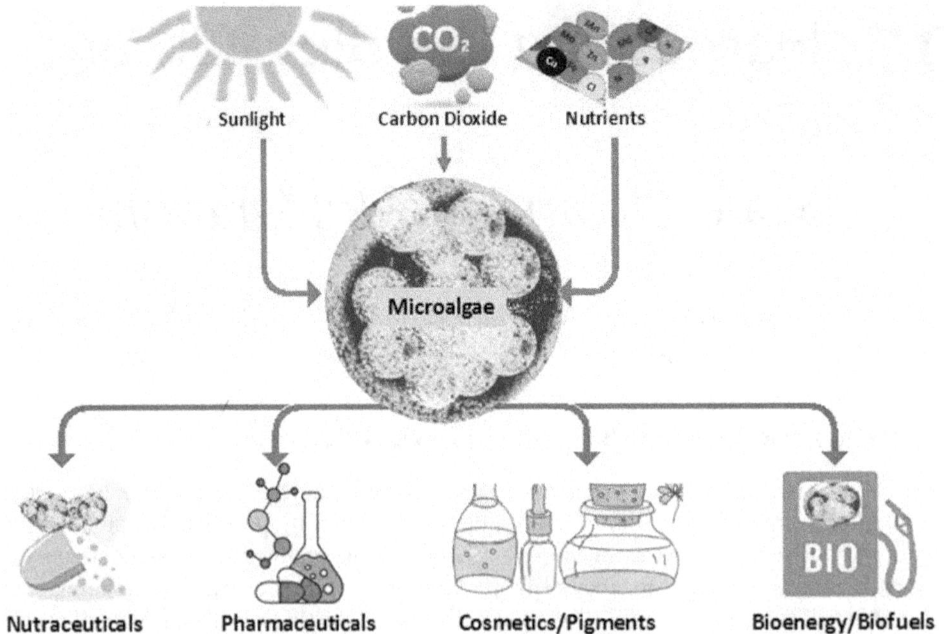

FIGURE 21.1 Microalgae potential for producing value-added products from CO_2 and sunlight.

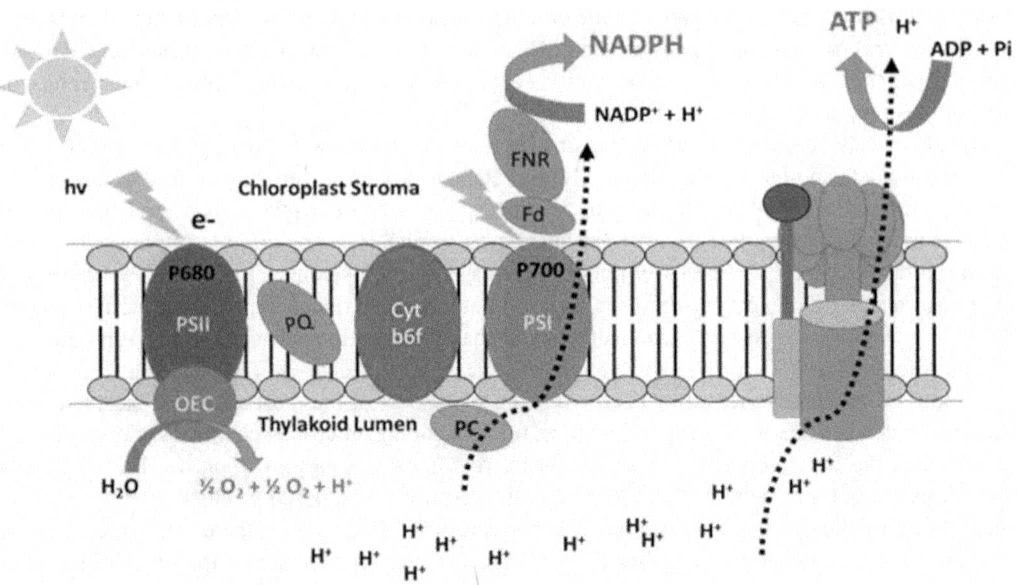

FIGURE 21.2A Electron transport chain that governs the light harvestation and reducing equivalents production in the form of ATP and NADPH.

metabolites of high commercial values. They can transform industrial flue gas and different types of wastewater into useful molecules which can further be converted into high-value biomass and value-added chemicals. They serve as an effective carbon sequestration platform and convert these inputs into biomass. Microalgae have developed to thrive in a variety of nutritional modes, including autotrophic, mixotrophic, and heterotrophic environments, exhibiting metabolic flexibility in response to light and carbon supply [5]. Systematic advances in strain selection, improvement, and

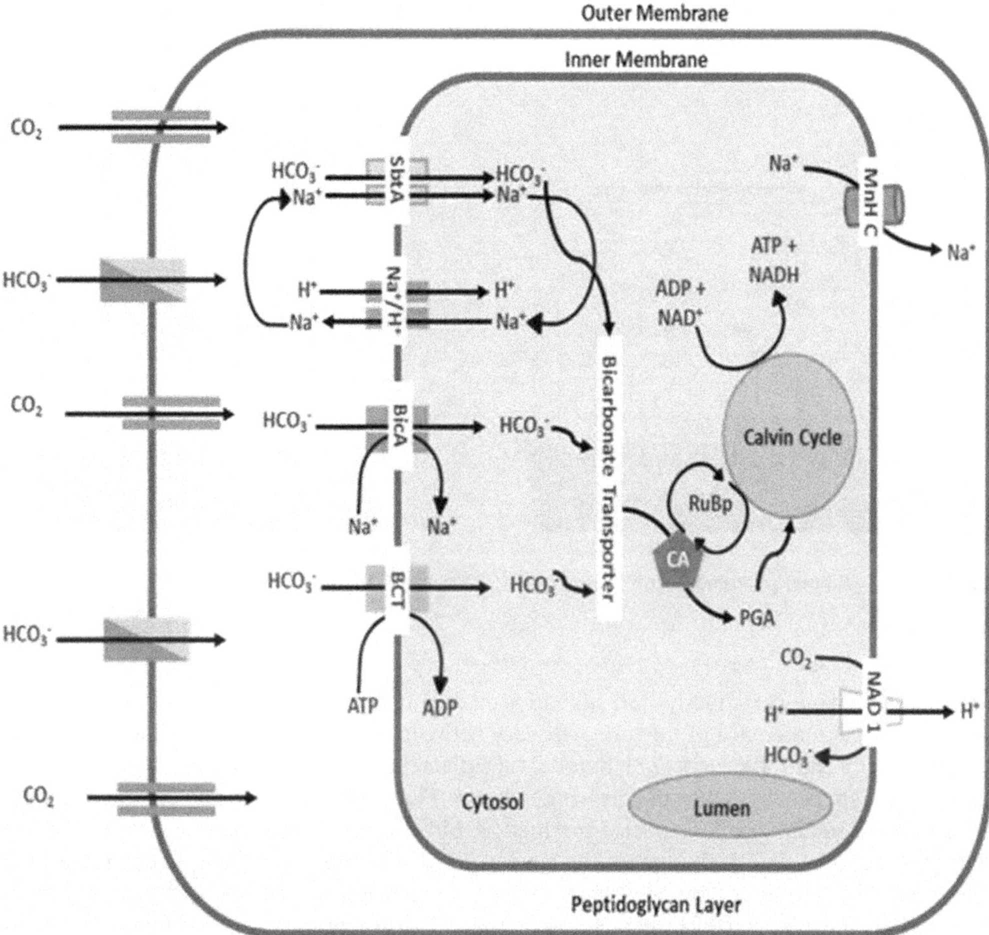

FIGURE 21.2B Carbon assimilation mechanism in microalgae including Calvin Benson cycle and reverse TCA cycle.

cultivation technologies have resulted in optimized strains with increased yields of biomass, lipids, carbohydrates, proteins, and other high-value molecules. Though the major objective of microalgae research is to increase biomass production, which may then be converted into long chain fatty acids (LCFAs) in response to various stress conditions, microalgal biomass that would be used as a source for biofuel production and to extract a spectrum of products [6]. The major biological components which are rich in energy storage reserves are proteins, polysaccharides, and lipids. The attained biomass can also be converted thermo-chemically into bio-oils and biogas based on their application. A biorefinery technique may be adopted to improve the techno-economic feasibility of algae cultivation and the net energy ratio. A unified biorefinery is composed of a vast network of highly integrated biological upstream and downstream processes that utilize photosynthetic machinery as the biocatalytic platform.

21.2 MICROALGAL PRODUCTION TECHNOLOGIES

21.2.1 Tubular Photobioreactors

These PBRs are cylindrical in shape and are made of glass, PVC, or plastic, and it is the most often utilized kind of PBR system today. The simplicity of this form of PBR, as well as its huge luminous

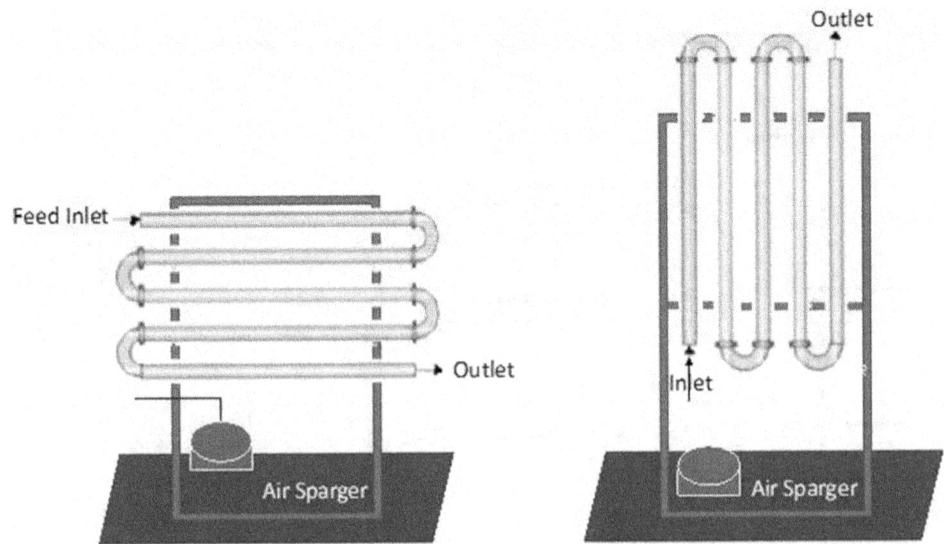

FIGURE 21.3 Tubular photobioreactor in horizontal and vertical alignment.

surface area, make it particularly well suited for outdoor applications (Figure 21.3). It is estimated that the outer circumference of the tube will vary between 10 and 60 mm, while the length would vary between 10 and 100 meters [7,8]. Smaller tube diameters are used in reactors to allow uniform and optimal light penetration to the growing cultures. Fluid velocities between 0.2 and 0.5 meters per second are suitable for larger-scale production. Most of the time, mixing is accomplished by using a sparger placed at the bottom of the reactor. To avoid excessive oxygen concentrations from accumulating in the PBR system, the tubes are linked to a degasser unit. It is known as horizontal, vertical, or inclined tubular PBR, depending on the orientation of the tube inside the tube [9]. The negatives include the buildup of dissolved oxygen, excessive power consumption, high temperature, high pH, CO_2 and O_2 gradients, expensive capital, and operational costs, as well as photo-limiting restrictions. A tubular photobioreactor is a device that uses airlift and bubble columns. Natural sunshine is a good source of illumination for this form of PBR.

21.2.2 Air Lift Photobioreactor

An airlift PBR has a riser and a descender, unlike a bubble column PBR (Figure 21.4). The riser function is employed, as in the bubble column technique. The down comer does not have a sparger since it is not sparged. To minimize structural damage to the PBR and mutual shadowing, this reactor's diameter could be 0.4 m, and its height could be up to 5 m. Increased tube diameter reduces light penetration into the tube core, resulting in reduced illumination [10]. This kind of reactor uses transparent material that allows light to pass through easily like glass and thermoplastics. The mixing in this kind of reactor gives a flashing light effect in the PBR. The bubbles transfer upward in the riser on the dark side of the PBR and downward in the down comer on the lit side of the PBR. The PBR has a high mass transfer rate, low power consumption, and consistent shear stress due to the uniform recirculation of air that promote mixing [11–13]. When designing an air lift PBR, one critical design parameter to consider is the top-to-down developed gas holdup difference. The disadvantages of this kind of PBR include expensive initial capital expenses and frequent cleaning and maintenance.

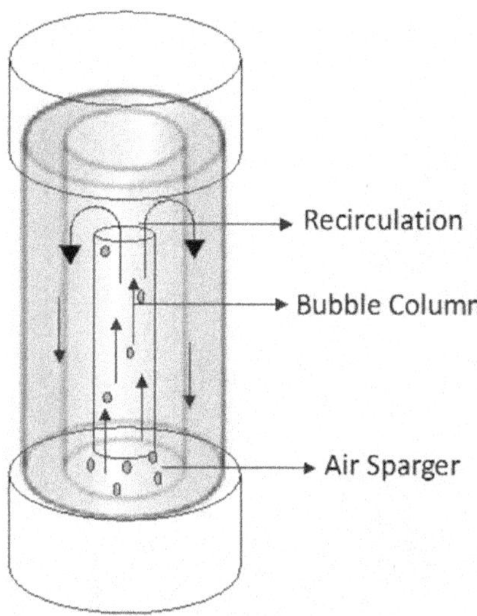

FIGURE 21.4 Air lift photobioreactor.

21.2.3 Flat Plate Photobioreactor

Flat plate PBRs are made of plastic or acrylic sheet and are used for a variety of applications. As seen in Figure 21.5, mixing may be accomplished either by sparging or by employing a motor. This reactor model offers one of the highest surface area to volume ratio values for microalgae production when compared to other reactor designs [14,15]. When it comes to PBR systems, the surface to volume ratio is defined as the amount of surface area exposed to the sunlight per unit volume of the PBR reactor. The high light conversion efficiency and minimal dissolved oxygen buildup are due to efficient heat exchange and long illuminated surface. Mixing in these reactors is usually performed through aeration or by a pump. There are a variety of height and width options available, but the recommended height and width are less than 2.0 m and 0.30 m broad, respectively. It is possible to angle a flat PBR, i.e., lit by sunshine in order to limit light loss. Among the elements influencing biomass production in this form of PBR are the direction and angle of the flat panels, as well as the quantity of flat panels per land unit [16,17]. Fouling is a concern with these reactors because cells stick on the plastic walls, decreasing the light exposure for the suspended microalgal cells, and contamination is also an issue with these PBR.

21.2.4 Bag Photobioreactor

Microalgae are grown in bag PBRs, which are plastic bags with holes in the bottom (Figure 21.6). An aeration system is used to prevent the algal biomass from sedimenting in the system. A plastic bag system for growing microalgae is used in conjunction with a frame system to maintain the structure. Bag PBR has design issues, including the size of the bag, the materials used in their construction, the technique of aeration, and the frame's structural integrity [18,19]. Bag PBRs have been tested for several species with volumes ranging from 3 L to 150 L for different species [20,21]. One of the most significant disadvantages of this form of PBR is the need to have lower cycle durations and establishing a standard disposal procedure with less environmental impact.

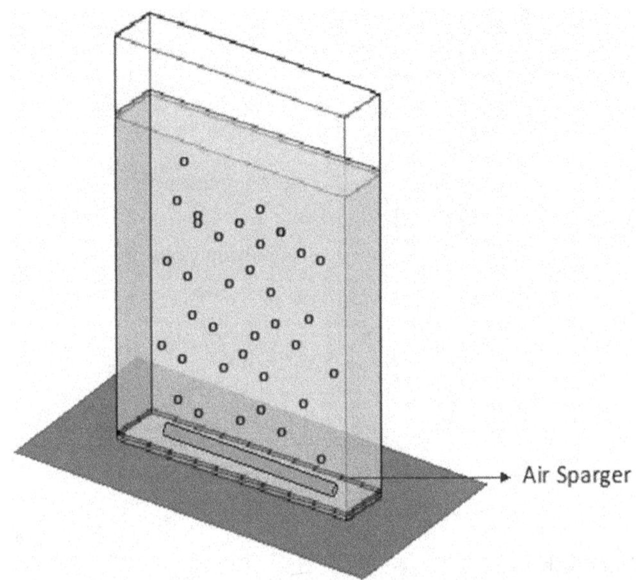

FIGURE 21.5 Flat plate photobioreactor.

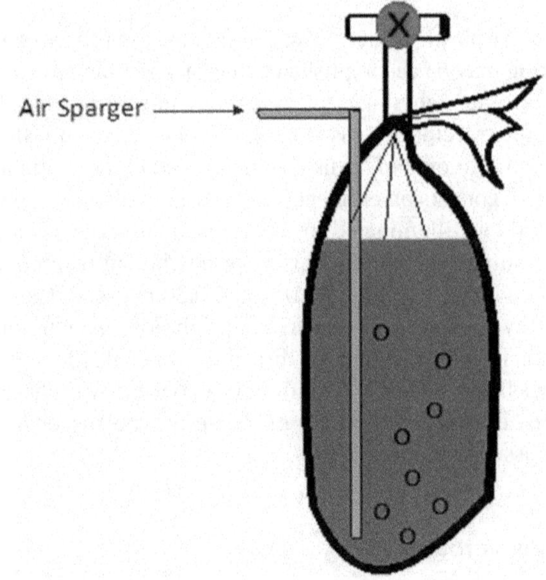

FIGURE 21.6 Bag photobioreactor.

21.2.5 Stirred Tank Photobioreactor

Stirred tank reactors (STR) were originally proposed to grow microalgae photo-autotrophically at both industrial scale and lab scales; however, it was found to be ineffective. Few studies have reported the growth of microalgae was achieved via the use of a stirred tank photobioreactor with a total capacity of 3.6 m^3. It was made up of two sealed turbidostats and a set of open, hydraulically coupled continuous flow stirred-tank reactors with a continuous flow [22]. According to the

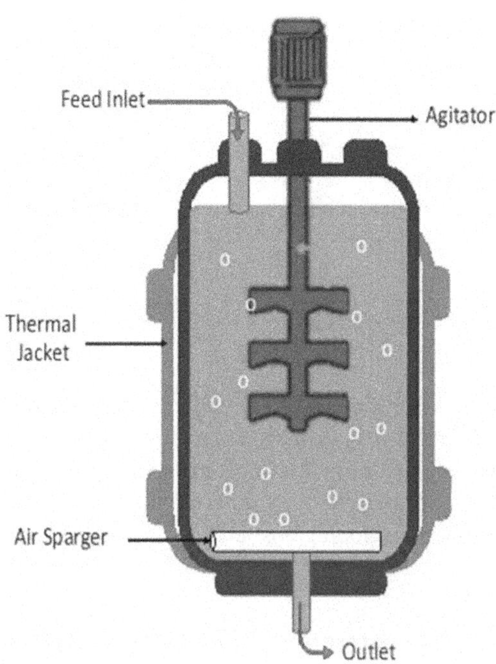

FIGURE 21.7 Stirred tank photobioreactor.

findings, the reactors served as a biomass amplifier for the injected culture. This approach was recently utilized to improve the deterministic model to anticipate microalgal productivity and evaluate its practical viability of using it on a wide scale; however, no reports of its subsequent deployment have been published [19,23,24] (Figure 21.7).

21.2.6 Open Pond Cultivation

Naturally occurring or man-made raceway ponds and inclined surface systems propelled by paddle wheels are all examples of open-air farming systems [23]. These systems typically operate at water depths ranging from 15 to 30 cm. They are representative of the traditional procedures that are utilized in the production of algal biomass [25,26]. Various kinds of open reactors have been put into operation by different organizations since the late 1950s; however, the most widely utilized systems are shallow large ponds, tanks (both cylindrical and circular), and raceway ponds (Figure 21.8). In an open pond cultivation system when nutrients are introduced, the mixing is done by a paddle wheel to ensure even distribution of nutrients [27,28]. The biggest raceway-based microalgae production facility is located in USA, which has a growing area of 440,000m^2, and it produces *Spirulina* [29]. The use of an inclined surface provides two benefits over the traditional flat pond in terms of having better flow and optimal light exposure as the mean depth of the reactor is less than 2 m [30]. Due to the culture's reduced thermal inertia, the temperature may be increased fast in the morning. An open pond with a flat panel of adequate size and orientation can be used to better regulate temperature and sunshine, improving algal productivity in an outdoor culture. Recent research found that elevating the CO_2 levels in the microalgal growth media through specialized carbon dioxide sparging units it has been noted that the carbon dioxide absorption capabilities have increased by 80% in an microalgal cultivation unit [31].

Unlike most closed systems, an open culture unit has instant access to sunlight and is easier to build and manage. Open culture units have been studied for 40 years. Some major drawbacks identified during the process include (i) only a few algal species can be grown successfully at scale;

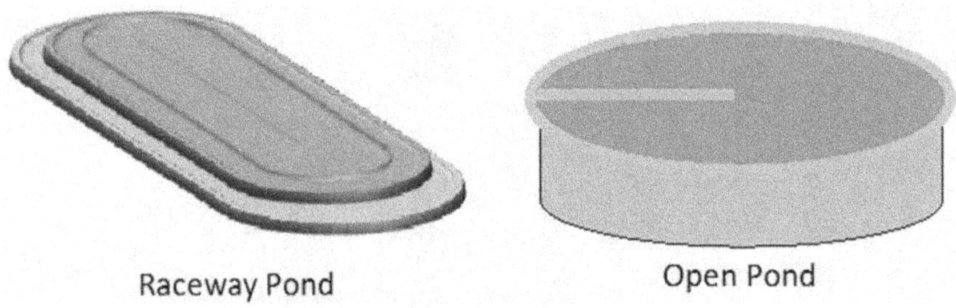

FIGURE 21.8 Open ponds and raceway ponds.

(ii) wild microbial predators are a problem; (iii) significant evaporation losses, making water conservation a concern; (iv) inefficient CO_2 use; and (v) a requires large area. Despite efforts to boost open pond productivity with temperature control, fertilizer delivery, pond level optimization, CO_2 injection systems, and other approaches, open pond productivity remains inferior to closed systems. Currently there is focus on developing cost-effective closed culture techniques, which will be detailed more later.

21.3 PHOTO-BIOFACTORIES PRODUCING VALUE-ADDED PRODUCTS

Microalgae form significant source of feedstock for synthesis of high-value chemicals, including carotenoids, polyunsaturated fatty acids, such as EPA, DHA, phycobilins, and astaxanthin (Figure 21.9).

21.3.1 Nutraceutical and Edible Oils

PUFAs, particularly those with a chain length of 20 carbons or more, are quickly gaining popularity in the pharmaceutical and nutraceutical industries because to their many potential health, nutritional, and therapeutic advantages. In microalgae, eicosapentaenoic acid (EPA) and docosahexaenoic acid (DHA) are the most prevalent fatty acids (DHA; 22:6 n-3). Numerous studies have shown that PUFAs may help prevent cardiovascular and inflammatory disorders, aid in brain growth, and protect mental health, among other advantages [32]. A restricted number of fish oils are used to make commercial PUFAs, although these oils are of varying quality and content. Fish get their EPA from microalgae, which are the principal producers of polyunsaturated fatty acids (PUFAs) [33]. Direct microalgal EPA generation is now the subject of a wide range of research and development initiatives. When glucose is used as a carbon source under heterotrophic circumstances, *Aurantiochytrium limacinum* and *Schizochytrium* sp. LU310 produce the greatest levels of DHA at rates of 4 and 5.7 g/l/d, respectively [34,35]. When compared to autotrophic systems, the omega-3 fatty acids production is increased by two to three times in heterotrophic production systems. In terms of PUFA generation, nitrogen and temperature are the most important elements, with the optimum sources of nitrogen for optimal cell development, and EPA content, being nitrate and urea. To boost EPA production, tryptone and yeast extract have been shown to be effective [36,37]. A drop in membrane fluidity of 10–15°C results in membrane lipid desaturation, which leads to the creation of PUFAs. It's been shown that employing glucose or glycerol as a carbon source, researchers have enhanced the DHA synthesis using heterotrophic batch or fed-batch culture methods [38]. DHA synthesis in *Phaeodactylum tricornutum* has been boosted by metabolic engineering. It is utilized in baby feeds, health supplements, and fortified food items that are made from algae. Food, beverage, and supplement manufacturers have begun using *Cryptecodinium* and *Schizochytrium* species' algal oils in their production processes in an effort to optimize the DHA/EPA ratios [39,40].

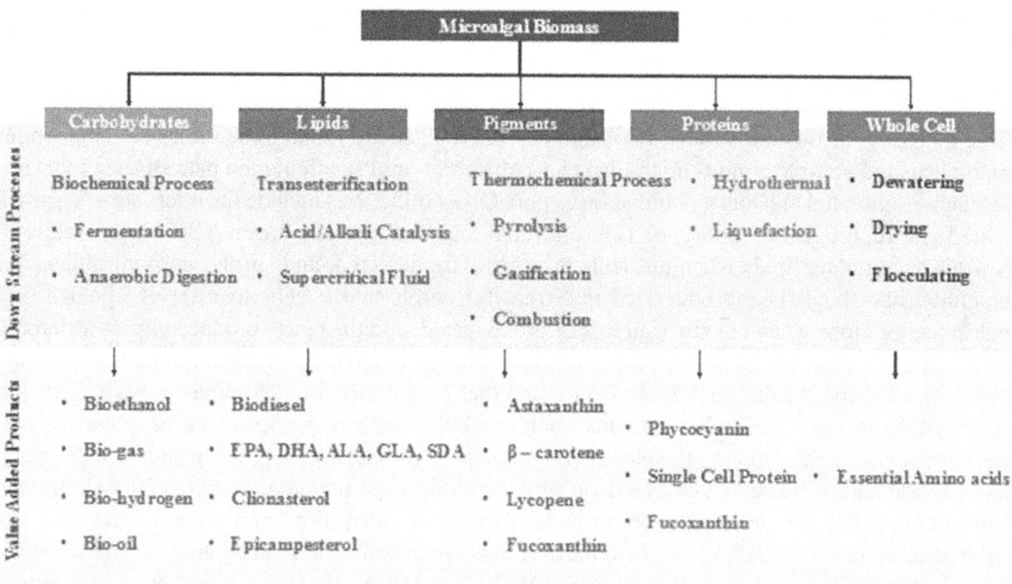

FIGURE 21.9 Multiple biobased products produced from microalgae and the processes involved.

21.3.2 Pigments

Novel carotenoids can be produced in microalgae, which are being studied as a cell factory. Pigment-accumulating microalgae species such as *Dunaliella, Haematococcus, Chlorella zofingiensis*, and *Chlorella sorokiniana* are widely recognized. *Dunaliella* sp., a kind of green microalgae, is a rich source of natural ß-carotene. They may collect enormous amounts of ß-carotene when exposed to certain harsh environmental circumstances, such as high salinity and high light intensity, extreme temperatures, and nutritional deprivation. Inter-thylakoid area of chloroplasts contains oil-containing globules, where ß-carotene is accumulated. Excessive light may be quenched by the globules of ß -carotene, which are located within the cells [41]. Known for its potent antioxidant capabilities, the commercial pigment astaxanthin has earned a prestigious place in the cosmetics and food industries. Carotenoid pigments hematochrome and xanthophyll belong to the xanthophyll group. Free radical scavenging action makes astaxanthin one of nature's most potent antioxidants. Astaxanthin is important in human metabolism because it protects the skin from photo-oxidation by UV radiation, generates antibodies, prevents cancer, and is used in cancer treatment. Due to the dramatic decrease in age-related illness symptoms, it is often referred to as the "fountain of youth". Due to its UV-protective properties, the cosmetics sector incorporates this pigment into sunscreen creams [42,43]. The *Haematococcus pluvialis* strain of microalgae is the most major source of astaxanthin synthesis, with astaxanthin concentration of up to 3% DCW observed. A kilo of astaxanthin now costs over $2,000, with synthetic astaxanthins accounting for the majority of the market value of over $240 million each year [44]. Numerous research on microalgal physiology, biochemistry, and cell biology are currently being linked with genetic and metabolic engineering to better understand how astaxanthin production works in the body under stress [45]. Additionally, other carotenoids, such as bixin, are utilized in dairy goods such as cheese, butter, and margarine, and also in cosmetics that produce a yellowish to peach-colored tone. To prevent eye illnesses such as macular degeneration and cataracts, lutein and zeaxanthin are utilized. Canthaxanthin and violaxanthin, both orange-colored pigments, are often used in the food industry as colorants for the yolk and skin of eggs, salmon aquaculture, and sausages. In the cosmetic industry, canthaxanthin is utilized in tanning tablets and is generated by *Nannochloropsis* sp. [46–48].

21.3.3 Lipids

More than 20% of the lipids found in oleaginous microalgae are found to be saturated. Oleaginous microalgae and non-oleaginous microalgae have different lipid accumulation patterns due to a variety of environmental and nutrient imbalances [49]. Other influences include the microalgae's growth mode (autotrophic, mixotrophic, or heterotrophic), culture age, and strain. Oleaginous microorganisms accumulate lipids when nutrients (nitrogen) are depleted, and surplus carbon (glucose) is integrated into the cells and converted into triacylglycerols by the cells themselves (TAGs). Cell multiplication stops when the nitrogen supply is exhausted, and the newly produced lipids are stored inside the cells themselves. It is possible to grow during periods of nutritional deficiency, though after 15 h in N-free media, the researchers found that *E. gracilis* could continue growth [50,51]. It was possible because chloroplast proteins, such as ribulose-1, 5-bisphosphate carboxylase/oxygenase, were reallocated, allowing the plant to continue growth (RuBisCO). During situations of nutritional deficit and decreased photosynthetic activity, chloroplast breakdown may offer a significant contribution to the nitrogen needed for growth. Microalgae cultivations may be more economically viable if nutrients and medium can be reused. Lipids are accumulated during nitrogen deprivation, which is a survival response until more favorable circumstances are restored, and stored molecules are consumed according to their energy content, beginning with lipids, and moving on to carbs and proteins. Studies using batch cultures of *Chlorella* and *Scenedesmus* sp. utilizing glucose or acetate as the carbon source have revisited the favorable influence on lipid formation following nitrogen restriction in these organisms. These findings also reveal that high cell density cultures and hence high lipid production can only be achieved via the use of cultivation methodologies such as fed-batch cultures. The lipid content of hydrolyzed molasses and de-oiled microalgal biomass leftovers was found to be 42% and 5.58 g/l, respectively [52,53]. Study showed that *Chlorella* sp. may effectively collect lipids by using carbohydrates and amino acids combined when 95% of the nitrogen in the wastewater was eliminated in heterotrophic batch cultures with 10 g/l glycerol [54]. Using glucose, glycerol, or acetate as growth media, researchers compared *Chlorella minutissima*'s ability to collect lipids in autotrophic, mixotrophic, and heterotrophic conditions using batch cultivations [55,56]. When glucose was used as a carbon source, the greatest biomass concentration and fat content was discovered. Having depleted nitrogen, the remaining carbon supply has a significant impact on the sort of lipids that build in the cells; pigments and antioxidants are all included in these lipids.

21.3.4 Biodiesel

Microalgae exhibit benefits as oil producers with greater photosynthetic efficiency and quicker growth rates. They have increased lipid synthesis, and microalgal strains have the potential to be genetically manipulated to produce a fatty acid composition, i.e., ideal for human consumption. Many microalgal strain cultures grown under phototrophic conditions have an oil content ranging between 20 and 50% of their dry weight, depending on the strain. As a result, microalgae with an oil content of around 50% have the potential to produce 86,515 kg biodiesel per hectare per year, compared to soy and canola, which produce 170 and 340 kg (per hectare per year), respectively [57,58]. Among microalgae, *Chlorella* strains exhibit considerable benefits, such as better photosynthetic efficiency over other photosynthetic organisms. Deficiencies in nutrients such as nitrogen and phosphorus in the culture medium have been shown to significantly enhance the production of cholesterol by *Chlorella*, with nitrogen having the greatest impact on lipid accumulation. *C. vulgaris* is known to store more lipids when the nitrogen-to-carbon ratio is lower, according to scientific evidence. An attempt has been undertaken to evaluate the biomass production of *Chlorella* strains in large-scale bioreactors as a result of the potential of this organism. Fermentation technology, i.e., now available for the heterotrophic production of oleaginous microalgae and subsequent processing into biofuels may be used to produce biofuels that are sustainable, cost-effective, and have minimal operating costs. Oleaginous microalgae grown under heterotrophic conditions produce oils that vary in concentration from 42% to 73% of their dry weight. Depending on the extraction technique used, lipid recovery from cells might range between 62% and 97% [49,53,59]. Despite the fact that algae biodiesel production is technically viable and ecologically desirable, a number of obstacles must be solved before the method can effectively compete with fossil fuel–derived goods on a global scale.

21.3.5 Hydrocarbons

Hydrocarbons are energy-dense molecules that may be collected from cell cultures and treated to catalysis in order to make various products such as crude oil, gasoline terpenoids, pharmaceuticals, polymer precursors, and specialty chemical products. In the microalgae kingdom, *Botryococcusbraunii* is well-known for its capacity to produce and accumulate a wide range of hydrocarbons. This hydrocarbon mixture is made up solely of carbon and hydrogen as components, as well as a number of ether lipids and other compounds. *B. braunii* strains that have been isolated from the wild and from the laboratory produce various forms of hydrocarbons. It has been proven that the exponential growth phase of *B. braunii* is the best stage for achieving maximum hydrocarbon output [60,61]. There have only been a few investigations on the generation of hydrocarbons by *Botryococcus* strains under heterotrophic conditions. The effects of the carbon source on the growth and morphology of *B. braunii* were investigated under autotrophic, mixotrophic, and heterotrophic conditions, respectively. Cells were shown to be able to grow in complete darkness when fed glucose or mannose [62]. Comparing cells cultivated with and without glucose, cells cultured with glucose had significantly bigger cell sizes, colony sizes, and intracellular oil granules, as compared to cells cultured without glucose.

21.4 CONSTRAINTS FOR PILOT-SCALE AND INDUSTRIAL-SCALE PRODUCTIONS

21.4.1 Metabolic/Biological Constraints

Open ponds have the potential to become polluted. It is possible for ponds to get infected due to a variety of factors, including infiltration from different algae strains and alternative species, dirt scraps, leaves, and other mobile items. Although the algal species grown for commercial purposes in outside ponds typically grow under extremely selective circumstances, contamination by means of a number of undesirable algal species is not uncommon. As a result, although the capital costs of establishing an associated open lake algal farm are minimal, the risk of contamination of the appropriate algal oil producing species by invading species remains considerable. The pollutants do not only originate from non-organic sources, but they also come from a wide range of biological and completely distinct algal species that may end up being a problem in the future as well. Environmental contaminants may have an influence on the hydrogen ion concentration and pH scale of the social group medium, among other things. The following are the primary pollutants that have an impact on algal culture in open ponds: protozoans, entirely different algae species, animals, insects, and other organic resources such as bacteria, viruses, and fungus. Because of this, the company has confirmed that rigorous tradition preservation and items that encourage the increase of desirable algal species over the establishment of contaminated species are necessary [63,64]. Open ponds, large impurities, and large amounts of water are often eliminated by means of a golf stroke correctly sized exhibit inside the water glide. Important pollutants that sink to the bottom of the pile are encircled in pits that are arranged at the proper angle to shepherd the flow and should then be removed from the sediment traps. They are regulated by the use of an effective batch-tradition and the restarting of social groups at regular intervals with modern, algal matter in those areas under control. A cautious approach to life preservation, offering items that favor the popular species over the contaminated species, as well as a long-term quantity of nonstop way of life, are all possible with caution. Raceway cultures of *P. tricornutum* have also been successfully maintained in laboratory for more than a year, demonstrating their long-term effectiveness. The contamination of *D. salina* ponds by several other algae species of *Dunaliella* is greatly reduced by the use of protective high salinities in the ponds [65]. Continual monitoring is required for tradition renovation, the most important type of monitoring is that the daily microscopic examination, which looks for any abnormal morphological changes and, as a result, the presence of contaminating organisms such as various algal and protozoa species that may be present. Even routine assessments of the pool's nutritional awareness should be carried out to avoid the occurrence of acute nutrient deficits. The routine examination of changes in pH and O_2 levels inside the pool throughout the day may even serve as a beneficial early warning system. The pursuit of O_2 and pH has the advantage of allowing for more computerization flexibility. Pulse amplitude modulated fluorometry (PAM) is a recent replacement strategy that looks to be quite sensitive when it comes to determining the physiological state of algal cells. A slow population buildup of the desired organism may be used to overcome contamination with alga that are completely different from the one being used [63]. It is possible to avoid this problem by covering ponds with transparent membranes or by using greenhouses, which allows the extra effective strains to be completely developed without becoming infected in the first place.

21.4.2 Physiochemical and Process Constraints

21.4.2.1 Evaporation from Open Ponds

The considerable evaporation from the open ponds in dry, tropical locations makes it difficult to grow microalgae. This is a challenge both in terms of increasing the salt concentration in the medium and acquiring enough water to make up for the water loss [66]. To compensate for evaporation losses, algal culture system sites should be chosen in a way heat exchange on surfaces is very less.

21.4.2.2 Light Availability and Odor Issues

Due to the fact that light is the only source of radiant vigor, insufficient light penetration into the pond becomes a nuisance. Algal growth is often regarded as one of the most important variables influencing the environment, particularly light. The most significant difficulties encountered are as follows: using the open pond system, light only reaches the top 8–10 cm of the water. During the course of the algae growth and reproduction, the way of life grows so thick that it prevents light from penetrating deeper into the water. Low intensities may not be sufficient to induce algae growth at the bottom of the pond. Direct sunlight is just too intense for the majority of algae, which need only approximately one-tenth the amount of light they get from direct sunshine to be successful. The use of too much intensity may also result in image obstruction and image oxidation. Long-term exposure to light may also result in modest damage to the algae series antenna, which may be quickly repaired by the cell if the cell is kept in the dark. Several studies have attempted to demonstrate that it should be possible to achieve high production in open ponds due to the fact that the temperature and light-weight intensity fluctuate at various times of the day and throughout the year. Furthermore, even during some of the brightest summer season days, outdoor algal cultures were shown to be light-limited; therefore, in order to achieve the highest possible productivity, it has been recommended that artificial sources of sunlight be available for use during the nights and cloudy days. It is because of the sunlight saturation effect that a 20-fold increase in incident energy results in only a four-fold increase in the quantity of energy used by the algae, as previously stated [67]. The development of algal culture in open ponds will result in foul odor emanating from the ponds. This disadvantage is mostly due to a scarcity of gasoline. Algae that are suspended below the surface of the water are unable to photosynthesize, and as a consequence, they disintegrate and decay.

The decomposition process consumes dissolved gas, and as a result, gas concentrations in the water column fall, resulting in the foul odor. Right designing refers to bringing the algae inside the house at the appropriate moment, before they die and decompose, in order to untangle this illustration.

21.4.3 Economical Constraints

The cost of PBRs is the most significant element influencing the cost of large-scale biomass. The fall in the PBR value resulted in a reduction in the cost of biomass. The methods for reducing value vary depending on the kind of algal strain, the type of PBRs, and the method of biomass postharvest processing [68]. The most important value determinants are irradiation conditions, mixing, the photosynthetic potency of the algae, the medium, and the price of carbon dioxide. The use of raw materials is considered while attempting to reduce the cost of PBR via connectivity [69]. Carbon dioxide (CO_2) is the most expensive resource that may be used in biomass production. It is possible to reduce the cost of greenhouse gas emissions to zero if industrial flue gases are easily available on the market by using flue gases from industrial sources [5]. The bottleneck for the low-cost production of microalgal can be overcome by the development of a large number of productive PBR systems.

21.5 BIO-INTEGRATIONS TO OVERCOME CONSTRAINTS

21.5.1 Biohydrogen Production from Microalgae

Biohydrogen is a potential energy carrier because of its renewable nature, lack of CO_2 emissions, and high energy density per unit weight. Hydrogen may be produced via electrolysis of water, thermo catalytic reformation of organic molecules containing hydrogen, or biological activities [70]. Currently, most hydrogen is produced through electrolysis of water or steam reformation of methane. These include photo-fermentations, dark fermentations, and direct bio-photolysis. Biological hydrogen production by photosynthetic microorganisms requires a low-energy solar reactor, but electrochemical hydrogen production by solar battery-based water splitting requires high-energy solar batteries. The prospect for clean, carbon-free renewable energy from abundant natural resources like sunlight and water is exciting [71]. Unicellular green algae, cyanobacteria, obligate anaerobic, and nitrogen-fixing bacteria are now known to have genes and proteins for hydrogen generation. Anoxygenic photosynthetic bacteria (APSB), as well as anaerobic and nitrogen-fixing bacteria, are examples of these microorganisms. Using cyanobacteria to synthesize molecular hydrogen has been reviewed many times in the previous decade [72]. These organisms may create H_2 in two ways: (1) as a by-product of nitrogen fixation by nitrogenases, and (2) directly via bidirectional hydrogenase. Incorporating hydrogen production across multiple species and enzyme systems is a novel concept with the potential to minimize feedstock utilization and associated costs while enhancing hydrogen yields. Creating PBRs and genetically modifying hydrogen-producing microorganisms are also crucial aspects in increasing overall process yield. Utilizing a photosynthetic technology platform, the technique utilizes viable, high-growth, high-oil-content algae strains.

21.5.2 Bioelectricity Production Using Microalgae

Due to resource scarcity, lipid-based algae biofuels have been proposed as an alternative. This is attributed to high algal production per hectare per year, as well as its ability to capture CO_2 from flue gases. As recently proved in life cycle assessment (LCA) or energy evaluations, the total interest in algae biofuel may overcome the costs associated with huge quantities of nitrogen and phosphate fertilizers, significant volumes of diluted cultures, and oil extraction methods. In actuality, harvesting costs may account for 20–30% of total production costs, and when combined with oil extraction, they may account for over 50% [73]. Alternative transformation techniques should be investigated to avoid the extraction stage. One approach is to perform direct anaerobic digestion (AD) of raw collected algae to create methane and CO_2. The biomethane generated may be used to create energy or heat, while the CO_2 produced can be used to grow microalgae [74]. Depending on the usage, the remaining biomass might be used to generate fiber or fertilizer that are reused in anaerobic digestion to produce biogas. It prevents harvesting, concentration, and oil extraction, which reduces costs and the total quantity of energy due.

21.5.3 Biofuel Production with Simultaneous Waste Remediation

The fact that microalgae flourish in conditions that need minimal freshwater and may be farmed on land typically used for agricultural plants makes them a prospective source of biofuel feedstock. They can even survive in municipal, agricultural, and industrial effluents. These wastes include high levels of nitrogen and phosphorus, which lead to eutrophication of rivers and lakes. A number of studies have shown that microalgae can thrive in diverse wastewater conditions, removing considerable quantities of nitrogen and phosphorus even though they need these elements for protein, nucleic acid, and phospholipid production. If demand is high enough, CO_2 may be acquired cheaply from nearby industries. Although microalgae have been used for over 40 years in combination with physical and chemical removal methods, their application in municipal wastewater treatment is relatively new. Using wastewater resources, which can be used to cultivate microalgae and cleanse sewage at cheap cost and with little environmental impact, may be a future option for enhancing algal biofuel production sustainability [75,76]. As a consequence of the high biomass and lipid productivities observed in several of the reviewed research, it seems that this dual cultivation technique has great promise as a feasible way for producing sustainable and renewable energy. The synergy between wastewater treatment and microalgae-based biofuel production may also have been accomplished due to the existing infrastructure. The algae produce the oxygen, while the bacteria provide the carbon dioxide. As a consequence of reaching an ecological equilibrium, the system becomes particularly resilient and resistant to oscillations common in bacteria-based wastewater treatment systems. The oxygen produced by algae also reduces the need for costly mechanical oxidation of wastewater, saving money. This technology also eliminates greenhouse gas emissions and removes nitrogen and phosphorus, all while preserving ecological balance. Another example is the use of enhanced ponds for wastewater treatment, CO_2 scrubbing, and microalgae-based biofuel production [77,78]. A number of challenges remain, such as lowering harvesting costs and harvesting in a way that allows for bio-product manufacturing, such as harvesting without the use of chemical coagulants. Harvesting might be done via pH-change flocculation-sedimentation or by customized electrical modulations.

21.6 INTEGRATED BIOREFINERY APPROACH

To build a biorefinery framework, the functional components and aspects of photosynthesis may be connected in a closed loop with the photosynthetic process. Approaches for integrating algae bioprocessing into a sustainable closed-loop system are many (Figure 21.10). Hydrothermal liquefaction (HTL), water recycling, nutrient recycling of acidogenic process effluents, wastewater treatment, and biomass valorization from cited processes are the most prominent technologies that have immediate applications in the biorefinery platform. As it can be integrated in situ with petroleum-based oil refineries and therefore, substantially reduce infrastructure costs, HTL has become a popular biorefinery technique for turning algal wet biomass into biocrude. Dewatering, biomass separation, and solvent-based lipid/oil extractions may all be reduced by using this method.

ETPs with microalgae for tertiary wastewater treatment might be another option for polishing ponds in future smart cities. Under mixotrophic circumstances, some microalgae may use organic carbon sources in addition to carbon dioxide and produce larger yields of biomass [5,79]. A high-value product may be synthesized via heterotrophy, which depends on organic carbon as a feedstock.

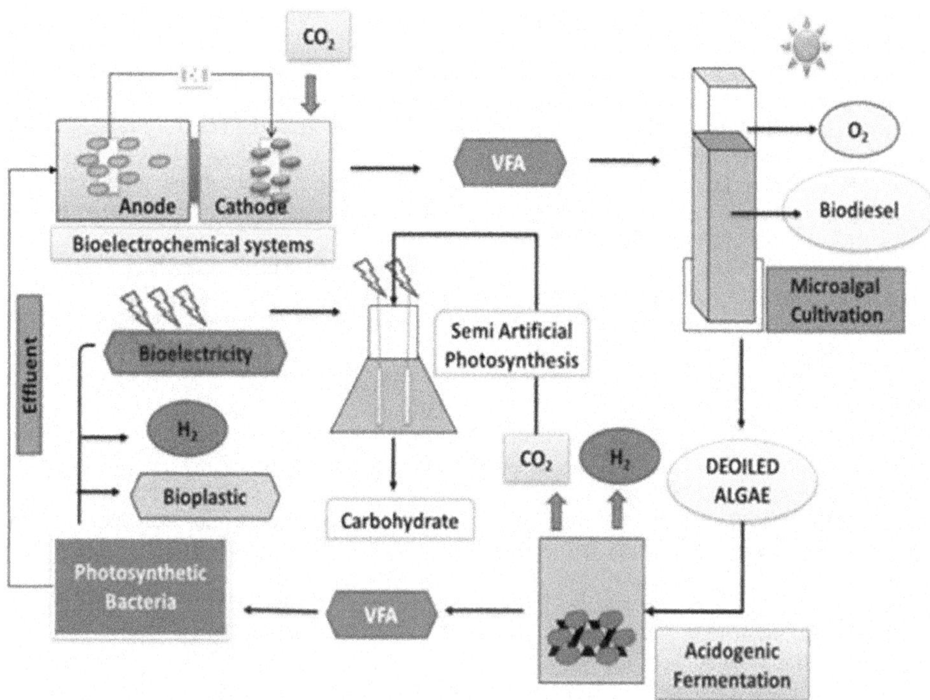

FIGURE 21.10 Photo-biorefinery through process integrations.

Biodiesel generation from microalgae relies heavily on both biomass output and lipid productivity, both of which may be increased by mixotrophic culturing in addition to autotrophy.

21.7 FUTURE PERSPECTIVES FOR PHOTO-BIOREFINERY

In contrast to terrestrial plants, significant development has been achieved in the photosynthetic platform for the production of biofuels and bioproducts since they have drawn considerable attention as instruments for the collection of light energy and CO_2. The major mechanisms by which photosynthetic organisms sequester CO_2 are oxygenic and anoxygenic photosynthesis. Using algae and photosynthetic bacteria, we'll examine two different approaches to photosynthetic sequestration in the sections that follow. Biodiesel, pharmaceuticals, and cosmetics/polymers with high added value may all be produced using microalgal culture in an integrated biorefinery. Waste treatment, value addition, and climate change mitigation all benefit from photo-symbiosis, a vital multidisciplinary bioprocess integration.

ACKNOWLEDGMENT

The authors thank the director of CSIR-IICT for the support. SKB acknowledges "Centre for Technological Excellence in Water Purification: project (GAP-789) for providing research fellowship.

REFERENCES

[1] Blankenship RE. Early evolution of photosynthesis. Plant Physiol 2010;154:434–8. https://doi.org/10.1104/pp.110.161687.

[2] Venkata Mohan S, Modestra JA, Amulya K, Butti SK, Velvizhi G. A circular bioeconomy with bio-based products from CO2 sequestration. Trends Biotechnol 2016;34:506–19. https://doi.org/10.1016/J.TIBTECH.2016.02.012.

[3] Singh J, Dhar DW. Overview of carbon capture technology: Microalgal biorefinery concept and state-of-the-art. Front Mar Sci 2019;6.

[4] Manisalidis I, Stavropoulou E, Stavropoulos A, Bezirtzoglou E. Environmental and health impacts of air pollution: A review. Front Public Heal 2020;8:1–13. https://doi.org/10.3389/fpubh.2020.00014.

[5] Butti SK, Mohan SV. Autotrophic biorefinery: Dawn of the gaseous carbon feedstock. FEMS Microbiol Lett 2017;364:fnx166–fnx166.

[6] Mutanda T, Naidoo D, Bwapwa JK, Anandraj A. Biotechnological applications of microalgal oleaginous compounds: Current trends on microalgal bioprocessing of products. Front Energy Res 2020;8. https://doi.org/10.3389/fenrg.2020.598803.

[7] Huang Q, Jiang F, Wang L, Yang C. Design of photobioreactors for mass cultivation of photosynthetic organisms. Eng 2017;3:318–29. https://doi.org/10.1016/J.ENG.2017.03.020.

[8] Posten C. Design principles of photo-bioreactors for cultivation of microalgae. Eng Life Sci 2009;9:165–77. https://doi.org/10.1002/elsc.200900003.

[9] Torzillo G, Chini Zittelli G. Tubular photobioreactors 2015:187–212. https://doi.org/10.1007/978-3-319-20200-6_5.

[10] Mirón AS, Gómez AC, Camacho FG, Grima EM, Chisti Y. Comparative evaluation of compact photobioreactors for large-scale monoculture of microalgae. Prog Ind Microbiol 1999;35:249–70. https://doi.org/10.1016/S0079-6352(99)80119-2.

[11] Ketheesan B, Nirmalakhandan N. Feasibility of microalgal cultivation in a pilot-scale airlift-driven raceway reactor. Bioresour Technol 2012;108:196–202. https://doi.org/10.1016/J.BIORTECH.2011.12.146.

[12] Sánchez Mirón A, Cerón García MC, García Camacho F, Molina Grima E, Chisti Y. Mixing in bubble column and airlift reactors. Chem Eng Res Des 2004;82:1367–74. https://doi.org/10.1205/CERD.82.10.1367.46742.

[13] Fernandes BD, Mota A, Ferreira A, Dragone G, Teixeira JA, Vicente AA. Characterization of split cylinder airlift photobioreactors for efficient microalgae cultivation. Chem Eng Sci 2014;117:445–54. https://doi.org/10.1016/J.CES.2014.06.043.

[14] Marsullo M, Mian A, Ensinas AV, Manente G, Lazzaretto A, Marechal F. Dynamic modeling of the microalgae cultivation phase for energy production in open raceway ponds and flat panel photobioreactors. Front Energy Res 2015;3. https://doi.org/10.3389/fenrg.2015.00041.

[15] Banerjee S, Ramaswamy S. Dynamic process model and economic analysis of microalgae cultivation in flat panel photobioreactors. Algal Res 2019;39:101445. https://doi.org/10.1016/J.ALGAL.2019.101445.

[16] Sierra E, Acién FG, Fernández JM, García JL, González C, Molina E. Characterization of a flat plate photobioreactor for the production of microalgae. Chem Eng J 2008;138:136–47. https://doi.org/10.1016/J.CEJ.2007.06.004.

[17] Slegers PM, Wijffels RH, van Straten G, van Boxtel AJB. Design scenarios for flat panel photobioreactors. Appl Energy 2011;88:3342–53. https://doi.org/10.1016/J.APENERGY.2010.12.037.

[18] Moheimani NR. Long-term outdoor growth and lipid productivity of Tetraselmissuecica, Dunaliellatertiolecta and Chlorella sp (Chlorophyta) in bag photobioreactors. J Appl Phycol 2013;25:167–76. https://doi.org/10.1007/s10811-012-9850-0.

[19] Wang B, Lan CQ, Horsman M. Closed photobioreactors for production of microalgal biomasses. Biotechnol Adv 2012;30:904–12. https://doi.org/10.1016/J.BIOTECHADV.2012.01.019.

[20] Cui J, Purton S, Baganz F. Characterisation of a simple 'hanging bag' photobioreactor for low-cost cultivation of microalgae. J Chem Technol Biotechnol 2022;97:608–19. https://doi.org/https://doi.org/10.1002/jctb.6985.

[21] Tang H, Chen M, Simon Ng KY, Salley SO. Continuous microalgae cultivation in a photobioreactor. Biotechnol Bioeng 2012;109:2468–74. https://doi.org/https://doi.org/10.1002/bit.24516.

[22] Rusch KA, Christensen JM. The hydraulically integrated serial turbidostat algal reactor (HISTAR) for microalgal production. Aquac Eng 2003;27:249–64. https://doi.org/10.1016/S0144-8609(02)00086-9.

[23] Singh RN, Sharma S. Development of suitable photobioreactor for algae production—A review. Renew Sustain Energy Rev 2012;16:2347–53. https://doi.org/10.1016/J.RSER.2012.01.026.

[24] Ogbonna JC, Yada H, Masui H, Tanaka H. A novel internally illuminated stirred tank photobioreactor for large-scale cultivation of photosynthetic cells. J Ferment Bioeng 1996;82:61–7. https://doi.org/10.1016/0922-338X(96)89456-6.

[25] Kumar K, Mishra SK, Shrivastav A, Park MS, Yang JW. Recent trends in the mass cultivation of algae in raceway ponds. Renew Sustain Energy Rev 2015;51:875–85. https://doi.org/10.1016/J.RSER.2015.06.033.

[26] Benemann JR, Tillett DM, Weissman JC. Microalgae biotechnology. Trends Biotechnol 1987;5:47–53. https://doi.org/10.1016/0167-7799(87)90037-0.

[27] Xu L, Weathers PJ, Xiong X-R, Liu C-Z. Microalgal bioreactors: Challenges and opportunities. Eng Life Sci 2009;9:178–89. https://doi.org/https://doi.org/10.1002/elsc.200800111.
[28] Grobbelaar JU. Microalgal biomass production: Challenges and realities. Photosynth Res 2010;106:135–44. https://doi.org/10.1007/s11120-010-9573-5.
[29] Chinnasamy S, Sood A, Renuka N, Prasanna R, Ratha SK, Bhaskar S, et al. Ecobiological aspects of algae cultivation in wastewaters for recycling of nutrients and biofuel applications. Biofuels 2014;5:141–58. https://doi.org/10.4155/bfs.13.78.
[30] Grivalský T, Ranglová K, da CâmaraManoel JA, Lakatos GE, Lhotský R, Masojídek J. Development of thin-layer cascades for microalgae cultivation: Milestones (review). Folia Microbiol (Praha) 2019;64:603–14. https://doi.org/10.1007/s12223-019-00739-7.
[31] Butti SK, Venkata Mohan S. Photosynthetic and lipogenic response under elevated CO2 and H2 conditions—high carbon uptake and fatty acids unsaturation. Front Energy Res 2018;6. https://doi.org/10.3389/fenrg.2018.00027.
[32] Gammone MA, Riccioni G, Parrinello G, D'Orazio N. Omega-3 polyunsaturated fatty acids: Benefits and endpoints in sport. Nutrients 2018;11. https://doi.org/10.3390/nu11010046.
[33] Barta DG, Coman V, Vodnar DC. Microalgae as sources of omega-3 polyunsaturated fatty acids: Biotechnological aspects. Algal Res 2021;58:102410. https://doi.org/10.1016/J.ALGAL.2021.102410.
[34] Grima EM, Pérez JAS, Camacho FG, Medina AR, Giménez AG, López Alonso D. The production of polyunsaturated fatty acids by microalgae: From strain selection to product purification. Process Biochem 1995;30:711–19. https://doi.org/10.1016/0032-9592(94)00047-6.
[35] Adarme-Vega TC, Lim DKY, Timmins M, Vernen F, Li Y, Schenk PM. Microalgal biofactories: A promising approach towards sustainable omega-3 fatty acid production. Microb Cell Fact 2012;11:1–10. https://doi.org/10.1186/1475-2859-11-96.
[36] Morales-Sánchez D, Martinez-Rodriguez OA, Martinez A. Heterotrophic cultivation of microalgae: Production of metabolites of commercial interest. J Chem Technol & Biotechnol 2017;92:925–36. https://doi.org/https://doi.org/10.1002/jctb.5115.
[37] Sahin D, Tas E, Altindag UH. Enhancement of docosahexaenoic acid (DHA) production from Schizochytrium sp. S31 using different growth medium conditions. AMB Express 2018;8. https://doi.org/10.1186/s13568-018-0540-4.
[38] Boelen P, van Dijk R, Damsté JSS, Rijpstra WIC, Buma AGJ. On the potential application of polar and temperate marine microalgae for EPA and DHA production. AMB Express 2013;3:1–9. https://doi.org/10.1186/2191-0855-3-26.
[39] Hejazi MA, Wijffels RH. Milking of microalgae. Trends Biotechnol 2004;22:189–94. https://doi.org/10.1016/J.TIBTECH.2004.02.009.
[40] Chauton MS, Reitan KI, Norsker NH, Tveterås R, Kleivdal HT. A techno-economic analysis of industrial production of marine microalgae as a source of EPA and DHA-rich raw material for aquafeed: Research challenges and possibilities. Aquaculture 2015;436:95–103. https://doi.org/10.1016/J.AQUACULTURE.2014.10.038.
[41] Ren Y, Sun H, Deng J, Huang J, Chen F. Carotenoid production from microalgae: Biosynthesis, salinity responses and novel biotechnologies. Mar Drugs 2021;19. https://doi.org/10.3390/md19120713.
[42] Kumar G, Shekh A, Jakhu S, Sharma Y, Kapoor R, Sharma TR. Bioengineering of microalgae: Recent advances, perspectives, and regulatory challenges for industrial application. Front Bioeng Biotechnol 2020;8. https://doi.org/10.3389/fbioe.2020.00914.
[43] Norsker N-H, Barbosa MJ, Wijffels R. Microalgal biotechnology in the production of nutraceuticals. Biotechnol Funct Foods Nutraceuticals 2010. https://doi.org/10.1201/9781420087123-c17.
[44] Nguyen KD. A Comparative Case of Synthetic vs. Natural, Tennessee Research and Creative Exchange; 2013.
[45] Shah MMR, Liang Y, Cheng JJ, Daroch M. Astaxanthin-producing green microalga haematococcus pluvialis: From single cell to high value commercial products. Front Plant Sci 2016;7. https://doi.org/10.3389/fpls.2016.00531.
[46] Kim SY, Kwon YM, Kim KW, Kim JYH. Exploring the potential of nannochloropsis sp. Extract for cosmeceutical applications. Mar Drugs 2021;19. https://doi.org/10.3390/md19120690.
[47] Rebelo BA, Farrona S, Rita Ventura M, Abranches R. Canthaxanthin, a red-hot carotenoid: Applications, synthesis, and biosynthetic evolution. Plants 2020;9:1–18. https://doi.org/10.3390/plants9081039.
[48] Lubián LM, Montero O, Moreno-Garrido I, Huertas IE, Sobrino C, González-del Valle M, et al. Nannochloropsis (Eustigmatophyceae) as source of commercially valuable pigments. J Appl Phycol 2000;12:249–55. https://doi.org/10.1023/A:1008170915932.

[49] Patel A, Karageorgou D, Rova E, Katapodis P, Rova U, Christakopoulos P, et al. An overview of potential oleaginous microorganisms and their role in biodiesel and omega-3 fatty acid-based industries. Microorganisms 2020;8. https://doi.org/10.3390/microorganisms8030434.

[50] Wang Y, Seppänen-Laakso T, Rischer H, Wiebe MG. Euglena gracilis growth and cell composition under different temperature, light and trophic conditions. PLoS One 2018;13:e0195329–e0195329. https://doi.org/10.1371/journal.pone.0195329.

[51] Gissibl A, Sun A, Care A, Nevalainen H, Sunna A. Bioproducts from euglena gracilis: Synthesis and applications. Front Bioeng Biotechnol 2019;7. https://doi.org/10.3389/fbioe.2019.00108.

[52] Maurya R, Paliwal C, Chokshi K, Pancha I, Ghosh T, Satpati G, et al. Hydrolysate of lipid extracted microalgal biomass residue: An algal growth promoter and enhancer. Bioresour Technol 2016;207. https://doi.org/10.1016/j.biortech.2016.02.018.

[53] Alishah Aratboni H, Rafiei N, Garcia-Granados R, Alemzadeh A, Morones-Ramírez JR. Biomass and lipid induction strategies in microalgae for biofuel production and other applications. Microb Cell Fact 2019;18:1–17. https://doi.org/10.1186/s12934-019-1228-4.

[54] Pereira I, Rangel A, Chagas B, de Moura B, Urbano S, Sassi R, et al. Microalgae growth under mixotrophic condition using agro-industrial waste: A review. In: Basso TP, Basso TO, Basso LC, editors. Biotechnological Applications of Biomass, Rijeka: IntechOpen; 2021. https://doi.org/10.5772/intechopen.93964.

[55] Sharma AK, Sahoo PK, Singhal S, Patel A. Impact of various media and organic carbon sources on biofuel production potential from Chlorella spp. 3 Biotech 2016;6:1–12. https://doi.org/10.1007/s13205-016-0434-6.

[56] Heredia-Arroyo T, Wei W, Ruan R, Hu B. Mixotrophic cultivation of Chlorella vulgaris and its potential application for the oil accumulation from non-sugar materials. Biomass Bioenergy 2011;35:2245–53. https://doi.org/10.1016/j.biombioe.2011.02.036.

[57] Bošnjaković M, Sinaga N. The perspective of large-scale production of algae biodiesel. Appl Sci 2020;10:1–26. https://doi.org/10.3390/app10228181.

[58] Schlagermann P, Göttlicher G, Dillschneider R, Rosello-Sastre R, Posten C. Composition of algal oil and its potential as biofuel. J Combust 2012;2012:285185. https://doi.org/10.1155/2012/285185.

[59] Ochsenreither K, Glück C, Stressler T, Fischer L, Syldatk C. Production strategies and applications of microbial single cell oils. Front Microbiol 2016;7:1539. https://doi.org/10.3389/fmicb.2016.01539.

[60] Eroglu E, Okada S, Melis A. Hydrocarbon productivities in different Botryococcus strains: Comparative methods in product quantification. J Appl Phycol 2011;23:763–75. https://doi.org/10.1007/s10811-010-9577-8.

[61] Hirano K, Hara T, Ardianor, Nugroho RA, Segah H, Takayama N, et al. Detection of the oil-producing microalga Botryococcus braunii in natural freshwater environments by targeting the hydrocarbon biosynthesis gene SSL-3. Sci Rep 2019;9:1–11. https://doi.org/10.1038/s41598-019-53619-y.

[62] Gouveia JD, Ruiz J, van den Broek LAM, Hesselink T, Peters S, Kleinegris DMM, et al. Botryococcus braunii strains compared for biomass productivity, hydrocarbon and carbohydrate content. J Biotechnol 2017;248:77–86. https://doi.org/https://doi.org/10.1016/j.jbiotec.2017.03.008.

[63] Di Caprio F. Methods to quantify biological contaminants in microalgae cultures. Algal Res 2020;49:101943. https://doi.org/https://doi.org/10.1016/j.algal.2020.101943.

[64] Fabris M, Abbriano RM, Pernice M, Sutherland DL, Commault AS, Hall CC, et al. Emerging technologies in algal biotechnology: Toward the establishment of a sustainable, algae-based bioeconomy. Front Plant Sci 2020;11. https://doi.org/10.3389/fpls.2020.00279.

[65] Baldev E, Ali DM, Sathya R, Thajuddin N. Chapter 21—Critical parameters affecting large-scale production of microalgal biomass in outdoor open raceway ponds. In: Gurunathan B, Sahadevan R, Zakaria ZA, editors. Biofuels and Bioenergy, Elsevier; 2022, pp. 463–78. https://doi.org/https://doi.org/10.1016/B978-0-323-85269-2.00006-X.

[66] Borowitzka MA, Vonshak A. Scaling up microalgal cultures to commercial scale. Eur J Phycol 2017;52:407–18. https://doi.org/10.1080/09670262.2017.1365177.

[67] Juneja A, Ceballos R, Murthy G. Effects of environmental factors and nutrient availability on the biochemical composition of algae for biofuels production: A review. Energies 2013;6:4607–38. https://doi.org/10.3390/en6094607.

[68] Norsker N-H, Barbosa MJ, Vermuë MH, Wijffels RH. Microalgal production—A close look at the economics. Biotechnol Adv 2011;29:24–7. https://doi.org/https://doi.org/10.1016/j.biotechadv.2010.08.005.

[69] Ummalyma SB, Sahoo D, Pandey A. Microalgal biorefineries for industrial products. Microalgae Cultiv Biofuels Prod 2019:187–95. https://doi.org/10.1016/B978-0-12-817536-1.00012-6.

[70] Nagarajan D, Lee DJ, Kondo A, Chang JS. Recent insights into biohydrogen production by microalgae—From biophotolysis to dark fermentation. Bioresour Technol 2017;227:373–87. https://doi.org/10.1016/J.BIORTECH.2016.12.104.

[71] Show KY, Yan Y, Zong C, Guo N, Chang JS, Lee DJ. State of the art and challenges of biohydrogen from microalgae. Bioresour Technol 2019;289:121747. https://doi.org/10.1016/J.BIORTECH.2019.121747.

[72] Limongi AR, Viviano E, De Luca M, Radice RP, Bianco G, Martelli G. Biohydrogen from microalgae: Production and applications. Appl Sci 2021;11:1–14. https://doi.org/10.3390/app11041616.

[73] Zabed HM, Akter S, Yun J, Zhang G, Zhang Y, Qi X. Biogas from microalgae: Technologies, challenges and opportunities. Renew Sustain Energy Rev 2020;117:109503. https://doi.org/10.1016/J.RSER.2019.109503.

[74] Mussgnug JH, Klassen V, Schlüter A, Kruse O. Microalgae as substrates for fermentative biogas production in a combined biorefinery concept. J Biotechnol 2010;150:51–6. https://doi.org/10.1016/J.JBIOTEC.2010.07.030.

[75] Hwang J-H, Church J, Lee S-J, Park J, Lee WH. Use of microalgae for advanced wastewater treatment and sustainable bioenergy generation. Environ Eng Sci 2016;33:882–97. https://doi.org/10.1089/ees.2016.0132.

[76] Khoo KS, Chia WY, Chew KW, Show PL. Microalgal-bacterial consortia as future prospect in wastewater bioremediation, environmental management and bioenergy production. Indian J Microbiol 2021;61:262–9. https://doi.org/10.1007/s12088-021-00924-8.

[77] Zhang C, Li S, Ho SH. Converting nitrogen and phosphorus wastewater into bioenergy using microalgae-bacteria consortia: A critical review. Bioresour Technol 2021;342:126056. https://doi.org/10.1016/J.BIORTECH.2021.126056.

[78] Cheng DL, Ngo HH, Guo WS, Chang SW, Nguyen DD, Kumar SM. Microalgae biomass from swine wastewater and its conversion to bioenergy. Bioresour Technol 2019;275:109–22. https://doi.org/10.1016/J.BIORTECH.2018.12.019.

[79] Mohan SV, Butti SK, Amulya K, Dahiya S, Modestra JA. Waste biorefinery: A new paradigm for a sustainable bioelectric economy. Trends Biotechnol 2016;34:852–5. https://doi.org/10.1016/j.tibtech.2016.06.006.

22 Algal Cultivation for Fuel Production

Saeed Fatima, Sajja S. Chandrasekhar, Bapanipally Govardhan, Nivedita Sahu, and S. Sridhar

22.1 INTRODUCTION

Immense research and development activities are going on to bring up projects focusing on renewable energy sources that may be actively explored to alleviate the concerns of global warming, exponentially rising oil prices, CO_2 emissions, and possibly creating new job possibilities (Singh et al. 2011). Biofuels have grown significantly as more than just a sustainable energy source, particularly for transportation. Biofuels and biorefineries have long been expected to help ameliorate some of these issues and create more steady and greener economies. Biofuels have been developed in three generations so far. Three generations of biofuel have been developed till now (Jeswani et al. 2020). Corn, soybean, sugarcane, and rapeseed were used to make first-generation biofuels. The rocket increase in food prices is one of the concerns as they are frequently used in biofuel generation. Second-generation biofuels have numerous advantages over first-generation biofuels. They are made from lignocellulosic biomass like miscanthus, switchgrass, or poplar. The main benefits are increased yields and reduced land needs (Biswas et al. 2018). Microalgae are considered as a suitable alternative and a renewable feedstock for third-generation biofuels. They are able to generate carbohydrates, lipids, hydrocarbons, and hydrogen, which can be transformed into different forms of energy. Furthermore, methane is obtained from the macroalgal biomass via anaerobic digestion, along with bio-oil and synthesis gas by pyrolysis/gasification (Randrianarison et al. 2010). Microalgae grow in a variety of protein (6–52 wt. %), carbohydrate (5–23 wt. %), and lipids (6–52 wt. %) proportions (7–23 wt. %). Microalgae with a high lipid content are typically thought to be a potential source of third-generation biofuels in the future. One of the most promising options for producing biofuels is Eustigmatophytes, which are high in one or both of the polyunsaturated fatty acids 20:5(n-3) and 22:6(n-3) (Kannah et al. 2021). Moreover, their cultivation is easy in saline/brackish, wastewater, and nonarable land. Under suitable conditions, algae are more productive than plants. The maximum specific growth rate of microalgal species possesses a median value of 1/day and for higher plants is ≤ 0.1/day (Chisti 2010). They come in various colors viz. green, red, orange, and yellow. There are about 100,000 different types of microalgae found in fresh water and saline water. *Chlamydomonas reinhardtii, Dunaliella salina*, and other *Chlorella* species, as well as *Botryococcus braunii*, are common species. *Phaeodactylum tricornutum, Thalassiosira pseudonana, Nannochloropsis*, and *Isochrysis* sp. are other prominent algae groups. Microalgae mainly have triglycerides, phospholipids, glycolipids, and free fatty acids. Due to the presence of all these fuel feedstocks, it has become competitive in the biodiesel and aviation fuel market (Fu et al. 2015). In this chapter, the strength and constraints associated with microalgae cultivation for biofuel production are presented. The main objective of this study was to get a brief overview of the advances and development of the techniques involved in algae cultivation along with fuels production. Among different fuels, hydrogen has the highest price, followed by biodiesel and then ethanol. While gasoline has the lowest price, its combustion contributes significantly to climate change.

 DOI: 10.1201/9781003219156-27

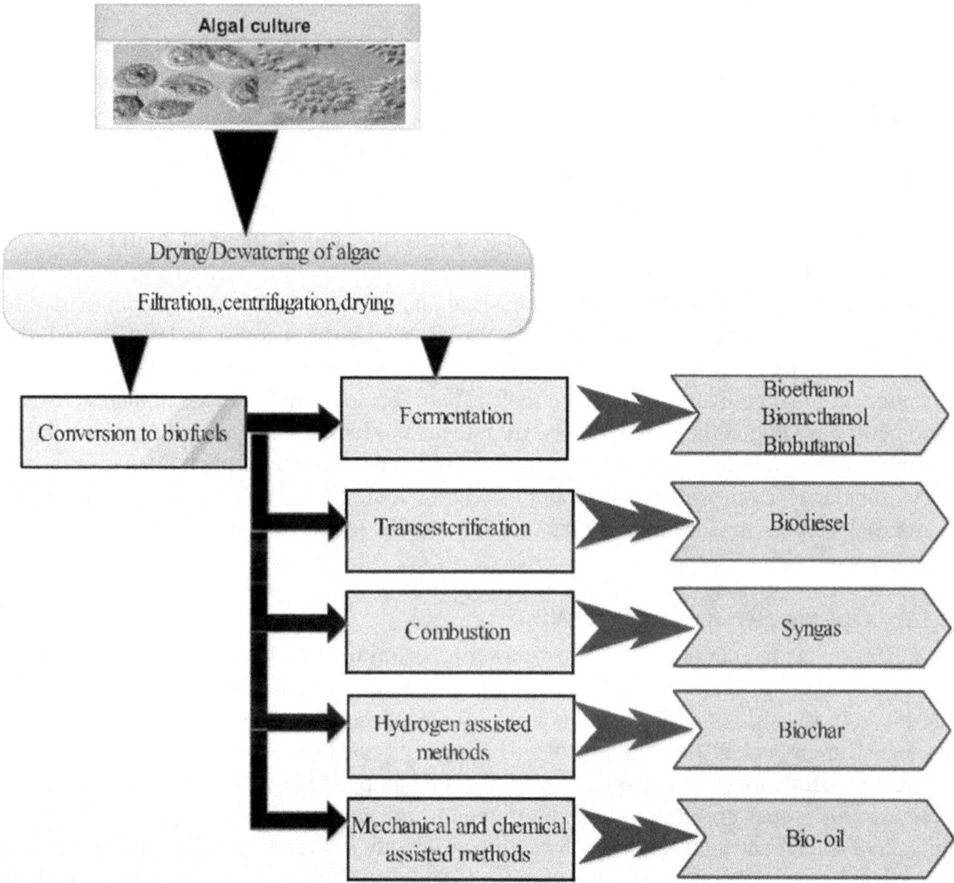

FIGURE 22.1 Conversion of algal biomass to different biofuels.

22.2 MICROALGAE CULTIVATION

Microalgal cultivation in wastewater has great potential for the production of biofuels. The main genera cultivated include *Laminaria, Porphyra, Undaria, Gracilaria, Euchema, Ulva,* and *Chondrus*. Also, the microalgae exhibit significant efficiency in removing contaminants from wastewater. Algae cultivation is relatively easy as it requires simple conditions, that is, light, water, carbon source, temperature, and macro/micronutrients (Suparmaniam et al. 2019). At present, there are different techniques available for algae cultivation on a large scale. Open raceway ponds are the commonly used systems due to their low cost and ease of operation, and in closed culture, systems called photobioreactors (PBRs) due to higher biomass yield and better control opportunities (Hossain and Mahlia 2019). Eutrophic lakes are a great source of cultivation to generate large microalgae biomass consisting of *Chlorella, Microcystis,* and other genera. Many researchers are focusing to integrate these two to obtain the best results. To promote the growth of microalgae, wastewater and flue gas can be used as a carbon and nutrient source. A wide variety of culture parameters like pH, salinity, temperature, light intensity dissolved oxygen, carbon dioxide, and nutrient availability (Veerabadhran et al. 2021). Light intensity, pH, salinity, nutrients availability, temperature, CO_2, and dissolved oxygen content are all factors that influence microalgal growth.

22.2.1 Types of Cultivation

Basically, two types of algal cultivation systems are common on a small and large scale. Photoautotrophic cultivation with photobioreactors (PBRs) and heterotrophic cultivation using open raceway ponds (ORPs).

22.2.1.1 Photoautotrophic Cultivation

It is done in closed cultivation systems in which carbon is supplied as carbon dioxide (CO_2). They need a relatively small area for operation, and they can fully utilize the light sources due to their artificial availability (Yen et al. 2019). The phototrophic cultivation of microalgae can be done on land unsuitable for agricultural use near to industrial and municipal sources on a commercial basis. These systems are designed to allow for perfect mixing with the increase in light accessibility. It helps for improved gas exchange and optimized cell growth (Pegallapati et al. 2014). They are a great source for algal cultivation on a large scale. These reactors can be varied in their construction and geometry based upon the requirements, from towers to tanks. They can be tubular or flat panel shapes and can be glass or plastic made. A rather high concentration of algal biomass can be produced by tubular reactors with bubble columns (Gupta et al. 2015). Because too much oxygen can harm microalgae growth, a supplementary tank is provided to remove the oxygen produced during photosynthesis.

22.2.1.2 Open Waterway Ponds (OWPs)

These algal ponds are economical and low-power consuming culture techniques that are made at depths of 20–50 cm to make sure sufficient light perforation amplifies biomass production. They can be made in several configurations, like raceway ponds, circular ponds, shallow ponds, closed ponds, etc. (Kumar et al. 2015). These systems have great potential to utilize natural carbon dioxide for microalgae cultivation. These systems can also be equipped with additional accessories like a rotating arm for proper mixing of algal culture and to prevent sedimentation of algal biomass. To optimize and spread nutrients and carbon dioxide uniformly to the algal cultures, the ponds can be made oval or race track in construction (Quinn and Davis 2015). Microalgae growing in conjunction with sewage or wastewater treatment plants is a new technique for large-scale output. For a satisfactory development rate of microalgae cells, raceway ponds are created to a shallow depth of 0.3 m. However, certain microalgae strains, such as *Scenedesmus* sp., *Chlorella* sp., *Dunaliella salina*, *Spirulina*, and *Nannochloropsis* sp., prefer the open pond system.

22.3 TECHNICAL CHALLENGES FOR ALGAL CULTIVATION

The commercial production of superior quality algal biomass for biofuel production is a great challenge for researchers. pH, micronutrients, salinity, nitrogen, phosphorous, carbon, temperature, and light are the major parameters that affect microalgal cultivation. Open pond cultivation can get easily contaminated with bacteria, viruses, fungi, etc. To avoid these problems in open pond systems prefilters like a sand filter, coagulation treatments can be done (McBride et al. 2014). Due to the evaporative cooling in open ponds, the culture temperature becomes difficult to control, which leads to CO_2 diffusion in the atmosphere. The microalgal culture can be suitably maintained with controllable agents of bacteria and viruses. Also, probiotic bacteria can be integrated with the cultivation system to avoid toxic bacterial tolerance during cultivation (Guldhe et al. 2017). Nitrogen and phosphorus are critical resources for large-scale agriculture. Due to variances in process integration methodologies, there is a lot of variability in the demand for these resources for large production.

Closed systems like PBRs have significant problems with temperature control, cleaning, maintenance, biofueling, etc. The algal biomass production can also be hindered by the accumulation of dissolved oxygen and the proliferation of benthic algae (Koller 2015). The cost economics related to its design and operation needs to be explored in-depth to overcome problems related to the system price. Microalgae cultivation requires optimum temperature conditions for a high yield of biofuels. The photosynthetic efficiency can be improved by the genetic engineering approach to enhance the biomass production rate (Ugwu et al. 2008). Many scientific communities are continuously trying to find out a wide variety of new and modified methods for microalgal cultivation, drying, and harvesting.

22.4 ECONOMIC CHALLENGES FOR ALGAL CULTIVATION

The production of high-value algal biomass on large scale is critical. One of the key causes of interest in using microalgae as a biofuel feedstock is the great productivity potential associated with it. The cost estimations based on the small-scale systems are sensitive and cannot be directly used for large-scale cultivation systems. The proper use of the pretreatment process like sedimentation, flotation, etc. can make up to 25% cost reduction in total costs for its concentration (Singh et al. 2011). Cost reduction can be obtained by increasing lipids content in algal cells and decreasing the investments in harvesting. Proper choice of natural mutant selection can also help to increase output yields. Increased demand for microalgae could lead to fertilizer shortages. Genetic modifications can be applied to mutants to improve production efficiency. One more approach that can increase algal growth is to transport nutrients to the sea by terrestrial sources. The algal tubular bioreactor can produce dry algal biomass for US$ 32.16 per kg. Algal may need about 14–21 kg of CO_2, 15×109 kg of nitrogen, and $1–2 \times 109$ kg of phosphorus per gallon of biodiesel. It requires 123 billion to 143 trillion of water to replace 5% of gasoline/diesel with algal biofuels. Colder night temperatures reduce the productivity of open ponds on a big scale (Shurin et al. 2016). The excess heat from power plants and other industrial sources can be merged to eliminate this problem. Lipid extraction before esterification is an area where more research is needed. The cost of biomass pretreatment will be greatly reduced as a result of this (Bhardwaj et al. 2020). Nowadays, major oil producers of the world like the UAE and USA are running 10–20% of their transportation on biofuel.

22.5 TYPES OF MICROALGAE FOR FUEL PRODUCTION

There are over 10 million microbial species on the earth, out of which 4,000 and more have been discovered so far (Wang et al. 2010). The type of microbial strain determines the quality of lipid content suitable for biofuel production. To produce the best quality biofuels, the primary objective is to produce strains with suitable lipid content and productivity (Mata et al. 2010). It would be great, for example, if the strains could significantly accumulate lipids despite food shortage. Table 22.1

TABLE 22.1
Different Algal Species to Produce Various Types of Biofuels

Microalgal Species	Target Biofuel	Reaction Conditions	References
C. protothecoides **[F]**	Biodiesel	Heterotrophy; nitrogen limitation	[Xu et al. 2006]
S. obliques *Chlamydomonas reinhardtii* **[F/M]**	Biohydrogen	Indirect process; light biophotolysis; (Fe–Fe) enzymes	[Appel and Schulz 1998]
Nannochloropsis salina **[F/M]**	Biogas	Photobioreactor, large scale, 35°C	[Quinn et al. 2014]
Chlorococum sp. **[F/M]**	Bioethanol	Fermentation	[Singh and Patidar 2018]
Spirulina sp. **[F/M]**	Bio methanol	Gasification/anaerobic fermentation	[Rodionova et al. 2017]
Microalgal consortium **[F/M]**	Biochar	Hydrothermal liquefaction	[Roberts et al. 2013]

gives Different algal species to produce various type of biofuels. Mainly there are two types of microalgae, that is, freshwater algae and marine algae.

22.5.1 Freshwater Algae

There are ten major phyla (divisions) of freshwater algae as per their microscopical occurrence. In general, the number of integral species for terrestrial and freshwater algae gives some hint of the ecological and taxonomic diversity of these groups. Among them, the green algae and diatoms exceed the other species, which indicates their extensive prevalence and capability to live in distinct habitats (Peng et al. 2020). Diatoms, in particular, are biologically successful as planktonic and benthic creatures (about 1,600 species). In addition to the foregoing, also list other phyla, such as *Raphidophyta* (two species), *Haptophyta* (five species), *Eustigmatophyta* (three species), *Prasinophyta* (13 species), and *Glaucophyta* (two species) (Miller and Wheeler 2012). Although these small phyla are interesting from taxonomic and phylogenetic interest, they have less effect on the freshwater environment.

22.5.2 Marine Algae

Marine microalgae proved to be the widespread source of primary biomass. It is gaining a lot of attention as a source of novel compounds and biotechnologically useful genes. A broad range of microalgae can be found in the diverse marine environment. There are at least 30,000 known species of marine microalgae (Rodolfi et al. 2009). Microalgae are photosynthetic cells that are generally unicellular, though certain complicated relationships result in bigger colonies. Marine microalgae are a diverse category consisting of prokaryotic and eukaryotic organisms, like bacteria (cyanobacteria, also called blue-green algae) and diatoms. The blue-green species is vast and presumably not fully explored (Maeda et al. 2018). The microalgae strains should be flexible and sturdy enough to tolerate the shear stress caused by vigorous mixing or due to other microbial interferences. Also, the strains should be strong enough to adjust to the changes in physicochemical parameters of the growing environment.

22.6 TYPES OF PROCESSES TO CONVERT MICROALGAE INTO BIOFUELS

The conversion of microalgal biomass to various valuable renewable biofuels can be done through various techniques.

(1) Transesterification
(2) Thermochemical conversion like combustion, pyrolysis, gasification, thermochemical liquefaction, etc.
(3) Biochemical/biological conversion in terms of anaerobic digestion, fermentation, and photobiological hydrogen production

22.6.1 Transesterification

Transesterification is a chemical reaction in which one alcohol replaces another alcohol from an ester. It employs alcohols like methanol, ethanol, propanol, butanol, and amyl alcohols are for transesterification. Ethanol and methanol are often used due to their low cost and chemical/physical advantages (Makareviciene et al. 2017). This method is extensively employed to reduce the viscosity of triglycerides, which upgrades the engine performance by improving the physical properties of renewable fuels. The transesterification method is typically used to produce fatty acid methyl esters (also known as biodiesel fuel). The transesterification of lipids is used in the generation of biofuel from microalgae. The algal substrate is primarily composed of neutral lipids (Adeniyi et al. 2018). It exists in the form of mono, di, and triglycerides or free fatty acids (FFA) inside the cells. They are more soluble in nonpolar solvents such as hexane and chloroform.

22.6.1.1 Case Study for Production of Biodiesel from Microalgae

In this study, the microalgal transesterification of lipids is done. The transesterification and esterification both are facilitated via acidic catalysis to produce biodiesel. Two sites were chosen, and their temperature during investigation ranged from 24?–33? and 25?–31?. The inorganic acids like sulfuric acid, hydrochloric acid were used as reaction catalysts due to their insensitivity to the free fatty acids content of the oil feedstock. Chloroform-based extraction processes employing chloroform and methanol as solvents have shown higher yield of biodiesel than the traditional extraction-transesterification process. The pH of samples during experiments was observed to be neutral to alkaline. In the experiment extraction with hexane performed in which hexane is used as a solvent and two separate layers of biodiesel and hexane will be obtained at the end. The concept used during experiments was the principle of DNA extraction. There are basic steps in biodiesel extraction:

- The chemical and physical methods like blending, grinding, or sonicating was done to break the algal cells—this exposes the lipids within the algal cells.
- Solvent extraction by hexane.
- Biodiesel purification from reagents used during cell lysis step.

22.6.1.1.1 Determination of Moisture in Algae

The algae used is directly obtained from the lake water that contains a high amount of moisture. For determining the water content, take 10 g of algae sample and kept it at 95°C for 24 h. Take the sample out and measure the weight dry sample.

$$\text{Moisture content}\left(\text{in }\%\right)=\frac{(\text{wet sample}-\text{dry sample})\times 100}{\text{Wet sample}}$$

22.6.1.1.2 Pretreatment of Algae

The microalgae used in the experiments were obtained from the open lake. The microalgal biomass was stored at –20°C, and then its temperature is increased till 25°C before transferring it into digestion reactor. The different compositions of methanol and sulfuric acid were added into the digestion reactor, and the obtained mixture is mixed in digester. Microwave irradiation are applied to the digestion reactor. The water content of the biomass was reduced to 65% of the initial water content after irradiation. The microalgae sample is now centrifuged in REMI R-24 centrifuge for eight minutes at 3000 rpm, excess water is removed, and the solid mass is kept for 24 h at –20°C. Three sample of solid mass each of 12 g is taken for the pretreatment of algae so as to break the cell walls of algae obtained after centrifugation.

Treatment with sulfuric acid: 50 mL of sulfuric acid (30% conc.) is added to the solid mass sample. The solution is kept for 120 h at room temperature.

Treatment with methanol: Algae sample is dipped inside methanol and kept for 24 h. The beaker is covered with aluminum foil in order to avoid the evaporation of methanol at room temperature. After 24 h the solution is sonicated for 30 minutes.

Treatment by sonication in the third sample: Distilled water is added and kept for 24 h, and then it is sonicated for 30 minutes.

22.6.1.2 Extraction of Biodiesel

From each pretreated sample, three different samples were collected and are extracted by adding solvent. By mixing partially soluble sample solvent in the treated sample, the biofuel compounds

TABLE 22.2
Yield of Biodiesel with Different Solvents Used

S. No.	Chemical Used	Solvent Used	Yield (%)
1	H_2SO_4	Chloroform	6.875
2	H_2SO_4	Hexane	11.25
3	H_2SO_4	Chloroform + hexane	25.33
4	Distilled water	Chloroform	45
5	Distilled water	Chloroform + hexane	3.075
6	Distilled water	Hexane	5.075
7	Methanol	Hexane	2.55
8	Methanol	Chloroform	23

will dissolve into the solvent and form a layer on the top that can be separated out. The rest biomass solution remains beneath. The samples collected from each pretreated solution are divided into three parts. Three different solvents are added to the samples. Hexane and chloroform are used as a solvent. Table 22.2 gives yield of biodiesel for different solvents.

22.6.1.3 Extraction with Hexane

The mixture obtained by each pretreatment technique underwent extraction with hexane as solvent, and 8 ml hexane was added, and the mixture is shaken vigorously for five minutes. After some retention time, the mixture was centrifuged, and the upper layer with hexane was transferred into a glass vial. Extraction is repeated three times to obtain complete biodiesel in the hexane layer. The cited steps are repeated using chloroform and hexane: chloroform as solvents for extraction.

However, lipid extraction before esterification can be improved to increase productivity. The cost of the pretreatment methods can be reduced if there are other alternatives to process algal biomass without drying. The existing biodiesel production processes require lipid material with no fatty acids content. This further increases the processing cost. The alternative esterification processes are still at research or bench scale. The cost of producing biofuel from algal biomass is around 50 €/L, making it appealing to begin commercial production as a single fuel (Ahrens and Sander 2010).

22.6.2 Combustion/Pyrolysis

Thermal techniques like combustion or pyrolysis are carried out to obtain products from microalgae that are compatible with diesel engines. The conversion of algal biomass to several useful forms of energy in the presence of air or oxygen is known as combustion (Ripoll et al. 2017). While pyrolysis takes place in the absence of oxygen and catalyst at very high (350–700?) temperatures. It transforms the algal biomass into bio-oil, bio alcohols, biochar, coke, biomethane, and higher gaseous hydrocarbons. Many researchers have studied the combustion and pyrolysis of microalgae like *N. gaditana*. Triglycerides form the majority of algal biomass, and its pyrolysis produces several products like alkanes, alkenes, alkadienes, aromatics, and carboxylic acids (Thangalazhy-Gopakumar et al. 2012). Because oxygen is removed simultaneously during thermal processing, the environmental benefits of using oxygenated fuels are lost. The differences in each of the components present in algal biomass show different thermal behavior due to their diverse degradation properties. The mineral content of microalgal biomass determines its ash content, which has a substantial impact on the decomposition behavior of the biomass due to its catalytic influence during thermal degradation (Gai et al. 2015). Pyrolysis is the most competitive technique for producing liquid biofuels because it provides fuels with a high fuel/feed ratio. However, quick pyrolysis of the lignocellulosic materials is the major focus of researchers. The combative potential of fuel production from microalgae biomass has been demonstrated by pyrolysis of *Emiliana huxleyi*, *Gephyrocapsa oceanica*, *Chlorella protothecoides*, and *Spirulina platensis*.

22.6.3 Gasification

The incomplete oxidation of biomass into a combustible gas at high temperatures (800–900°C) is known as gasification. The biomass combines with oxygen and water (steam) to produce syngas, a mixture of CO, H_2, CO_2, N_2, and CH_4 in the standard gasification process. Gasification as a biomass-to-energy pathway has the advantage of being able to produce syngas from a wide range of potential feedstocks (Stucki et al. 2009). Syngas is a low-calorie gas (typically 4–6 MJ m^{-3}) that can be directly burned or utilized as a fuel for gas engines or gas turbines.

22.6.4 Thermochemical Liquefaction

The algal biomass can be converted to liquid fuel by a thermochemical liquefication process. It is a high pressure (5–20 MPa), high temperature (300–350?) process supported by a catalyst to yield liquid biofuels (Arvindnarayan et al. 2017). The thermochemical liquefication of algal biomass with high moisture content (80–98%) in hot compressed water is one of the most suitable processes for biofuel production. The recovery of the inorganic nutrients (sulfates, nitrates, and phosphates) in biomass is accelerated by thermochemical liquefication. This process produces completely sterilized products, that is, free from possible pathogens including bacteria, fungi, and viruses (Dote et al. 1994).

22.6.5 Fermentation

Fermentation is the process of converting biomass to alcohol. On a commercial basis, sugarcane and starch-containing materials are used to make alcohol. Vegetable-derived ethanol is considered the cleanest fuel (Xia et al. 2016). Fermentation is the process of converting biomass to ethanol. On a commercial basis, sugar cane and starch-containing components are used to make ethanol (Li et al. 2014). This process entails (a) starch solubilization (liquefication), (b) soluble starch conversion to glucose (hydrolysis), and (c) glucose conversion to ethanol (fermentation). Crushing biomass, adding water and yeast, and fermenting it in big tanks called fermenters are all part of the process. Hydrolysis can be done by different methods like acidic, alkaline, enzymatic, autoclaving, microwave, etc.

22.6.6 Anaerobic Digestion

In this process, the biomass is degraded in the absence of oxygen with the help of microbes. Biomethane, biohydrogen, and digestate (solid-liquid mixture) are the main outcomes of this process. A large number of microbes are involved in this process, several biological interactions occur between the microbes and substrate during this process (Sialve et al. 2009). In the anaerobic digestion biological process, the anaerobic bacteria lead the biomass conversion to biogas, that is, biomethane and carbon dioxide. The biogas production efficiency is mainly determined by temperature, pH, and C/N ratio. They are the key factors affecting the process.

Because of the high cellulose percentage in the feedstock, anaerobic digestion efficiency is low, resulting in poor economics. High cellulose-containing bio feedstock is typically retreated to degrade the bio feedstock into simpler molecules in order to improve biogas generation efficiency (Zamalloa et al. 2011). Mechanical grinding, fungal and enzymatic hydrolysis, acid and alkali chemical treatment, and steam explosion are all common pretreatment procedures. The addition of exogenous catalysts to the fermentation process has also been suggested as a way to boost the digestion rate and biogas yield.

22.7 BIOFUELS FROM MICROALGAE

The fuels obtained from biomass in liquid and gaseous form are known as biofuels. Some of the common biofuels are bio-alcohol, biodiesel, bio-oil, biodiesel/kerosene/diesel, biohydrogen biomethane, etc. These can be easily blended with the existing fuels. To deliver the alternative fuels, three generations of biomass are under the study of researchers.

22.7.1 Biodiesel

Biodiesel possesses combustion characteristics similar to those of diesel and is produced by a mono-alcoholic (methanol or ethanol) transesterification process aided by alkali, acids, or enzymes. In order to optimize the transesterification process, remarkable technological developments have been made. Algal oil: diesel in 20% ratio is believed to be the best replacement for diesel engines because it entirely reduced hydrocarbon emissions. Most of the biodiesel was produced via acid catalysis in the traditional process (Hossain et al. 2008). However, this process lowers the activity over that of base catalysis. The ultrasound-aided transesterification can be employed which is superior to batch reaction in traditional methods. Due to the insolubility and stability, hydrophobic catalysts can be used to achieve > 90% yields of free fatty acids under moderate temperature and pressure. Some examples include MgO, CaO, and Al_2O_3 (Silva et al. 2014).

22.7.2 Biohydrogen

Hydrogen is a valuable fuel with numerous applications in fuel cells, coal liquefaction, and heavy oil upgrading. Apart from the existing process for hydrogen production, it can also be obtained by special microalgal species. For the synthesis of hydrogen from microalgae, indirect water photolysis is used (Lam et al. 2019). *Chlamydomonas reinhardtii* proved to be emerging green energy resource for bio-hydrogen production. It is receiving a lot of attention due to its flexibility in the photobiological production of hydrogen. Solar energy is turned into chemical energy in the form of carbohydrates for the creation of hydrogen, which is then employed as substrates for the subsequent reaction (Kumar et al. 2013).

22.7.3 Bioethanol

Bioethanol is one of the environmentally benign biofuels with qualities similar to gasoline that may be produced through a fermentation process from the biomass containing large amounts of starch (Özçimen et al. 2020). The raw material used in bioethanol manufacturing has an impact on the final product. It consists of three steps: carbohydrate hydrolysis, sugar fermentation into ethanol, and ethanol separation and purification (Lee et al. 2015).

22.7.4 Syngas

Depending on the biomass feedstock, the gas product formed by pyrolysis is referred to as syngas or bio-syngas. Bio-syngas is produced by converting biomass into carbon monoxide, hydrogen, methane, water, other hydrocarbons, and ashes in the presence of oxygen, water vapor, or air (Lee et al. 2020). High temperature (800–1,200°C) is required for gasification, and the feedstock must include no more than 20% water in the biomass. Syngas is a prospective fuel source or raw material for the manufacturing of chemicals, especially for the direct synthesis of methanol and ethylene glycol-based on the appealing hydrogen to carbon monoxide ratio (Azadi et al. 2014). It has an energy density of approximately 50% that of natural gas. H_2, O_2, CO, CO_2, N_2, CH_4, and water vapor make up the majority of the syngas composition. Microalgae like *Chlorella* and *Enteromorpha*, as well as macroalgae like *Ulva lactuca*, have high energy content. Unfortunately, due to their high ash, salt, and moisture content, they cannot be used directly. Instead, they can be easily employed as a co-feedstock for combustion with traditional boiler fuels (Raheem et al. 2021). Higher moisture content in feedstock will favor the production of H_2 and CO_2 in terms of syngas composition. It will reduce the carbon monoxide generation depending upon the composition of the algal feedstock used along with a decrease in the possible conversion of H_2O to H_2 by water-gas shift reaction.

22.8 CONCLUSIONS

Due to the depletion of nonrenewable energy sources, different energy sources must be investigated. Sequential growing of microalgae in a microalgal farming facility considerably improves the economic viability of microalgal biofuel production. Extraction of high-value products from harvested algal biomass, thermal processing, and the use of leftover biomass for valuable chemicals are all changing rapidly. The use of a synergistic approach to biofuel production can make the technology more viable and cost-effective. Under altered growing circumstances of alkalinity, temperature, and pH, metabolic/genetic engineering aids in the production of high-quality biofuel. However, more study is needed into less expensive harvesting, conversion, disrupting, and extraction methods. The results of this study reveal that microalgae are particularly efficient at producing a variety of biorefinery products, which enhances the process's economics. It can contribute to total society development by providing nourishment, health, jobs, and general socioeconomic well-being, in addition to satisfying energy needs.

ACKNOWLEDGMENT

The author of the manuscript acknowledges the (AcSIR) Ghaziabad, India, and (CSIR), New Delhi, India, to provide senior research fellowship to carry out PhD work. The authors acknowledge the Knowledge and Information Management (KIM) Division of IICT for plagiarism check and providing manuscript communication number IICT/Pubs./2022/097.

REFERENCES

Adeniyi OM, Azimov U, Burluka A. Algae biofuel: Current status and future applications. Renewable and Sustainable Energy Reviews. 2018 Jul 1;90:316–35.

Ahrens T, Sander H. Microalgae in wastewater treatment: Green gold from sludge. Bio Forum Europe. 2010;14:16–18.

Appel J, Schulz R. Hydrogen metabolism in organisms with oxygenic photosynthesis: Hydrogenases as important regulatory devices for a proper redox poising? Journal of Photochemistry and Photobiology B: Biology. 1998 Nov 1;47(1):1–1.

Arvindnarayan S, Prabhu KK, Shobana S, Kumar G, Dharmaraja J. Upgrading of micro algal derived bio-fuels in thermochemical liquefaction path and its perspectives: A review. International Biodeterioration & Biodegradation. 2017 Apr 1;119:260–72.

Azadi P, Brownbridge GP, Mosbach S, Inderwildi OR, Kraft M. Production of biorenewable hydrogen and syngas via algae gasification: A sensitivity analysis. Energy Procedia. 2014 Jan 1;61:2767–70.

Bhardwaj N, Agrawal K, Verma P. Algal biofuels: An economic and effective alternative of fossil fuels. In Microbial strategies for techno-economic biofuel production, 2020 (pp. 207–27). Springer, Singapore.

Biswas B, Singh R, Kumar J, Singh R, Gupta P, Krishna BB, Bhaskar T. Pyrolysis behavior of rice straw under carbon dioxide for production of bio-oil. Renewable Energy. 2018 Dec 1;129:686–94.

Chisti Y. Fuels from microalgae. Biofuels. 2010 Mar 1;1(2):233–5.

Didem O, Anil TK, Benan I, Tugba O. Bioethanol production from microalgae. In Handbook of microalgae-based processes and products (pp. 373–89).

Fu J, Yang C, Wu J, Zhuang J, Hou Z, Lu X. Direct production of aviation fuels from microalgae lipids in water. Fuel. 2015 Jan 1;139:678–83.

Gai C, Liu Z, Han G, Peng N, Fan A. Combustion behavior and kinetics of low-lipid microalgae via thermogravimetric analysis. Bioresource Technology. 2015;181:148–54.

Guldhe A, Kumari S, Ramanna L, Ramsundar P, Singh P, Rawat I, Bux F. Prospects, recent advancements and challenges of different wastewater streams for microalgal cultivation. Journal of Environmental Management. 2017 Dec 1;203:299–315.

Gupta PL, Lee SM, Choi HJ. A mini review: Photobioreactors for large scale algal cultivation. World Journal of Microbiology and Biotechnology. 2015 Sep;31(9):1409–17.

Hossain AS, Salleh A, Boyce AN, Chowdhury P, Naqiuddin M. Biodiesel fuel production from algae as renewable energy. American Journal of Biochemistry and Biotechnology. 2008;4(3):250–4.

Hossain N, Mahlia TM. Progress in physicochemical parameters of microalgae cultivation for biofuel production. Critical Reviews in Biotechnology. 2019 Aug 18;39(6):835–59.

Jeswani HK, Chilvers A, Azapagic A. Environmental sustainability of biofuels: A review. Proceedings of the Royal Society A. 2020 Nov 25;476(2243):20200351.

Kannah RY, Kavitha S, Banu JR, Sivashanmugam P, Gunasekaran M, Kumar G. A mini-review of biochemical conversion of algal biorefinery. Energy & Fuels. 2021.

Koller M. Design of closed photobioreactors for algal cultivation. In Algal biorefineries, 2015 (pp. 133–86). Springer, Cham.

Kumar Gupta S, Kumari S, Reddy K, Bux F. Trends in biohydrogen production: Major challenges and state-of-the-art developments. Environmental Technology. 2013 Jul 1;34(13–14):1653–70.

Kumar K, Mishra SK, Shrivastav A, Park MS, Yang JW. Recent trends in the mass cultivation of algae in raceway ponds. Renewable and Sustainable Energy Reviews. 2015 Nov 1;51:875–85.

Lam MK, Loy AC, Yusup S, Lee KT. Biohydrogen production from algae. In Biohydrogen, 2019 Jan 1 (pp. 219–45). Elsevier.

Lee OK, Oh Y-K, Lee EY. Bioethanol production from carbohydrate-enriched residual biomass obtained after lipid extraction of Chlorella sp. KR-1. Bioresource Technology. 2015;196:22–7.

Lee XJ, Ong HC, Gan YY, Chen WH, Mahlia TM. State of art review on conventional and advanced pyrolysis of macroalgae and microalgae for biochar, bio-oil and bio-syngas production. Energy Conversion and Management. 2020 Apr 15;210:112707.

Li K, Liu S, Liu X. An overview of algae bioethanol production. International Journal of Energy Research. 2014 Jun 25;38(8):965–77.

Maeda Y, Yoshino T, Matsunaga T, Matsumoto M, Tanaka T. Marine microalgae for production of biofuels and chemicals. Current Opinion in Biotechnology. 2018 Apr 1;50:111–20.

Makareviciene V, Gumbyte M, Skorupskaite V, Sendzikiene E. Biodiesel fuel production by enzymatic microalgae oil transesterification with ethanol. Journal of Renewable and Sustainable Energy. 2017 Mar 9;9(2):023101.

Mata TM, Martins AA, Caetano NS. Microalgae for biodiesel production and other applications: A review. Renewable and Sustainable Energy Reviews. 2010 Jan 1;14(1):217–32.

McBride RC, Lopez S, Meenach C, Burnett M, Lee PA, Nohilly F, Behnke C. Contamination management in low cost open algae ponds for biofuels production. Industrial Biotechnology. 2014 Jun 1;10(3):221–7.

Miller CB, Wheeler PA. Biological oceanography, 2012 Apr 11. John Wiley & Sons.

Özçimen D, Koçer AT, İnan B, Özer T. Bioethanol production from microalgae. In Handbook of microalgae-based processes and products, 2020 Jan 1 (pp. 373–89). Academic Press.

Pegallapati AK, Arudchelvam Y, Nirmalakhandan N. Energetic performance of photobioreactors for algal cultivation: Brief review. Environmental Science & Technology Letters. 2014 Jan 14;1(1):2–7.

Peng L, Fu D, Chu H, Wang Z, Qi H. Biofuel production from microalgae: A review. Environmental Chemistry Letters. 2020 Mar;18(2):285–97.

Quinn JC, Davis R. The potentials and challenges of algae based biofuels: A review of the techno-economic, life cycle, and resource assessment modeling. Bioresource Technology. 2015 May 1;184:444–52.

Quinn JC, Hanif A, Sharvelle S, Bradley TH. Microalgae to biofuels: Life cycle impacts of methane production of anaerobically digested lipid extracted algae. Bioresource Technology. 2014 Nov 1;171:37–43.

Raheem A, Abbasi SA, Mangi FH, Ahmed S, He Q, Ding L, Memon AA, Zhao M, Yu G. Gasification of algal residue for synthesis gas production. Algal Research. 2021 Oct 1;58:102411.

Randrianarison G, Ashraf MA. Microalgae: A potential plant for energy production. Geology, Ecology, and Landscapes. 2017 Apr 3;1(2):104–20.

Ripoll N, Silvestre C, Paredes E, Toledo M. Hydrogen production from algae biomass in rich natural gas-air filtration combustion. International Journal of Hydrogen Energy. 2017 Feb 23;42(8):5513–22.

Roberts GW, Fortier MO, Sturm BS, Stagg-Williams SM. Promising pathway for algal biofuels through wastewater cultivation and hydrothermal conversion. Energy & Fuels. 2013 Feb 21;27(2):857–67.

Rodionova MV, Poudyal RS, Tiwari I, Voloshin RA, Zharmukhamedov SK, Nam HG, Zayadan BK, Bruce BD, Hou HJ, Allakhverdiev SI. Biofuel production: Challenges and opportunities. International Journal of Hydrogen Energy. 2017 Mar 23;42(12):8450–61.

Rodolfi L, ChiniZittelli G, Bassi N, Padovani G, Biondi N, Bonini G, Tredici MR. Microalgae for oil: Strain selection, induction of lipid synthesis and outdoor mass cultivation in a low-cost photobioreactor. Biotechnology and Bioengineering. 2009 Jan 1;102(1):100–12.

Shurin JB, Burkart MD, Mayfield SP, Smith VH. Recent progress and future challenges in algal biofuel production. F1000 Research. 2016;5.

Sialve B, Bernet N, Bernard O. Anaerobic digestion of microalgae as a necessary step to make microalgal biodiesel sustainable. Biotechnology Advances. 2009 Jul 20;27(4):409–16.

Silva C, Soliman E, Cameron G, Fabiano LA, Seider WD, Dunlop EH, Coaldrake AK. Commercial-scale biodiesel production from algae. Industrial & Engineering Chemistry Research. 2014 Apr 2;53(13):5311–24.

Singh A, Nigam PS, Murphy JD. Mechanism and challenges in commercialisation of algal biofuels. Bioresource Technology. 2011 Jan 1;102(1):26–34.

Singh G, Patidar SK. Microalgae harvesting techniques: A review. Journal of Environmental Management. 2018 Jul 1;217:499–508.

Stucki S, Vogel F, Ludwig C, Haiduc AG, Brandenberger M. Catalytic gasification of algae in supercritical water for biofuel production and carbon capture. Energy & Environmental Science. 2009;2(5):535–41.

Suparmaniam U, Lam MK, Uemura Y, Lim JW, Lee KT, Shuit SH. Insights into the microalgae cultivation technology and harvesting process for biofuel production: A review. Renewable and Sustainable Energy Reviews. 2019 Nov 1;115:109361.

Thangalazhy-Gopakumar S, Adhikari S, Chattanathan SA, Gupta RB. Catalytic pyrolysis of green algae for hydrocarbon production using H+ ZSM-5 catalyst. Bioresource Technology. 2012 Aug 1;118:150–7.

Ugwu CU, Aoyagi H, Uchiyama H. Photobioreactors for mass cultivation of algae. Bioresource Technology. 2008 Jul 1;99(10):4021–8.

Veerabadhran M, Gnanasekaran D, Wei J, Yang F. Anaerobic digestion of microalgal biomass for bioenergy production, removal of nutrients and microcystin: Current status. Journal of Applied Microbiology. 2021 Oct;131(4):1639–51.

Wang L, Min M, Li Y, Chen P, Chen Y, Liu Y, Wang Y, Ruan R. Cultivation of green algae Chlorella sp. in different wastewaters from municipal wastewater treatment plant. Applied Biochemistry and Biotechnology. 2010 Oct;162(4):1174–86.

Xia A, Jacob A, Tabassum MR, Herrmann C, Murphy JD. Production of hydrogen, ethanol and volatile fatty acids through co-fermentation of macro-and micro-algae. Bioresource Technology. 2016 Apr 1;205:118–25.

Xu H, Miao X, Wu Q. High quality biodiesel production from a microalga Chlorella protothecoides by heterotrophic growth in fermenters. Journal of Biotechnology. 2006 Dec 1;126(4):499–507.

Yen HW, Hu IC, Chen CY, Nagarajan D, Chang JS. Design of photobioreactors for algal cultivation. In Biofuels from algae, 2019 Jan 1 (pp. 225–56). Elsevier.

Zamalloa C, Vulsteke E, Albrecht J, Verstraete W. The techno-economic potential of renewable energy through the anaerobic digestion of microalgae. Bioresource Technology. 2011 Jan 1;102(2):1149–58.

Index

A

B

C

D

E

G

H

I

J

K

L

M

N

O

Q

R

S

T

U

V

W

www.ingramcontent.com/pod-product-compliance
Lightning Source LLC
LaVergne TN
LVHW081313110826
845149LV00006B/1498

* 9 7 8 1 0 3 2 1 1 2 6 9 5 *